Handbook of Climate Change Impacts on River Basin Management

Climate change not only involves rising temperatures but it can also alter the hydro-meteorological parameters of a region and the corresponding changes emerging in the various biotic or abiotic environmental features.

One of the results of climate change has been the impact on sediment yield and its transport. These changes have implications for various other environmental components, particularly soils, water bodies, water quality, land productivity, sedimentation processes, glacier dynamics, and risk management strategies to name a few.

This volume presents a diverse collection of case studies from researchers across the globe examining the impacts of climate change on river basin management in various geographical, hydrological, and socioeconomic contexts. The case studies yield important insights that can help formulate strategies to build resilience and adapt river basins to a changing climate.

Handbook of Climate Change Impacts on River Basin Management

Case Studies

Edited by
Saeid Eslamian, Mir Bintul Huda, Nasir Ahmad Rather,
and Faezeh Eslamian

CRC Press
Taylor & Francis Group
Boca Raton London New York

CRC Press is an imprint of the
Taylor & Francis Group, an **informa** business

Designed cover image: Shutterstock

MATLAB® and Simulink® are trademarks of The MathWorks, Inc. and are used with permission. The MathWorks does not warrant the accuracy of the text or exercises in this book. This book's use or discussion of MATLAB® or Simulink® software or related products does not constitute endorsement or sponsorship by The MathWorks of a particular pedagogical approach or particular use of the MATLAB® and Simulink® software.

First edition published 2024
by CRC Press
2385 NW Executive Center Drive, Suite 320, Boca Raton FL 33431

and by CRC Press
4 Park Square, Milton Park, Abingdon, Oxon, OX14 4RN

CRC Press is an imprint of Taylor & Francis Group, LLC

© 2024 selection and editorial matter, Saeid Eslamian, Mir Bintul Huda, Nasir Ahmad Rather, and Faezeh Eslamian; individual chapters, the contributors

ISBN: 978-1-032-04182-7 (hbk)
ISBN: 978-1-032-75314-0 (pbk)
ISBN: 978-1-003-47339-8 (ebk)

DOI: 10.1201/9781003473398

Typeset in Times
by Deanta Global Publishing Services, Chennai, India

Dedicated to

My loving parents, my dear Mom, Nighat Nasreen, and my dear Dad, Gh. Mohi-ud-Din Rather for supporting me throughout. Their affection, love, encouragement, and patience have helped me achieve success. They are the driving force in my life and career. They have taught me to believe in myself and my dreams.

My love for them can never be quantified. May Allah bless them always!

Contents

PART I South and North America

Chapter 1 Watershed Management Case Studies in the United States and Canada 3

Saeid Eslamian and Mousa Maleki

Chapter 2 Sediment Yield in Extreme Cold Environments: The Case of the
Sagavanirktok River in Arctic Alaska, USA 13

Horacio Toniolo, Emily Youcha, Eric LaMesjerant,
John Keech, and Joel Bailey

Chapter 3 Regional Precipitation and Streamflow Trends in the Bermejo River Basin in
the Context of Global Change in Argentina 43

Sabrina N. Ayala, Marcela H. González, and Saeid Eslamian

Chapter 4 Statistical Modeling of Summer Precipitation in the Pampa Region of Argentina 62

Alfredo L. Rolla, Paula C. Oliveri, Sabrina N. Ayala, Eugenia M. Garbarini,
Marcela H. González, and Saeid Eslamian

PART II Africa

Chapter 5 Climate Change Impact on Sediment Yield: Case Studies in Africa 89

Faiza Hallouz and Mohamed Meddi

Chapter 6 Basin Management Case Studies in North Africa 112

Youssef Brouziyne, Fouad Moudden, Lahcen Benaabidate, Lhoussaine Bouchaou,
Younes Hamed, Oumaima Attar, Mohammed El Hafyani, Abdelhadi El ouali,
Ali Essahlaoui, and Abdelghani Chehbouni

Chapter 7 Global Climate Change and Free Market Economy: Double Exposure
Influence on Co-operatives in Tanzania: Hopes and Fears 121

Luka Sabas Njau and Neema Kumburu

PART III   South Asia

PART IV   East Asia

PART V   The Middle East

Preface to Volume III: Case Studies

This volume presents a diverse collection of case studies from researchers across the globe examining the impacts of climate change on river basin management in various geographical, hydrological, and socioeconomic contexts. The case studies yield important insights that can help formulate strategies to build resilience and adapt river basins to a changing climate.

Part I begins with watershed management studies in North America highlighting challenges and best practices from the United States and Canada. Climate models project increasing precipitation extremes for the region, threatening water security. The Arctic Alaska case study on sediment yield from the Sagavanirktok River demonstrates the complex interactions between climate change, permafrost thaw, and hydrological processes in cold environments. Studies from Argentina analyse trends in precipitation, streamflow, and sediment yield in the Bermejo River Basin, finding increased risks of flooding and erosion due to climate change. Together, these case studies illustrate the multifaceted impacts already being felt across the Americas.

Part II then shifts to the African continent, where vulnerability to climate change impacts is high due to low adaptive capacity. The sections on sediment yield and basin management in North African countries demonstrate complex interactions between land use, management practices, and a changing hydrological regime due to increasing temperatures and extremes. The studies from Tanzania and Nigeria provide a key socioeconomic context, highlighting the disproportionate risks faced by rural communities dependent on the land for livelihoods. The findings point to the need for climate-sensitive management approaches and adaptation strategies tailored to local contexts.

Moving into Asia, Part III examines research from South and East Asian river basins experiencing substantial human development pressures alongside threats from climate change. The Bangladesh case studies reveal impacts on aquatic biodiversity, sediment loads, land surface temperatures, and coastal zones. The research from India and China likewise demonstrates changing hydrological patterns, sediment dynamics, and stresses on water resources. Large dams are identified as exacerbating local climate change impacts in the Three Gorges case study. Overall, these studies underscore the urgency of integrated management to balance development needs with sustainability in fast-growing Asian economies.

Finally, Part IV features case studies from the Middle East focusing on impacts in Turkey, Iran, and the Gulf. The climate modeling research for Turkey demonstrates the value of high-resolution data for adapting to extremes at local scales. The Iranian studies investigate hydrological responses to drought stress and evaporation losses from irrigation, illustrating regional vulnerabilities and technological solutions. Together, these studies highlight the unique challenges facing Middle Eastern river basins characterized by water scarcity, high reliance on irrigated agriculture, and socioeconomic imbalances.

In summary, the collective findings presented across these case studies point to the multifaceted nature of climate change impacts on river basins, as well as the urgent need for holistic, integrated management approaches that consider the complex interactions between physical, biological, and human systems. This volume provides a valuable foundation for decision makers seeking to formulate policies and measures to enhance the resilience of river basins worldwide in the face of a changing climate.

The aim of this book is to cater to the needs of researchers, graduate students, and academicians. It is expected to help the reader with all of the basic principles and processes. Climate change being a burning issue has been put forth in many manuscripts, but this book helps in building it up right from the basics up to the integrated case studies level. It can rightly be summed up as being the A–Z of the stated title.

The book has a great scope in being incorporated into the curriculum of engineering students, especially in the fields of water resources, environmental engineering, physical geography, and climate change studies. Technically the book is sound enough to find its way as a reference for major hydrology and soil water courses.

The book discusses the topics in detail and has been weaved in a manner to maintain a rhythm and a continuous flow in the topics discussed. Effort has been put in to deal with almost all of the related topics and a comprehensive approach has been taken. One of the main highlights of this book is the incorporation of case studies from around the globe.

Saeid Eslamian
Isfahan University of Technology, Isfahan, Iran

Mir Bintul Huda
National Institute of Technology, Srinagar, India

Nasir Ahmad Rather
Baba Ghulam Shah Badshah University, Rajouri, India

Faezeh Eslamian
McGill University, Montreal, Canada

About the Editors

Saeid Eslamian is a full professor of environmental hydrology and water resources engineering in the Department of Water Engineering at Isfahan University of Technology, Isfahan, Iran, where he has been since 1995. His research focuses mainly on statistical and environmental hydrology in a changing climate. In recent years, he has worked on modeling natural hazards, including floods, severe storms, wind, drought, pollution, water reuses, sustainable development, resiliency, etc. Formerly, he was a visiting professor at Princeton University, New Jersey, and the University of ETH Zurich, Switzerland. On the research side, he started a research partnership in 2014 with McGill University, Canada. He has contributed to more than 600 publications in journals, books, and technical reports. He is the founder and chief editor of the *International Journal of Hydrology Science and Technology* (IJHST). Eslamian is now associate editor of three important publications: *Journal of Hydrology* (Elsevier), *Eco-Hydrology and Hydrobiology* (Elsevier), *Journal of Water Reuse and Desalination* (IWA), and *Journal of the Saudi Society of Agricultural Sciences* (Elsevier). Professor Eslamian is the author of approximately 40 books and 370 book chapters.

Dr Eslamian's professional experience includes membership on editorial boards, and he is a reviewer of approximately 150 Web of Science (ISI) journals, including *ASCE Journal of Hydrologic Engineering, ASCE Journal of Water Resources Planning and Management, ASCE Journal of Irrigation and Drainage Engineering, Advances in Water Resources, Groundwater, Hydrological Processes, Hydrological Sciences Journal, Global Planetary Changes, Water Resources Management, Water Science and Technology, Eco-Hydrology, Journal of American Water Resources Association, American Water Works Association Journal*, etc. UNESCO has also nominated him for a special issue of the *Eco-Hydrology and Hydrobiology Journal* in 2015.

Professor Eslamian was selected as an outstanding reviewer for the *Journal of Hydrologic Engineering* in 2009 and received the EWRI/ASCE Visiting International Fellowship in Rhode Island (2010). He was also awarded outstanding prizes from the Iranian Hydraulics Association in 2005 and the Iranian Petroleum and Oil Industry in 2011. Professor Eslamian was chosen as a distinguished researcher of Isfahan University of Technology (IUT) and Isfahan Province in 2012 and 2014, respectively. In 2016, he was a candidate for national distinguished researcher in Iran.

He has also been the referee of many international organizations and universities, including the US Civilian Research and Development Foundation (USCRDF), the Swiss Network for International Studies, the Majesty Research Trust Fund of Sultan Qaboos University of Oman, the Royal Jordanian Geography Center College, and the Research Department of Swinburne University of Technology of Australia. He is also a member of the following associations: American Society of Civil Engineers (ASCE), International Association of Hydrologic Science (IAHS), World Conservation Union (IUCN), GC Network for Drylands Research and Development (NDRD), International Association for Urban Climate (IAUC), International Society for Agricultural Meteorology (ISAM), Association of Water and Environment Modeling (AWEM), International Hydrological Association (STAHS), and UK Drought National Center (UKDNC).

Professor Eslamian finished Hakimsanaei High School in Isfahan in 1979. After the Islamic Revolution, he was admitted to IUT for a BS in Water Engineering and graduated in 1986. After graduation, he was offered a scholarship for a master's degree program at Tarbiat Modares University, Tehran, Iran. He finished his studies in hydrology and water resources engineering in 1989. In 1991, he was awarded a scholarship for a PhD in Civil Engineering at the University of New South Wales, Australia. His supervisor was Professor David H. Pilgrim, who encouraged him to work on "Regional Flood Frequency Analysis Using a New Region of Influence Approach". He earned a PhD in 1995 and returned to his home country and IUT. In 2001, he was promoted to associate professor and in 2014 to full professor. For the past 28 years, he has been nominated for different positions at IUT, including university president consultant, faculty deputy of education, and head of department. Eslamian is now the director of the Center of Excellence in Risk Management and Natural Hazards (RiMaNaH). Professor Eslamian has made three scientific visits to the United States, Switzerland, and Canada in 2006, 2008, and 2015, respectively. In the first, he was offered the position of visiting professor by Princeton University and worked jointly with the late Professor Eric F. Wood at the School of Engineering and Applied Sciences for one year. The outcome was a contribution to hydrological and agricultural drought interaction knowledge by developing multivariate L-moments between soil moisture and low flows for northeastern US streams.

Recently, Professor Eslamian has published 14 handbooks with Taylor & Francis (CRC Press): *Handbook of Engineering Hydrology* (2014), *Urban Water Reuse Handbook* (2016), *Underground Aqueducts Handbook* (2017), *Handbook of Drought and Water Scarcity* (2017), *Constructed Wetlands: Hydraulic Design* (2019), *Handbook of Irrigation System Selection for Semi-Arid Regions* (2020), *Urban and Industrial Water Conservation Methods* (2020), *Flood Handbook* (2022), *Handbook of Irrigation Hydrology and Management* (2022); and several other books with other renowned publishers such as *An Evaluation of Groundwater Storage Potentials in a Semiarid Climate*, *Advances in Hydro Geochemistry Research*, *Handbook of Eurasian Forecasting*, *Water Scarcity Global Perspectives, Issues and Challenges*, *Handbook of Water Harvesting and Conservation* (Wiley) and *Handbook of Disaster Risk Reduction and Resilience*, *Earth Systems Protection and Sustainability*, *Handbook of Hydroinformatics* (Elsevier), and many others.

Professor Eslamian has been appointed as World's Top 2% Researcher by Stanford University, California, in 2019, 2020, and 2022. He has also been a Grant Assessor/Report Referee/Award Jury/Invited Researcher for international organizations such as the United States Civilian Research and Development Foundation (2006), Intergovernmental Panel on Climate Change (2012), World Bank Policy and Human Resources Development Fund (2021), Stockholm International Peace Research Institute (2022) respectively.

Mir Bintul Huda presently works as a Consultant for the Sheikh-ul-Alam Sheikh Nuruddin Water Resources Chair under the Ministry of Water Resources, Government of India at the National Institute of Technology Srinagar. The position is aimed at carrying out studies in Water Resources with emphasis on the impact of Climate Change on it and the adaptation of strategies with respect to the Planning and Management of Water Resources Systems of the Indus Basin of the Himalayas. In addition to carrying out the research, she is actively involved in guiding the M.Tech and PhD dissertations in this field at the Institute.

She has a PhD in Civil Engineering, a Master's in Water Resources Engineering, and a B.Tech in Agricultural Engineering. She has expertise in Water Resources and Management Studies, especially stream flow and its hydraulics, hydraulic structures and their interaction, and climate change and related studies.

In addition, she is a corporate member of the Institute of Engineers (India) Kolkata. Dr Mir has numerous publications in reputed international journals and several book chapters with renowned publishing houses. In addition to being a reviewer in many journals, she is also a part of many academic houses and journals.

Nasir Ahmad Rather (Ph.D.) is a senior Assistant Professor of Civil Engineering at College of Engineering and Technology, Baba Ghulam Shah Badshah University, Rajouri, Jammu & Kashmir, India. Dr. Rather did his Ph.D. from National Institute of Technology, Srinagar. His research interests include Water Resources and Geo-technical Engineering, Protective Filters, Climate Change, Hydraulic Structures and River Engineering. Dr. Rather is actively involved in student engagement through his meticulous teaching and learning skills. Besides classroom teaching, he loves to impart practical knowledge to his students through projects and provides research guidance to many scholars. He has made a significant contribution in the field of water resources engineering which is evident through his publications in high impact-factor journals. Dr. Rather has also received fellowship from the Ministry of Education (MoE), Government of India. Further, he is a member of several prestigious academic bodies and societies of international repute. He is the editorial board member and reviewer of numerous international journals. For his pioneering contribution, he is recognized as a subject expert in the said field across many institutes. He is also a member of various academic and administrative bodies.

Faezeh Eslamian has a PhD in Bioresource Engineering from McGill University, Montreal, Canada. Her research focuses on the development of a novel lime-based product to mitigate phosphorus loss from agricultural fields. Faezeh completed her Bachelor's and Master's Degrees in Civil and Environmental Engineering from Isfahan University of Technology, Iran, where she evaluated natural and low-cost absorbents for the removal of pollutants such as textile dyes and heavy metals. Furthermore, she has conducted research on the worldwide water quality standards and wastewater reuse guidelines. Faezeh is an experienced multidisciplinary researcher with an interest in soil and water quality, environmental remediation, water reuse, and drought management.

Contributors

Ismail Abd-Elaty
Water & Water Structures Engineering
    Department
Faculty of Engineering
Zagazig University
Zagazig, Egypt

Zohra Abdelkrim
Institute of Urban Techniques Management
University of M'sila
M'sila, Algeria

Merve Açar
Istanbul Technical University
Türkiye

Mahmoud M. Afify
Civil Engineering Department
Benha Faculty of Engineering
Benha University
Benha, Egypt

Aziza Ahsan
Department of Environmental Management
Glasgow Caledonian University
Scotland, UK

Oumaima Attar
International Water Research Institute
Mohammed VI Polytechnic University (UM6P)
Benguerir, Morocco

Sabrina N. Ayala
Centro de Investigaciones del Mar y la
    Atmósfera (CIMA)
Universidad de Buenos Aires
Buenos Aires, Argentina

Joel Bailey
Water and Environmental Research Center
University of Alaska
Fairbanks, USA

Prabal Barua
Department of Knowledge Management for
    Development
Young Power in Social Action
Chittagong, Bangladesh

Lahcen Benaabidate
Laboratory of Functional Ecology and
    Environmental Engineering
Faculty of Sciences & Techniques
Fez, Morocco

Lhoussaine Bouchaou
Sciences Faculty of Gafsa, Earth sciences
    Department
University of Gafsa
Tunisia

Youssef Brouziyne
International Water Management Institute
    (IWMI)
MENA Office
Giza, Egypt

Ferat Çağlar
Istanbul Technical University
Türkiye

Abdelghani Chehbouni
Unité mixte de recherche (UMR) Centre
    d'études spatiales de la biosphère (Cesbio)
Institut de recherche pour le développement
    (IRD)
Toulouse, France

Davoud Khodadadi Dehkordi
Department of Water Engineering and
    Sciences
Ahvaz Branch
Islamic Azad University
Ahvaz, Iran

Adebayo Oluwole Eludoyin
Department of Geography
Obafemi Awolowo University
Ile-Ife, Nigeria

Saeid Eslamian
Department of Water Science and Engineering
College of Agriculture, Isfahan University of
    Technology
Isfahan, Iran

Ali Essahlaoui
Department of Geology
Laboratory of Geoengineering and
 environment
Faculty of Sciences, Moulay Ismail University
Meknès, Morocco

Bachir Faid
Institute of Urban Techniques Management
Mohamed Boudiaf University
M'sila, Algeria

Jihui Fan
Institute of Mountain Hazards and Environment
Chinese Academy of Sciences
Sichuan, China

Majid Galoie
Civil Engineering Department
Imam Khomeini International University
Qazvin, Iran

Eugenia M. Garbarini
Centro de Investigaciones del Mar y la
 Atmósfera (CIMA)
Universidad de Buenos Aires
Buenos Aires, Argentina

Albrecht Gnauck
Ecosystem and Environmental Informatics
Brandenburg, Germany

Marcela H. González
Centro de Investigaciones del Mar y la
 Atmósfera (CIMA)
Universidad de Buenos Aires
Buenos Aires, Argentina

Erdem Görgün
Istanbul Technical University
Türkiye

Ceren Ballı Gözen
Istanbul Technical University
Türkiye

Mohammed El Hafyani
Department of Geology
Laboratory of Geoengineering and
 Environment
Faculty of Sciences, Moulay Ismail University
Meknès, Morocco

Faiza Hallouz
Department of Geography and Territorial
 Planning
Faculty of Earth Sciences, Geography and
 Territorial Planning
University of Science and Technology
Houari BOUMEDIENE, Algiers

Younes Hamed
Unité mixte de recherche (UMR)
 Centre d'études spatiales de la biosphère
 (Cesbio)
Institut de recherche pour le
 développement (IRD)
Toulouse, France

Asude Hanedar
Namik Kemal University
Türkiye

Md Hasibul Hasan
School of Human Evolution and Social
 Change
Arizona State University
AZ, USA

Mir Bintul Huda
Water Resources Management Center
National Institute of Technology (NIT)
Srinagar, India

Amir S. Ibrahim
Civil Engineering Department
Benha Faculty of Engineering
Benha University
Benha, Egypt

Azlyana Syahirah binti Haji Ibrahim
Department of Geography, Environment and
 Development Studies
Faculty of Arts and Social Sciences, University
 of Brunei Darussalam
Bandar Seri Begawan, Brunei Darussalam

Arshib Imtiaj
Department of Hydro Sciences
Faculty of Environmental Sciences
Technische Universität Dresden
Dresden, Germany

Selahattin İncecik
Istanbul Technical University
Türkiye

Shafi Noor Islam
Geography, Environment and Development
 Studies
Faculty of Arts and Social Sciences (FASS)
Universiti Brunei Darussalam
Bandar Seri Begawan, Brunei Darussalam

Faezeh Jannesary
Department of Water Science and Engineering
College of Agriculture, Isfahan University of
 Technology
Isfahan, Iran

Fatemeh Jannesary
Department of Nanotechnology Engineering
Faculty of Chemistry, University of Isfahan
Isfahan, Iran

John Keech
Water and Environmental Research Center,
University of Alaska
Fairbanks, USA

Neema Kumburu
Department of Management
Moshi Co-operative University
Moshi, Tanzania

Eric LaMesjerant
Water and Environmental Research Center
University of Alaska
Fairbanks, USA

Mousa Maleki
Department of Civil, Architectural, and
 Environmental Engineering
Illinois Institute of Technology
Chicago, United States

Arindan Mandal
Jawaharlal Nehru University
New Delhi, India

Mohamed Meddi
Ecole Nationale Supérieure d'Hydraulique de
 Blida
Blida, Alegria

Artemis Motamedi
Civil Engineering Department
Technical University of Buein Zahra
Qazvin, Iran

Fouad Moudden
Department of Geology, Laboratory of
 Geoengineering and environment
Faculty of Sciences
Moulay Ismail University
Meknès, Morocco

Mahdi Moudi
College of Management
Chengdu University of Information Technology
Chengdu, China

Abolfazl Nasseri
Agricultural Engineering Research
 Department
East Azarbaijan Agricultural and Natural
 Resources Research and Education Centre,
 AREEO
Tabriz, Iran

Luka Sabas Njau
Department of Gender and Community
 Development
Moshi Co-operative University
Moshi, Tanzania

Paula C. Oliveri
Centro de Investigaciones del Mar y la
 Atmósfera (CIMA)
Universidad de Buenos Aires
Buenos Aires, Argentina

Victor Otti
Civil Engineering Department
Federal Polytechnic
Oko, Nijeria

Abdelhadi El ouali
Department of Geology, Laboratory of
 Geoengineering and environment
Faculty of Sciences, Moulay Ismail University
Meknès, Morocco

Alagappan Ramanathan
Jawaharlal Nehru University
New Delhi, India

Md. Mostafizur Rahman
Department of Urban and Regional Planning
Rajshahi University of Engineering &
 Technology
Rajshahi, Bangladesh

Shahriar Rahman
School of Natural Sciences
Macquarie University
Australia

Sanzida Rahman
Department of Environmental Management
Glasgow Caledonian University
Scotland, UK

Sandra Reinstädtler
Faculty of Environment and Natural
   Sciences
Brandenburg University of Technology (BTU)
Cottbus-Senftenberg, Germany

Alfredo L. Rolla
Centro de Investigaciones del Mar y la
   Atmósfera (CIMA)
Universidad de Buenos Aires
Buenos Aires, Argentina

Ahmad Lotfi-Sadigh
Department of Water Engineering
University of Tabriz
Tabriz, Iran

Monica Sharma Shamurailatpam
Jawaharlal Nehru University
New Delhi, India

Zekiye Şengul
Department of Agricultural Economics
Faculty of Agriculture
Siirt University
Siirt, Turkey

Abdelaziz El Shinawi
Environmental Geophysics Lab (ZEGL),
   Geology Department
Faculty of Science, Zagazig University
Zagazig, Egypt

Horacio Toniolo
Water and Environmental Research
   Center
University of Alaska
Fairbanks, USA

Md. Salah Uddin
Sonali Bank Ltd. Chapai Nawabganj
Rajshahi, Bangladesh

Yurdanur S. Ünal
Istanbul Technical University
Türkiye

Emily Youcha
Water and Environmental Research Center
University of Alaska
Fairbanks, USA

Part I

South and North America

1 Watershed Management Case Studies in the United States and Canada

Saeid Eslamian and Mousa Maleki

1.1 INTRODUCTION

Rivers no longer flow directly across the landscape without barriers to infrastructure (Surian and Rinaldi 2003; Pinter and Heine 2005), and the majority of rivers no longer show the historical extent of their change (Poff et al. 2007; Postel and Richter 2012). However, flow variability over time and space is a key feature of natural rivers, and river "flow regime," in coordination with sediment inlets, determines not only the geomorphic settings but also the composition and rate of key ecosystem processes, such as primary production (Poff et al. 1997). These settings, which include lateral canal migration and dynamic interactions between the riverbed, the floodplain, and the coastal area, are part of a healthy river response to changes in the surrounding landscape and changes in discharge. Indeed, these settings allow rivers to absorb disturbances and protect the surrounding ecosystem and landslides from floods and human impacts. However, the ability of rivers and their organisms to respond to altered flow regimes is not unlimited. Changes due to urbanization, over-harvesting of water, or climate change, that occur rapidly and lead to flows outside the natural range of diversification, will have important consequences for river ecosystems and their dependents (IJC 2001; Maleki et al. 2022; Norman and Melious 2004; Pinter and Heine 2005).

The biodiversity and productivity of native rivers may decline, water quality for human consumption may be compromised, and, in some areas, the risk of flooding, along with damage to property and community, may grow (Bunn and Arthington 2002; Maleki et al. 2022; Allan et al. 2005; Mir et al. 2022).

While the effects of global change on water access and sustainability of river ecosystem services have been considered (Shaikh et al. 2022; Alcamo et al. 2003; Lettenmaier et al. 1999; RRBC 2010; Mir et al. 2021a), the overlapping effects of water change and atmosphere and the effects of dams and other human infrastructure have not been studied on a global scale (Poff and Hart 2002).

Anticipating and planning adaptive strategies for many under-stressed rivers might be critical due to over-water abstraction or land development, and this stress may be exacerbated by climate change (Huitema et al. 2016). Identifying and prioritizing actions that can now be taken to increase the resilience of river ecosystems in the face of disturbances may minimize effects such as biodiversity loss or severe flooding (Palmer et al. 2007). Even if no action is taken at this time, identifying coping options should help prepare communities to deal with climate-related problems at the time of emergence.

The rule of climate adaptation shows that theoretical explanations for how and why barriers to adaptation, continuity, and change are created over time have not been developed (Biesbroek et al. 2013; Eisenack et al. 2014; GNEB 2010; Paavola 2007).

This is surprising because climate change is likely to alter interdependencies between actors, such as interdependencies between upstream and downstream water users or between water service providers, spatial planners, and water users (Naiman et al. 2005; Postel and Richter 2012; Mir et al. 2021b).

The river basin is also known as a catchment or a drainage basin. A river basin can have smaller sub-basins that combine to form a larger water basin. When it rains or ice and snow melt, the water that comes from it flows into the river basin, lakes, oceans, or the sea before leaving.

Rocks, mountains, and hills often separate river basins. Water that comes from a variety of sources, such as rivers, ponds, streams, rain, or melting snow and ice, flows through them and into a basin of water that enters another water, which is usually larger.

These types of lands are called watersheds, which are fissures or elevations that separate a river basin or catchment area. They are also known as drainage divisions because they separate the river system or river basin from other river systems. Watershed is a definition that refers to a basin that is smaller in size and flows into a smaller outlet such as a creek or lagoon. Watersheds are part of a watershed.

River basins and watersheds have different functions in ecology, although they are both landforms. One collects water from different sources such as house drainage, rainfall, and other surface waters and moisture. Another one divides the collection point or river basin where all the water from different sources is collected (Eslamian et al. 2023).

1.2 RIVER BASINS OF NORTH AMERICA

River basins play a vital role in shaping the hydrological and ecological systems of a region, providing essential resources for both human and natural communities. North America, with its diverse landscapes and extensive river networks, harbors numerous river basins that contribute to the continent's water availability, biodiversity, and socioeconomic activities. This scientific text provides an overview of the major river basins in North America, highlighting their characteristics, ecological significance, and human impacts.

1.2.1 MEXICO

1.2.1.1 Rio Grande Basin

The Rio Grande Basin forms part of the border between Mexico and the United States. It originates in the southern Rocky Mountains and flows through the arid landscapes of New Mexico and Texas before reaching the Gulf of Mexico. The basin faces challenges such as water scarcity, overuse, and conflicts between water demands from agriculture, urban areas, and environmental needs.

1.2.1.2 Balsas River Basin

The Balsas River Basin is located in south-central Mexico. It is the third-largest river basin in the country and supports diverse ecosystems. The basin is essential for agriculture and provides water for irrigation and hydroelectric power generation.

1.2.1.3 Lerma–Chapala Basin

The Lerma–Chapala Basin is situated in central Mexico and includes the Lerma River and Lake Chapala, the largest freshwater lake in Mexico. The basin is important for water supply, agriculture, and biodiversity conservation.

1.2.1.4 Grijalva–Usumacinta Basin

The Grijalva–Usumacinta Basin is in southern Mexico and includes the Grijalva and Usumacinta Rivers. It is characterized by tropical rainforests and supports a rich variety of flora and fauna. The basin is significant for hydroelectric power generation and provides water for irrigation and industrial use.

1.2.2 United States

1.2.2.1 Mississippi River Basin

The Mississippi River Basin is the largest river basin in North America, covering approximately 40% of the United States. It stretches from the Rocky Mountains in the west to the Appalachian Mountains in the east and drains into the Gulf of Mexico. The basin is of immense ecological importance, supporting diverse ecosystems and playing a vital role in transportation, agriculture, and hydroelectric power generation.

1.2.2.2 Colorado River Basin

The Colorado River Basin is located in the southwestern United States and northwestern Mexico. It spans seven US states and two Mexican states. The basin supplies water for irrigation, municipal use, and hydropower generation. It faces challenges such as water scarcity, over allocation, and ecological degradation.

1.2.2.3 Columbia River Basin

The Columbia River Basin extends from the Canadian Rockies in British Columbia, Canada, to the Pacific Ocean through the states of Washington and Oregon in the United States. It is one of the largest river basins in North America and is known for its hydroelectric power production. The basin supports diverse ecosystems and plays a vital role in transportation and agriculture.

1.2.3 Canada

1.2.3.1 Mackenzie River Basin

The Mackenzie River Basin is the largest river basin in Canada, covering an area of about 1.8 million km^2. It originates in the western part of the country and flows into the Arctic Ocean. The basin is characterized by vast expanses of boreal forests, tundra, and wetlands. It supports a rich diversity of wildlife and is an essential source of freshwater.

1.2.3.2 Fraser River Basin

The Fraser River Basin is located in British Columbia, Canada. It is the largest river basin in the province and supports diverse ecosystems, including temperate rainforests and salmon-bearing rivers. The basin is important for fisheries, transportation, and hydroelectric power generation.

1.2.3.3 Saint Lawrence River Basin

The Saint Lawrence River Basin spans parts of Canada and the United States. It connects the Great Lakes to the Atlantic Ocean and is a vital transportation route. The basin supports various economic activities, including shipping, tourism, and hydroelectric power generation.

The river basins in Mexico, the United States, and Canada are diverse and significant ecosystems that provide water resources, support biodiversity, and sustain various human activities. These basins face challenges such as water scarcity, pollution, and the impacts of climate change. It is crucial to implement sustainable water management practices and promote conservation efforts to protect and preserve these vital river basins for the benefit of present and future generations.

1.3 RIVER BASINS MANAGEMENT IN NORTH AMERICA

A useful field for applied environmental analysis is the watershed, which is described by the US Environmental Protection Agency as a "geographic area where water, sediment, and solutes are discharged to a common outlet – one point at a time. A larger river, a lake, an underwater aquifer, or an ocean" (RRBC 2010). More scientists are doing research on both border regions concerning bi-national watershed management, offering both a strong regional context and a valid science-based approach to the challenges introduced earlier in the paper.

Environment Canada has been instrumental in promoting integrated watershed management (Environment Canada 2010). Before examining the management of bi-national water resources, a brief overview of the existing internal management frameworks in the United States and Canada is helpful. The Clean Water Act states that water quality management is of national interest and responsibility and that the law provides a framework of regulations and measures that control pollution and protect water quality. In Table 1.1, the characteristics of the US and Canada's watersheds are summarized.

The internal management framework for water resources in Canada is similar to the United States, but also significantly different. As a federation of provinces and regions, Canada promotes the management of water resources through the actions of federal, provincial, and regional governments. Active management of water resources is within the formal authority of the province's constitution, and management in the territories is shared between the territorial. Special responsibilities managed by provincial and regional governments include the development of water management, hydroelectric infrastructure, and pollution control. The federal role in Canada is largely limited to ocean management and fisheries resources, navigation, First Nations issues, and the international context (Environment Canada 2010). The impacts of watershed management systems in the United States and Canada are mentioned in Table 1.2.

TABLE 1.1

Watershed Characteristics in the United States and Canada

Watershed Name	Area (km²)	Major Rivers	Land Use	Population
Columbia River	668,000	Columbia	Agriculture, forestry, mining, urban	2.5 million
Mississippi River	3,200,000	Mississippi, Missouri	Agriculture, urban, mining	50 million
Fraser River	220,000	Fraser	Agriculture, forestry, urban	2.5 million
Great Lakes	244,000	St. Lawrence, Niagara, Detroit, etc.	Fishing, recreation, shipping	35 million

Source: Author.

TABLE 1.2

Impacts of Watershed Management Systems in the United States and Canada

Watershed Name	Management System	Benefits	Challenges
Columbia River	Watershed Councils	Improved water quality, stakeholder collaboration	Inadequate funding, slow decision-making
Mississippi River	Nutrient Management Strategies	Reduced nutrient pollution, improved fish habitat	Opposition from agricultural industry, limited enforcement
Fraser River	Habitat Restoration Projects	Increased salmon populations, improved riparian habitat	Limited success in urban areas, lack of long-term monitoring
Great Lakes	International Joint Commission	Improved water quality, reduced toxic pollutants	Limited authority, insufficient resources

Source: Author.

1.4 THE US–MEXICO EXPERIENCE

The most important resource management issues occur in the US–Mexico border region, specifically twin cities within or adjacent to a bi-national watershed. Accordingly, the watershed approach offers significant application in the US–Mexico border areas. Challenges facing the region include adequate water supply, wastewater management, shared water resources management, and lack of groundwater management mechanisms. An extensive water resources management framework has been established over the past 125 years, and significant progress has been made on many of the issues of water quantity and quality that the two countries face. Examining some recent bi-national efforts helps to understand what efforts have been effective and why.

The Southwest Consortium for Environmental Research and Policy (SCERP) is an effort by the US Environmental Protection Agency (USEPA), between US and Mexican universities to support bi-national environmental research efforts and is a prime example of the "science-based policies" discussed earlier in this chapter. SCERP has supported a number of water resources research efforts over the past 20 years that have used the basin or watershed landscape in the US–Mexico joint watershed. In the Tijuana River Basin, several research projects have examined the vulnerability of water resources and related indicators within the basin, which demonstrates the usefulness of bi-national research teams and the watershed landscape.

The US Geological Survey (USGS), in partnership with other US and Mexican agencies, established the US–Mexico Border Environmental Health Initiative (BEHI) to study the interactions between human health, environmental factors, population growth, and economic development (Norman et al. 2010).

BEHI researchers have created a geographic database of human health and environmental variables in the border area and present this data on the web. BEHI research also includes the interaction between pollutant levels, human health, and wildlife species in transboundary watersheds (Norman et al. 2010). The successful results of this study demonstrate the long-term commitment of USGS staff to conduct joint research.

The International Boundary and Water Commission (IBWC) meets with stakeholders and shares information with border residents through the IBWC Citizens Forums in five sub regions; topics covered include water quality concerns related to IBWC wastewater treatment plants, flood risk, and operation of regional water management projects. To date, these efforts have been successful in sharing information, but the author has long argued that they also have two major drawbacks. First, they are merely US domestic efforts within the IBWC, with little formal involvement from the Mexican sector, Community Integrated Living Arrangement (CILA), or other Mexican agencies. Second, the forums are established solely as information-sharing mechanisms; little opportunity is afforded for regional collaboration or decision-making, efforts that are consistent with the idea of watershed councils. The other IBWC efforts in area-specific technical cooperation have been more successful, but cross-border efforts by Citizens' Halls have had limited success.

The latest example of joint US–Mexican environmental work under discussion is the Good Neighbor Environmental Board (GNEB). The GNEB was established through the Enterprise for the Americas Initiative Act, which was passed in 1992 to provide technical and input advice to the US Congress and President on cross-border environmental infrastructure issues (Eslamian et al. 2023). Although this effort is a US internal consultation mechanism, the GNEB has partnered with Mexican citizens to conduct research in support of annual reports and comments sent to the US President and senior members of Congress.

Annual reports on water resource management discuss human health, air pollution, border security, and environmental protection concerns in the post-911 periods, and provide research and thematic data on members of Congress, the government, and their staff. The comments focus on more urgent and sometimes contentious issues, and discuss the management and financing of the Commission on Border Environmental Cooperation and the North American Development Bank, the critical evaluation of border security infrastructure, and the increased Mexican involvement in the delegation.

What lessons can be learned from the experience of the United States and Mexico that can shed light on the North American perspective? First, successful endeavors require an initial and ongoing commitment to working together and bi-nationally. The results of the SCERP and USGS efforts clearly illustrate this finding, which the GNEB and IBWC are working hard to advance. Second, many members of bi-national research teams have worked together for years, and these connections and personal relationships are key to advancing this. Third, effective bi-national environmental research and management requires significant and sustained financial commitments. The USGS BEHI effort has been well received for several years. It supports the collection and integration of spatial data and collaborative research. Conversely, SCERP has received less sustainable funding over the past five years, which has negatively affected the ability of SCERP researchers to continue their work.

The last point to draw from the US–Mexico experience is the productive nature of university researchers. Work done by SCERP is largely conducted by researchers at member universities, and the GNEB, IBWC Citizen Forums, and the USGS efforts have benefited from the support, interest, and hard work of university partners.

1.5   THE US–CANADA EXPERIENCE

Although Canada is very well endowed with water resources, the northern US and Canada face similar challenges as Mexico in meeting the demand for adequate and safe water resources. With only 5.5% of the world's population, Canada contains 7% of the world's freshwater supply (Environment Canada 2010), but the lack of space coordination between the place of population and water resources will produce challenges for Canada in the demand for water slow down. This situation sees considerable regional variability, but various sub-regions of the US–Canada border face challenges in this regard. Water quality issues are also evident, as urbanization, population increase, and industrial, municipal, and agricultural demands generate impacts on water quality.

A brief review of the history of bi-national water resource management between the United States and Canada is a useful starting point for this discussion. The Boundary Waters Treaty of 1909 established the International Joint Commission (IJC) as a bi-national institution with primary responsibility for managing the shared waters in the Great Lakes and boundary river systems. Each country has three commissioners and its own section with technical staff, who are responsible for interpreting and following the treaty and resolving relevant disputes, not simply representing the interests of their own countries (Hulme 2005), a departure from many international affairs posts and institutions.

In 1972, the United States and Canada signed the Great Lakes Water Quality Agreement, with the goal of renewing each country's commitment to work together to protect water quality through an ecosystem perspective. In particular, the agreement sets out a series of general and specific objectives; sets research programs and standards; and describes the duties, powers, and responsibilities of the IJC. The agreement was amended by a protocol in 1987 that set timeframes under which programs were implemented, changing technologies were considered, and accountability was strengthened (IJC 1909). The coordinated agreement and protocol guide most IJC-based ecosystem work.

An ecosystem-based approach to water quality work was linked to integrated water resources management through the preparation and publication of IJC reference letters in 1998. In these letters, the governments of Canada and the United States requested the IJC to conduct research to establish international watershed delegations with an emphasis on operational details of structure, membership, and conditions. These letters explicitly endorse the proposal to "establish international watershed delegations to adopt an integrated ecosystem approach to transboundary environmental issues". In doing so, the US and Canadian federal governments supported a science-based bi-national policy to manage water resources.

The IJC is not the only bi-national organization in the United States and Canada to support such an approach. In 2002, the Commission for Environmental Cooperation (CEC) hosted a workshop entitled "Public Workshop on Freshwater Issues in North America" in which the CEC issued recommendations in line with the IJC References of 1998 (CEC 2002). Among the research options discussed were structures for effective integrated transboundary water management. The debate at the CEC also extended to the question of how the IJC and IBWC might contribute to such an outcome. All parties to the conference supported efforts to "collaborate on common approaches or lessons learned" that are at the heart of the transnational management's cross-border management efforts (CEC 2002).

What has happened on Earth that might provide lessons learned for the North American context? The International Committee of the Red Cross and Red Crescent Commission are making significant efforts to coordinate bi-national and integrated water management on the US–Canadian border. The Red River International Board is an IJC board that combines current activities with the International Engineering Board of Soris–Red Rivers and the International Red River Pollution Board to promote pollution control and water-sharing activities (IJC 1972). In particular, the Board should assist the IJC in addressing water quality conflicts in the basin, "by using the best available knowledge about the aquatic ecosystem and being aware of the needs, expectations and capabilities of the residents. Red River Basin" (IJC 1972; Milly et al. 2005). This is highly consistent with the science-based policy introduced earlier in this chapter.

The other organization which is working in the basin is the Red River Basin Commission (RRBC), a non-profit group that seeks to bring together state, local, provincial, and county governments to better manage the basin's natural resources. The RRBC was formed in 2002 when the Red River Basin Board, the International Coalition, and the Red River Water Resources Council joined forces to advance common goals. In 2005, the commission drafted the Red River Basin Natural Resources Framework, which set 13 goals for water quality protection, flood management, water supply, and wildlife protection (Poff et al. 2007). This local effort is a bi-national watershed approach that is consistent with the 1998 reference letters calling for an integrated watershed.

What experiences can be learned from the experience of the United States and Canada that can reflect the North American perspective? An important insight is the proactive approach by which the IJC has developed a science-based ecosystem approach to shared water management on the US–Canada border. This view was incorporated into the Great Lakes Water Quality Agreement in early 1972 and reinforced by reference letters in 1998 calling for the development of international watershed symbols. As introduced at the CEC workshop in 2002, such a view may have significant application in the US–Mexico border region, which faces somewhat similar challenges. The discussion now turns to other ideas for future research.

Table 1.3 provides an overview of the different funding sources available to support watershed management programs in the United States and Canada. These funding sources include federal and

TABLE 1.3

Funding Sources for Watershed Management Programs in the United States and Canada

Funding Source	Description	Examples
Federal Grants	Funds from the federal government to support watershed management	Clean Water Act Section 319, Environmental Quality Incentives Program
State/Province Grants	Funds from state or provincial governments to support watershed management	Watershed Restoration and Protection Program, Mississippi River Basin Healthy Watersheds Initiative
Private Foundations	Funds from private organizations to support watershed management	National Fish and Wildlife Foundation, The Nature Conservancy

state/provincial grants, private foundations, and corporate partnerships. Each of these sources has its own strengths and limitations, and understanding the funding landscape is essential for developing effective watershed management strategies. For example, federal grants can provide significant funding to support large-scale projects but may be subject to changes in political priorities. Private foundations can provide more flexible funding but may have narrower focus areas. Corporate partnerships can provide significant funding and access to expertise but may raise concerns about conflicts of interest. By understanding the range of funding sources available, watershed managers can develop more effective funding strategies and better leverage available resources to support their programs.

1.6 FIELDS OF FURTHER RESEARCH

In the context of watershed management, there are several key areas that could benefit from further research. These might include:

- **The impacts of climate change on watershed health and management:** Climate change is expected to have significant effects on water availability, quality, and flow patterns in watersheds. Further research is needed to better understand these impacts and develop effective management strategies to mitigate them.
- **The effectiveness of different management approaches:** There are many different strategies for managing watersheds, including regulatory frameworks, market-based approaches, and community-based initiatives. Further research is needed to evaluate the effectiveness of these approaches and identify best practices for achieving desired outcomes.
- **The role of stakeholder engagement in watershed management:** Effective stakeholder engagement is critical for successful watershed management, but there is limited understanding of how to engage stakeholders in ways that promote collaboration and achieve consensus. Further research is needed to develop effective strategies for stakeholder engagement in watershed management.
- **The socioeconomic impacts of watershed management:** Watershed management strategies can have significant impacts on local communities, including impacts on livelihoods, access to natural resources, and quality of life. Further research is needed to better understand these impacts and develop strategies for mitigating negative effects.
- **The integration of traditional ecological knowledge into watershed management:** Traditional ecological knowledge (TEK) can provide valuable insights into the ecological and cultural significance of watersheds. Further research is needed to better understand how TEK can be integrated with scientific knowledge to inform watershed management decisions and promote greater recognition of indigenous knowledge systems.
- **The use of emerging technologies in watershed management:** New technologies, such as remote sensing, machine learning, and sensor networks, are increasingly being used in watershed management. Further research is needed to evaluate the effectiveness of these technologies and identify best practices for their use.

1.7 CONCLUSIONS

Although the human and physical geography of North America is very different between the three countries on the continent, a comparative study of the management and policy of bi-national water resources offers interesting and useful insights. Important elements of regional management frameworks existing in the US–Mexico and US–Canada border areas have been identified and explored, with some basic ideas on how lessons learned in one area may be used and of interest in another. Ideas for future research efforts that can build on the work described in this chapter will serve as a basis for such future research agenda.

REFERENCES

Alcamo, J., Döll, P., Henrichs, T., Kaspar, F., Lehner, B., Rösch, T., and Siebert, S. (2003). Development and testing of the WaterGAP 2 global model of water use and availability. *Hydrological Sciences Journal*, 48(3), 317–337.

Allan, S. M., Tyrrell, P. J., and Rothwell, N. J. (2005). Interleukin-1 and neuronal injury. *Nature Reviews Immunology*, 5(8), 629–640.

Biesbroek, G. R., Klostermann, J. E., Termeer, C. J., and Kabat, P. (2013). On the nature of barriers to climate change adaptation. *Regional Environmental Change*, 13(5), 1119–1129.

Bunn, S. E., and Arthington, A. H. (2002). Basic principles and ecological consequences of altered flow regimes for aquatic biodiversity. *Environmental Management*, 30(4), 492–507.

Commission on Environmental Cooperation (CEC). (2002). Summary of a workshop hosted by the CEC entitled, "Public Workshop on Freshwater issues in North America" on 3 October 2002 in Albuquerque, NM.

Eisenack, K., Moser, S. C., Hoffmann, E., Klein, R. J., Oberlack, C., Pechan, A., Rotter, M., and Termeer, C. J. (2014). Explaining and overcoming barriers to climate change adaptation. *Nature Climate Change*, 4(10), 867–872.

Environment Canada. (2010). Government of Canada, Ottawa. http://www.ec.gc.ca/eau–water/default.asp ?lang=En&n=FDF30C16–1.

Eslamian, S., Roknizadeh, H., Maleki, M., and Koupai, J. A. (2023) Surface runoff water harvesting for irrigation. In *Handbook of Irrigation Hydrology and Management*, Ed. S. Eslamian and F. Eslamian. CRC Press, Boca Raton, 323–334.

Good Neighbor Environmental Board (GNEB). (2010). Website for the Good Neighbor Environmental Board, maintained by the U.S. Environmental Protection Agency, [On line]. http://www.epa.gov/ocem/gneb -page.htm, site. Accessed 29 December 2010.

Huitema, D., Adger, W. N., Berkhout, F., Massey, E., Mazmanian, D., Munaretto, S., Plummer, R., and Termeer, C. C. (2016). The governance of adaptation: Choices, reasons, and effects. Introduction to the Special Feature. *Ecology and Society*, 21(3), 165–181.

Hulme, P. E. (2005). Adapting to climate change: Is there scope for ecological management in the face of a global threat?. *Journal of Applied Ecology*, 42(5), 784–794.

International Joint Commission (IJC). (1909). Boundary waters treaty of 1909, [On line]. http://www.ijc.org/ en/home/main_accueil.htm. Accessed 30 December 2009.

International Joint Commission (IJC). (1972). IJC great lakes water quality agreement of 1972 [On line]. http:// www.ijc.org/rel/agree/quality.html.

International Joint Commission (IJC). (2001). IJC Directive of 7 February 2001 calling for the establishment of the International Red River Board [On line]. http://www.ijc.org/conseil_board/red_river/en/ irrb_mandate_mandat.htm. Accessed 29 December 2009.

Lettenmaier, D. P., Wood, A. W., Palmer, R. N., Wood, E. F., and Stakhiv, E. Z. (1999). Water resources implications of global warming: A US regional perspective. *Climatic Change*, 43(3), 537–579.

Maleki, M., Eslamian, S., Mustafa, F. B., and Madadi, M. (2022). Flood and building damages. In *Flood Handbook*, Eds. S. Eslamian and F. Eslamian. CRC Press, Boca Raton, 303–322.

Milly, P. C., Dunne, K. A., and Vecchia, A. V. (2005). Global pattern of trends in streamflow and water availability in a changing climate. *Nature*, 438(7066), 347–350.

Mir, B., Kumar, R., Lone, M. A., and Tantray, F. A. (2021a). Climate change and water resources of Himalayan region-review of impacts and implication. *Arab J Geosci*, 14, 1088. https://doi.org/10.1007/ s12517-021-07438-z.

Mir, B. H., Lone, M. A., Kumar, R., and Khoshouei, S. R. (2021b). A review on the implications of changing climate on the water productivity of Himalayan glaciers. *Water Productivity Journal*, 1(3), pp. 25–36. doi: 10.22034/wpj.2021.259848.1017.

Mir, B. H., Rather, N. A., and Eslamian, S. (2022). Social aspects of flooding. In *Flood Handbook*, Volume 1, 1st Edition, Ch. 6. CRC Press, Taylor & Francis Group, Boca Raton. https://doi.org/10.1201 /9781003262640-8, ISBN: 9781138584938.

Naiman, R. J., Bechtold, J. S., Drake, D. C., Latterell, J. J., O'Keefe, T. C., and Balian, E. V. (2005). Origins, patterns, and importance of heterogeneity in riparian systems. In Lovett, G. M., Turner, M. G., Jones, C. G., and Weathers, K. C. (Eds.), *Ecosystem Function in Heterogeneous Landscapes*. Springer, New York, 279–309. https://doi.org/10.1007/0-387-24091-8_14.

Norman, E. S., and Melious, J. O. (2004). Transboundary environmental management: A study of the Abbotsford-Sumas aquifer in British Columbia and western Washington. *Journal of Borderlands Studies*, 19(2), 101–119.

Norman, L. M., Callegary, J., van Riper, C., III, and Gray, F. (2010). The border environmental health initiative; investigating the transboundary Santa Cruz watershed. U.S. Geological Survey Fact Sheet 2010–3097, 2p.

Paavola, J. (2007). Institutions and environmental governance: A reconceptualization. *Ecological Economics*, 63(1), 93–103.

Palmer, M., Allan, J. D., Meyer, J., and Bernhardt, E. S. (2007). River restoration in the twenty-first century: Data and experiential knowledge to inform future efforts. *Restoration Ecology*, 15(3), 472–481.

Pinter, N., and Heine, R. A. (2005). Hydrodynamic and morphodynamic response to river engineering documented by fixed-discharge analysis, Lower Missouri River, USA. *Journal of Hydrology*, 302(1–4), 70–91.

Poff, N. L., Allan, J. D., Bain, M. B., Karr, J. R., Prestegaard, K. L., Richter, B. D., Sparks, R. E., and Stromberg, J. C. (1997). The natural flow regime. *BioScience*, 47(11), 769–784.

Poff, N. L., and Hart, D. D. (2002). How dams vary and why it matters for the emerging science of dam removal: An ecological classification of dams is needed to characterize how the tremendous variation in the size, operational mode, age, and number of dams in a river basin influences the potential for restoring regulated rivers via dam removal. *BioScience*, 52(8), 659–668.

Poff, N. L., Olden, J. D., Merritt, D. M., and Pepin, D. M. (2007). Homogenization of regional river dynamics by dams and global biodiversity implications. *Proceedings of the National Academy of Sciences*, 104(14), 5732–5737.

Postel, S., and Richter, B. (2012). *Rivers for Life: Managing Water for People and Nature*. Island Press, Washington, DC.

Red River Basin Commission (RRBC). (2010). RRBC website detailing the history and development of the Red River Basin Commission [On line]. http://www.redriverbasincommission.org/. Accessed 6 January 2010.

Shaikh, M., Lodha, P., and Eslamian, S. (2022). A framework for climate change impact assessment and mitigation of hydrological parameters of the watershed through trend analysis and hydrological modeling, Ch. 6. In Eslamian et al. (Eds.), *Handbook of Eurasian Forecasting*. Nova Science Publishers, Inc., 79–110.

Surian, N., and Rinaldi, M. (2003). Morphological response to river engineering and management in alluvial channels in Italy. *Geomorphology*, 50(4), 307–326.

2 Sediment Yield in Extreme Cold Environments

The Case of the Sagavanirktok River in Arctic Alaska, USA

Horacio Toniolo, Emily Youcha, Eric LaMesjerant, John Keech, and Joel Bailey

2.1 INTRODUCTION

The hydrologic regime of rivers in the Arctic is different from that of rivers in temperate environments. In the Arctic, severely cold weather throughout winter and the presence of permafrost (perennially frozen soils) create conditions of little to no baseflow, and in spring during breakup, surface runoff is considerable (Lamb and Toniolo, 2016).

However, the sedimentologic regimes of rivers in arctic and temperate regions have features in common that influence sediment availability and sediment transport capacity (Amamra et al., 2018). Such features include the area, relief, geology, and soil cover of the watershed, and the precipitation amount, duration, and intensity (Gordeev, 2006).

In Arctic Alaska, permafrost plays a key role in hydrologic processes; its distribution is continuous and its thickness ranges from 250 to 600 m (Osterkamp and Payne, 1981). Permafrost deepens northward and limits the storage capacity of water in the soils. The active layer (i.e., the zone below the ground surface and above the permafrost table that thaws seasonally) is usually 50 cm deep, but this depth may vary from 25 to 100 cm. As air temperature increases, so does permafrost temperature, which leads to permafrost thaw (Osterkamp and Romanovsky, 1999; Jorgenson et al., 2006; Slater and Lawrence, 2013; Liljedahl et al., 2016). Permafrost thawing coupled with a predicted increase in precipitation (ACIA, 2004; IPCC, 2007) will have an impact on water and sediment fluxes (Jorgenson and Shur, 2007; Gooseff et al., 2009; Grosse et al., 2013; Stuefer et al., 2017).

Many descriptions of the basic hydraulic and sedimentological characteristics of major rivers and tributaries in temperate regions can be found in the literature (see case studies presented in books, such as Chang, 2002; Julien, 2002; Bridge, 2003; ASCE, 2008; and see research articles by Hicks and Mason, 1991; Navratil et al., 2006; Ji et al., 2011, to name a few). Literature that provides data on streams located in extreme cold environments is somewhat limited (i.e., Ashton, 1986; Prowse, 1993; Beltaos and Burrel, 2000; Shiklomanov et al., 2006; ASCE, 2008). Literature that provides data on streams in Arctic Alaska is very limited. In fact, sediment transport processes in extreme cold settings have not been studied as extensively as sediment transport processes in temperate regions. Many reasons account for the difference in research interest. In Arctic Alaska, those reasons include travel distance to rivers, harsh weather conditions, and difficult logistical challenges of working in remote locations (Lamb and Toniolo, 2016; Toniolo, 2020).

Regional changes in climate around the world have been widely reported (see for instance IPCC, 2007; Milliman et al., 2008; Hamilton, 2010; Walsh et al., 2011; Lenton, 2012; Su et al., 2016, and many others). Expected changes in Arctic regions include an increase in precipitation and discharge, permafrost degradation, and increased sediment loads in streams (Piliouras and Rowland, 2020).

DOI: 10.1201/9781003473398-3

The focus of this chapter is a multiyear hydro-sedimentological study of a reach approximately 160 km long on the Sagavanirktok River, located in Arctic Alaska. A review of relevant literature is presented first. A description of the background, study area, fieldwork (including applied methodology), and obtained results follows, and the chapter ends with a discussion and the main conclusions of this study.

2.2 RELEVANT LITERATURE

The review of relevant literature is grouped into two main subjects: sediment transport and climate change in extreme cold regions, and previous sediment transport studies in Arctic Alaska.

2.2.1 SEDIMENT TRANSPORT AND CLIMATE CHANGE IN EXTREME COLD REGIONS

Syvitski (2002) published a seminal article linking climate and sediment flux in Arctic rivers. Syvitski demonstrated that the sediment loads in Arctic rivers are controlled by the watershed's surface temperature. He developed a model capable of predicting sediment load discharged by arctic rivers that are ungauged. The model was applied to nearly 50 Arctic and Subarctic streams. Two rivers in Alaska—the Colville and the Yukon rivers—were considered in the analysis. Syvitski's model predicted a 22% increase in sediment load for every 2°C of warming.

Holmes et al. (2002) reported results from a study that included 19 Arctic and Subarctic rivers, most located in Russia and Canada. A single Alaskan river, the Yukon, was considered in their work. They did not find a long-term trend in the amount of sediment delivered to the Arctic Ocean, but they found stepwise changes in two of the studied rivers. The authors speculated that an increase in sediment flux would be expected in response to an increase in discharge due to climate change.

Gordeev (2006) provided a summary of available literature related to suspended sediment flux from rivers located in Canada and Russia into the Arctic Ocean. The author applied the Morehead et al. (2003) sediment transport model to future climatic scenarios (in terms of increasing temperature) and reported that for every 2°C of warming, an increase of 30% in sediment flux could be expected.

Stuefer et al. (2017) conducted a study of water flux from coastal Arctic watersheds in Alaska. The authors analyzed historical river discharge from 11 streams but focused on 6 small and low-gradient watersheds (Putuligayuk River, Upper Fish Creek, Ublutuoch River, Blackfish Creek, Crea Creek, and SnowNet Creek) located on the Alaska North Slope. All 6 studied watersheds are located west of the Sagavanirktok River watershed. The last year considered in their analysis was 2015. The authors reported an increase in annual discharge and interannual variability. Thus, an increase in sediment load could be expected.

In the context of a warming Arctic, Piliouras and Rowland (2020) conducted a remote sensing analysis of Arctic deltas to characterize their morphologies. Six deltas were studied: the Colville, Kolyma, Lena, Mackenzie, Yenisei, and Yukon. In terms of watershed area and cumulative discharge volume, the Yenisei and Lena deltas encompass the two largest systems in the Arctic. The other deltas studied represent intermediate and small systems. The authors found large variability in channel width, which could indicate different channel dynamics and different mechanisms in delivering sediment to the ocean.

2.2.2 PREVIOUS SEDIMENT TRANSPORT STUDIES IN ARCTIC ALASKA

As indicated by Lamb and Toniolo (2016), one of the first studies of suspended sediment transport in Arctic Alaska was done in 1962 on the Colville River by Arnborg et al. (1967). Total suspended load and dissolved organic material were measured, and the authors determined that 75% of sediment transport occurred during breakup (i.e., initial open water flow in a stream previously frozen). Subsequent studies confirmed the main results from that study. For instance, Braun et al. (2000) reported that 88% of the annual sediment load flowing into Lake Sophia (in Nunavut, Canada) occurred during snowmelt. Forbes and Lamoureux (2005) indicated that 93% of suspended sediment transport occurred during snowmelt on the Lord Lindsay River (in Nunavut, Canada).

Hodel (1986) presented a somewhat detailed hydraulic characterization of the Sagavanirktok River. While bed sediment diameters were presented in this work, no measurements of bed load were made. Hodel reported that the suspended sediment load, measured after the spring breakup discharge peak, was approximately 2500 t/day and 7000 t/day at Sagwon (station located mid-basin, nearly 70 miles (113 km) south of Prudhoe Bay) and West Channel (located in Prudhoe Bay, near the river outlet), respectively.

McNamara et al. (2008) focused on the effects on bed sediment transport conditions of an extreme summer rainfall event (Kane et al. 2003) recorded in August 2002 on the Upper Kuparuk River watershed. The authors reported an estimated volume of 870 m³ of bed sediment moved by that single, and unusual (at the time), precipitation event and indicated that the high flow that resulted moved all sediment sizes up to particles of 0.5 m. As a result of this flood event, the river cross sections showed significant changes.

Dornblaser and Striegl (2009) studied suspended sediment flux to the Bering Sea through the Yukon River. They studied three locations on the Yukon, Tanana, and Porcupine Rivers in the Subarctic and Arctic regions of Alaska from 2001 to 2005. These authors found that approximately 50% of the annual sediment yield happened during spring breakup.

Toniolo et al. (2010) characterized the basic hydraulic conditions of a reach of the Anaktuvuk River near Umiat during spring breakup and reported suspended sediment concentrations. The headwaters of this braided, gravel-bed river are located in the Brooks Range. The vertical relief from the river's headwaters to its confluence with the Colville River is 2063 m. The contributing area upstream of the study reach is 7058 km². Measured suspended sediment concentration (SSC) ranged from nearly 50 mg/L to approximately 380 mg/L.

Toniolo et al. (2013) conducted an extensive fieldwork campaign during spring breakup in 2011 on seven pristine rivers in two topographic regions: the northern foothills of the Brooks Range and the coastal plain within the National Petroleum Reserve in Alaska (NPR-A), a vast area of approximately 92,000 km². The reported SSC values were relatively low for all seven rivers, the maximum value being around 190 mg/L.

The results of a multiyear study that focused on the initial quantification of suspended sediment transport in three pristine rivers (i.e., the Anaktuvuk, Chandler, and Itkillik Rivers) were reported by Lamb and Toniolo (2016). The study was conducted during the open-water season (i.e., from spring breakup to the beginning of freeze-up). The maximum SSC measured was approximately 600 mg/L (Chandler River), 800 mg/L (Itkillik River), and 1600 mg/L (Anaktuvuk River). Calculated monthly sediment yields showed high temporal variability. The authors demonstrated the importance of widespread rainfall events on suspended sediment loads.

Toniolo et al. (2017a) reported discharge and SSC data collected during the 2015 spring breakup on the Sagavanirktok River, approximately 20 miles (32 km) south of Prudhoe Bay. The SSC ranged from a few milligrams per liter (i.e., clear water at the beginning of breakup) to nearly 4500 mg/L.

2.3 BACKGROUND

This multiyear hydro-sedimentological study along an approximately 160 km reach of the Sagavanirktok River began in 2015, in response to unprecedented winter and spring flooding of the river that extensively damaged the Dalton Highway, the only terrestrial route that connects Prudhoe Bay, an oil support town in Arctic Alaska, with the city of Fairbanks in Interior Alaska. As a result of the flooding, the Dalton Highway was closed for nearly 3 weeks. In addition, the extreme event threatened the infrastructure of the Trans-Alaska Pipeline System (TAPS). To restore the roadway, the Alaska Department of Transportation and Public Facilities (ADOT&PF) made extensive repairs along several kilometers of the highway, extracting gravel from the active floodplain to use as road-building material. The ADOT&PF also funded a study, along with the Alyeska Pipeline Service Company (the entity that owns and operates the TAPS), to monitor the Sagavanirktok River.

The study was undertaken by researchers at the University of Alaska Fairbanks (UAF), who installed six hydro-meteorological stations (named consecutively DSS1 to DSS5, and ASS1) in the spring and summer of 2015 along the Sagavanirktok River reach. Water level, runoff, and basic meteorological data are collected at these stations. Five additional meteorological stations were installed at different watershed elevations in the summer of 2015 to improve understanding of the overall hydrology and water balance of the basin. Information from each of the stations is transmitted in near real-time by a complex telemetry system.

Seven pits were excavated in September 2015 near stations DSS1 to DSS4 to examine the river's bed sediment transport processes and natural ability to refill its excavated areas in the river channel and adjacent floodplain. Bathymetric surveys were conducted after the excavations and on a monthly basis during the summers of 2016–2019 to determine bed sediment deposition within the pits.

2.4 STUDY AREA

The Sagavanirktok River originates in the mountains of the Brooks Range and flows north to the Beaufort Sea east of Deadhorse (Figure 2.1). The river's total basin area at DSS5 (the UAF observation station at Franklin Bluffs) is approximately 13,500 km² and lies mostly in the central Brooks Range (>50%). Less than 20% of the basin area is located on the Coastal Plain, and the remaining basin area is in the northern foothills of the Brooks Range. The basin has a low hydraulic gradient near the Arctic Ocean (the Coastal Plain area) and a high hydraulic gradient in the headwaters (the Brooks Range). The basin length is approximately 250 km, and the stream length is at least 300 km. The elevation of the Sagavanirktok River basin ranges from sea level to nearly 2500 m above mean sea level (AMSL), with a mean elevation of 785 m AMSL. The middle to upper Sagavanirktok River basin contains the Ivishak River to the east and the upper Sagavanirktok River to the west. The Ivishak River basin area is nearly equivalent to the upper Sagavanirktok River basin area. According to the National Land Cover Database (Homer et al., 2015), the Sagavanirktok River basin land cover consists of shrubs (43.3%), sedge (14.0%), and wetlands (2.4%); the remaining land cover is classified as barren (37.3%), open water (2.0%), and glacier/perennial snow (1.0%). The Sagavanirktok River is fed by snowmelt, rain, and summer glacial melt. The floodplain varies in width, from narrow at the river's source to several kilometers wide where the river discharges into the Arctic Ocean (Toniolo et al., 2016a). The Sagavanirktok River, which is adjacent to the Dalton Highway and the trans-Alaska pipeline, is characterized by extensive braiding, with some bars vegetated by shrubs (Toniolo et al., 2016b). Since 1983, river discharge has been measured upstream of the Ivishak River confluence at the US Geological Survey (USGS) Sagavanirktok River station (15908000), located near Pump Station 3 (Dalton Highway at Milepost [MP] 325).

The map in Figure 2.1 shows the study area and location of the UAF hydro-sedimentological and meteorological stations. These stations collectively are the observation network. The first station (DSS5—East Bank) was installed during the spring breakup in 2015. In addition to measuring water level, discharge, and meteorology at the stations, suspended sediment samples were collected. Later that summer, four sediment research sites (DSS1, DSS2, DSS3, and DSS4) were installed. All are located on the west bank of the Sagavanirktok River. Happy Valley (DSS3) and MP318 (DSS4) are located upstream of the confluence of the Sagavanirktok and Ivishak Rivers. The contributing areas of DSS3 and DSS4 are 5800 km² and 4725 km², respectively. The other two sediment research sites (DSS1 and DSS2) are located downstream of the confluence of the two rivers: DSS1 is located in the westernmost channel of the lower Sagavanirktok River, approximately 13 km south of Deadhorse at MP405; DSS2 is located approximately 13 km downstream of the confluence with the Ivishak River. The basin area upstream of DSS1 is larger than 13,500 km² (this area was determined

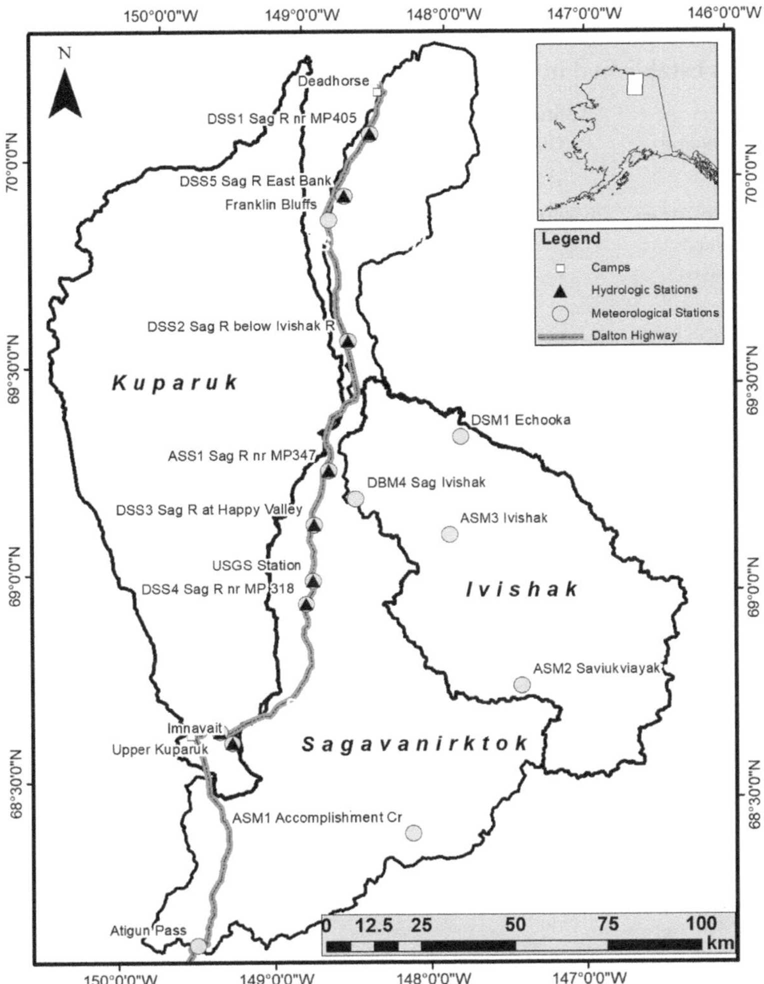

FIGURE 2.1 Sagavanirktok River basin and the hydro-sedimentological and meteorological observation network (from Toniolo et al., 2017b). The black line indicates the Sagavanirktok and Ivishak watershed boundaries.

around 20 km south of the station, near DSS5, where the river bifurcates into east and west channels, with multiple small channels in between). The basin area upstream of DSS2 is 11,790 km^2. The site locations were selected to improve our understanding of the sediment transport conditions along the reach. Observations at the four sediment research sites include meteorological variables, bathymetric surveys, discharge measurements, and specific measurements to quantify bed and suspended sediment. Table 2.1 provides the location of the stations along the Sagavanirktok River and within its watershed. In the summer of 2015, a hydro-meteorological observation station was installed for Alyeska Pipeline Service Company near MP347 (ASS1), 29 km upstream of the Ivishak River confluence. The contributing area of this station is approximately 6375 km^2.

2.5 FIELDWORK

The main field activities, including pit excavation, bathymetric surveys, snow surveys, and discharge measurements, are briefly described in the following paragraphs.

TABLE 2.1

Station Locations Established in the Sagavanirktok River Watershed

Site ID	Station Name	Latitude (WGS84)	Longitude (WGS84)	Elevation (m) (NAVD88)	Data Type
		Hydro-Sedimentological Research Sites			
DSS1	Sagavanirktok River near Deadhorse MP405 site 042	70.09912	−148.50892	26	Air temperature (AT), Relative Humidity (RH), Wind Speed (WS), Wind Direction (WD), Water Levels (WL), Water Temperature (WT) Discharge (Q), Suspended Sediment (SS), Turbidity (TU)
DSS2	Sagavanirktok River downstream of the Ivishak River	69.59580	−148.62600	137	AT, RH, WS, WD, WL, WT, Q, SS, TU
DSS3	Sagavanirktok River at Happy Valley site 005	69.15065	−148.82323	291	AT, RH, WS, WD, WL, WT, Q, SS, TU
DSS4	Sagavanirktok River near MP318 site 066	68.95835	−148.85997	371	AT, RH, WS, WD, WL, WT, Q, SS, TU, Rainfall (P)
		General Hydrological or Meteorological Monitoring			
DSS5	Sagavanirktok River East Bank	69.9461	−148.6714	60	AT, RH, WS, WD, WL, WT, Q
	Franklin Bluffs Repeater at Sagavanirktok River East Bank (DSS5)	69.9459	−148.6538	100	WS, WD
DSM1	Echooka Met	69.3709	−147.8546	335	AT, RH, WS, WD, Soil Temperature (ST), Barometric Pressure (PB), P, Net Radiation (NR), Snow Depth (SD),
ASM3	Ivishak Met (Alyeska)	69.13382	−147.92338	401	AT, RH, WS, WD, PB, NR, SD, P
DBM4	Sag–Ivishak Met	69.2156	−148.5519	431	AT, RH, WS, WD, ST, PB, NR, SD, P
ASM2	Saviukviayak Met (Alyeska)	68.7705	−147.4323	945	AT, RH, WS, WD, PB, NR, SD, P
ASM1	Accomplishment Creek Met, formerly DBM1 (Alyeska)	68.4116	−148.1365	1458	AT, RH, WS, WD, ST, NR, SD, P

2.5.1 EXCAVATION

As mentioned earlier, in September 2015 pits were excavated at four locations in the Sagavanirktok River channel and active floodplain. The pits were 2–3 m in depth and 10–20 m in diameter. Pits were excavated in shallow water near the edge of the active river channel, as well as in gravel bars near the active main channel. Approximately 80 to 350 m^3 of sediment was extracted from the river channel at each pit.

2.5.2 SURVEYING

At each of the four sediment research sites, survey control was established using standard static GPS practices. At least 4 hours of observation data were collected and then processed using the National Geodetic Survey (NGS) OPUS (Online Positioning User Service). The OPUS solution formed the basis for horizontal and vertical control at each site. The horizontal coordinates used were calculated as Alaska State Plane Zone IV, and the vertical datum used was NAVD88, Geoid 12B.

Immediately after the excavation of each test pit was completed in September 2015, the pit and the adjacent area were surveyed to record the initial state. This survey was performed using the real-time kinematic (RTK) GPS survey technique, employing the control established above. An echo sounder was incorporated with the RTK GPS rover, and this was attached to a kayak or jet powerboat to record the bathymetry in and around the pits. Oftentimes, during periods of low flow, pits, and adjacent areas could be surveyed on foot using the rover alone.

The initial surveys of September 2015 provided pit volume calculations and surrounding bathymetry as a basis to track deposition and erosion at the pits and river channels that had been excavated. Beginning in June 2016, the pits were resurveyed at approximately 1-month intervals throughout the summer (three surveys per season) to monitor changes.

2.5.3 SNOW SURVEYS

Snow surveys were conducted every April (2016–2019) at the locations of historical and newly established snow survey sites within the Sagavanirktok River basin (Kane et al., 2006; Berezovskaya et al., 2007, 2008, 2010a, 2010b; Stuefer et al., 2011, 2012, 2014) to estimate the end-of-winter snow water equivalent (SWE), which is defined as follows:

$$SWE = \left(\frac{h_s * \rho_s}{\rho_w} \right) \tag{2.1}$$

where h_s denotes snow depth and ρ_s and ρ_w denote snow and water density, respectively.

To measure SWE, the double sampling technique was used (Rovansek et al., 1993). In this sampling technique, 5 snow density and 50 snow depth measurements are collected. Snow cores for density measurements were collected using an Adirondack tube, and snow depths were measured along an L-shaped transect, with samples approximately 1 m apart (Derry et al., 2009).

2.5.4 WATER LEVEL MEASUREMENTS

Vented and non-vented pressure transducers were used to measure water levels (also referred to as river stage) every 15 minutes at each station. Errors associated with the pressure transducer itself are typically less than 1 cm. Point measurements of the stage were collected using traditional surveying equipment. Benchmarks at each station were established; the vertical datum is NAVD88 (GEOID12A). Automated cameras installed at the stations took hourly photographs of the river at each site. The images were used to qualitatively evaluate the river stage, corroborate pressure transducer data, and observe river ice conditions.

2.5.5 DISCHARGE MEASUREMENTS

Acoustic Doppler current profiler (ADCP) techniques were used to measure discharge at each site. Both ADCP bottom tracking and either an RTK or satellite-based augmentation system differential GPS were used as the reference to measure river velocity. To calculate river discharge and determine any directional bias, numerous transects were made from both the left-to-right-bank and the right-to-left-bank directions when possible.

Measurements during spring breakup in any braided river located in extreme cold environments involve challenges ranging from river access and personnel safety to the quality of any given measurement. Iced banks usually prevent boat launches when attempting to carry out measurements, and floating blocks of ice moving downstream pose a safety concern for members of the field crew in boats. In the event that a boat can be launched, there is no certainty that all channels in a given river cross-section can be measured. The Sagavanirktok River is no exception in this regard. This task (i.e., measuring discharge during spring breakup) was successfully accomplished by performing the measurements from a helicopter, as reported by Keech et al. (2018). The ADCP was located in a high-speed trimaran which was connected to the helicopter by a 15 m line. The aircraft crew consisted of three people: the pilot, one spotter, and one person in charge of the software used to run the ADCP. This approach allowed the team to obtain a complete view of the entire river cross-section. Multiple channels (even small ones) could be measured since the helicopter could easily lift the trimaran out of the water and reposition it. This method of measuring discharge during spring breakup when river travel is hazardous due to floating ice resulted in high-quality measurements, which are needed for any further hydrological analyses.

During the summer, a jet powerboat or a kayak, depending on the flow conditions, was used to tow the trimaran containing the ADCP to take discharge measurements.

2.5.6 Suspended Sediment

Automated portable ISCO 3700 samplers were deployed at the four sediment sites on the Sagavanirktok River (DSS1–DSS4) to monitor SSC on a daily basis during spring and summer breakup. Two samplers were placed at each site and programmed to collect water samples every 48 hours, 24 hours apart. Thus, if one sampler failed, a partial record was still available. Sampler intake tubes were located approximately 12 cm above the streambed. Each intake tube was attached to a rebar tripod mount that kept the intake tube off the streambed.

Grab samples were collected at the river's edge (near the autosamplers) and in the main channel, and depth-integrated samples were collected in the main channel. The collection of grab and depth-integrated samples involved taking in 1 L volume of river water. Depth-integrated samples were compared with ISCO samples collected simultaneously. The goal of this comparison was to establish a relationship between SSC from the main channel and SSC at the edge of the river near the ISCO autosampler.

The field samples were analyzed at the Water and Environmental Research Center at UAF to determine the SSC of the Sagavanirktok River at each study site. Following ASTM Standard 3977-97, the samples were vacuum-filtered using Whatman GF/C glass microfiber filters that had a particle retention size of 1.2 μm. Then the percentage of organic matter in each sample was determined using ASTM Standard 2974-17 (Test Method C), in which samples were placed in a muffle furnace at 440°C for 12 hours.

2.5.7 Bed Sediment

To characterize the bed sediment transport rate between bathymetric surveys at the excavated pits, pit volume change was calculated between consecutive surveys. In addition, bed sediment at different locations (inside and upstream of the pits) was sampled and grain size distributions were calculated.

2.6 RESULTS

The following paragraphs present a summary of the main results obtained during this study. The findings on precipitation for the entire watershed are reported first and then the findings from two stations in particular (DSS2 and DSS3), for which rating curves and suspended sediment relationships are included. Finally, the findings on bed sediment transport conditions along the study reach of the Sagavanirktok River are presented.

2.6.1 PRECIPITATION

Measurements of cold season and warm season precipitation were taken throughout the Sagavanirktok River basin. Sonic sensors were used at each meteorological station to observe cold season precipitation (in the form of snow depth), and end-of-winter snow depth and density were measured in mid-April to calculate snow water equivalent (SWE) at various snow survey sites. In the water balance of the basin, end-of-winter SWE represents total cold-season precipitation. After snowmelt, during the warm season (in general, summer is June to September in Arctic Alaska), tipping bucket rain gauges (which are shielded to reduce undercatch) are used to measure rainfall at these meteorological stations. The sum of the warm season rainfall and the end-of-winter SWE approximates total annual precipitation.

2.6.1.1 Cold-Season Precipitation

Beginning in April 2016, snow survey sites previously established in the Sagavanirktok River watershed, as described in Kane et al. (2014) and Stuefer et al. (2014), were visited along with several newly established sites. Snow depth and density measurements were made to calculate SWE at 32 locations in the basin. Snow survey sites are located in all three regions of the basin, namely the mountains, foothills, and the coastal plain. Maps in Figures 2.2 and 2.3 show examples of

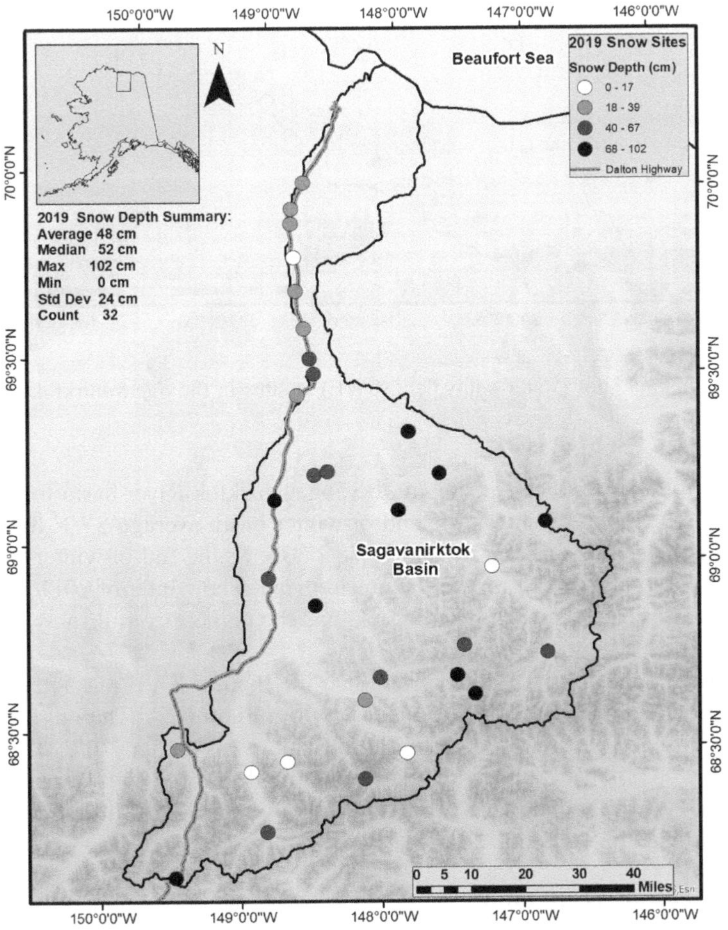

FIGURE 2.2 Example of snow depth at sites in the Sagavanirktok River basin in mid-April 2019.

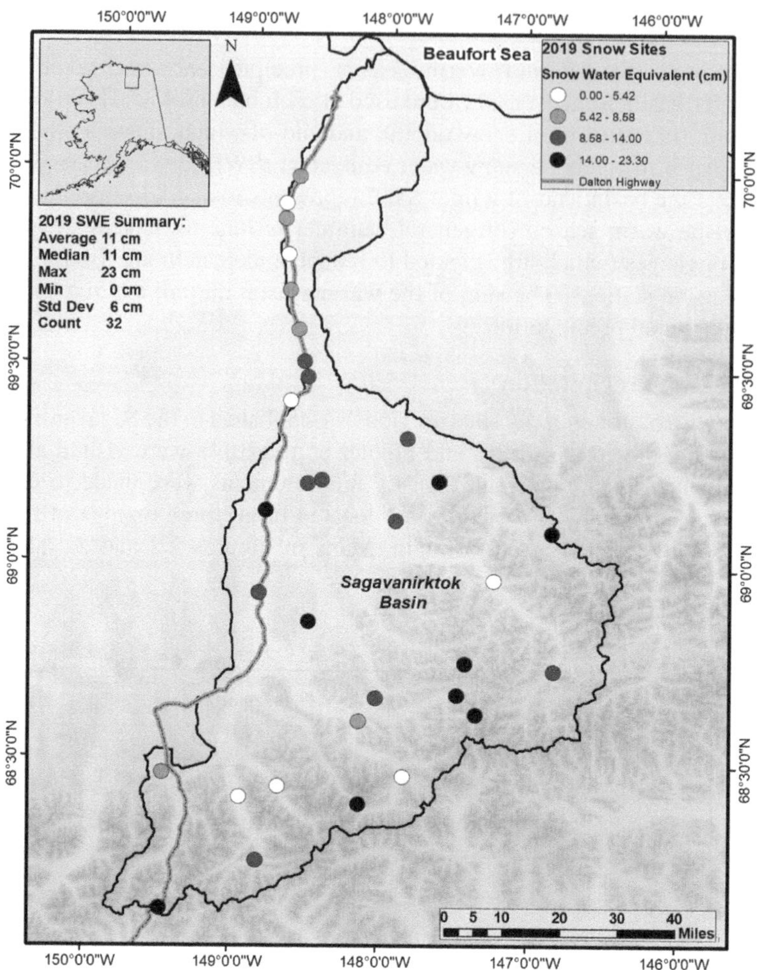

FIGURE 2.3 Example of snow water equivalent (SWE) at sites in the Sagavanirktok River basin in mid-April 2019.

snow depth and SWE, respectively, at sites in the Sagavanirktok River basin for winter 2018/2019 (Toniolo et al., 2020). Table 2.2 shows the end-of-winter basin average SWE for each year of the study compared with the long-term mean, and Table 2.3 shows the end-of-winter SWE by region for each year of the study, compared with the long-term mean. The winter of 2017/2018, which Sturm et al. (2018) describe as having a record snowpack, had the highest overall snow depth and SWE of the study period.

All regions of the Sagavanirktok River basin had the highest SWE in winter 2017/2018. Snow survey site elevation and SWE were plotted, as shown in Figure 2.4. The only year in the 4-year period with a clear trend of increasing SWE with elevation was winter 2017/2018, when the snowpack was unusually deep. The results from other years roughly agree with Homan and Kane (2015) and Kane et al. (2014), who found that SWE at the end of winter varied little from the coastal plain to the continental divide in the Brooks Range.

2.6.1.2 Warm-Season Precipitation

In the Sagavanirktok River basin, rainfall increases with elevation (southward). Since approximately 75% of the basin lies in the foothill and mountain regions, several meteorological stations were

TABLE 2.2

End-of-Winter Basin Average Snow Water Equivalent, Sagavanirktok Basin

Year	Number of Snow Survey Sites	Basin Average SWE (cm)
2015/2016	32	9.7
2016/2017	32	12.2
2017/2018	32	19.2
2018/2019	32	10.6
Long term	12 (2006-2019)	11.5

(Long-Term Data from Kane et al., 2012)

TABLE 2.3

End-of-Winter Snow Water Equivalent by Region, Sagavanirktok Basin

Year	Coastal Plain (n_{sites}=up to 8)	Foothills (n_{sites}=up to 8)	Mountains (n_{sites}=up to 16)
	Average SWE (cm)		
2015/2016	7	13	10
2016/2017	9	12	14
2017/2018	12	19	23
2018/2019	8	13	11
Long term	10	11	11
	(n_{years}=18)	(n_{years}=12)	(n_{years}=12)

(Long-Term Data from Kane et al 2012)

established there to better characterize precipitation in this data-sparse area. Rainfall data are collected using tipping buckets at the following Sagavanirktok River basin stations: Accomplishment Creek (ASM1/DBM1), Saviukviayak (ASM2), Sag–Ivishak (DBM4), Ivishak (ASM3), Sagavanirktok River near MP318 (DSS4), Echooka (DSM1), and Sagavanirktok River near MP347 (ASS1). An attempt is made to measure total annual rainfall; however, during mid-May and mid-September, the area may receive snow or rain, or a mixture of the two. Table 2.4 presents the cumulative rainfall from 2007 to 2019 at each UAF station. Figures 2.5 and 2.6 show plots of cumulative rainfall at each station during the summers of 2016 to 2019. Additional nearby stations are also included in the plots (Arp and Stuefer, 2019; Stuefer and Youcha, 2020; Kane et al., 2021a, 2021b). The Sagavanirktok River basin's average rainfall is not estimated due to the limited number of stations with tipping buckets installed at high elevations. Only two tipping buckets are located in the mountain region (ASM1 and ASM2), and only one is at high elevation (1458 m).

In 2019, rainfall totals within the basin varied from 50 mm on the coastal plain (Franklin Bluffs) to 550 mm at Ivishak station in the foothill region (Figure 2.6), similar to 2018. In both 2018 and 2019, cumulative rainfall was much greater than what was observed in 2016 and 2017, indicating wet conditions in the watershed during the last 2 years.

Table 2.5 shows the cumulative monthly rainfall totals for 2016 to 2019 at each station. If a rain gauge malfunctioned, or was not yet installed, data are not presented for the monthly total shown in the table. Significant rain can occur in any summer month, but historically, August is the rainiest

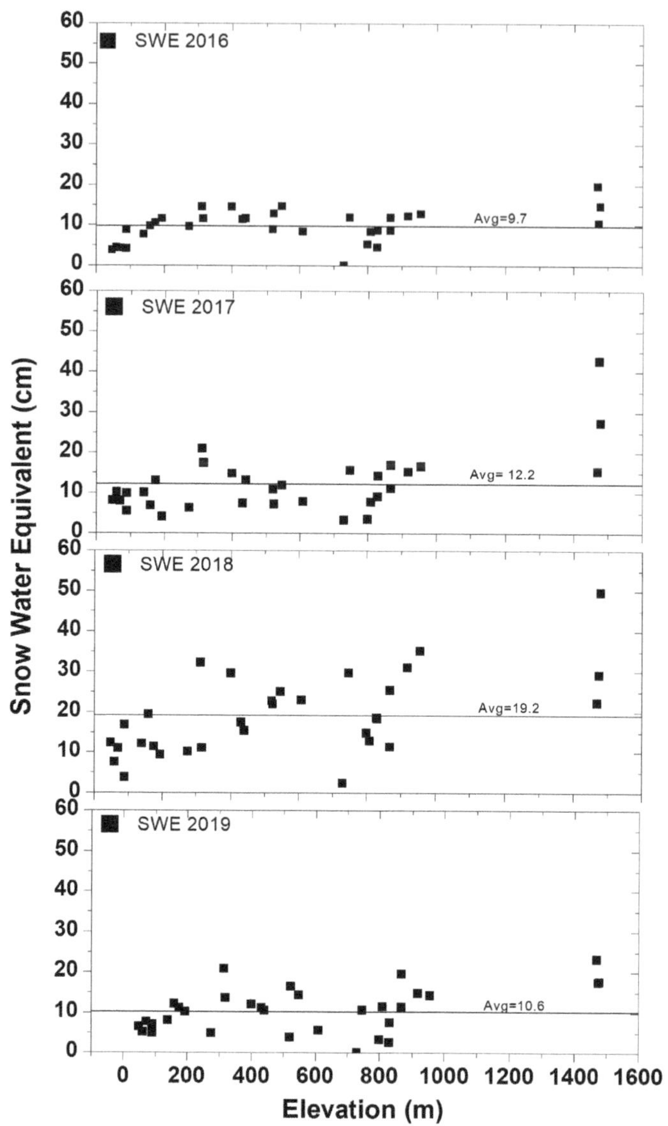

FIGURE 2.4 End-of-winter SWE as a function of elevations within the Sagavanirktok River watershed during the last four winter seasons (2016–2019). The horizontal line in each plot indicates the basin average for each year.

month in the region (Kane et al., 2008). In 2019, the month of August was particularly wet; in 2018, unusual rain events continued into September. The increased amount of rainfall during these 2 years resulted in several very high-flow events in the Sagavanirktok River.

Cumulative rainfall was examined at Accomplishment Creek and at Sag–Ivishak stations from 2006 to 2019. Rainfall at Accomplishment Creek (n=8), located in the mountain region, was variable, ranging from 150 mm to 420 mm depending on the year (Figure 2.7). Rainfall at Sag–Ivishak (n=6) was also variable year-to-year, ranging from 50 mm to 300 mm (Figure 2.7). At both stations, the most rainfall in the period of record occurred in 2018. The last widespread drought across the central North Slope region occurred in 2007; both stations recorded low rainfall that year.

TABLE 2.4

Cumulative Rainfall from 2007 to 2019 at UAF Stations within the Sagavanirktok Basin

Station	Elev. (m)	Region	2007 (mm)	2008 (mm)	2009 (mm)	2010 (mm)	2012 (mm)	2016 (mm)	2017 (mm)	2018 (mm)	2019 (mm)
Accomplishment Creek (ASM1)	1458	Mountains	183	228	275	160	179			412	331
Saviukviayak (ASM2)	945	Mountains							188		556
Sag–Ivishak (DBM4)	431	Foothills	49	212	157				220	303	250
Ivishak (ASM3)	401	Foothills							225	556	445
Sagavanirktok River near MP318 (DSS4)	366	Foothills								331	250
Echooka (DSM1)	335	Foothills							208	356	310
Sagavanirktok River near MP347 (ASS1)	241	Foothills						270	222	275	203

(2007 to 2012 data from Kane et al., 2012 and Kane et al., 2014)

2.6.2 Discharge Measurements

During spring breakup, river discharge was measured several times, when possible, near the stations. Individual measurements of summer discharge and water levels (i.e., during ice-free conditions) were used to develop rating curves for each site. Figure 2.8 shows an example of this type of curve. The relationships between water level and discharge (defined by the rating curves) were applied to the nearly continuous water level data to estimate continuous river discharge. Figure 2.9 shows the estimated continuous discharge in the Sagavanirktok River, along with all point measurements, at a site upstream (DSS3) and downstream (DSS2) of the Ivishak River confluence.

2.6.3 Cumulative Volumetric Runoff

The temporal discharge variation (Figure 2.9) and estimated cumulative volumetric runoff each year were examined for the Sagavanirktok River downstream of the Ivishak (DSS2) station and at the Happy Valley (DSS3) station (Figure 2.10). Cumulative volumetric runoff is calculated using estimated continuous discharge, and it is normalized to the basin area at the gauge site being analyzed (DSS2 and DSS3) in order to compare different basins. Cumulative volumetric runoff yields insights into the basin water balance and can validate precipitation measurements. In years with high precipitation, cumulative volumetric runoff may also be high. Cumulative volumetric runoff and precipitation data could be used to calculate the runoff ratio for a river (the percentage of precipitation that becomes runoff). Examination of cumulative volumetric runoff at stations with long-term records may allow us to see changes to the basin water balance over time.

Although calculating cumulative volumetric runoff is straightforward for the summer period (continuous discharge is estimated using a rating curve), a calculation for the spring breakup period is less straightforward due to the presence of ice in the river, which prevents the use of a rating curve to estimate continuous discharge. Continuous discharge was estimated for parts of spring breakup by interpolating frequent manual discharge measurements. For this reason, there is greater uncertainty about cumulative volumetric runoff during the spring breakup period, particularly for years

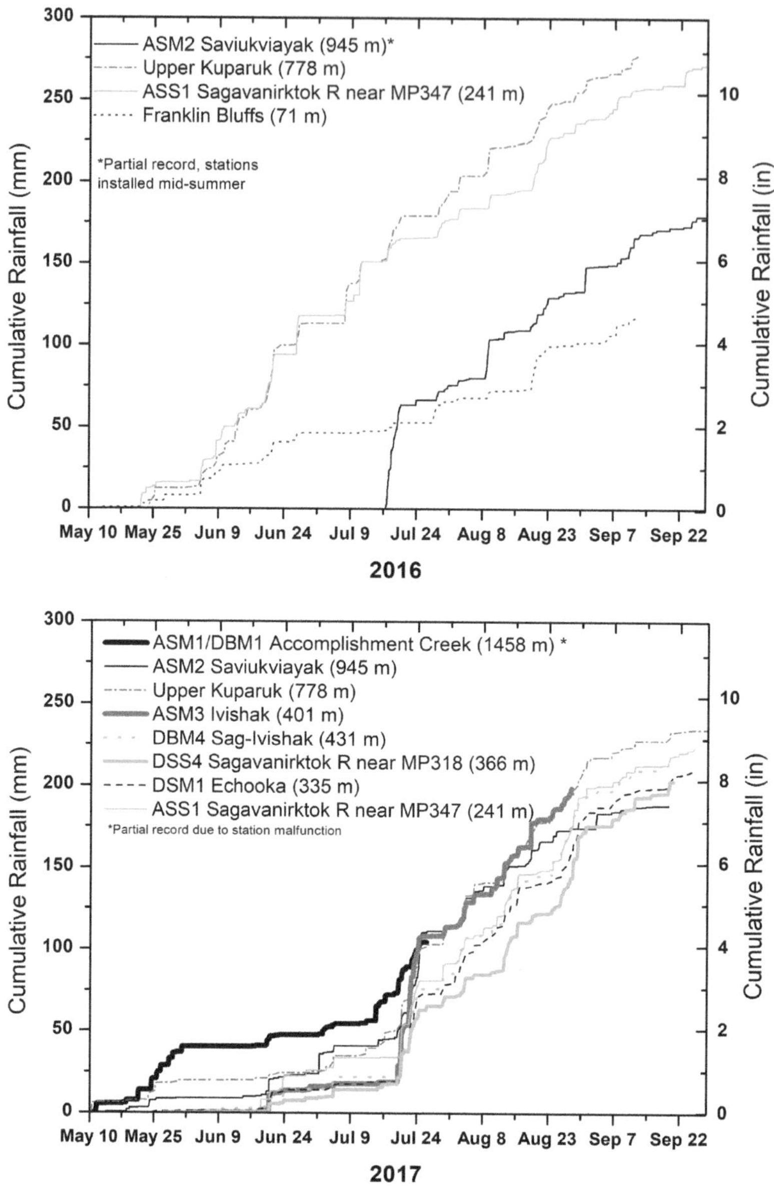

FIGURE 2.5 Cumulative rainfall at stations in or near the Sagavanirktok River basin, 2016 and 2017. For 2016, the total summer rainfall plotted is lower than the actual total because several stations were installed mid-summer (Upper Kuparuk, Franklin Bluffs data from Kane et al., 2021a and Kane et al., 2021b).

when discharge measurements were less frequent. As spring breakup progresses, ice disappears from the river channel and the rating curve can then be used to estimate the later breakup discharge.

Stuefer et al. (2017) reported the long-term mean cumulative volumetric runoff for the Sagavanirktok River USGS station, located upstream of DSS2 and DSS3, as 295 mm, which compares with the present study's calculation for DSS2 and DSS3 in 2016 and 2017. However, the estimated cumulative volumetric runoff in the Sagavanirktok River during the study period was much higher in 2018 and 2019 (Figure 2.10; Tables 2.6 and 2.7) compared with 2016 and 2017 for both the DSS2 and DSS3 stations. The cumulative volumetric runoff during the spring breakup of 2018 was

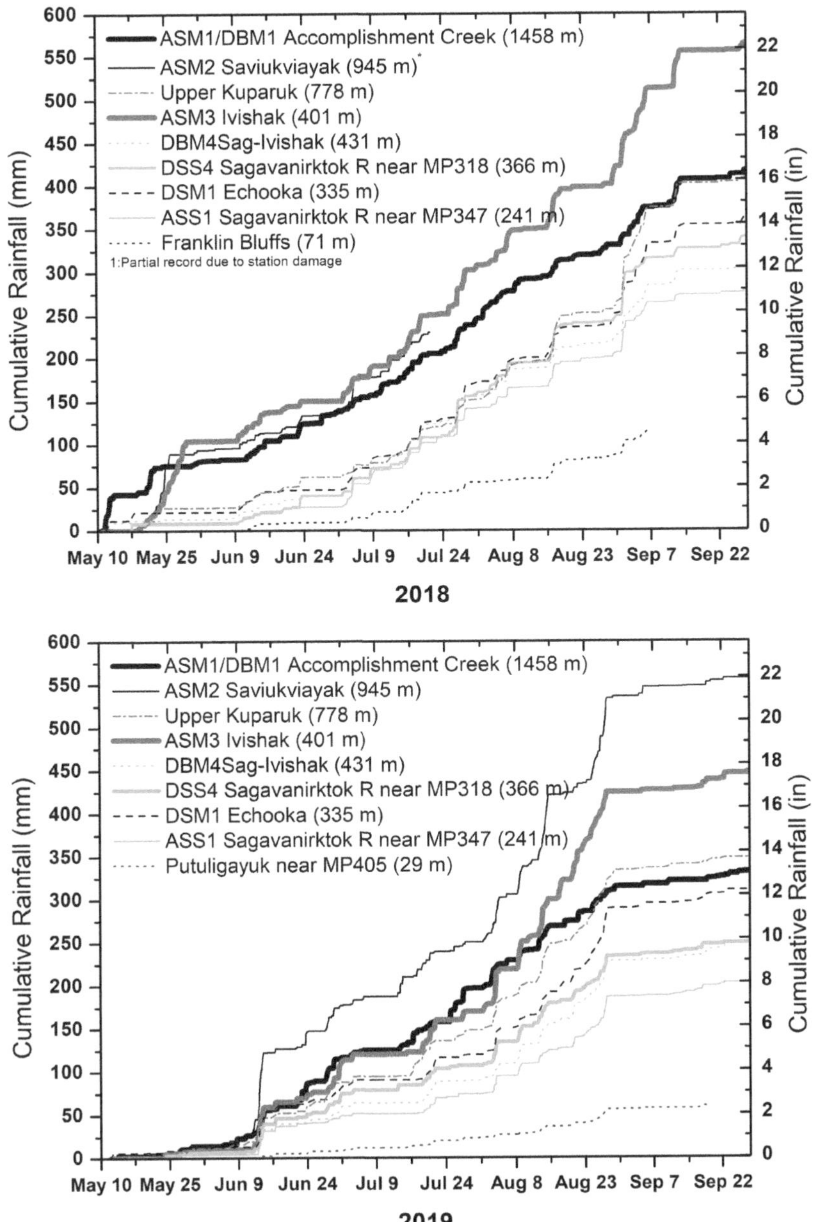

FIGURE 2.6 Cumulative rainfall at stations in or near the Sagavanirktok River basin, 2018 and 2019. Note the change in *y*-axis scale compared with previous plots (Upper Kuparuk, Franklin Bluffs, Putuligayuk data from Kane et al., 2021a, Stuefer and Youcha, 2020, Kane et al., 2021b, and Arp and Stuefer, 2019).

high due to the record-deep snowpack. Additionally, 2018 and 2019 had higher summer cumulative volumetric runoff from numerous summer and fall rain events.

2.6.4 Sediment Grain Size

Toniolo et al. (2016b) found that the average grain-size distribution (D_{50}) of suspended sediment at each Sagavanirktok River sediment research site (DSS1 to DSS4) in 2016 ranged from 20 to 50

TABLE 2.5

Cumulative Monthly Rainfall at Each Station for the Study Period 2016 to 2019

Month and Year	ASM1 Accom-plishment Creek	ASM2 Saviuk-viayak	DBM4 Sag-Ivishak	ASM3 Ivishak	DSS4 Sagava-nirktok River MP318	DSM1 Echooka	ASS1 Sagava-nirktok River MP347
				Total Rainfall (mm)			
May-16							16
Jun-16							102
Jul-16		76		44			59
Aug-16	41	72		33			61
Sep-16	22	25		0			36
May-17		9	8	3	0	10	8
Jun-17		15	14	16	9	13	26
Jul-17		90	71	98	62	65	66
Aug-17		96	108	120	104	105	107
Sep-17		14	16	40	29	27	34
May-18	79	94	2	87	9	24	0
Jun-18	57	42	27	46	31	25	20
Jul-18	109		116	157	120	126	115
Aug-18	93		82	129	110	96	81
Sep-18	67	59	63	119	60	85	50
May-19	13	17	5	9	7	13	4
Jun-19	102	159	53	98	60	70	45
Jul-19	84	76	34	64	41	38	26
Aug-19	117	284	138	254	127	169	112
Sep-19	6	12	6	10	15	13	18

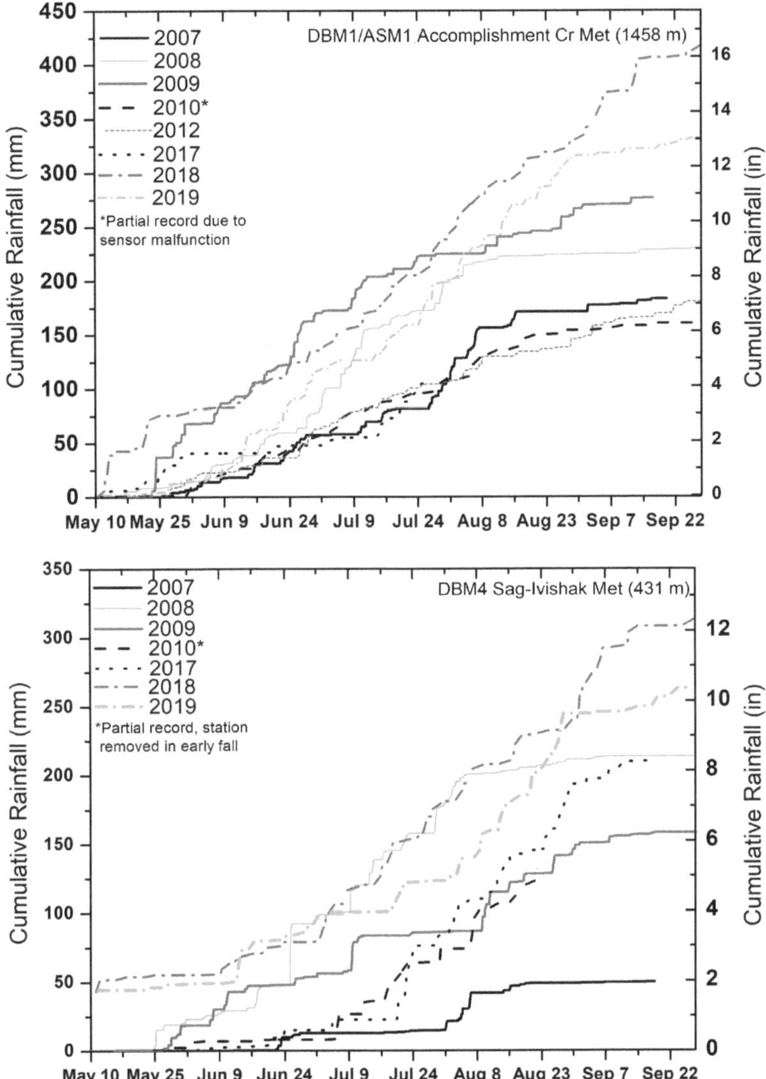

FIGURE 2.7   Cumulative rainfall at Accomplishment Creek (ASM1) and Sag–Ivishak (DBM4) stations for the period of record, 2006–2019 (2006–2012 data from Kane et al., 2012 and Kane et al., 2014). The years 2018 and 2019 had the highest summer rainfall during the study period.

microns, which corresponds to silt-sized particles (ranging from medium to coarse silt). Toniolo et al. (2017b) reported suspended sediment grain sizes of 16 to 55 microns at Happy Valley (DSS3), upstream of the Ivishak River confluence, in early summer 2017.

Toniolo (2020) reported the following average grain-size distribution of bed sediment samples collected upstream of the excavated pits: 0.03 m, 0.03 m, 0.06 m, and 0.075 m for DSS1, DSS2, DSS3, and DSS4, respectively.

2.6.5   Suspended Sediment

The findings presented in this section are from the analysis of SSC in the Sagavanirktok River at three stations: Happy Valley, located upstream of the Ivishak River confluence (DSS3); a station

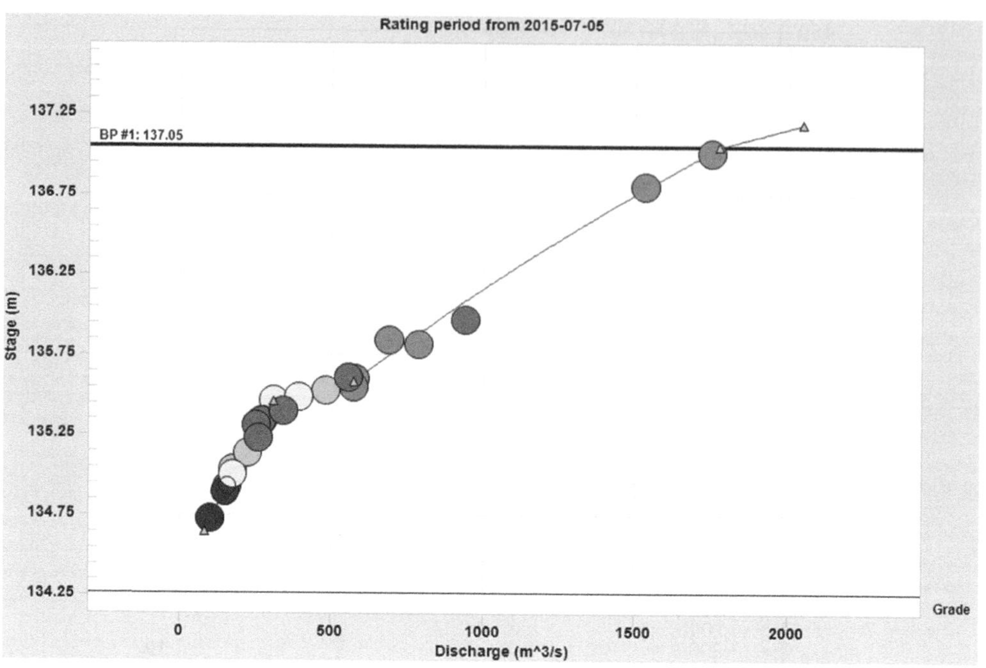

FIGURE 2.8   Rating curve for the Sagavanirktok River downstream of the Ivishak River confluence (station DSS2). Measurements of discharge and river stage occurred between 2015 and 2019.

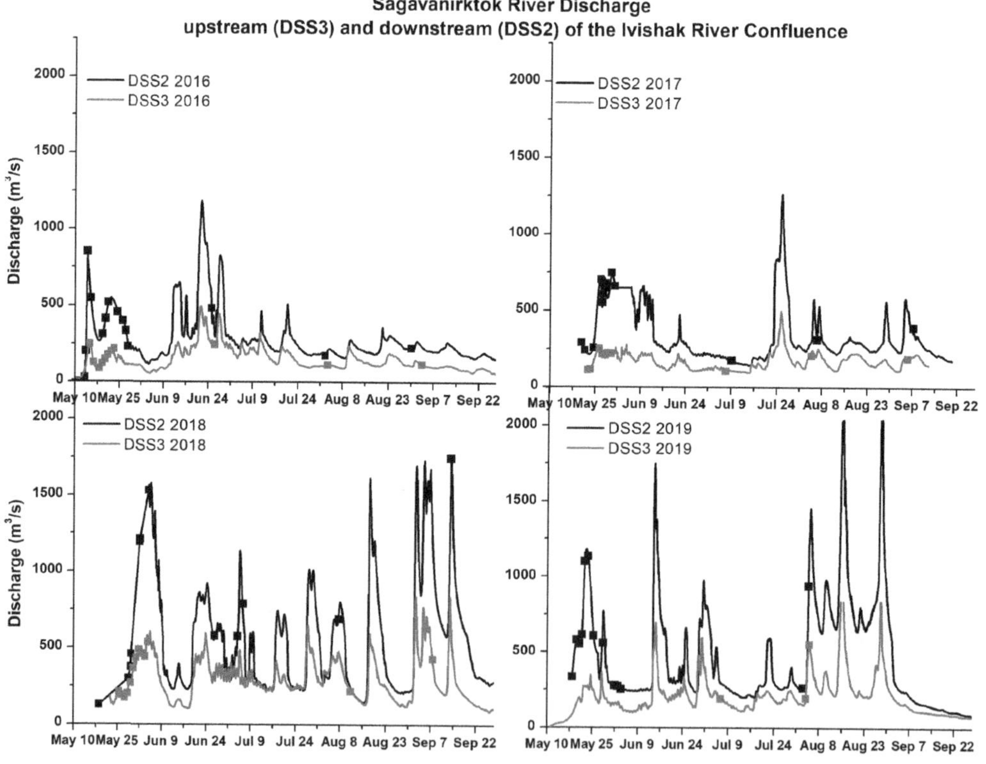

FIGURE 2.9   Hydrographs for the Sagavanirktok River upstream (DSS3) and downstream (DSS2) of the Ivishak River confluence, 2016–2019.

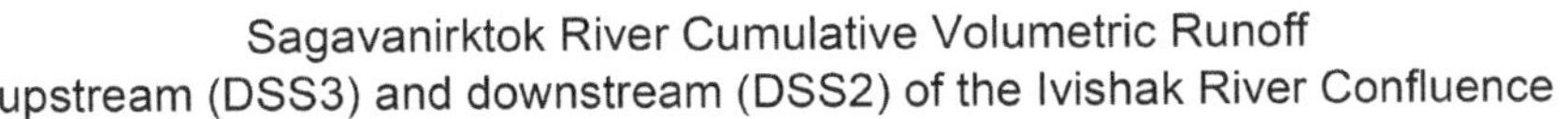

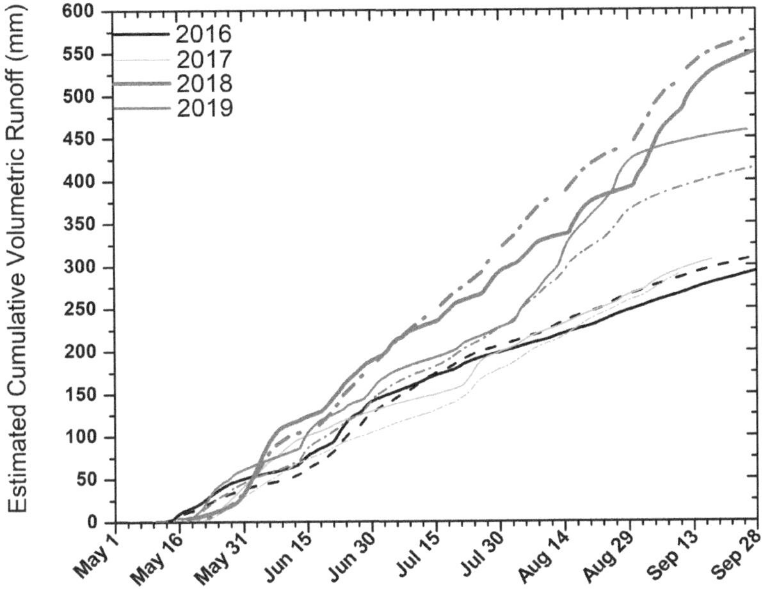

FIGURE 2.10   Estimated cumulative volumetric runoff at the Sagavanirktok River upstream (DSS3) and downstream (DSS2) of the Ivishak confluence. DSS3: solid lines; DSS2: dashed lines.

TABLE 2.6
Estimated Cumulative Volumetric Runoff for the Sagavanirktok River Downstream of the Ivishak River Confluence (DSS2) Normalized to the Watershed Area

Year	Estimated Total Cumulative Volumetric Runoff (mm)	Estimated Spring Cumulative Volumetric Runoff (mm)	Estimated Summer Cumulative Volumetric Runoff (mm)
2016	306	47	259
2017	307	62	245
2018	550	108	442
2019	460	53	407

TABLE 2.7
Estimated Cumulative Volumetric Runoff for the Sagavanirktok River Upstream of the Ivishak River Confluence (DSS3) Normalized to Watershed Area

Year	Estimated Total Cumulative Volumetric Runoff (mm)	Estimated Spring Cumulative Volumetric Runoff (mm)	Estimated Summer Cumulative Volumetric Runoff (mm)
2016	315	32	283
2017	289	46	243
2018	567	82	485
2019	416	31	385

located downstream of the Ivishak River confluence (DSS2); and East Bank, located upstream of where the river splits into the east and west channels just south of Prudhoe Bay (DSS5).

Grab samples, collected from shore next to the ISCO intake, were used to verify the ISCO SSC data. Water samples were also collected from the main channel, using the depth-integrated sampler or a grab sample. These samples were taken from fast-flowing water in the middle of the river.

As previously mentioned, a study of SSC during the 2015 spring breakup flood found concentrations of up to 4500 mg/L at DSS5 (Toniolo et al., 2017a). The 2015 breakup was one of the earliest on record; the entire watershed warmed quickly, resulting in a fast mechanical breakup of river ice. The early, rapid removal of anchor ice from the streambed resulted in more sediment available for transport. In 2016, the SSC from ISCO samplers during spring breakup was significantly lower (less than 250 mg/L) than the SSC reported for the spring breakup of 2015 (Toniolo et al., 2016a). Breakup flows in 2016 were lower, and breakup was slower; consequently, the flows had lower energy and sediment was not available (i.e., it was protected by ice). The SSC from ISCO samplers during spring breakup 2017 was near 1000 mg/L at Happy Valley (DSS3) despite similar breakup discharges in 2016 and 2017.

Usually during the summer months, SSC was low (ranging from <3 to 200 mg/L) at each station during periods of lower flows and increased greatly during periods of high flows. However, in 2015 and 2016, summer SSC remained less than 250 mg/L, while high flow events in 2018 and 2019 resulted in SSC greater than 1000 mg/L.

One depth-integrated sample was collected at DSS2 during a high-flow event in September 2018; SSC measured 1140 mg/L (river discharge of 1780 m³/s). Another sample was collected at DSS2 in August 2019, with an SSC of 980 mg/L (river discharge of nearly 1000 m³/s). The highest depth-integrated sample collected from Happy Valley (DSS3) was 700 mg/L during an event on 4 August 2019 (river discharge of 550 m³/s).

The plots in Figure 2.11 show the relationship between SSC collected from shore and SSC collected from the main channel. The suspended sediment samples collected from the main channel were plotted against river discharge to obtain the SSC rating curves (Figure 2.12). These relationships, along with the continuous river discharge dataset, can be used to determine the suspended sediment yield for the river. The station downstream of the Ivishak confluence (DSS2) characterizes the suspended sediment yield from both the upper Sagavanirktok and the Ivishak tributary. The corresponding sediment rating curve in Figure 2.12 can be applied to the cumulative discharge for DSS2 to estimate the Sagavanirktok River suspended sediment yield into the Beaufort Sea. Figure 2.13 shows the estimated suspended sediment yield for 2016 through 2019, based on estimated cumulative discharge data from the stations downstream (DSS2) and upstream (DSS3) of the Ivishak River confluence. Uncertainty exists in the estimated suspended sediment yield shown in Figure 2.13. This uncertainty is associated with the rating curves in Figure 2.12, which are less defined at higher flow/SSC, and with SSC samples from the main channel, which were only collected during the summer period. Also, continuous discharge for parts of the spring breakup period is estimated using interpolation methods (as mentioned). The suspended sediment yield was estimated as soon as the river appeared turbid, usually during the rising limb just prior to the spring peak discharge. Early breakup flows have lower amounts of suspended sediment because anchor ice prohibits sediment transport.

The suspended sediment yields in 2018 and 2019 were much greater than the suspended sediment yields in 2016 and 2017 due to higher flows (Figure 2.13), which resulted from several large rain events in August and September of 2018 and 2019. Note that the 2018 spring breakup period also generated a large volume of sediment that was transported by the streams. Spring breakup in 2018 was slow due to basin-wide cold air temperatures (Toniolo et al., 2018), which resulted in low breakup flows throughout May and slow thermal degradation of river ice. However, by early June, with the record-deep snowpack melting, discharge increased, rapidly removing any remaining river ice and transporting large amounts of suspended sediment. In 2019, suspended sediment was also transported during spring breakup (although less than in 2018) due to the more mechanical breakup

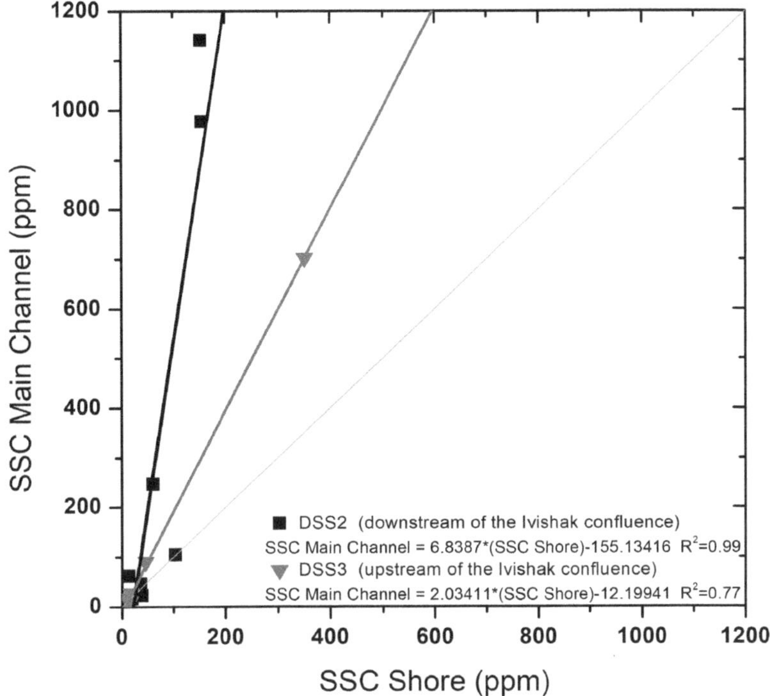

FIGURE 2.11 Relationship between SSC of Sagavanirktok River water samples collected from shore and the main channel.

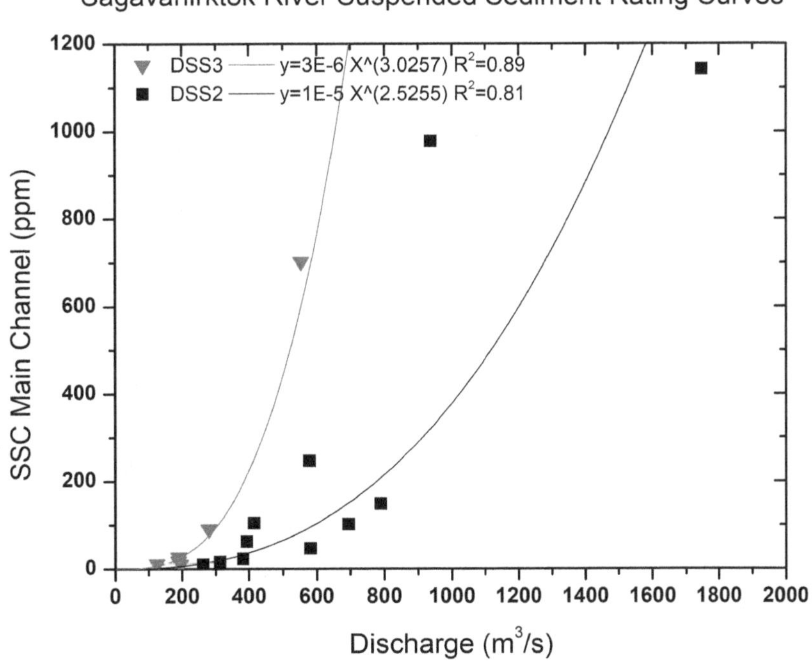

FIGURE 2.12 Relationship between Sagavanirktok River discharge and SSC at each station.

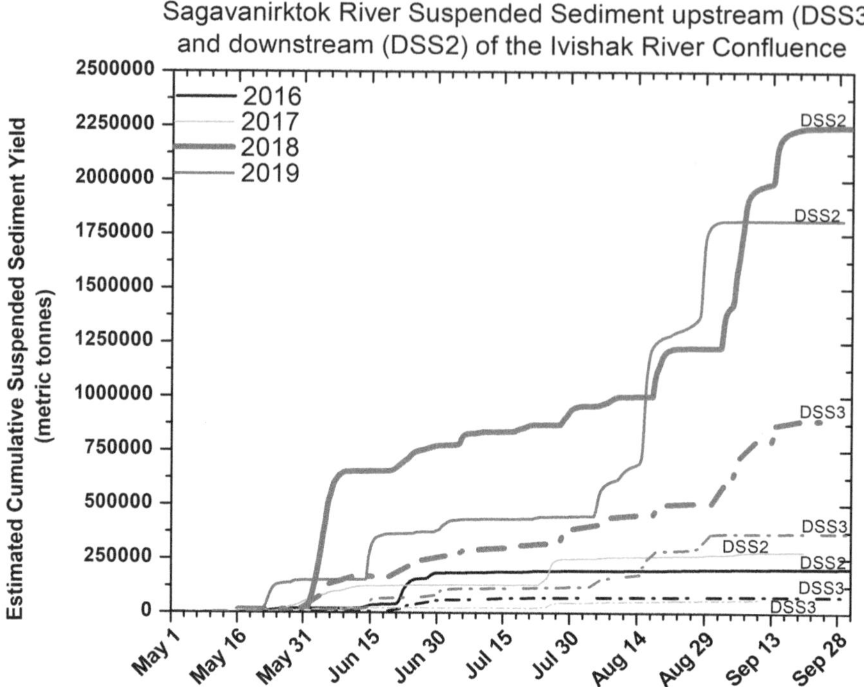

FIGURE 2.13 Estimated suspended sediment yield on the Sagavanirktok River downstream (DSS2) and upstream (DSS3) of the Ivishak River confluence. The years 2018 and 2019 had higher suspended sediment yields than 2016 and 2017. DSS2: solid lines; DSS3: dashed lines.

of river ice (Toniolo et al., 2020). In both 2016 and 2017, the amount of suspended sediment in the river during spring breakup was much less than in 2018 and 2019 due to longer-lasting bottom ice and/or lower breakup flows.

Plots in Figure 2.13 show a well-defined increase in suspended sediment yield at DSS3 and DSS2 during 2018 and 2019, compared with the values for 2016 and 2017. Further inspection of the plots indicates a difference between upstream (DSS3) and downstream (DSS2) of the Ivishak River confluence, in that suspended sediment yield significantly increased during the second half of summer and early fall at DSS2. It is speculated that either glaciers (more are located within the Ivishak River watershed) release extra water and sediment as summer progresses, or there is increased rainfall in the Ivishak watershed compared with the upper Sagavanirktok watershed.

2.6.6 BED SEDIMENT TRANSPORT CONDITIONS

The available literature on Siberian streams (Alekseev and Lisitsyna, 1974) indicates that sediment transport loads are very small, while research on a river in Canada (Milliman and Meade, 1983) indicates that sediment loads on the Mackenzie River can be considered normal when compared with temperate streams. According to Morehead et al. (2003), if the watershed temperature is low, relatively low sediment transport loads can be expected.

In the case under study, the Sagavanirktok River, Hodel (1986) stated that "many workers have observed that little or no bed load is currently being carried onto the Beaufort Sea shelf", which could be an indication that the river is aggradational for a longer distance than that previously reported by Boothroyd and Timson (1983), who suggested the river was aggradational the last 20 km of its course.

Recently, using data (i.e., pit volume change, discharge, sediment characteristics, and water slope) collected at DSS1–DSS4 during the multiyear effort on the Sagavanirktok River, Toniolo (2020) published an article on the river's bed-sediment transport conditions, applying a simple, basic structure commonly used for bed sediment transport calculations. The structure takes the following form:

$$q^* = \alpha \left(\tau^* - \tau_c^* \right)^{1.5} \tag{2.2}$$

where q^* is the dimensionless sediment transport rate per unit width, which is defined as

$$q^* = \frac{q_s}{\left(RgD^3 \right)^{\frac{1}{2}}} \tag{2.3}$$

where q_s is the bed load transport rate per unit width, and τ^* (Equation 2.2) is the Shields parameter, which is defined as

$$\tau^* = \frac{\tau_0}{\rho RgD} \tag{2.4}$$

where ρ is the density of water, R is the submerged specific gravity of natural sediments ($R = 1.65$ for quartz), g is gravity, D is the sediment diameter, and τ_0 is the bed shear stress, which is given by the following equation

$$\tau_0 = \gamma HS \tag{2.5}$$

where γ is the specific gravity of water, H is the average water depth, and S is the water slope.

Additionally, α (Equation 2.2) is a constant coefficient. Several values were reported for this coefficient. For instance, the value was 8 in the pioneering work of Meyer-Peter and Muller (1948), 5.7 in Fernandez Luque and Van Beek (1976), and 3.97 in a reanalysis done by Wong and Parker (2006).

Shields (1936) determined that a minimum value is needed to initiate the motion of non-cohesive bed-sediment particles. This value is known as the critical Shields number, τ^*_c (Equation 2.2), which ranges from 0.03 to 0.06 for gravel-bed rivers, as reported in the published literature (ASCE, 2008).

Finally, the total bed load transport rate, Q_s, can be calculated as

$$Q_s = q_s B \tag{2.6}$$

where B is the channel width.

Toniolo (2020) reported that α values for the study locations along the river (DSS1–DSS4) ranged from 0.012 to 0.045, which are considerably smaller values than those previously reported. His findings corroborate the statement made by Hodel (1986). Toniolo speculates that spring breakup plays a key role in the armoring conditions of the riverbed, which in turn directly affects the river's bed-sediment transport load. Warmer air temperatures could change breakup conditions, meaning a more rapid breakup could occur. Such circumstances could allow the partial or total destruction of the riverbed's armored layer, facilitating bed sediment transport processes.

2.7 DISCUSSION

Results from this multiyear hydro-sedimentological study of the Sagavanirktok River indicate that the last 2 years (2018 and 2019) had extremely high flow and high SSC. Consequently, the sediment yield was also quite high. Though one could argue that the study period is relatively short, an analysis of available long-term records for the Sagavanirktok River USGS station (located upstream of DSS2 and DSS3) and for the Kuparuk River USGS station (located near Deadhorse) provides additional supporting information.

The Sagavanirktok River USGS station has a 38-year record (1982–2019); however, only one or two discharge measurements are made during the spring breakup period, resulting in increased uncertainty in the data during breakup. Thus, only the discharge record in the summer months was examined. The mean monthly discharge during the months of July, August, and September was examined (Figure 2.14) to look for changes in the long-term discharge record. Compared with the summer discharge values recorded in previous years, the ones recorded in 2018 and 2019 were higher.

The Kuparuk River is adjacent to and west of the Sagavanirktok River. The USGS reports daily discharge for the Kuparuk River from a station located near Deadhorse (USGS station 15896000). This station has a 49-year record (1971–2019). Because of their proximity, the Kuparuk and the Sagavanirktok watersheds are influenced by similar climatic patterns. The Kuparuk River's largest yearly event is spring breakup. Rain events during the summer months contribute to the river's total runoff. The cumulative volumetric runoff (normalized to the watershed area) was calculated for each year of the 49-year record (Table 2.8) to look for changes in runoff patterns in recent years. Stuefer et al. (2017) reported a gradual increase in cumulative volumetric runoff in the Kuparuk River between 1971 and 2015, with the maximum cumulative volumetric runoff occurring (at the time) in the year 2014. This chapter expanded upon that finding by including the 2016 to 2019 study period. The highest cumulative volumetric runoff for the Kuparuk River basin occurred in 2018, due to record-deep snowpack and a wet summer. Both the Sagavanirktok and Kuparuk watersheds had numerous rain events in September 2018, resulting in above-normal streamflow. It is apparent from the yearly ranks shown in the last column of Table 2.8 that the past 7 years have ranked

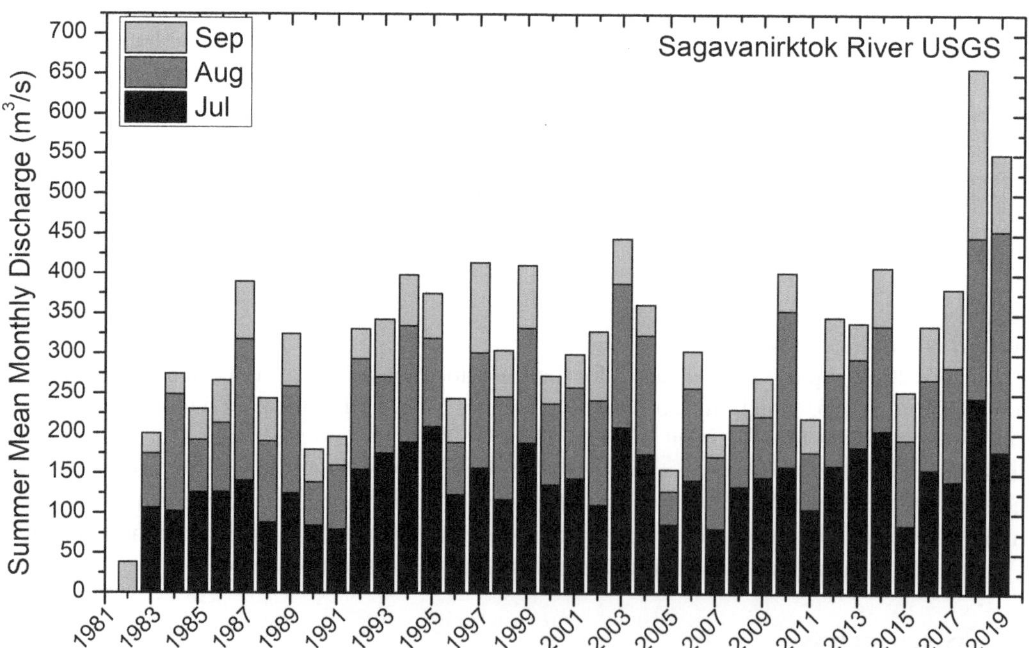

FIGURE 2.14 Mean monthly discharge in July, August, and September at USGS Sagavanirktok River station (USGS, 2019).

TABLE 2.8

Kuparuk River near Deadhorse, Cumulative Volumetric Runoff 1971–2019

Year	Cumulative Volumetric Runoff (mm)	Rank, n=49 yr
1971	171	17
1972	170	18
1973	184	13
1974	71	49
1975	90	46
1976	157	25
1977	169	19
1978	147	29
1979	112	42
1980	174	16
1981	156	27
1982	252	3
1983	115	41
1984	159	24
1985	80	47
1986	167	21
1987	128	39
1988	137	34
1989	236	5
1990	92	45
1991	111	43
1992	136	35
1993	160	23
1994	129	37
1995	165	22
1996	168	20
1997	227	6
1998	124	40
1999	139	31
2000	180	14
2001	150	28
2002	185	12
2003	157	26
2004	195	10
2005	129	38
2006	136	36
2007	79	48
2008	100	44
2009	175	15
2010	144	30
2011	138	32
2012	138	33
2013	252	4
2014	297	2
2015	220	9
2016	222	8
2017	193	11
2018	306	1
2019	225	7
Mean	**162**	

(USGS, 2019)

in the top 11 on record for the highest cumulative volumetric runoff for the Kuparuk River basin. Additionally, one of the earliest spring breakups on record occurred in 2015 in both the Kuparuk and the Sagavanirktok rivers. Rapid spring breakup, as observed in 2015 and 2019, may lead to increased sediment transport during spring breakup.

Higher discharge implies high suspended sediment loads. Additionally, a warming Arctic could mean an increase in the duration of summer, resulting in more opportunities for sediment transport. Thus, the combined analyses indicate that sediment yield is increasing in these two large arctic rivers.

2.8 CONCLUSIONS

A multiyear hydro-sedimentological study has been done along a roughly 160 km reach of the Sagavanirktok River, located in Arctic Alaska, United States. Several monitoring stations were installed along the river and within the watershed. Fieldwork included snow surveys at the end of winter, and water level and discharge measurements during breakup and summer. For suspended sediment analysis, water samples collected nearshore by autosamplers deployed onshore and depth-integrated samples collected from the main channel were used. These samples were collected in the summer. Discharge and sediment rating curves have been developed for the study sites in order to estimate suspended sediment yield. Data recorded indicate that 2018 and 2019 had extremely high flows. A similar conclusion is drawn from the analysis of the Kuparuk River watershed, which is adjacent to the Sagavanirktok River watershed and has a long hydrologic record.

A comparison was made of two stations installed along the Sagavanirktok River: one located upstream of the river's confluence with the Ivishak River (DSS3) and one located downstream of the confluence (DSS2). The stations showed important differences in sediment yield during the second half of summer and early fall. A noticeable increase in sediment yield was observed at the station located downstream of the Ivishak confluence. This finding may be interpreted as the "extra supply" from melting glaciers located in the Ivishak River watershed or from increased precipitation in the Ivishak River watershed compared with the upper Sagavanirktok River watershed.

As the Arctic moves toward a wetter environment due to climate warming (Piliouras and Rowland, 2020), an increase in sediment fluxes to the oceans can be expected.

ACKNOWLEDGMENTS

Funding to conduct this research was provided by public and private entities: the Alaska Department of Transportation and Public Facilities, and the Alyeska Pipeline Service Company. The authors acknowledge Grace Pedersen for editing assistance during the preparation of this chapter.

NOTATION

B:	channel width
D:	sediment diameter
g:	acceleration of gravity
h_s:	snow depth
H:	average water depth
Q_s:	total bed load transport rate
q_s:	bed load transport rate per unit width
q^*:	dimensionless sediment transport rate per unit width
R:	submerged specific gravity of sediment ($R = 1.65$ for quartz)

S: water slope

SWE: snow water equivalent

α: *parameter* in the bed sediment transport relation

γ: specific gravity of water

ρ_s: snow density

ρ_w: density of water

τ_o: boundary shear stress

τ^*:   dimensionless Shields parameter

τ^*c: critical Shields number

REFERENCES

ACIA. (2004). *Impacts of a Warming Arctic.* Arctic Climate Impact Assessment. Cambridge University Press, Cambridge, 146pp.

Alekseev, V., and Lisitsyna, K. (1974). Stok vzveshennyk nanosov. (Discharge of suspended sediment.) In *Mirovoi Vodnyi Balans I Vodnye Resursy Zemli (World Water Balance and Earth's Water Resources).* Gidrometeoizdat, Leningrad, 510–516.

Amamra, A., Khanchoul, K., Eslamian, S., and Hadj Zobir, S. (2018). Suspended sediment estimation using regression and artificial neural network models: Kebir watershed, northeast of Algeria, North Africa. *International Journal of Hydrology Science and Technology*, 8(4), 352–371.

Arnborg, L., Walker, H. J., and Peippo, J. (1967). Suspended load in the Colville River, Alaska, 1962. *Geografiska Annaler: Series A, Physical Geography*, 49, 131–144.

Arp, C., and Stuefer, S. (2019). Terrestrial ecological observation network: Putuligayuk Met station datasets. https://ine.uaf.edu/werc/teon.

ASCE. (2008). Sedimentation engineering: Processes, measurements, modeling and practice. Manual 110. In Garcia, M. (Ed.). ASCE Publication, 1132pp.

Ashton, G. (Ed.). (1986). *River and Lake Ice Engineering.* Water Resources Publications, Littleton, CO, 485pp.

Beltaos, S., and Burrel, B. (2000). Suspended sediment concentration in the S. John River during ice breakup. *Proceedings of the Canadian Society of Civil Engineers Annual Conference*, Toronto, 235–242.

Berezovskaya, S. L., Derry, J. E., Kane, D. L., Gieck, R. E., Lilly, M. R., and White, D. M. (2007). Snow survey data for the Sagavanirktok river Bullen point hydrology study: Spring 2007. University of Alaska Fairbanks, Water and Environmental Research Center, Report INE/WERC 07.18, Fairbanks, AK, 17pp.

Berezovskaya, S. L., Derry, J. E., Kane, D. L., Lilly, M. R., and White, D. M. (2008). Snow survey data for the Sagavanirktok River Bullen point hydrology study: Spring 2008. University of Alaska Fairbanks, Water and Environmental Research Center, Report INE/WERC 08.15, Fairbanks, AK, 30pp. [Amended Figures 3 and 4, Aug. 26, 2008].

Berezovskaya, S. L., Derry, J. E., Kane, D. L., Gieck, R. E., and Lilly, M. R. (2010a). Snow survey data for the central north slope watersheds: Spring 2009. University of Alaska Fairbanks, Water and Environmental Research Center, Report INE/WERC 09.01, Fairbanks, AK, 45pp.

Berezovskaya, S., Hilton, K., Derry, J., Youcha, E., Kane, D., Gieck, R., Homan, J., and Lilly, M. (2010b). Snow survey data for the central north slope watersheds: Spring 2010. University of Alaska Fairbanks, Water and Environmental Research Center, Report INE/WERC 10.01, Fairbanks, AK, 50pp.

Boothroyd, J. C., and Timson, B. S. (1983). The Sagavanirktok and adjacent river systems, eastern North Slope, Alaska: An analog for ancient fluvial terrain on Mars. In *Permafrost: Proceedings of Fourth International Conference.* National Academy Press, Washington, DC, 74–79.

Braun, C., Hardy, D. R., Bradley, R. S., and Retelle, M. J. (2000). Streamflow and suspended sediment transfer to Lake Sophia, Cornwallis Island, Nunavut, Canada. *Arctic, Antarctic, and Alpine Research*, 32, 456–465.

Bridge, J. (2003). *Rivers and Floodplains: Form, Processes, and Sedimentary Record.* Blackwell, Oxford, 491pp.

Chang, H. (2002). *Fluvial Processes in River Engineering.* Krieger, Malabar, FL, 432pp.

Derry, J. E., Kane, D. L., Lilly, M. R., and Toniolo, H. (2009). Snow-course measurement methods, north slope, Alaska. December 2009, University of Alaska Fairbanks, Water and Environmental Research Center, Report INE/WERC 2009.07, Fairbanks, AK, 11pp.

Dornblaser, M. M., and Striegl, R. G. (2009). Suspended sediment and carbonate transport in the Yukon River Basin, Alaska: Fluxes and potential future responses to climate change. *Water Resources Research*, 45, W06411. https://doi.org/10.1029/2008WR007546.

Fernandez Luque, R., and Van Beek, R. (1976). Erosion and transport of bed load sediment. *Journal of Hydraulic Research*, 14(2), 127–144. https://doi.org/10.1080/00221687609499677.

Forbes, A. C., and Lamoureux, S. F. (2005). Climatic controls on streamflow and suspended sediment transport in three large Middle Arctic catchments, Boothia Peninsula, Nunavut, Canada. *Arctic, Antarctic, and Alpine Research*, 37, 304–315.

Gooseff, M. N., Balser, A., Bowden, W. B., and Jones, J. B. (2009). Effects of hillslope thermokarst in Northern Alaska. *Eos, Transactions American Geophysical Union*, 90(4), 29–36.

Gordeev, V. V. (2006). Fluvial sediment flux to the Arctic Ocean. *Geomorphology*, 80, 94–104.

Grosse, G., Jones, B., and Arp, C. (2013). Thermokarst lakes, drainage, and drained basins. In Shroder, J. F. (Ed.), *Treatise on Geomorphology* (Vol. 8). Academic Press, San Diego, CA, 325–353.

Hamilton, S. K. (2010). Biogeochemical implications of climate change for tropical rivers and floodplains. *Hydrobiologia*, 657(1), 19–35. https://doi.org/10.1007/s10750-009-0086-1.

Hicks, D., and Mason, P. (1991). *Roughness Characteristics of New Zealand Rivers*. National Institute of Water and Atmospheric Research, New Zealand, 329pp.

Hodel, K. (1986). The Sagavanirktok river, north slope Alaska: Characterization of an Arctic Stream. USGS Open-file Report 86-267, 29pp.

Holmes, R. M., McClelland, J. W., Peterson, B. J., Shiklomanov, I. A., Shiklomanov, A. I., Zhulidov, A. V., Gordeev, V. V., and Bobrovitskaya, N. N. (2002). A circumpolar perspective on fluvial sediment flux to the Arctic Ocean. *Global Biogeochemical Cycles*, 16(4), 1098. https://doi.org/10.1029/2001GB001849.

Homan, J. W., and Kane, D. L. (2015). Arctic snow distribution patterns at the watershed scale. *Hydrology Research*, 46(4), 507–520.

Homer, C., Dewitz, J., Yang, L., Jin, S., Danielson, P., Xian, G., Coulston, J., Herold, N., Wickham, J., and Megown, K. (2015). Completion of the 2011 National Land Cover Database for the conterminous United States – Representing a decade of land cover change information. *Photogrammetric Engineering and Remote Sensing*, 81(5), 345–354.

IPCC. (2007). Climate change 2007: The physical science basis. Contribution of working group I to the fourth assessment report of the intergovernmental panel on climate change. Solomon, S., Qin, D., Manning, M., Chen, Z., Marquis, M., Averyt, K. B., Tignor, M., and Miller, H. L. (Eds.). Cambridge University Press, Cambridge, and New York, NY, 996pp.

Ji, U., Julien, P., and Park, S. (2011). Sediment flushing at the Nakdong River estuary barrage. *Journal of Hydraulic Engineering, ASCE*, 137(11), 1522–1535.

Jorgenson, M. T., and Shur, Y. (2007). Evolution of lakes and basins in northern Alaska and discussion of the thaw lake cycle. *Journal of Geophysical Research*, 112, F02S17. https://doi.org/10.1029/2006JF000531.

Jorgenson, M. T., Shur, Y. L., and Pullman, E. R. (2006). Abrupt increase in permafrost degradation in Arctic Alaska. *Geophysical Research Letters*, 33, L02503. https://doi.org/10.1029/2005GL024960.

Julien, P. (2002). *River Mechanics*. Cambridge University Press, Cambridge, 433pp.

Kane, D. L., McNamara, J. P., Yang, D., Olsson, P. Q., and Gieck, R. E. (2003). An extreme rainfall/runoff event in Arctic Alaska. *Journal of Hydrometeorology*, 4, 1220–1228.

Kane, D. L., Berezovskaya, S., Irving, K., Busey, R., Chambers, M., Blackburn, A. J., and Lilly, M. R. (2006). Snow survey data for the Sagavanirktok River/Bullen Point hydrology study: Spring 2006. University of Alaska Fairbanks, Water and Environmental Research Center, Report INE/WERC 06-03, Fairbanks, AK, 10pp.

Kane, D. L., Hinzman, L. D., Gieck, R. E., McNamara, J. P., Youcha, E. K., and Oatley, J. A. (2008). Contrasting extreme runoff events in areas of continuous permafrost, Arctic Alaska. *Hydrology Research*, 39, 287–298.

Kane, D. L., Youcha, E. K., Stuefer, S. L., Toniolo, H., Schnabel, W. E., Gieck, R. E., Myerchin-Tape, G., Homan, J., Lamb, E., and Tape, K. (2012). Meteorological and hydrological data and analysis report for the Foothills/Umiat corridor and Bullen projects: 2006–2011. University of Alaska Fairbanks, Water and Environmental Research Center, Report INE/WERC 12.01, Fairbanks, AK, 260pp.

Kane, D. L., Youcha, E. K., Stuefer, S. L., Myerchin-Tape, G., Lamb, E., Homan, J. W., Gieck, R. E., Schnabel, W. E., and Toniolo, H. (2014). Hydrology and meteorology of the Central Alaskan Arctic: Data collection and analysis, Final Report. University of Alaska Fairbanks, Water and Environmental Research Center, Report INE/WERC 14.05, Fairbanks, AK, 168pp.

Kane, D., Hinzman, L., Stuefer, S., and Arp, C. (2021a). Updated meteorological, radiation, and soil temperature data, Kuparuk River and nearby watersheds: Upper Kuparuk Station, 1994–2018. Arctic Data Center. https://doi.org/10.18739/A29P2W70H.

Kane, D., Hinzman, L., Stuefer, S., and Arp, C. (2021b). Updated meteorological, radiation, and soil temperature data, Kuparuk River and nearby watersheds: Franklin Bluffs Station, 1985–2018. Arctic Data Center. https://doi.org/10.18739/A2R20RX59.

Keech, J., Terwilliger, M., Bailey, J., and Toniolo, H. (2018). Using a helicopter to measure river discharge under extreme environmental conditions: The case of the Sagavanirktok River, Alaska. *Water*, 10(6), 697.

Lamb, E., and Toniolo, H. (2016). Initial quantification of suspended sediment loads for three Alaska North Slope rivers. *Water*, 8(10), 419.

Lenton, T. M. (2012). Arctic climate tipping points. *Ambio*, 41(1), 10–22. https://doi.org/10.1007/s13280-011 -0221-x.

Liljedahl, A. K., Boike, J., Daanen, R. P., Fedorov, A. N., Frost, G. V., Grosse, G., and Zona, D. (2016). Pan-Arctic ice-wedge degradation in warming permafrost and influence on tundra hydrology. *Nature Geoscience*, 9, 312–318. https://doi.org/10.1038/ngeo2674.

McNamara, J., Oatley, J., Kane, D., and Hinzman, L. (2008). Case study of a large summer flood on the North Slope of Alaska: Bedload transport. *Hydrology Research*, 39(4), 299–308.

Meyer-Peter, E., and Muller, R. (1948). Formulas for bed load transport. *International Association for Hydraulic Structures Research*, Stockholm.

Milliman, J., and Meade, R. (1983). Worldwide delivery of river sediment to the oceans. *Journal of Geology*, 91, 1–21.

Milliman, J., Farnsworth, K., Jones, P., Xu, K., and Smith, L. (2008). Climatic and anthropogenic factors affecting river discharge to the global ocean, 1951–2000. *Global and Planetary Change*, 62(3–4), 187–194. https://doi.org/10.1016/j.gloplacha.2008.03.001.

Morehead, M. D., Syvitski, J. P., Hutton, E. W. H., and Peckham, S. D. (2003). Modeling the temporal variability in the flux of sediment from ungauged river basins. *Global and Planetary Change*, 39, 95–110.

Navratil, O., Albert, M., Herouin, E., and Gresillon, J. (2006). Determination of bankfull discharge magnitude and frequency: Comparison of methods on 16 gravel-bed river reaches. *Earth Surface Processes and Landforms*, 31, 1345–1363.

Osterkamp, T. E., and Payne, M. W. (1981). Estimates of permafrost thickness from well logs in northern Alaska. *Cold Regions Science and Technology*, 5, 13–27.

Osterkamp, T. E., and Romanovsky, V. E. (1999). Evidence for warming and thawing of discontinuous permafrost in Alaska. *Permafrost and Periglacial Processes*, 10(1), 17–37. https://doi.org/10.1002/(sici)1099-1530(199901/03)10:1<17::aid-ppp303>3.0.co;2-4.

Piliouras, A., and Rowland, J. C. (2020). Arctic river delta morphologic variability and implications for riverine fluxes to the coast. *Journal of Geophysical Research: Earth Surface*, 125, e2019JF005250. https:// doi.org/ 10.1029/2019JF005250.

Prowse, T. (1993). Suspended sediment concentration during river ice breakup. *Canadian Journal of Civil Engineering*, 20, 872–875.

Rovansek, R. J., Kane, D. L., and Hinzman, L. D. (1993). Improving estimates of snowpack water equivalent using double sampling. In *Proceedings of the 61st Western Snow Conference*, Quebec City, Quebec, 157–163.

Shields, I.-A. (1936). *Anwendung der ahnlichkeitmechanik und der turbulenzforschung auf die gescheibebewegung*. Mitt. Preuss Ver.-Anst., 26, Berlin.

Shiklomanov, A., Yakovlera, T., Lammers, R., Karasev, I., Vorosmarty, C., and Linder, E. (2006). Cold region river discharge uncertainty – Estimates from large Russian rivers. *Journal of Hydrology*, 326, 231–256.

Slater, A. G., and Lawrence, D. M. (2013). Diagnosing present and future permafrost from climate models. *Journal of Climate*, 26(15), 5608–5623. https://doi.org/10.1175/jcli-d-12-00341.1.

Stuefer, S. L., and Youcha, E. K. (2020). Hydrology of Imnavait Creek and Kuparuk River in Alaska. Upper Kuparuk dataset. University of Alaska Fairbanks, Water and Environmental Research Center. https:// ine.uaf.edu/werc/werc-projects/imnavait/.

Stuefer, S. L., Youcha, E. K., Homan, J. W., Kane, D. L., and Gieck, R. E. (2011). Snow survey data for the central north slope watersheds: Spring 2011. University of Alaska Fairbanks, Water and Environmental Research Center, Report INE/WERC 11.02, Fairbanks, AK, 47pp.

Stuefer, S. L., Homan, J. W., Youcha, E. K, Kane, D. L., and Gieck, R. E. (2012). Snow survey data for the central north slope watersheds: Spring 2012. University of Alaska Fairbanks, Water and Environmental Research Center, Report INE/WERC 12.22, Fairbanks, AK, 38pp.

Stuefer, S. L., Homan, J. W., Kane, D. L., Gieck, R. E., and Youcha, E. K. (2014). Snow survey results for the Central Alaskan Arctic, Arctic Circle to Arctic Ocean. University of Alaska Fairbanks, Water and Environmental Research Center, Report INE/WERC 14.01, Fairbanks, AK, 96pp.

Stuefer, S. L., Arp, C. D., Kane, D. L., and Liljedahl, A. K. (2017). Recent extreme runoff observations from coastal arctic watersheds in Alaska. *Water Resources Research*, 53, 9145–9163. https://doi.org/10.1002/2017WR020567.

Sturm, M., Parr, C., Pedersen, S., Apr, C., Urban, F., Welker, J., and Serreze, M. (2018). A report on the snow conditions of the North Slope and Brooks Range of Alaska during Winter of 2018. University of Alaska Fairbanks, Geophysical Institute – UAF Occasional Report 2018, August 20, 2018, Fairbanks, AK, 27pp.

Su, F., Zhang, L., Ou, T., Chen, D., Yao, T., Tong, K., and Qi, Y. (2016). Hydrological response to future climate changes for the major upstream river basins in the Tibetan Plateau. *Global and Planetary Change*, 136, 82–95. https://doi.org/10.1016/j.gloplacha.2015.10.012.

Syvitski, J. (2002). Sediment discharge variability in Arctic rivers: Implications for a warmer future. *Polar Research*, 21(2), 323–330.

Toniolo, H. (2020). Bed-sediment transport conditions along the Sagavanirktok River in Northern Alaska, USA. *Water*, 12(3), 774.

Toniolo, H., Derry, J., Irving, K., and Schnabel, W. (2010). Hydraulic and sedimentological characterization of a reach on the Anaktuvuk River, Alaska. *Journal of Hydraulic Engineering*, 136(11), 935–939.

Toniolo, H., Vas, D., Prokein, P., Kenmitz, R., Lamb, E., and Brailey, D. (2013). Hydraulic characteristics and suspended sediment loads during spring breakup in several streams located on the National Petroleum Reserve in Alaska. *Natural Resources*, 4(2), 220–228.

Toniolo, H., Tape, K. D., Tschetter, J. W., Homan, J. W., Youcha, E. K., Vas, D., Gieck, R. E., Keech, J., and Upton, G. (2016a). Sagavanirktok river spring breakup 2016: Final report. University of Alaska Fairbanks, Water and Environmental Research Center, Report INE/WERC 16.08, Fairbanks, AK.

Toniolo, H., Tschetter, T., Tape, K. D., Cristobal, J., Youcha, E. K., Schnabel, W. E., Vas, D., and Keech, J. (2016b). Hydro-sedimentological monitoring and analysis for material sites on the Sagavanirktok river: 2015–2016 data report. University of Alaska Fairbanks, Water and Environmental Research Center, Report INE/WERC 16.02, Fairbanks, AK.

Toniolo, H., Stutzke, J., Lai, A., Youcha, E., Tschetter, T., Vas, D., Keech, J., and Irving, K. (2017a). Antecedent conditions and damage caused by 2015 spring flooding on the Sagavanirktok River, Alaska. *Journal of Cold Regions Engineering*, 31(2). https://doi.org/10.1061/(ASCE)CR.1943-5495.0000127.

Toniolo, H., Youcha, E. K., Tape, K. D., Paturi, R., Homan, J., Bondurant, A., Ladines, I., Laurio, J., Vas, D., Keech, J., Tschetter, T., and LaMesjerant, E. (2017b). Hydrological, sedimentological, and meteorological observations and analysis on the Sagavanirktok river: 2017 interim report. University of Alaska Fairbanks, Water and Environmental Research Center, Report INE/WERC 17.18, Fairbanks, AK.

Toniolo, H., Youcha, E. K., Tape, K. D., Bondurant, A., LaMesjerant, E., Ladines, I., Vas, D., Keech, J., and Laurio, J. (2018). Hydrological, sedimentological, and meteorological observations and analysis of the Sagavanirktok river: 2018 interim report. University of Alaska Fairbanks, Water and Environmental Research Center, Report INE/WERC 18.16, Fairbanks, AK.

Toniolo, H., Youcha, E. K., Bondurant, A., Ladines, I., Keech, J., Vas, D., LaMesjerant, E., and Bailey, J. (2020). Hydrological, sedimentological, and meteorological observations and analysis of the Sagavanirktok river: 2019 final report. University of Alaska Fairbanks, Water and Environmental Research Center, Report INE/WERC 20.01, Fairbanks, AK.

USGS (U.S. Geological Survey). (2019). Water-resources data for the United States. http://waterdata.usgs.gov/nwis/.

Walsh, J. E., Overland, J. E., Groisman, P. Y., and Rudolf, B. (2011). Ongoing climate change in the Arctic. *Ambio*, 40, 6–16. https://doi.org/10.1007/s13280-011-0211-z.

Wong, M., and Parker, G. (2006). Reanalysis and correction of bed load relation of Meyer-Peter and Muller using their own database. *Journal of Hydraulic Engineering*, 132. https://doi.org/10.1061/(ASCE)0733-9429(2006)132:11(1159).

3 Regional Precipitation and Streamflow Trends in the Bermejo River Basin in the Context of Global Change in Argentina

Sabrina N. Ayala, Marcela H. González, and Saeid Eslamian

3.1 INTRODUCTION

Trend detection studies of hydro-meteorological variables are particularly relevant in the context of climate change. In particular, the Intergovernmental Panel on Climate Change (IPCC) through its Fifth Assessment Report AR5 [1] expressed that

> it is likely that anthropogenic influences have affected the global water cycle since 1960. Anthropogenic influences have contributed to observed increases in atmospheric moisture content in the atmosphere (medium confidence), to global-scale changes in precipitation patterns over land (medium confidence), to intensification of heavy precipitation over land regions where data are sufficient (medium confidence), and to the changes in surface and sub-surface ocean salinity (very likely).

However, rainfall observed changes present great spatial variability in the last decades. For instance, Gemmer et al. [2] studied monthly trends at gauging stations in China in 1951–2002 and found the presence of both increments and reductions in precipitation at all of the months. Longobardi and Villardi [3] detected a predominance of negative trends in the last 30 years of the 20th century in annual and seasonal (particularly boreal winter) precipitation in southern Italy. A decrease in annual precipitation was also found by de Luis et al. [4] in Slovenia and da Silva et al. [5] in southern Portugal in the periods 1951–2007 and 1960–2000, respectively. On the contrary, changes towards more humid conditions were detected in south-eastern South America since the mid-20th century. Barros et al. [6] documented the primarily positive trends in annual rainfall for 1956–1991 in sub-tropical Argentina east of the Andes, and related this with changes in the mean latitudinal position of circulation systems. Liebmann et al. [7] found increments in austral summer precipitation in southern Brazil for 1976–1999, associated with increases in the number of rainy days and the amount of rainfall that occurred at each event. Similarly, Haylock et al. [8] detected positive trends in annual and extreme rainfall in southern Brazil, Paraguay, Uruguay, and north-eastern Argentina in the last 40 years of the 20th century, which they partially explained through a change towards more El Niño–dominating conditions using canonical correlation analysis.

Located within the La Plata River Basin in south-eastern South America, the Bermejo River Basin covers approximately 123,000 km^2, with 90% of its area in northern Argentina and the remaining 10% in southern Bolivia. The Argentinean sector of the basin (hereinafter, BRB) is placed in a region known as the Gran Chaco, eastward of the Andes Mountain range. This area is characterized by the

DOI: 10.1201/9781003473398-4

presence of the Wet Chaco to the east, the Dry Chaco in the central region, and the Yungas at the western extreme [9]. The climate in Gran Chaco is subtropical, and highly influenced by the combination of the Andes topographic barrier [10, 11], and the regional circulation associated with the Southern Atlantic high-pressure system, the intermittent thermal-orographic north-western Argentinean Low [12], and the summer Chaco low-level jet. As a result, precipitation in this area presents a pronounced annual cycle, with the lowest accumulated rainfall values during the austral winter season.

The BRB can be divided in three sub-basins: Upper Bermejo and San Francisco, located in the western area, and Lower Bermejo in the center and eastern sectors. The Bermejo River flow has a regime highly influenced by the annual cycle of precipitation, and runs in a NW–SE direction along the Argentinean–Bolivian border, meeting the Grande de Tarija River at the Junta de San Antonio. Then, it continues its NW–SE path within the Upper Bermejo until it receives the contribution of the San Francisco River on its right margin at the Junta de San Francisco. After this confluence point, the river course continues within the Lower Bermejo sub-basin, and eventually drains into the Paraguay River with little contribution from other tributaries. The BRB is also characterized by a pronounced variation in terrain elevation, particularly in the western sector: Upper Bermejo has a maximum elevation close to 5000 m, while minimum altitude is 280 m. According to the Binational Commission for the Development of the Upper Bermejo and Grande de Tarija River Basins (Comisión Binacional de la Alta Cuenca del Río Bermejo y Río Grande de Tarija, COBINABE) high sediment production rates through soil erosion are present in the region of greatest elevation of the Upper Bermejo, owing to steep terrain and intense summer precipitation events. Flooding risk in the Lower Bermejo (flat terrain) increases when incoming maximum streamflow from the upriver sub-basins surpasses the 2000 m^3 s^{-1} threshold, particularly during the rainy summer period [13].

The BRB is also the part of a region that has been through a westward expansion of the agricultural border in the second half of the 20th century, partially as consequence of the increment in annual precipitation in south-eastern South America [14, 15], which resulted in deforestation and land use changes. Dominant economic activities are focused on ranching, forestry, horticulture, and rainfed agricultural production of soybean, cotton, corn, and beans [16].

Since precipitation is a key variable in the hydrological cycle given its implications on soil moisture, surface runoff, flood risk, crop yield, human settlements, and others, the aim of this study is to detect and evaluate the spatial distribution of annual and seasonal precipitation trends using the linear regression and Mann–Kendall statistical techniques. For this purpose, the work is divided in two inter-related analyses. First, long-term trends in Argentinean territory will be studied considering the periods 1961–2018 and 1985–2018, to explore possible trend changes towards the last decades. Second, focus will be put into regional trends in the BRB and surrounding areas of northern Argentina in 1985–2018, to examine the spatial variability of changes when increasing station density and to analyse the observed trends at basin scale in the context of change at country scale. Additionally, streamflow trends in hydrological stations within the BRB will be examined and compared with the distribution of rainfall trends in the basin.

3.2 METHODOLOGY

3.2.1 TREND DETECTION

3.2.1.1 Linear Regression

Linear regression through least squares (LR) is a widely used statistical technique for the characterization and quantification of a gradual change in a hydro-meteorological variable [2, 4, 17]. The linear slope obtained through LR expresses the amount of change of the variable by unit of time. A t-test can be applied to determine if the variable presents an actual statistically significant linear slope, which is a parametric test that assumes normality and independency of data. The test statistic

t_s is calculated as the quotient between the estimate of the slope (b), obtained through least squares estimation, and the standard error of b. In a two-tailed t-test for a time series comprised of n observations, and considering a level of significance $\alpha = 0.05$, under the null hypothesis $Ho : b = 0$ the linear slope is said to be statistically significant if the absolute value of t_s surpasses the threshold that accumulates 97.5% probability in the student's t probability distribution with $n-2$ degrees of freedom.

3.2.1.2 Mann–Kendall Trend Test

The Mann–Kendall trend test (MK) is another statistical technique used for trend detection in hydro-meteorological time series [18–20]. Unlike LR, MK is a non-parametric test. Therefore, it does not require any assumptions on the distribution of the data and it can be used with asymmetrical variables. Also, MK is insensitive to the presence of extreme values in the time series.

MK allows the detection of a monotonous increase/decrease of the variable in time, that does not necessarily imply a linear trend. For a time series comprised of n observations, the parameter S counts the number of times that a certain value x_i measured at instant i is exceeded at a subsequent time j $(j > i)$ by x_j:

$$S = \sum_{i=1}^{n-1} \sum_{j=i+1}^{n} sgn\left(x_j - x_i\right) \tag{3.1}$$

with $sgn(x)$ the mathematical sign function. The variance of S is computed as

$$Var(S) = \frac{n(n-1)(2n+5) - \sum_{k=1}^{P} t_k(t_k-1)(2t_k+5)}{18} \tag{3.2}$$

where P denotes the number of groups with repeated observations and t_k is the number of repeated values belonging to the k-th group; this term is omitted in the absence of repeated observations in the time series. Combining (3.1) and (3.2), the MK's Z statistic is calculated as

$$Z = \begin{cases} \dfrac{S-1}{\sqrt{Var(S)}}, & \text{if } S > 0 \\[2mm] 0, & \text{if } S = 0 \\[2mm] \dfrac{S+1}{\sqrt{Var(S)}}, & \text{if } S < 0 \end{cases} \tag{3.3}$$

The Z statistic follows a standard Gaussian distribution and takes positive (negative) values if there is an increment (reduction) of the variable in time. In a two-tailed test considering a level of significance $\alpha = 0.05$ under the null hypothesis $Ho : Z = 0$ a statistically significant trend exists in the time series if the absolute value of Z surpasses the threshold that accumulates 97.5% probability in the standard normal distribution.

As in the case of LR, MK was designed to detect trends in time series with serial independence. If this requirement is not met, type I error of the test is increased [21]. Some authors proposed different approaches to avoid the serial dependence issue [21, 22]. In this study the block-bootstrap method was added to the MK test scheme only when the time series presents significant autocorrelation in the first lag at 95% confidence [23]. A block of longitude 5 along with 2000 simulations were considered for the block-bootstrap resampling [24], where for each simulation (sample) the MK's Z' statistic was obtained. In these particular cases, for a two-tailed test under the null hypothesis $Ho : Z = 0$ a statistically significant increasing trend with a significance level $\alpha = 0.05$ exists in the original time series if the positive Z is greater than the threshold given by the 50-th greatest values of Z' obtained from the 2,000 samples. Similarly, a statistically significant decreasing trend with a significance level $\alpha = 0.05$ exists in the original time series if the negative Z is less than the threshold given by the 50-th smallest values of Z' obtained from the 2000 simulations.

3.2.2 DATA

Monthly precipitation data from meteorological stations was used. For the change analysis at country scale (referred to as "global change"), the Argentinean National Weather Service provided data starting in 1961. The Water Administration of Chaco province and the Ministry of Water Resources of Argentina were added as sources of information for the trend study focused on the Bermejo River Basin (BRB) located in northern Argentina (hereinafter known as "regional change"). In order to increase station density in the latter case, records starting in 1985 were considered. All of the data were consisted, allowing up to 10% missing records completed with the climatological monthly mean. For trend detection, the temporal evolution of rainfall at annual (ANN) and seasonal scale was obtained for each meteorological station by accumulating monthly values. The following definition for the seasons was considered: summer from December to February (DEF), autumn from March to May (MAM), winter from June to August (JJA) and spring from September to November (SON). The spatial distribution of meteorological stations for both types of analysis is detailed in Figure 3.1.

As part of the regional study, monthly streamflow data in the BRB provided by the Ministry of Water Resources of Argentina was employed to perform hydrological trend detection analysis. The

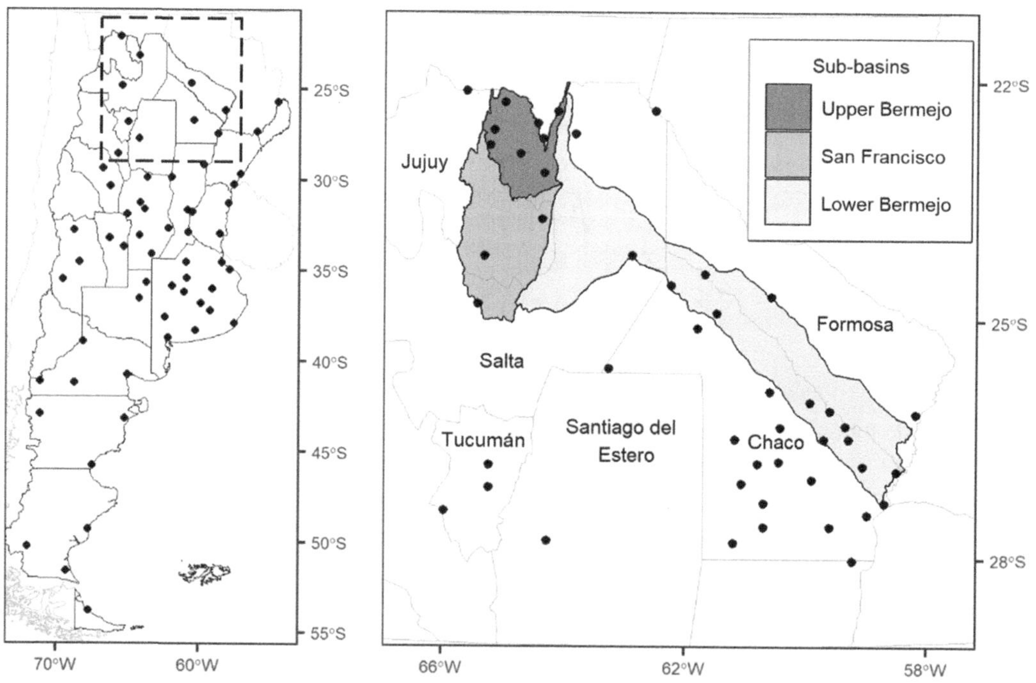

FIGURE 3.1 Left: rainfall stations used in the global trend analysis (dots). Political national and state ("provinces") boundaries are noted in black. The box indicates the relative location of the area considered in the regional analysis. Right: rainfall stations used in the regional trend analysis (dots). Black contour marks the limits of the BRB and its sub-basins. Province names are indicated.

hydrological stations considered for this purpose are Aguas Blancas and Pozo Sarmiento (both situated on the Bermejo River within the Upper Bermejo sub-basin), Caimancito (on the San Francisco River course, within the namesake sub-basin), and El Colorado (placed on the Bermejo River on the eastern sector of the Lower Bermejo sub-basin). The beginning of the hydrological year of each station was defined as the month with minimum mean monthly streamflow, not necessarily coincidental with the beginning of the calendar year. Then, the mean annual streamflow and the maximum monthly streamflow were determined for each station and for each hydrological year between 1984/1985 and 2017/2018. The relative location of the hydrological stations on the drainage system of the BRB is detailed in Figure 3.2.

The global and regional change analysis were performed by calculating the LR slope of annual and seasonal rainfall for each meteorological station, and performing t-tests of slope considering $\alpha = 0.05$. The results obtained through LR were compared to the Z statistic of MK, applying block-bootstrap in the presence of serial correlation at the first lag. The same procedure was followed to study trends in the mean annual ($Qann$) and maximum monthly ($Qmax$) streamflow of the BRB as part of the regional change analysis. It is important to note that although MK does not provide slope values as LR does, it allows studying the behavior of precipitation and streamflow along time without assuming the linearity or normality of the variables. Global change was analyzed considering two time periods: 1961–2018 (58 years) and 1985–2018 (34 years). Precipitation trend detection in the regional case was performed only in the latter period because of data availability and spatial coverage. Similarly, the analysis on $Qann$ and $Qmax$ for stations of the BRB was performed for the hydrological years comprised in 1984/1985–2017/2018.

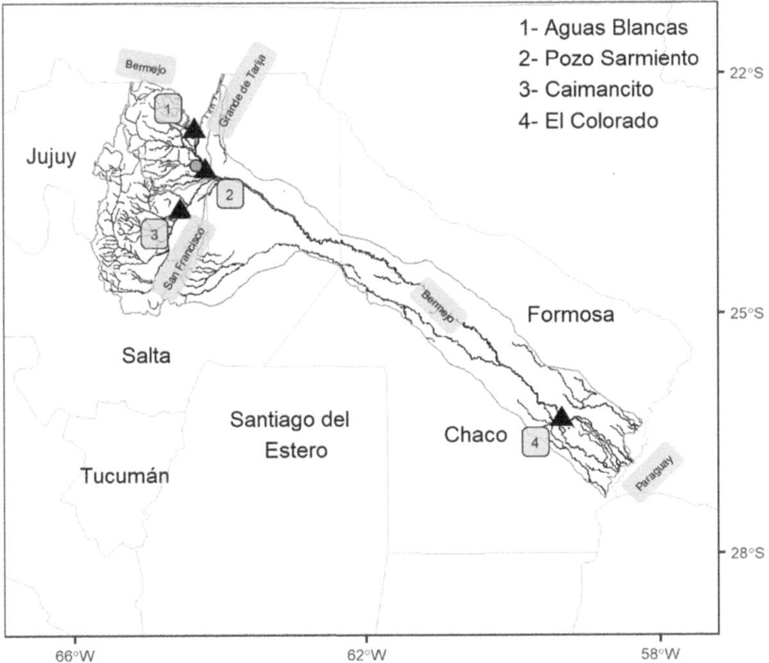

FIGURE 3.2 Streamflow stations used in the regional trend analysis (triangles) in the BRB drainage system (black contours). Names of the hydrological stations are indicated in the top right corner. Additionally, the gray dot symbolizes the position of Oran's meteorological station. The limits of the BRB and its sub-basins, as well as the names of the main river courses in the BRB are included.

3.3 RESULTS

3.3.1 Global Change Analysis

To frame the trend analysis in the climatological context of the region, mean annual and seasonal precipitation in the period 1985–2018 was obtained for the stations in the Argentinean territory (Figure 3.3). The ANN rainfall is highest (near 1800 mm) in the north-eastern extreme, and diminishes towards the west and south of the country. The lowest ANN accumulated values are found in the western sector between 30°S and 40°S, and in the Patagonia region in austral Argentina (south of 40°S) with the exception of a relative maximum in north-western Patagonia east of the Andes (40–45°S, 70–72°W). Most stations located in the center and north of Argentina show minimum accumulated seasonal rainfall during winter. At the north-western extreme of the country, summer is the season in which most of the ANN precipitation occurs, and spring constitutes the transitional season between both precipitation regimes. Compared to lower latitude stations, rainfall in southern Argentina has little variation throughout the year. The north-western Patagonia region presents maximum seasonal accumulated values between MAM and JJA.

The LR slope and MK test results for 1961–2018 are shown in Figure 3.4, alongside the ratio between the linear trend in ANN precipitation per decade and the mean ANN precipitation (named "LR quotient") considering 1985–2018 as the base period, for better comparison with the 1985–2018 global and regional analyses. There is a substantial spatial agreement between the sign of the LR slope and the Z statistic, suggesting that a monotonous change of precipitation in time can be properly represented by linear behavior. In the global analysis, most stations whose p-value of the MK test is lower than 5% coincide with the stations that present a significant linear trend with 95% confidence. The LR slope in Figure 3.4 shows that, while the change of ANN precipitation in time for the period 1961–2018 has some spatial variability in the global framework, the amount of stations with positive significant linear trends is greater than the number of cases with negative significant slope. The eastern sector of Argentina north of 35°S presents an increment in the ANN rainfall of approximately +5 mm year^{-1}, whereas a less intense trend (< +3 mm year^{-1}) is detected in the western sector in the latitude band 30°–35°S. The positive changes in precipitation are also present

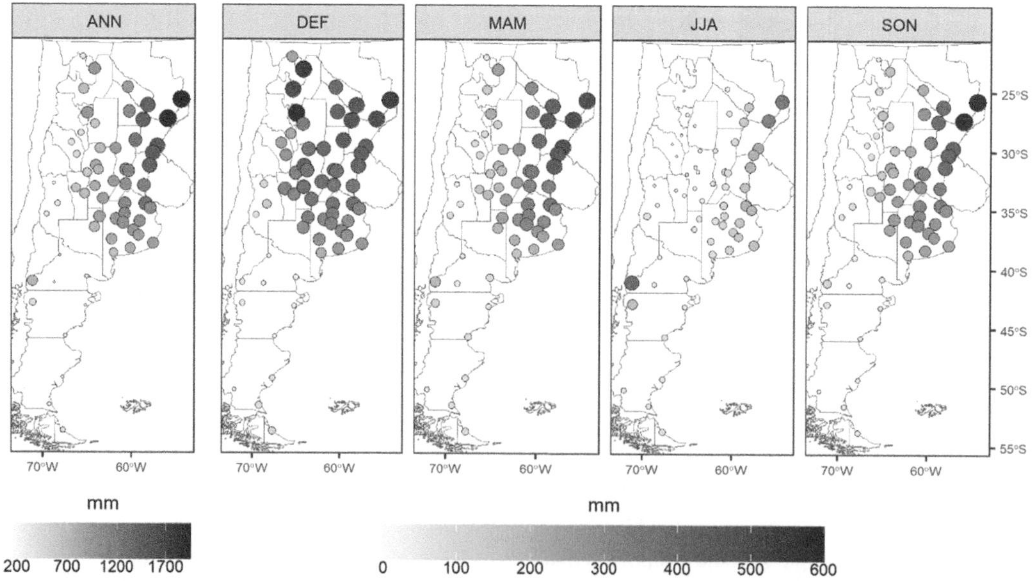

FIGURE 3.3 Mean annual (ANN) and seasonal precipitation (DEF, MAM, JJA, SON) at the meteorological stations considered in the global change analysis (1985–2018 base).

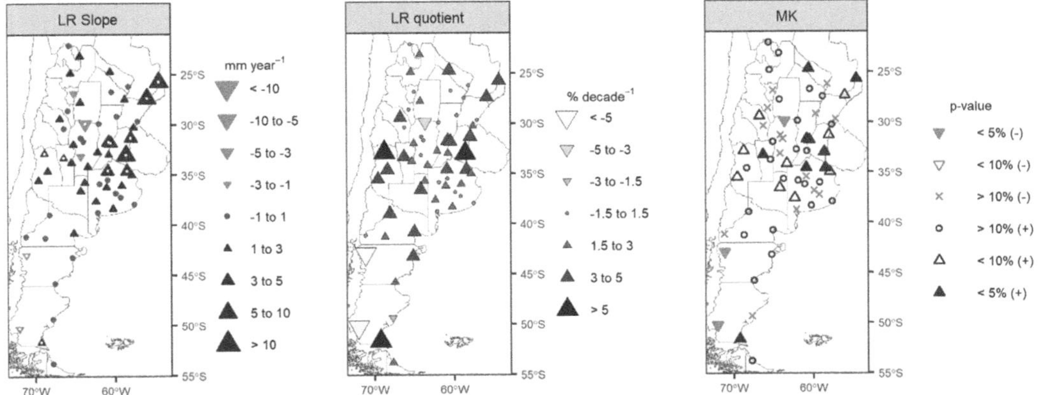

FIGURE 3.4 Global change fields for ANN precipitation in 1961-2018. Left column: LR slope. Gray (black) markers indicate a negative (positive) linear slope. Markers with a white dot indicate a significant slope at 95% confidence (t-test). Middle column: LR quotient, defined as the linear trend per decade in ANN precipitation relative to ANN mean precipitation (1985–2018). Downwards (upwards) triangles indicate negative (positive) change. Right column: p-value of MK's Z statistic. Gray (black) markers indicate negative (positive) Z.

in north-western Argentina, though not significant. In the central region, specific stations show weak decreases in the ANN rainfall, only statistically relevant in the north of Cordoba province (near –3.4 mm year^{-1}). Between 35°S and 40°S, there is a predominance, in both the number of cases and magnitude of linear trend, of increasing ANN rainfall. Station density in the Patagonia region is lower than in the rest of the territory. However, negative trends are present in the Andean stations (significant ones near –2 mm year^{-1}) while positive slopes prevail on the eastern Patagonian coast, only significant in the southern extreme (near +1.6 mm year^{-1}). The LR quotient in Figure 3.4 shows the areas where the change measured by the LR slope is most relevant relative to the climatological ANN precipitation. The highest positive values in the LR quotient are mainly concentrated in the central region of Argentina (between 30°S and 37°S), though increments in the ANN rainfall over +3% decade^{-1} are also present in north-eastern subtropical latitudes and in the Patagonian coast. The greatest decreases in the ANN precipitation relative to the climatological mean are maximum in western Patagonia (between –5.0% decade^{-1} and –12.2% decade^{-1}) where LR slope proved to be significant at 95% confidence according to the t-test, and are followed in intensity by a decrease in northern Cordoba province (near –4.4% decade^{-1}).

Since precipitation has a noticeable seasonality, with time of occurrence and intensity of peaks dependent upon sub-regions (Figure 3.3), the seasonal trends were studied in the global framework. Figure 3.5 shows the LR slope and MK test results for seasonal rainfall in the years 1961–2018. Once again, the direction of the change represented by the Z statistic agrees with the sign of the LR slope, with the exception of a few stations in which signs differ but the magnitudes of Z and linear slope are small. Under the t-test at 95% confidence, stations with significant trends are fewer and more scattered in each of the seasonal fields than in the ANN counterpart, meaning that the magnitude of the change seen in Figure 3.4 is probably due to smaller changes taking place along the year, rather than a single season concentrating most of the variation. The spatial pattern of trends in DEF resembles the pattern in the ANN case, although there is a contribution from other seasons to obtain the final ANN trend magnitudes. Changes in rainfall during MAM and SON strengthen the summer trends, particularly in subtropical eastern Argentina. During JJA, the season with the lowest mean accumulated precipitation for most subtropical stations (Figure 3.3), negative trends prevail in the north and center of the territory, being statistically significant in the latter region. On the other hand, in the southern Andes the ANN negative trend (Figure 3.4) is mostly due to a decrease in winter precipitation (Figure 3.5), the season in which climatological values are highest. Stations located on

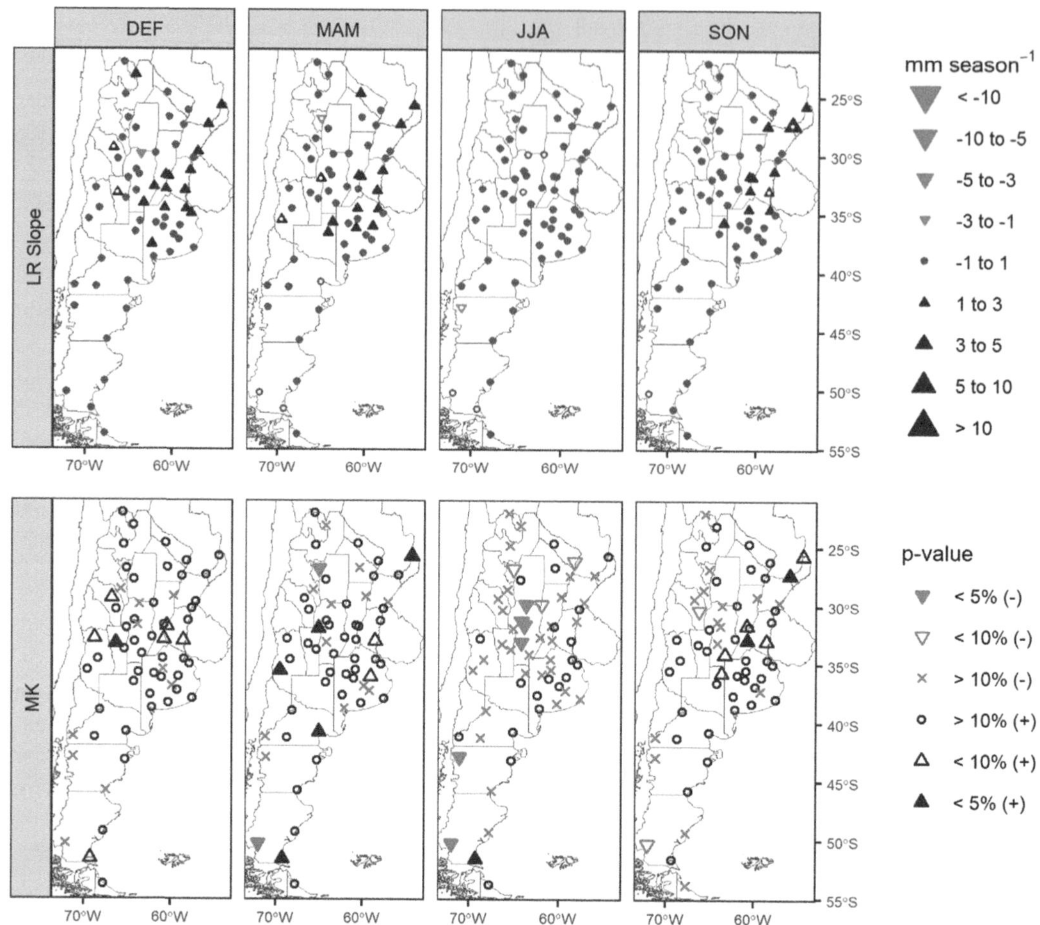

FIGURE 3.5 Global change fields for DEF, MAM, JJA, and SON precipitation in the 1961–2018 period. First row: LR slope. Gray (black) markers indicate a negative (positive) linear slope. Markers with a white dot indicate a significant slope at 95% confidence (T-test). Second row: p-value of MK's Z statistic. Gray (black) markers indicate negative (positive) Z.

the Patagonian eastern coast show little variation in the direction of change from season to season, reinforcing the behavior detected in the ANN LR slope field (Figure 3.4).

Since trends in time series are strongly dependent on the period considered, values of LR slope and MK's Z statistic were obtained for ANN (Figure 3.6) and seasonal (Figure 3.7) rainfall in 1985–2018 (last 34 years of the time period analyzed above), along with the LR quotient for the ANN trend. As in the study of the previous period, for this quotient, the ANN slope was divided by the 1985–2018 mean ANN rainfall.

Once again, in the 34-year period the LR slope serves as a good estimator of the direction and significance of change of the ANN and seasonal rainfall compared to the non-linear non-parametric MK test (Figure 3.6 and 3.7). An evident difference between the two time periods is a greater proportion of stations with a decrease in ANN and seasonal rainfall in 1985–2018 compared to the 58-year period (Figure 3.4 and Figure 3.5). The spatial distribution of changes in the sign of the LR slope between both periods is detailed in Figure 3.8.

The change in the sign of the linear trends of ANN precipitation predominates north of 28°S (region where the BRB is located) (Figure 3.8) indicating a decrease in annual accumulated values in the second period (Figure 3.6), whose magnitude surpasses the rate of increase observed in the

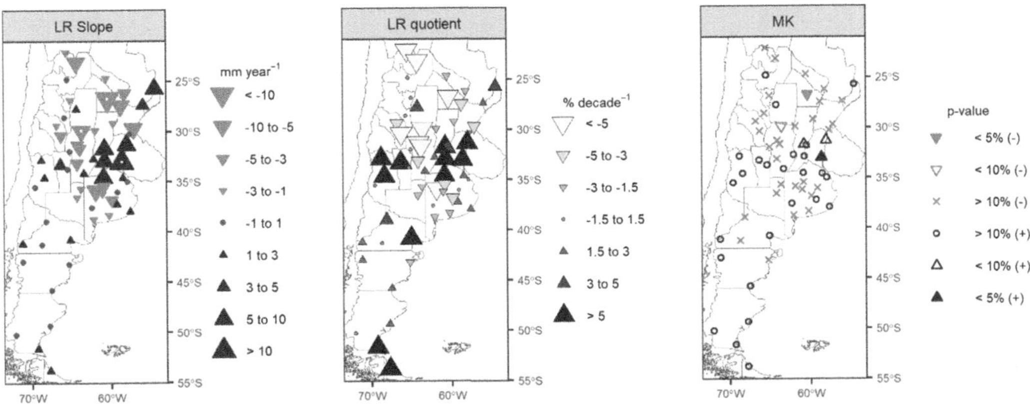

FIGURE 3.6   Same as in Figure 3.4, for ANN precipitation in 1985–2018.

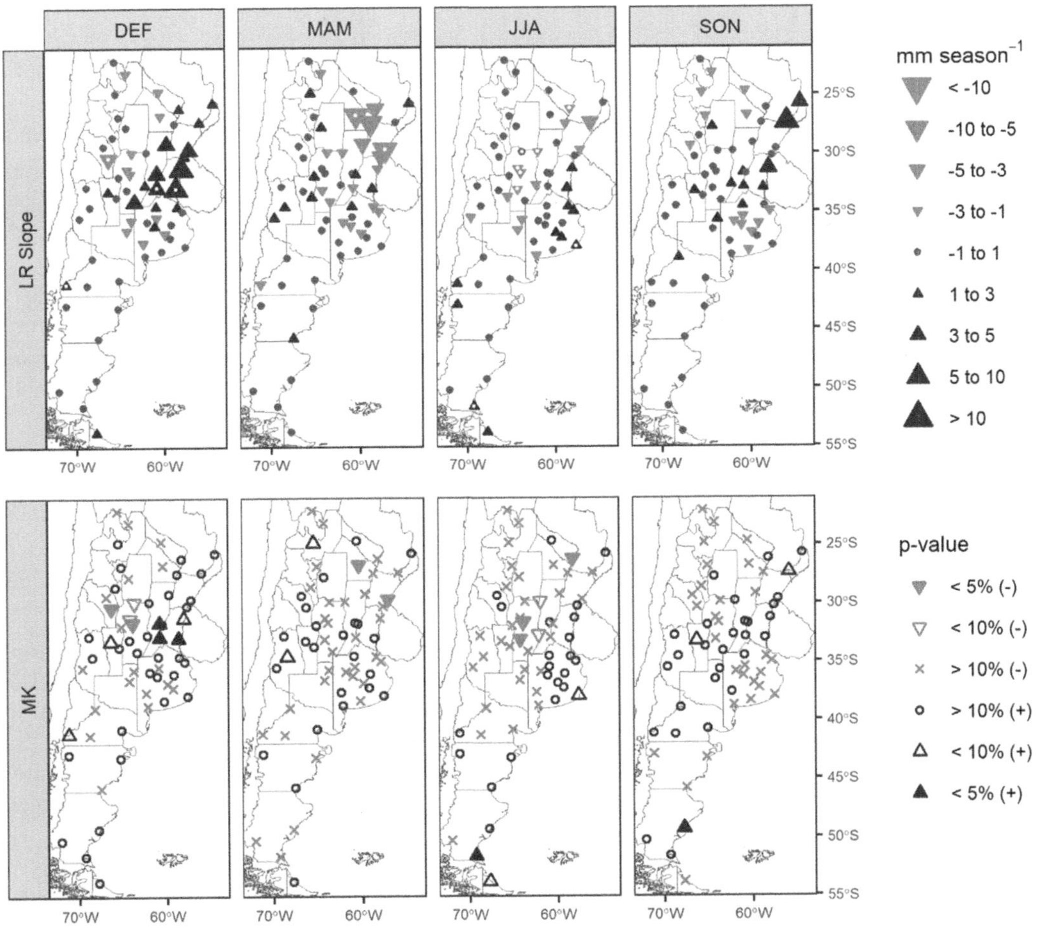

FIGURE 3.7   Same as in Figure 3.5, for seasonal precipitation in 1985–2018.

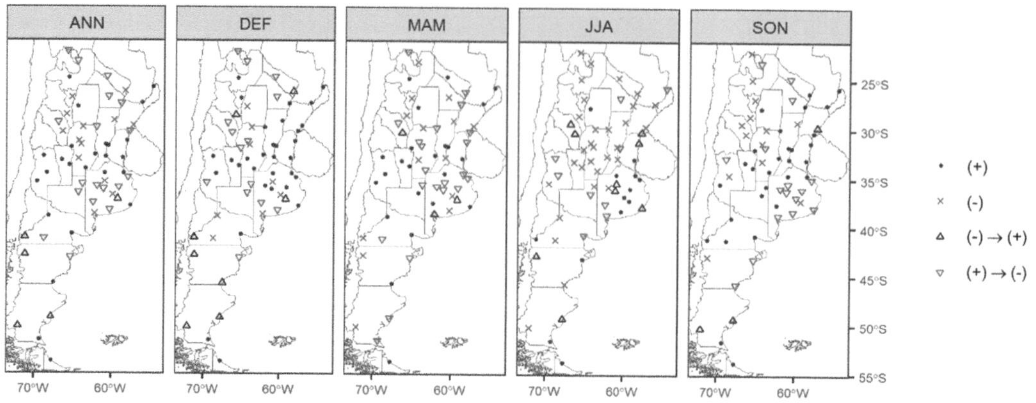

FIGURE 3.8   Variation in LR slope between 1961–2018 and 1985–2018. Gray crosses (black dots) indicate the presence of a negative (positive) slope in both periods. Gray downwards (black upwards) triangles mark a change in the sign of the slope, passing from positive (negative) in 1961–2018 to negative (positive) in 1985–2018.

first period considered (Figure 3.4). In north-western Argentina, the LR slope in the years 1985–2018 reached as high as –5 mm year^{-1} and represents a negative LR quotient with a magnitude of 5% decrease in ANN rainfall per decade relative to the climatological value (Figure 3.6), a much stronger rate of change than the one observed in the 1961–2018 quotient (Figure 3.4). The significant negative trends are found in the south-eastern sector of the subtropical latitude band in Chaco province (26°S–60°W), with an LR slope of –11 mm year^{-1} and an LR quotient of –10% decade^{-1} (Figure 3.5).

Stations that maintain the same increasing/decreasing behavior in the ANN rainfall in the last 34 years dominate the eastern and central regions of Argentina approximately between 28°S and 35°S (Figure 3.6 and Figure 3.8). In this sector the intensification of the trends is noticeable, accompanied by an increase in the magnitude of the LR quotient compared to the longer time period (Figure 3.4), indicating swifter changes relative to the mean ANN rainfall in the shorter period. However, no changes in statistical significance are detected in the stations with a negative LR slope, while a loss of significance according to both the t-test and MK test is identified in the eastern sector containing positive trends.

As in the case of northern Argentina, the 35°S–40°S latitude band shows an increment in the number of stations with negative linear trends of ANN rainfall in the 1985–2018 period (Figure 3.6 and Figure 3.8), though there is no statistical significance according to the t-test under 5% significance and the p-values associated with MK's Z are over 10%. The opposite variation in the sign of the LR slope and the Z statistic is observed in the Patagonian Andes south of 40°S, indicating a shift from decreasing (1961–2018) to increasing (1985–2018) ANN rainfall (Figure 3.6 and Figure 3.8). Nonetheless, the rate of relative growth in the latest years is less intense than the rate of relative reduction in the 54-year period, as shown by the LR quotient. Additionally, Figure 3.6 shows that the rise in precipitation is non-significant according to the t-test (with $\alpha = 0.05$) and the MK test (p-value > 10%). On the Patagonian eastern coast, most stations during the 1985–2018 period observed the positive LR slopes in ANN precipitation detected in the 54-year period (Figure 3.6 and Figure 3.8), albeit with some increases in the relative importance of the trends (> +5% decade^{-1}) compared to the 1961–2018 LR quotient field in Figure 3.4 (< +5% decade^{-1}).

Most stations located north of 28°S, where the BRB is situated, exhibit negative LR slopes in at least two seasons and contribute to the decrease observed in northern Argentina in 1985–2018 (Figure 3.7). In particular, the change of the sign of the LR slope when considering the last 34 years occurs primarily in DEF, MAM, and SON while the negative slope is maintained during winter

(Figure 3.8). The significance of the decrease in seasonal rainfall is mainly concentrated on the eastern sector near the 28°S parallel during MAM and JJA, and the absolute value of the linear slope in this area is maximum during autumn (between –3 mm season^{-1} and –7 mm season^{-1}) which represents a great proportion of the change in the ANN rainfall of the area (Figure 3.6). On the latitude band comprised in 28°S–35°S the sign reversal of the LR slope throughout the year is mostly towards decreasing seasonal precipitation. However, the spatial distribution of the sign reversals and the magnitude of the LR trends are such that the ANN rainfall trends are mainly maintained in this sector when considering the 34-year period (Figure 3.8). Decreasing precipitation trends in the 1985–2018 period are significant at 95% confidence and intensified with respect to the longer time period in the central region of the 28°S–35°S latitude band, particularly during winter. On the other hand, positive LR slopes in the summer are higher in the more recent period on the eastern side of the same latitude band, reaching between +4.5 mm season^{-1} and +8.1 mm season^{-1} in the significant cases according to the t-test (Figure 3.7). Between 35°S and 40°S the sign conversion observed towards the last 34 years is spatially scattered according to the season, with a predominance of stations whose LR slope becomes negative in the 1985–2018 period, giving as a result the sign conversion field of ANN rainfall (Figure 3.8). Seasonal LR slopes in this sector are less than 3 mm season^{-1} in magnitude, and trends are predominantly non-significant following the t-test and the MK test (Figure 3.7). Unlike in the 1961–2018 period, in the last 34 years the stations located in the Patagonian Andes show positive trends during DEF (Figure 3.8), significant in north-western Patagonia at 95% and 90% confidence according to the t-test and the MK test respectively (Figure 3.7), with a LR slope near +1.2 mm season^{-1}. During JJA, the most precipitating season of the year (Figure 3.3), positive trends are also present in north-western Patagonia in the 34-year period though not significant (Figure 3.7 and Figure 3.8). On the south-eastern Patagonian coast seasonal trends in the 1985–2018 period are mostly positive throughout the year. In north-eastern Patagonia, the LR slopes' magnitude is less than 1 mm season^{-1} (Figure 3.7), albeit the Argentinean eastern coast south of 40°S has the climatological low values of ANN rainfall compared to northern stations (Figure 3.3).

3.3.2 Regional Change Analysis

After studying the behavior of trends in ANN and seasonal rainfall in the global framework, considering its variations over the last 34 years, the regional analysis was addressed for the Argentinean north with a focus on the BRB. Precipitation and streamflow trend detection in the regional case was performed only in the 1985–2018 period because of data availability and spatial coverage north of 28°S. The change analysis of both variables was performed considering LR and MK techniques, as in the global framework.

3.3.2.1 Precipitation Trends

Values of LR slope, LR quotient, and significance of MK's Z for ANN and seasonal rainfall are depicted in Figure 3.9 and Figure 3.10 in 47 stations distributed in the provinces of Jujuy, Salta, Formosa, Chaco, Tucuman, and Santiago del Estero. Part of these stations are located inside the limits of the BRB. Additionally, Table 3.1 details the number of stations in the region with positive or negative trends according to each technique.

As before, there is good spatial agreement between the distribution of LR trends in the ANN rainfall and the sign of MK's Z (Figure 3.9). Table 3.1 indicates that only in one out of 47 stations (located in 24.9°S–61.5°W) there is a contradiction between the direction of change of ANN precipitation suggested by both techniques, with LR showing a positive slope of +3.0 mm year^{-1}. Nonetheless, the change is non-significant according to the t-test (95% confidence) and the MK test (p-value > 10%). Another small difference in the results derived from these methods lies in the number of stations with significantly decreasing ANN rainfall in the 1985–2018 period (5 for LR versus 3 for MK), though both methods coincide in detecting noteworthy

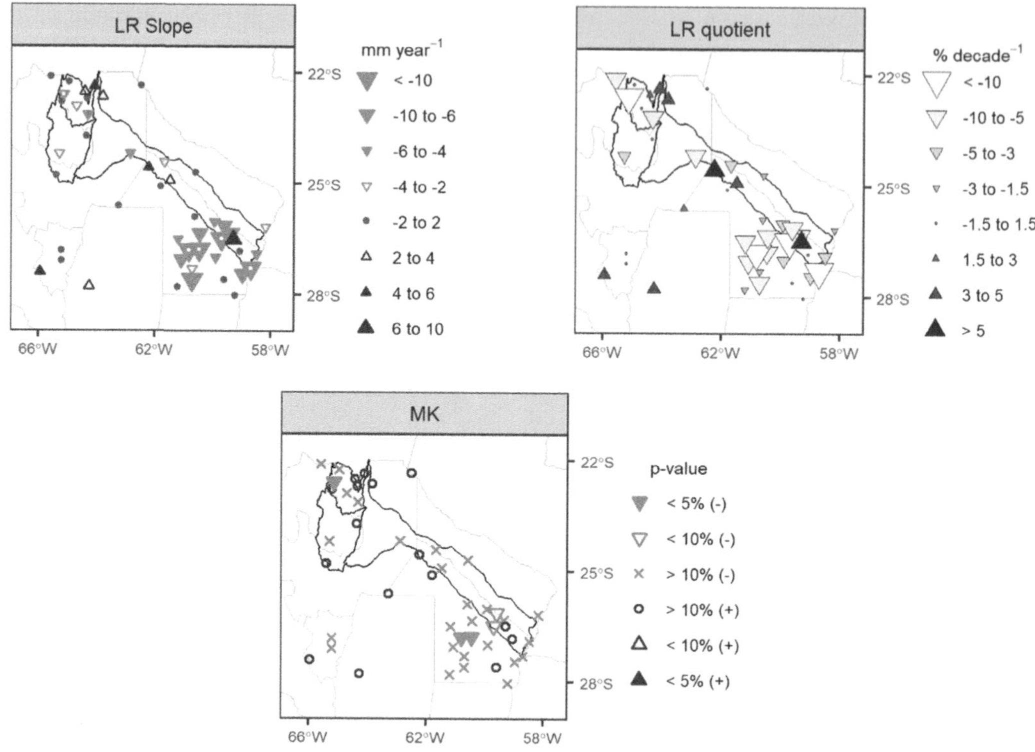

FIGURE 3.9 Regional change fields for ANN precipitation in 1985–2018. First row (left): LR slope. Gray (black) markers indicate a negative (positive) linear slope. Markers with a white dot indicate a significant slope at 95% confidence (t-test). First row (right): LR quotient, defined as the percent of change in ANN precipitation per decade relative to ANN mean precipitation (1985–2018). Downwards (upwards) triangles indicate negative (positive) change. Second row: p-value of MK's Z statistic. Gray (black) markers indicate negative (positive) Z. Black contour marks the BRB and its sub-basins borders.

reductions in ANN rainfall in the southern sector of the province of Chaco, and southern Lower Bermejo (Figure 3.9).

By increasing the density of stations in the regional analysis, the spatial variability in trends can be addressed. Decreasing the ANN precipitation noticed in the global analysis (Figure 3.6) spreads throughout most stations located in southern Chaco when considering the regional framework (Figure 3.9). Statistically significant LR trends according to the t-test oscillate between –10.8 mm year^{-1} and –17.3 mm year^{-1} in this area, representing a reduction relative to the ANN climatological precipitation of –9.9% decade^{-1} to –15.5% decade^{-1} following the LR quotient. These constitute the highest rates of reduction (as seen through both the LR slope and the LR quotient) of ANN rainfall in the Argentinian territory when compared to the global analysis (Figure 3.6).

West of 62°W the variability in the sign of the trends increases, although in the San Francisco and Upper Bermejo sub-basins the stations with the most relevant changes according to the LR quotient present negative slopes (Figure 3.9). In north-western Salta, on the western sector of the Upper Bermejo, there is a significant reduction in ANN rainfall (at 95% confidence through the t-test and MK test) of –4.7 mm year^{-1} which represents over 10% reduction per decade in ANN precipitation relative to the climatological values. The significance of this change was not detected in the previous global analysis.

In the 1985–2018 period the number of stations with negative seasonal LR slopes in northern Argentina doubled the stations with increasing seasonal precipitation, with the exception of DEF

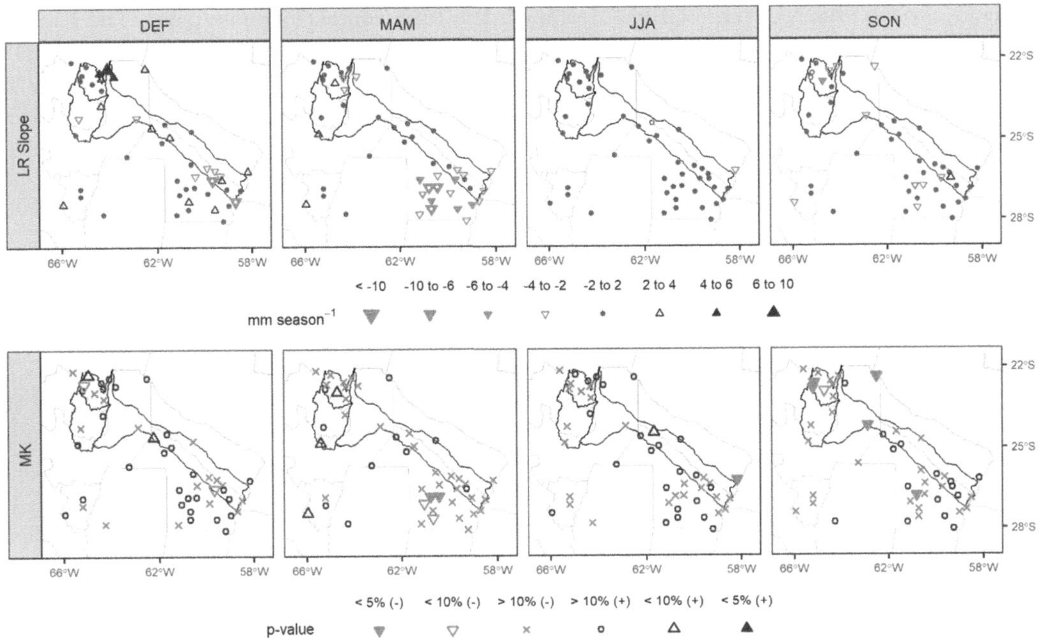

FIGURE 3.10 Regional change fields for DEF, MAM, JJA, and SON precipitation in 1985-2018. First row: LR slope. Gray (black) markers indicate a negative (positive) linear slope. Markers with a white dot indicate a significant slope at 95% confidence (t-test). Second row: p-value of MK's Z statistic. Gray (black) markers indicate negative (positive) Z. Black contour marks the BRB and its sub-basins borders.

TABLE 3.1

Number of stations in regional analysis that present increasing or decreasing rainfall trends according to the LR and MK techniques. Between parenthesis is the number of stations whose trend is statistically significant in each technique (p-value < 5%)

	Increasing trend		Decreasing trend	
1985-2018	**LR**	**MK**	**LR**	**MK**
ANN	15 *(0)*	16 *(0)*	32 *(5)*	31 *(3)*
DEF	25 *(0)*	29 *(0)*	22 *(2)*	18 *(0)*
MAM	15 *(0)*	13 *(0)*	32 *(4)*	34 *(2)*
JJA	15 *(1)*	23 *(0)*	32 *(1)*	24 *(1)*
SON	14 *(0)*	15 *(0)*	33 *(4)*	32 *(5)*

(Table 3.1). Additionally, winter is the season with the largest differences between the LR slope and the MK test results, although JJA is the one with less climatological rainfall in the region (Figure 3.3) and the magnitude of the linear trends and MK's Z is small (Figure 3.10).

Figure 3.10 shows that the significant reduction in the ANN precipitation in the south-eastern sector, south of the Lower Bermejo, is mainly due to the decrease in seasonal rainfall during MAM, and partially during DEF and SON. Negative LR slopes in autumn are over 2 mm season^{-1} in magnitude, with significant trends between –5.4 mm season^{-1} and –7.5 mm season^{-1}. In this area, the total reduction in MAM precipitation during the period 1985–2018 represents over 60% of the mean autumn rainfall. MAM precipitation is mostly decreasing at stations located within the Lower

Bermejo, though with weaker LR slopes compared to their southern counterparts. The strongest significant trends on the eastern extreme of the Lower Bermejo are negative and are found in DEF (near −6 mm season^{-1}). The middle section of the sub-basin (west of 60°W) presents relatively low LR slopes throughout the year, though in this area the MK test detects a significant increasing trend in DEF and JJA (90% confidence; the latter one also identified through the t-test with 95% confidence), and a significant reduction in the SON precipitation (95% confidence).

Figure 3.10 shows that the San Francisco sub-basin also presents negative LR slopes in spring, when the climatological dry season ends (Figure 3.3), although with no statistical significance according to the t-test and the MK test. In summer and autumn, the sign of the trends varies within the sub-basin, with MK detecting a significant rise in MAM rainfall (90% confidence) in the southern San Francisco basin. This increment (< +4 mm season^{-1}) is not statistically relevant according to the t-test.

Finally, the Upper Bermejo sub-basin presents a negative LR slope and Z statistic at every station considered in the regional analysis in spring. In particular, this decrease is significant through both the statistical tests (95% confidence for the t-test; p-value < 10% for the MK test) on the western sector of the sub-basin; though the magnitude of these trends is less than 2 mm season^{-1}, the total reduction in the SON precipitation during the period 1985–2018 represents over 20% more than the mean spring rainfall in the area. During DEF, the climatologically most precipitating season (Figure 3.3), the highest LR slopes are positive and higher than +4 mm season^{-1} in north-eastern Upper Bermejo. However, these trends are not statistically significant according to the t-test and the MK test.

3.3.2.2　Streamflow Trends

Figure 3.11 shows the mean annual cycle of streamflow for the hydrological stations depicted in Figure 3.2, together with the mean annual cycle of rainfall for two locations within the BRB: Oran, situated in the western sector of the BRB, and El Colorado, located in the eastern extreme (notice this station measures both rainfall and streamflow). Streamflow in all stations presents seasonality, with a peak in the warm season and a marked descent towards the winter. The hydrological year begins in September for the Upper Bermejo stations, and a month later in Caimancito and El Colorado. Lower streamflows are found in Aguas Blancas, located upriver and close to the confluence between the Bermejo and the Grande de Tarija Rivers, and in Caimancito in the San Francisco sub-basin. On the other hand, after receiving the contribution from its upriver tributaries, Pozo Sarmiento and El Colorado have the largest streamflow values with peaks over 1000 m^3 s^{-1}. In particular, El Colorado presents a slower decrease in streamflow towards the winter compared to the other stations. This behavior could be due to the combination of a weaker slope in the terrain of the Lower Bermejo, the contribution from the precipitation on the Wet Chaco region in the east, and the sum of streamflows from the Upper Bermejo and the San Francisco sub-basins.

The rainfall cycles depicted in Figure 3.11 serve as an approximation of the precipitation regimes typical of the BRB. As seen in Section 3.3.1, higher accumulated values are found in the warm season, with the annual range of rainfall increasing towards north-western Argentina. The simultaneity of the precipitation and streamflow cycles suggests that precipitation highly controls the flow regime, particularly in the Upper Bermejo and San Francisco sub-basins where a swift reduction in rainfall is accompanied by rapidly decreasing streamflows. Therefore, it is interesting to analyse if the negative trends found in ANN rainfall in western Upper Bermejo and eastern Lower Bermejo in the 1985–2018 period translate into changes in streamflow for the hydrological years spanning from 1984/1985 to 2017/2018. Thus, values of LR slope and MK's Z were obtained for the time series of the mean annual ($Qann$) and maximum monthly ($Qmax$) streamflow of the hydrological stations considered; the results are depicted in Figure 3.12.

Despite the presence of negative LR trends of ANN rainfall in the western Upper Bermejo and San Francisco sub-basins (Figure 3.8), LR and MK suggest that increasing $Qann$ is present in Aguas Blancas, Caimancito, and Pozo Sarmiento (+1.5, +1.2, +10.2 m^3 s^{-1} decade^{-1}, respectively).

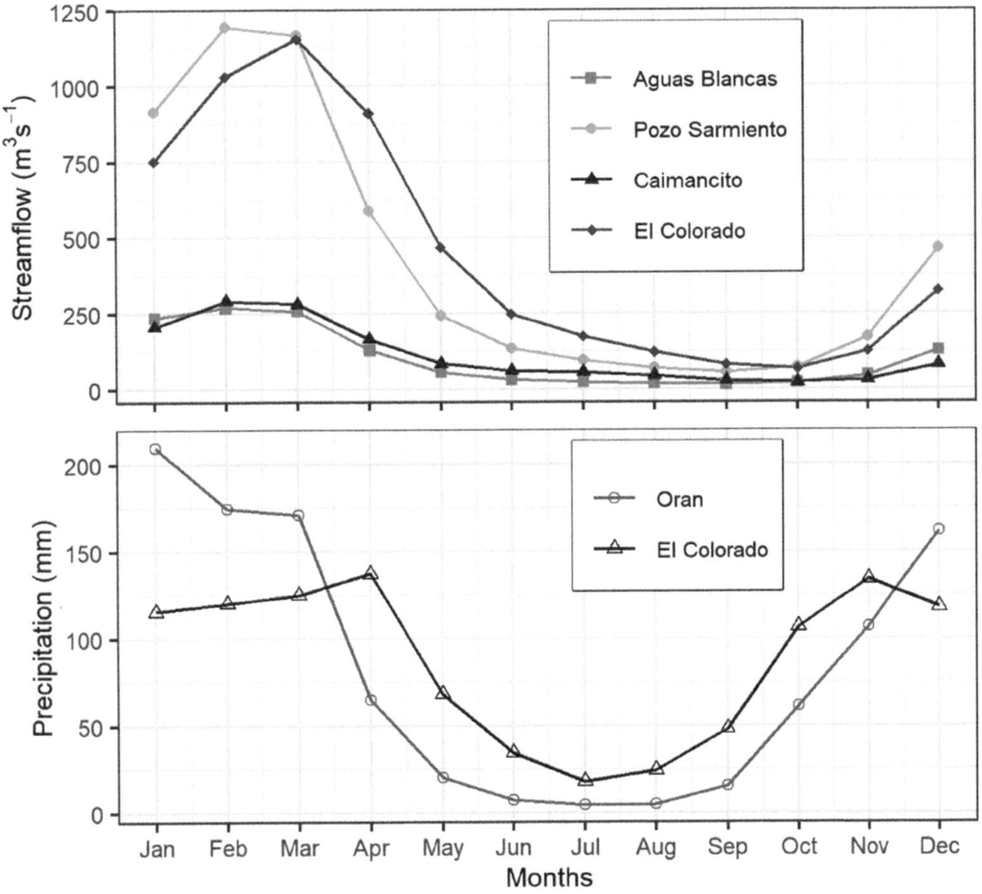

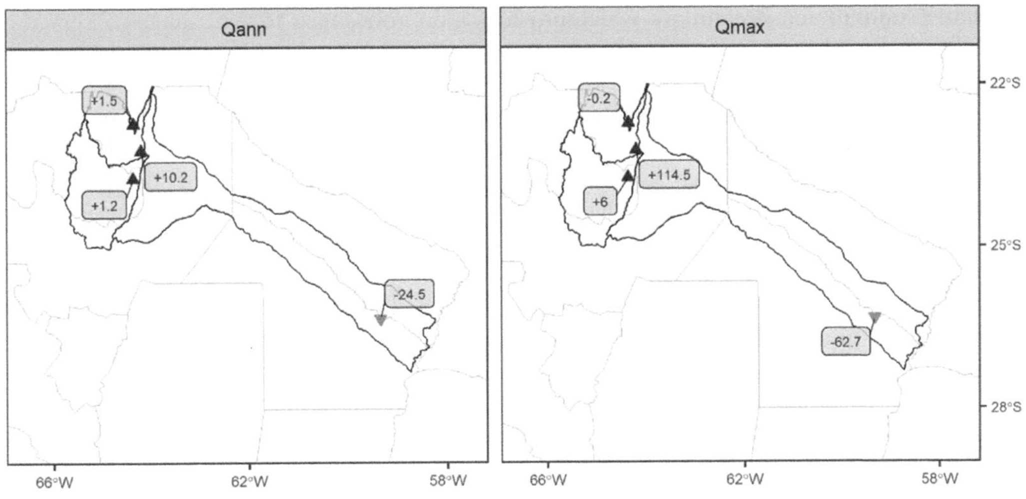

FIGURE 3.11   Top: mean annual cycle of streamflow for the hydrological stations in the BRB. Bottom: mean annual cycle of rainfall for two stations located within the BRB.

FIGURE 3.12   Trend analysis for the hydrological year 1984/1985–2017/2018 in the BRB, considering $Qann$ (left) and $Qmax$ (right). In both cases, black (gray) triangles indicate a positive (negative) Z statistic. Numbers indicate the value of the LR slope in $m^3\ s^{-1}\ decade^{-1}$.

However, these trends are weak when divided by the long-term (1985–2018) mean annual streamflow $\overline{Qann}$: $Qann$ in Pozo Sarmiento, the station with the highest relative trend, increases only +2.4% per decade. On the other hand, negative LR slope and MK's Z for $Qann$ is present in El Colorado, situated near the region with the strongest reduction in ANN rainfall when compared to other meteorological stations of the global and regional framework. However, the relative decrease associated with this LR slope (-24.5 m^3 s^{-1} decade^{-1}) only represents a 5.4% reduction of Qann per decade when compared to the long-term mean annual streamflow $\overline{Qann}$. When performing the t-test and the MK test, none of the stations present a significant $Qann$ trend in the period 1984/1985–2017/2018 (95% confidence).

A similar pattern of change is present in the $Qmax$ case, with mainly positive trends in the north-western sub-basins while the opposite is observed in the eastern sector of the Lower Bermejo. Rates of change of $Qmax$ for Pozo Sarmiento and El Colorado are the strongest in the BRB according to the LR slope (+114.5 m^3 s^{-1} decade^{-1} and -62.7 m^3 s^{-1} decade^{-1}, respectively) and also relative to the long-term mean of maximum streamflow $\overline{Qmax}$ (+7.6% decade^{-1} and -5.0 % decade^{-1}, respectively). As in the case of $Qann$, none of the stations present a significant trend of $Qmax$ according to the t-test and the MK test (95% confidence). However, it is noteworthy that in DEF and MAM, the seasons when the highest monthly streamflow in El Colorado typically occurs (Figure 3.11), rainfall trends are mainly negative in the eastern Lower Bermejo as well as south of the sub-basin.

3.4 CONCLUSIONS

The purpose of this work was to identify and study the spatial distribution of trends in annual (ANN) and seasonal rainfall in Argentina through Linear Regression (LR) and Mann–Kendall (MK) trend detection techniques, with emphasis on the Bermejo River Basin (BRB) situated in northern Argentina. The periods 1961–2018 and 1985–2010 were considered in the analysis of changes at the country scale (global framework), while the latter was the focus of study when addressing trends in the BRB (regional framework). Streamflow trend detections for the hydrological stations located within the BRB were also performed through the LR and MK techniques, considering the hydrological areas spanning from 1984/1985 to 2017/2018. A good spatial agreement was found between the LR and MK results in terms of the spatial distribution and significance of trends in rainfall and streamflow, which provides more robustness to the changes observed and quantified through LR and suggests that a great proportion of the changes in the variables can be explained through a linear behavior.

During the 1961–2018 period, the positive trends in ANN rainfall predominate in the global framework, with the highest significant increments mainly located at the subtropical latitudes in eastern Argentina ($>$ +3 mm year^{-1}), as was found in previous studies during the second half of the 20th century. The highest significant reductions in the ANN precipitation relative to the climatological mean were present in the Patagonian Andes (between -5.0% decade^{-1} and -12.2% decade^{-1}), in concordance with the decreasing trends detected by Castañeda and González [25] in the 1950–1999 period. Despite the inter-seasonal variability of the sign and magnitude of the trends, winter is the season with the greatest predominance of stations that present precipitation reduction rates. The rainfall trend spatial patterns changed noticeably when considering the most recent 34-year period, since the number of stations with the negative slopes in ANN rainfall increased. The sign reversal towards decreasing rates was mainly focused in northern Argentina and in the region within the 35°S–40°S latitude band, with seasonal rainfall showing trend sign reversal mostly in DEF, MAM, and SON.

When increasing station density in the regional change analysis for the 1985–2018 period the spatial variability of slopes becomes more pronounced. However, as seen in the global framework, there is a predominance of negative trends in the ANN rainfall in northern Argentina. The strongest generalized reductions in both regional and global frameworks are concentrated in the eastern extreme of the Lower Bermejo and south of Chaco province, reaching up to a 15.5% decrease per decade in the ANN rainfall relative to the climatological mean. Negative significant trends are also found on the western side of the Upper Bermejo. These reductions are mostly due to decreases in seasonal precipitation in summer and autumn for the eastern Lower Bermejo sector, and in spring for the Upper sub-basin.

González et al. [26] found a similar spatial pattern in the inversion of trends in northern Argentina when considering time periods beginning closer to the 21st century. Barros and Camilloni [15] observed an alike behavior in the stations located in the center and north-eastern Argentina, and suggested that the multi-decadal variability could be a factor responsible for such trend reversal. On the other hand, Zak et al. [27], Grau et al. [28], Piquer-Rodríguez et al. [29], and Torrella et al. [30] quantified the native vegetation area lost to agricultural and grazing activities in different sectors of the Dry Chaco in the second half of the 20th century, finding accelerated rates of deforestation since the 1990s (with variations according to the area considered). Transformations in land cover can lead to modifications in the water balance of ecosystems, altering water quality and fertility of the soil [31]. Moreover, land use change in the Dry Chaco can impact the precipitation regime through changes in surface albedo and in latent and sensible heat fluxes in the soil-atmosphere interface [32]. The transition from the native forests to croplands can potentially cause drier conditions that could serve as a possible explanation for the trends observed in rainfall in the Argentinean north in the last 34 years.

The streamflow trends in the BRB for the hydrological year spanning the periods 1984/1985–2017/2018 were non-significant. However, El Colorado station in eastern Lower Bermejo presents reductions in the mean annual and maximum monthly streamflow which are opposite to the behavior seen in upriver hydrological stations, and coincides with the sign of the ANN rainfall trend found in south-eastern Chaco province. Since precipitation controls the flow regime in the BRB, the decrease in streamflow in El Colorado could partially be due to the reduction in precipitation in the area rather than to the flow changes occurring upriver. Nonetheless, the other variables involved in the hydrological cycle must be taken into consideration to fully comprehend the long-term streamflow trends found in the BRB, such as changes in air temperature, evapotranspiration, soil infiltration, and land use and/or coverage, as well as multi-decadal atmospheric variability. In this sense, recently Lauro et al. [33] found a direct association between the Pacific Decadal Oscillation and mean annual streamflow in the western BRB sub-basins. The effective causes of trends, and the attribution of the anthropogenic influence on rainfall and streamflow variability in the BRB require further future investigation.

ACKNOWLEDGMENTS

Data was provided by the Argentinean National Weather Service (SMN), the Water Administration of Chaco province, the Ministry of Water Resources of Argentina, and the Regional Commission of the Bermejo River (COREBE). This research was supported by 2020-2022 UBACyT 20020190100090BA, 2018–2020 UBACyT 20620170100012BA, and UBACYT 2017–2019 20020160100009ba projects.

REFERENCES

1. IPCC. (2013). Summary for policymakers. In Stocker, T. F., Qin, D., Plattner, G.-K., Tignor, M., Allen, S. K., Boschung, J., Nauels, A., Xia, Y., Bex, V., and Midgley, P. M. (Eds.), *Climate Change 2013: The Physical Science Basis. Contribution of Working Group I to the Fifth Assessment Report of the Intergovernmental Panel on Climate Change.* Cambridge University Press, Cambridge and New York, NY, 17.
2. Gemmer, M., Becker, S., and Jiang, T. (2004). Observed monthly precipitation trends in China 1951–2002. *Theoretical and Applied Climatology*, 77, 39–45. https://doi.org/10.1007/s00704-003-0018-3.

3. Longobardi, A., and Villardi, P. (2010). Trend analysis of annual and seasonal rainfall time series in the Mediterranean area. *International Journal of Climatology*, 30, 1538–1546. https://doi.org/10.1002/joc.2001.

4. de Luis, M., Čufar, K., Saz, M. A., Longares, L. A., Ceglar, A., and Kajfež-Bogataj, L. (2014). Trends in seasonal precipitation and temperature in Slovenia during 1951–2007. *Reg Environ Change*, 14, 1801–1810. https://doi.org/10.1007/s10113-012-0365-7.

5. da Silva, R. M., Santos, C. A. G., Moreira, M., Corte-Real, J., Silva, V. C. L., and Medeiros, I. (2015). Rainfall and river flow trends using Mann–Kendall and Sen's slope estimator statistical tests in the Cobres River basin. *Natural Hazards*, 77, 1205–1221. https://doi.org/10.1007/s11069-015-1644-7.

6. Barros, V., Castañeda, M. E., and Doyle, M. (2000). Recent precipitation trends in Southern South America to the east of the Andes: An indication of a mode of climatic variability. In Smolka, P., and Volkheimer, W. (Eds.), *Southern Hemisphere Paleo and Neoclimates Key*. Springer, Berlin. https://doi.org/10.1007/978-3-642-59694-0_13.

7. Liebmann, B., Vera, C., Carvalho, L., Camilloni, I., Hoerling, M., Allured, D., Barros, V., Baez, J., and Bidegain, M. (2004). An observed trend in central south American precipitation. *Journal of Climate*, 17, 4357–4367.

8. Haylock, M. R., Peterson, T. C., Alves, L. M., Ambrizzi, T., Anunciação, Y. M. T., Baez, J., Barros, V. R., Berlato, M. A., Bidegain, M., Coronel, G., Corradi, V., Garcia, V. J., Grimm, A. M., Karoly, D., Marengo, J. A., Marino, M. B., Moncunill, D. F., Nechet, D., Quintana, J., Rebello, E., Rusticucci, M., Santos, J. L., Trebejo, I., and Vincent, L. A. (2006). Trends in total and extreme south American rainfall in 1960–2000 and links with sea surface temperature. *Journal of Climate*, 19(8), 1490–1512.

9. Morello, J., Matteucci, S., Rodríguez, A., and Silva, M. (2012). *Ecorregiones y complejos ecosistémicos argentinos*. Orientación Gráfica Editora, Buenos Aires.

10. Seluchi, M., Saulo, C., Nicolini, M., and Satyamurty, P. (2003). The Northwestern Argentinean low: A study of two typical events. *Monthly Weather Review*, 131, 2361–2378.

11. Garreaud, R. (2009). The Andes climate and weather. *Advances in Geosciences*, 22, 3–11.

12. Ferreira, L., Saulo, C., and Seluchi, M. (2010). Características de la depresión del noroeste argentino en el período 1997–2003: criterios de selección y análisis estadístico. *Meteorologica*, 35(1), 17–28.

13. COBINABE. (2010). *Generación y transporte de sedimentos en la Cuenca Binacional del Río Bermejo. Caracterización y análisis de los procesos intervinientes*. 1a ed., COBINABE, Buenos Aires, 230. ISBN 978-987-25793-7-1.

14. Murgida, A., González, M., and Tiessen, H. (2014). Rainfall trends, land use change and adaptation in the Chaco salteño region of Argentina. *Regional Environmental Change*, 14, 1387–1394.

15. Barros, V., and Camilloni, I. (2016). *La Argentina y el cambio climático: de la física a la política*. Eudeba, Ciudad Autónoma de Buenos Aires, Argentina, 286. ISBN 978-950-23-2655-9.

16. Arrieta, J., and Pastor, C. (1998). Relevamiento Socioeconómico y Ambiental de las Comunidades del Tramo Medio e Inferior de la Cuenca del Río Bermejo. *Informe y Anexos. Elemento: 2.5: Uso del Suelo en la Cuenca del Río Inferior*. Buenos Aires. Document extracted from the Regional Comission of the Bermejo River (COREBE) web page: http://corebe.org.ar/web2015/documentacion/ in January 2020.

17. Doyle, M., Saurral, R., and Barros, V. (2012). Trends in the distributions of aggregated monthly precipitation over the La Plata Basin. *International Journal of Climatology*, 32, 2149–2162.

18. Hannaford, J., and Buys, G. (2012). Trends in seasonal river flow regimes in the UK. *Journal of Hydrology*, 475, 158–174. https://doi.org/10.1016/j.jhydrol.2012.09.044.

19. Chen, J., Wu, X., Finlayson, B., Webber, M., Wei, T., Li, M., and Chen, Z. (2014). Variability and trend in the hydrology of the Yangtze River, China: Annual precipitation and runoff. *Journal of Hydrology*, 513, 403–412. http://dx.doi.org/10.1016/j.jhydrol.2014.03.044.

20. Ay, M., and Kisi, O. (2015). Investigation of trend analysis of monthly total precipitation by an innovative method. *Theoretical and Applied Climatology*, 120, 617–629. https://doi.org/10.1007/s00704-014-1198-8.

21. von Storch, V. H. (1995). Misuses of statistical analysis in climate research. In von Storch, V. H., and Navarra, A. (Eds.), *Analysis of Climate Variability: Applications of Statistical Techniques*. Springer-Verlag, Berlin, 11–26.

22. Yue, S., Pilon, P., Phinney, B., and Cavadias, G. (2002). The influence of autocorrelation on the ability to detect trend in hydrological series. *Hydrological Processes*, 16, 1807–1829.

23. Anderson, R. L. (1942). Distribution of the serial correlation coefficients. *Annals of Mathematical Statistics*, 13(1), 1–13.

24. Önöz, B., and Bayazit, M. (2012). Block bootstrap for Mann-Kendall trend test for serially dependent data. *Hydrological Processes*, 26(23), 3552–3560.

25. Castañeda, M., and González, M. (2008). Statistical analysis of the precipitation trends in the Patagonia region in southern South America. *Atmósfera*, 21(3), 303–317.
26. González, M., Dominguez, D., and Nuñez, M. (2012). Long term and interannual rainfall variability in argentinean Chaco Plain region. In Martín, O., and Roberts, T. (Eds.), *Rainfall: Behavior, Forecasting and Distribution*. Nova Science Publishers, 68–89. ISBN: 978-1-62081-5519.
27. Zak, M., Cabido, M., and Hodgson, J. (2004). Do subtropical seasonal forests in the Gran Chaco, Argentina, have a future? *Biological Conservation*, 120, 589–598.
28. Grau, H., Gasparri, N., and Aide, T. (2005). Agriculture expansión and deforestation in seasonally dry forests of north-west Argentina. *Environmental Conservation*, 32(2), 140–148.
29. Piquer-Rodríguez, M., Torella, S., Gavier-Pizarro, G., Volante, J., Somma, D., Ginzburg, R., and Kuemmerle, T. (2015). Effects of past and future land conversions on forest connectivity in the Argentine Chaco. *Landscape Ecology*, 30, 817–833.
30. Torrella, S., Ginzburg, R., and Galetto, L. (2015). Forest fragmentation in the Argentine Chaco: Recruitment and population patterns of dominant tree species. *Plant Ecology*, 216, 1499–1510.
31. Jobbágy, E., Nosetto, M., Santoni, C., and Baldi, G. (2008). El desafío ecohidrológico de las transiciones entre sistemas leñosos y herbáceos en la llanura Chaco-Pampeana. *Ecología Austral*, 18, 305–322.
32. Salazar, A., Baldi, G., Hirota, M., Syktus, J., and McAlpine, C. (2015). Land use and land cover change impacts on the regional climate of non-Amazonian South America: A review. *Global and Planetary Change*, 128, 103–119.
33. Lauro, C., Vich, A. I. J., and Moreiras, S. M. (2019). Streamflow variability and its relationship with climate indices in western rivers of Argentina. *Hydrological Sciences Journal*. https://doi.org/10.1080/02626667.2019.1594820.

4 Statistical Modeling of Summer Precipitation in the Pampa Region of Argentina

Alfredo L. Rolla, Paula C. Oliveri, Sabrina N. Ayala,
Eugenia M. Garbarini, Marcela H. González,
and Saeid Eslamian

4.1 INTRODUCTION

Argentina is located in southeastern South America occupying a total area of 2,791,810 km². In northern Argentina, the Andes mountains act blocking humid air from the Pacific Ocean, and the flow is governed by the South Atlantic High, winds prevail from the northeast advecting water vapor at low levels from tropical regions of the Atlantic Ocean and South America. South of 38°S, humid air can flow from the Pacific Ocean as the Andes height is reduced, thus establishing a flow from the west during the whole year. This generates an intense contrast between dense vegetation areas near the Andes mountains and the dry plains extending towards the Atlantic coast. The climatic adversities faced by the agricultural sector during the process of production generate a high degree of uncertainty about the result of the activity performed, which bears a high level of risk associated with the productive units. Given the diversity of climates and soils in Argentina, all farmers face the risk of losses due to climatic factors, whether as a consequence of drought, frost, hail, excess water, strong winds, floods, and many other adversities.

Results in agricultural production depend on forecasts and strategies adopted, which generally take into account the normal behavior of the climatic variables. These strategies are usually developed by the farmer himself in his region of interest and fundamentally aim to reduce vulnerability associated with adverse weather conditions. An example of these strategies is the protection of active crops, for example, irrigation by sprinkling as a method to reduce the impact of frost, supplemental irrigation to reduce water deficit, the assessment of ideal climatic conditions for sowing, fertilizing, knowing when to use proper equipment for harvesting crops (due to excess of precipitation) among others. Therefore, having a good understanding of the variability of regional precipitation is key for decision makers, particularly for the implementation of these strategies on crops.

Previous studies have detected a westward shift of the isohyets of annual accumulated precipitation in subtropical Argentina since the second half of the 20th century [1–3]. However, data from the last decades show some evidence of change in this behavior in specific regions, causing economic losses and major social problems. For example, negative trends have been observed in precipitation in some areas located in the Argentine Chaco region [4]. Knowledge of the interannual variability of precipitation is crucial to detect the effect produced by large-scale forcing on precipitation, especially in South America.

It is well known that the El Niño–Southern Oscillation (ENSO) has a great influence on precipitation in Southeast South America (SESA) [5–11]. Ashok et al. [12] defined a special type of El Niño, called El Niño Modoki (EMI), where the warming zone is limited to the tropical Central Pacific, while cooling zones are found to the east and west of this central zone. These events do not produce exactly the same effects on precipitation as those recorded by traditional El Niño events.

DOI: 10.1201/9781003473398-5

Another global climate forcing related to precipitation and sea surface temperature (SST) is the Indian Ocean Dipole (IOD) [13], whose positive phase is defined by the warming of the Southwest Indian Ocean and the cooling of the Northeast Indian Ocean. Chan et al. [14] found that in South America the positive phase of the IOD manifests itself as a dipole of precipitation anomalies, with increases in the La Plata basin and decreases in the central region of Brazil.

Another factor that influences the region is the Antarctic Oscillation (AAO), which is associated with wind variability [15]. Its positive phase is defined by negative pressure anomalies around the South Pole combined with a wavelike pattern at middle latitudes. The index is defined as the normalized pressure difference between 40°S and 70°S [16]. The positive phase is associated with an intensification of the zonal wind and therefore with a lower energetic exchange between high and middle latitudes preventing fronts from displacing meridionally along the South Pacific towards north-eastern Argentina. There are several papers investigating the effects of AAO on the South American climate. Reboita et al. [17] found that the frontogenesis function is strong during the negative phase of AAO and the trajectory of cyclonic activity shifts south during the positive phase of AAO. González and Vera [18], González et al. [19], González and Domínguez [20], and González [21] found a significant relationship between the negative phases of the AAO with the winter rains in the area of Comahue in the Andes, in north-western Patagonia. In the Atlantic Ocean, the South Atlantic Dipole (SAODI) [22] is an oscillation that has been defined in terms of the variability of its SST. Its positive phase is defined as the warming of the northeast (coast of Africa) and the cooling of the southwest of the tropical ocean (coast of Brazil). This effect is related to the position and intensity of the South Atlantic semi-permanent anticyclone and generates anomalies in the advection of humid air that generates precipitation anomalies.

For this study, the area selected is known as the Pampa Region, which represents Argentina's main agricultural production area. From an economic point of view, three-quarters of the total value of agricultural production corresponds to this area, covering 5 million hectares. Only the provinces of Buenos Aires, Santa Fe, and Córdoba generate 70% of the country's agricultural production. Therefore, this work aims to develop statistical forecasts of seasonal summer precipitation for subregions of the study area, defined on the basis of clusterization techniques. For this purpose, the study is organized as follows: Section 4.2 describes the data employed and the clusterization and predictor-selection techniques considered, Section 4.3 presents the main results, and Section 4.4 includes the major conclusions.

4.2 DATA AND METHODOLOGY

Daily precipitation data from 37 weather stations with complete records covering the study area belonging to the measurement network of the Argentine National Meteorological Service (SMN) in the Pampa Region for the period 1979–2016 were used (Figure 4.1). The data were consistent and only those with high quality were used. The missing data did not exceed 1% of the total daily data. Based on the series of daily-observed precipitation from every meteorological station considered, the seasonal accumulated precipitation series were generated. In this case, summer (DJF) comprises data from December to February, autumn (MAM) from March to May, winter (JJA) from June to August, and spring (SON) from September to November. Only seasonal summer precipitation (DJF) was used for this study. An exploratory analysis of these data series was performed using statistical methodologies.

Homogeneous subregions of weather stations were defined to deepen the study of distinctive precipitation characteristics. For this purpose, different hierarchical agglomerative grouping methods [23] were evaluated using a couple of distance metrics that were coherent with the problem to be solved, such as the Euclidean distance and the correlation between stations. In addition to this, non-hierarchical grouping methods such as K-means were also tested [23–25].

For each group, the forcings of the interannual variability of summer precipitation were studied by calculating the correlations between summer precipitation and meteorological and ocean

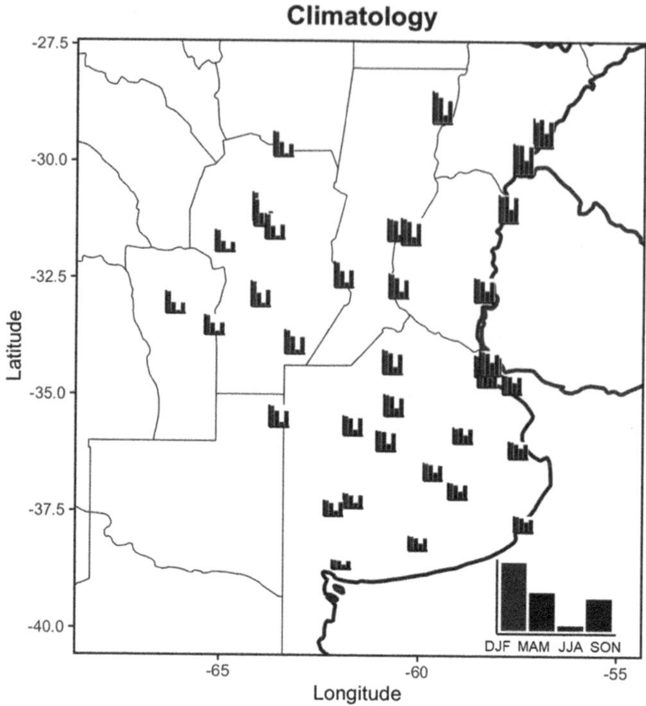

FIGURE 4.1   Weather station locations and mean seasonal cumulative precipitation of the period 1979-2015, for summer (DJF), autumn (MAM), winter (JJA), and spring (SON).

variables such as geopotential height (hgt) at 1000, 500, and 200 hPa, zonal (u) and meridional wind (v) at 850 hPa, precipitable water from surface to 700 hPa (tcw), and sea surface temperature (sst) obtained from NCEP/NCAR reanalyses [26]. The NCEP / NCAR Reanalysis Dataset is a continuously updated global gridded data set (1948–present) representing the state of the atmosphere. These reanalyses are built incorporating observations and forecasts from numerical meteorological models (NWP). It is a joint product of the National Centers for Environmental Prediction (NCEP) and the National Center for Atmospheric Research (NCAR). These reanalyses were used in the same period as the observations (1979–2016), because it is considered that since 1979 the global observation network incorporated satellite information, significantly improving the quality and veracity of the reanalyses at the regional level (southern hemisphere), which were used as predictors in the regression models. These datasets/reanalyses have a spatial resolution of 2.5° (approximately 250 km) and in this case a monthly temporal resolution, in particular, the month of November was selected. The dimension of each global variable's matrix is 144 (longitudes) × 73 (latitudes) × 37 (months of November).

Correlation maps were calculated between seasonal time series (DJF) precipitation observations and large-scale variables from the previously described reanalyses (hgt, u, v, tcw, and sst) in the month prior to the beginning of summer (November). A normal distribution test was performed to test statistical significance. The correlations were considered significant with 95% confidence when they exceeded the threshold of 0.35. The result of these correlations was global maps. Those areas with significant correlations that had a physical significance associated with known effects on precipitation in the region regarding the dynamics of the global atmosphere and its influence in the southern hemisphere were selected. In these regions with significant correlations, the predictor variables were defined as the temporal series corresponding to the month of November.

After the preselection of areas and predictors, the LASSO methodology (Least Absolute Shrinkage and Selection Operator) [27] was employed to select the best set of independent predictors. LASSO permits discarding predictors that explain the small variance of summer precipitation for the building of regression models.

Finally, multiple linear regression models were built with those predictors for each of the homogeneous regions found in the Argentine Pampas (hereinafter, the Pampa Region) with the aim of forecasting summer precipitation. The resulting models were validated through cross-validation [28], particularly leave-one-out cross-validation (LOOCV). This technique was chosen due to the low number of observations available, and it consists of using all of the observations but one for model building while the remaining observation is used for prediction. This process was repeated as many times as the number of years comprising the summer precipitation time series, which in this case spanned from 1979 to 2015. LOOCV allows simultaneously validating the forecast and verifying the stability of the models.

Contingency tables of observed and forecasted precipitation were elaborated to test the efficiency of the models to detect the rainfall categories. Observed precipitation tertiles were used to discriminate the categories: below-normal, normal, and above-normal precipitation (driest to wettest events, respectively). Additionally, certain indices were calculated to evaluate the performance of the models: the probability of detection (recall), false alarm relation (FPR), and percentage of hits (accuracy). Furthermore, the models' fit was analyzed by comparing the cumulative probability functions of observed and forecasted precipitation through the Chi-Square test.

4.3 RESULTS

4.3.1 DATA EXPLORATORY ANALYSIS

A precipitation database was created using daily data from weather stations which belong to the National Meteorological Service of Argentina and that report at operating time to the World Meteorological Organization's (WMO) network, covering the Pampa Region. All of the stations were selected to have the most complete precipitation series of the 1976–2016 period as was possible. In this selection process, those stations which had more than 10% of missing data in summer daily records were dismissed, in order not to affect the climate forcing analysis and therefore the forecast models that were the aim of future work. Following that criterion, five weather stations were eliminated. Their locations were not critical to represent the study area and their exclusion did not affect the regional clustering. The remaining stations had a relatively uniform distribution and they reasonably covered the region. The average separation between them was about 100 km. Employing the daily database, a series of accumulated precipitation of austral summer (DJF), autumn (MAM), winter (JJA), and spring (SON) were calculated. Further work was made only for the summer precipitation.

Figure 4.1 shows the climatology of seasonal precipitation for summer (DJF), autumn (MAM), winter (JJA), and spring (SON) by the meteorological station. At each location, a histogram is shown, where each bar represents seasonal precipitation of summer (first bar), autumn, winter, and spring (second, third, and fourth bars) over the 1979–2015 period. It can be seen that there is a pattern of increasing rainfall towards the north in summer. Contrasting values are observed between the south of Buenos Aires Province and Corrientes Province. On the other hand, there is a pattern of increasing rainfall towards the east in winter. Contrasting values can be noticed between the north-east of Buenos Aires Province and San Luis Province. Those patterns were observed in the representative series of the stations' clusters.

Figure 4.2 shows boxplots of summer accumulated precipitation observations for each of the 37 weather stations considered. Variability ranges can be observed. For a better visualization, observation data points are shown inside each boxplot. Each scale of gray identifies one cluster. The employed clustering will be explained in Section 4.3.1.1.

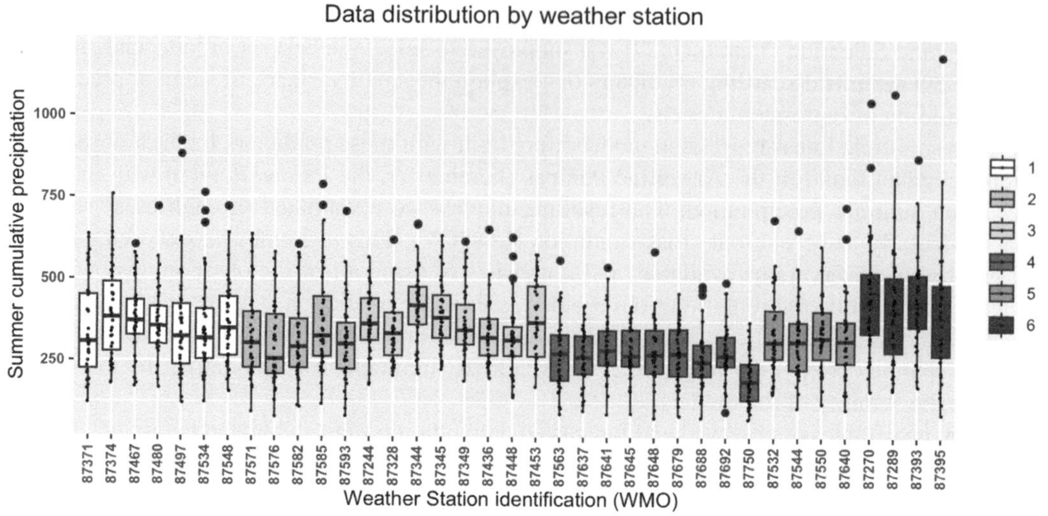

FIGURE 4.2 Boxplots of summer cumulative precipitation observations for all the weather stations employed. Each scale of gray corresponds to a different cluster identified with a number.

Some interesting details observed in Figure 4.2 are:

- Stations in cluster 4 are located in the south of the study region. They showed the lowest average observed rainfall in the considered period.
- Stations in cluster 6 are located in the north of the study region. They showed the largest average observed rainfall in the considered period.
- It can be observed that in general there were more cases of above-normal precipitation than below-normal precipitation.
- The highest rainfall events predominated among stations in cluster 6, which are located in the north of the Pampa Region. This can be explained because of the important influence that ENSO has in that region, leading to higher-than-normal summer precipitation in some years.
- It should be noted that summer precipitation showed an important variability over the whole study region.

4.3.1.1 Weather Stations Clustering

Different techniques of hierarchical and non-hierarchical clustering that were considered suitable for the present research, were applied. They were combined as follows:

4.3.1.1.1 Hierarchical Clustering

Distance matrices:

- Euclidean
- Correlation

Agglomerative methods:

- Single linkage: all of the distances between pairs of elements from two different groups are calculated and the minimum distance is considered a linkage criterion. It tends to produce the elongated and scattered clusters.

- Complete linkage: all the distances between pairs of elements from two different groups are calculated and the maximum distance is considered the distance between two clusters. It tends to produce more compact clusters.
- Average: all of the distances between pairs of elements from two different groups are calculated and the average distance is considered the distance between two groups. It tends to produce more compact clusters.
- Ward.D: it minimizes the total variance within a cluster.

Different types of hierarchical clusterings of 4 to 7 classes were analyzed in order to reasonably discriminate between the different summer precipitation regimes which exist in the Pampa Region. The best result was obtained by combining an Euclidean distance matrix and the minimum variance agglomerative method. This experiment's result is shown in Figure 4.3. By considering 6 subregions, a satisfactory grouping of stations was achieved through hierarchical clustering, both in terms of geographical distribution of stations and of the spatial variation of climatological summer rainfall.

4.3.1.1.2 Nonhierarchical Clustering: K-Means

K-Means was the partitioning method chosen to find hydrologically homogeneous subgroups of weather stations in terms of their summer precipitation regime. This methodology seeks to identify distinct non-overlapping clusters of data, given the number K of clusters desired. The K-Means algorithm works iteratively by assigning each meteorological station to one of the K groups according to the similarity between the observations. An initial random assignment of the observations into K clusters is performed, which allows the computation of the clusters' centroids as the feature mean for the observations within each cluster. Then, each observation is re-assigned to the cluster whose centroid lies closest, using the squared Euclidean distance as the measure of closeness

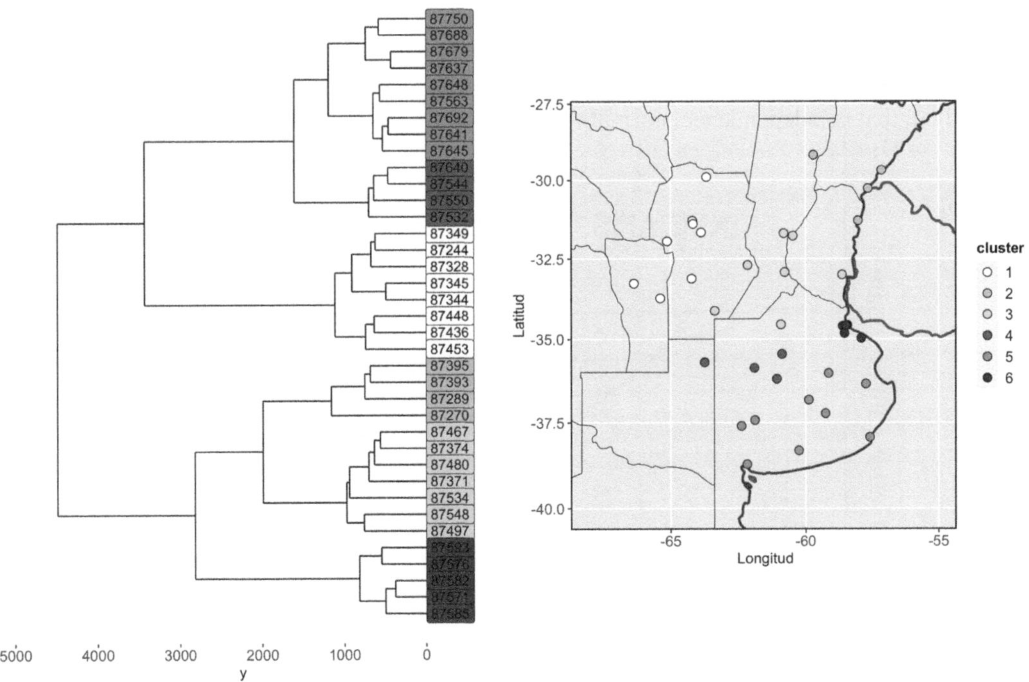

FIGURE 4.3 Best of the hierarchical clusterings considered. It employed Euclidean distance and the Ward.D agglomerative method of variance minimization within a group.

or similarity. The clusters' centroids are updated, and the iteration continues until no further re-assignment of observations minimizes the squared Euclidean distance. The Elbow technique was considered to evaluate the cohesion of the groups and the optimal number of clusters. This method computes the total within the sum of squares (WSS), defined as the sum of all the pairwise squared Euclidean distances between the observations in each subgroup summed over all k clusters, with k =1, 2, …, 10. The optimal number K of clusters marks the point after which the decrease in WSS is not meaningful despite increasing the number of groups.

Figure 4.4 shows the results derived from the Elbow technique, which suggests that the optimal number of groups oscillates between 5 and 6. Therefore, the latter was chosen for the clustering of the weather stations. Figure 4.5 presents the silhouette plot, used as a tool for cluster validation. Silhouette widths are computed for each weather station, allowing for detection when stations are properly clustered (high values of silhouette width), when stations lie between two clusters (silhouette width close to zero), or when stations are probably misclustered (negative silhouette widths). The results derived from the silhouette plot when considering K = 6 suggest that groups are satisfactorily clustered, with only one weather station (corresponding to cluster 5) presenting a negative silhouette width.

The spatial distribution of the 6 clusters is exhibited in Figure 4.6. The WMO identifier of each weather station is depicted on the map. The subgroups define areas that respect the features of summer precipitation previously noted, particularly in terms of the precipitation gradient from south to north and from east to west. The clusters delimit the central zone approximately in the region known as the "core" Pampa Region. It can be seen that this clustering resulted in consistent, both geographically and climatologically as well as the hierarchical, clustering shown in Figure 4.3. Clusters obtained from both methods were the same except for one weather station. The rest of the present work was carried out considering the K-means clustering.

4.3.1.1.3 Analysis of variance of clusters

An analysis of variance (ANOVA) was performed on the achieved regionalization to verify if the differences between the clusters were significant. The F statistic obtained through ANOVA is high enough to reject the null hypothesis of equal mean among the clusters (p-value < 0.05), therefore suggesting that the means of summer precipitation among regions are not statistically similar.

To analyse if the ANOVA result is valid, it is important to verify if the requirements of the method are fulfilled; that is independence, normality, and homoscedasticity. Normality of distributions was

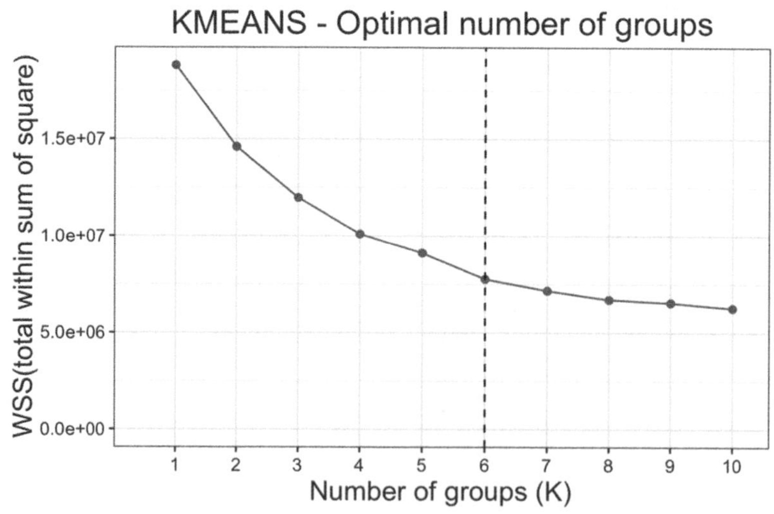

FIGURE 4.4 "Elbow" method for optimal number of groups (K) determination.

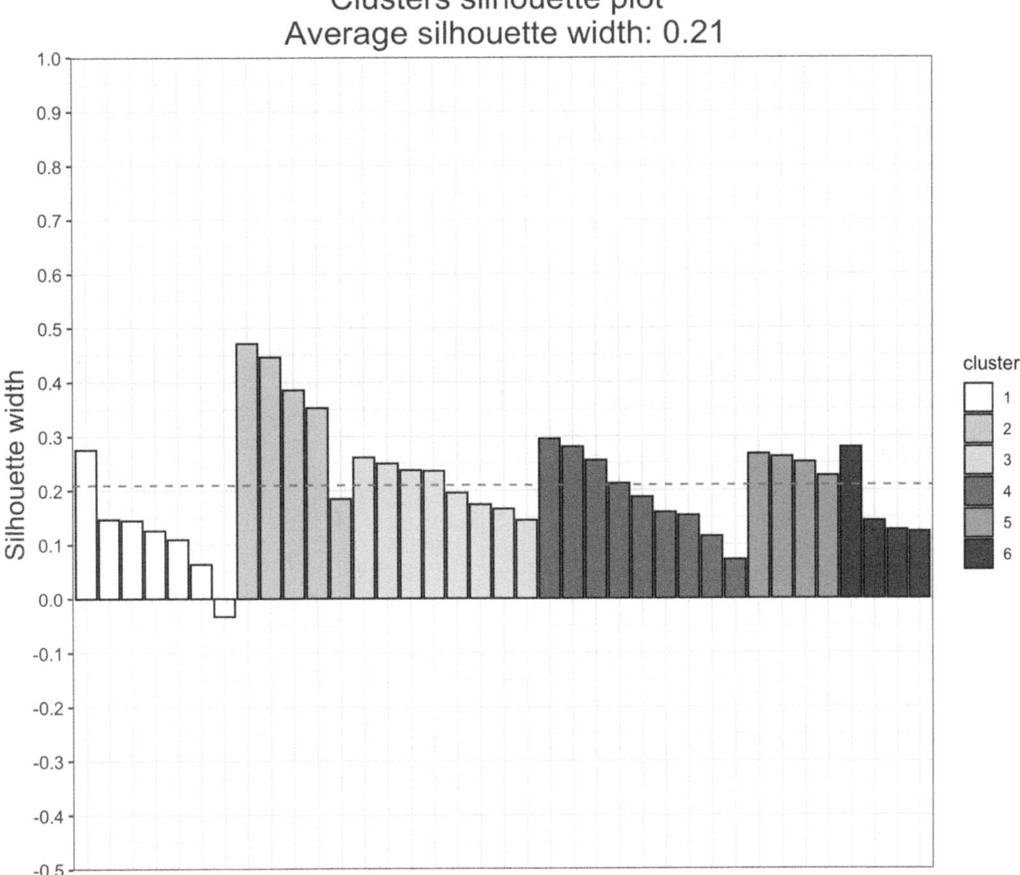

FIGURE 4.5 Clusters silhouette plot for K=6. Each bar indicates the performance of the clustering method for one of the 37 weather stations.

examined using the Shapiro–Wilk test. The statistic derived from the test has a p-value lower than 0.05, which suggests that the samples are not normally distributed. Therefore, a Box–Cox power transformation was applied to meet the normality assumption of the data. By considering a transformation power constant equal to 0.26, the summer rainfall data acquire a normal distribution according to the Shapiro–Wilk test.

Afterwards, the Levene test was applied to the transformed sample to check the requirement of homogenous variances among clusters (homoscedasticity). However, the null hypothesis of equal variances was rejected (p-value < 0.05) and therefore, the homoscedasticity assumption is not met and the parametric ANOVA is not valid.

Alternatively, the non-parametric ANOVA test for independent samples, also known as the Kruskal–Wallis (Wilcoxon) test, was applied to the transformed data. This version is similar to the traditional ANOVA, except that observations are replaced by their ranks which adds robustness to the test. According to the non-parametric ANOVA there are significant differences among the clusters (p-value < 0.05). Therefore, the Kruskall–Wallis test was applied again considering a p-value of 0.05 to perform pairwise multi comparisons between clusters. The results from this test show that, in general, most pairs of groups are statistically different, with the only exceptions being the pairs of clusters 1–2, 2–5, 3–5, and 3–6. Despite not considering the geographical position or spatial distance between clusters, groups geographically distant from one another present a difference. On

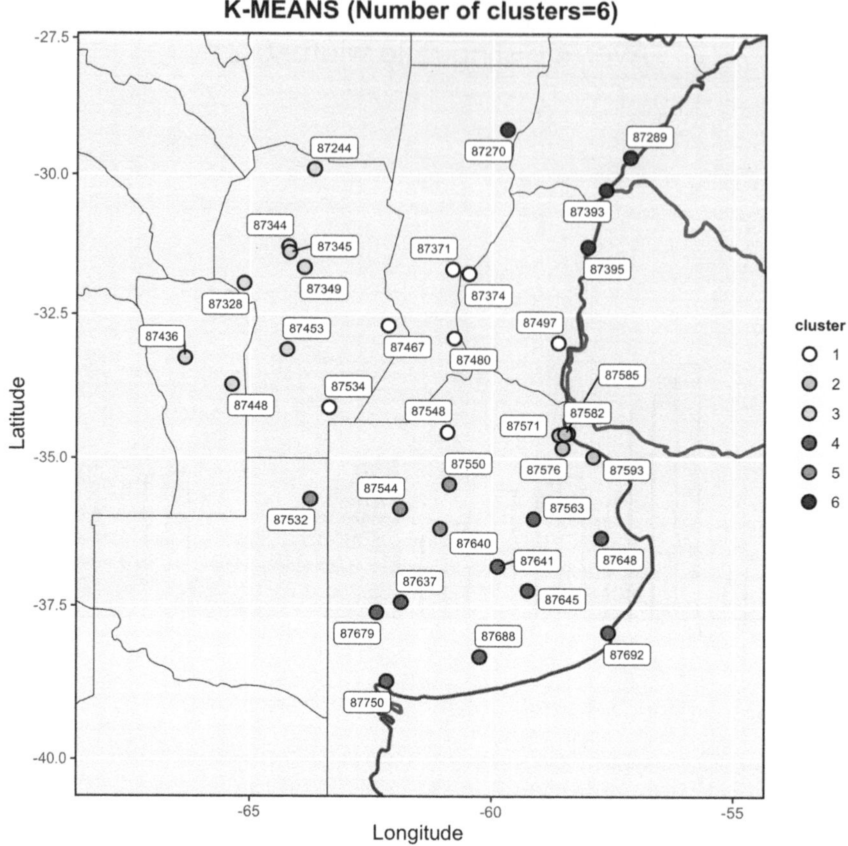

FIGURE 4.6　K-means clustering of the weather stations for K=6. Stations' WMO identifiers are shown.

the contrary, groups positioned geographically closer (that is with nearby borders) appeared to be similar in terms of summer precipitation, although from the perspective of the influence of climate forcings these groups could be different regions.

4.3.1.1.4　Regionalized Summer Precipitation

Figure 4.7 shows the boxplots of the six clusters obtained through the K-means method. These boxplots contain and summarize the information of summer precipitation already noted in Figure 4.2 for the individual weather stations. Group 6, which includes the northernmost stations, presents the highest mean summer precipitation as well as the most pronounced accumulated extremes of the entire Pampa Region. On the contrary, group 4 represents the area of lowest mean summer rainfall in the southern Pampa Region. Groups 5 and 6 (north of 35°S) show greater accumulated values than the ones observed for group 3 (northwesternmost cluster of stations).

Once the regionalization of the Pampa Region was achieved, the mean summer precipitation time series was constructed for each of the clusters defined. These series are shown in Figure 4.8 and correspond to the groups depicted in Figure 4.6, and serve as representatives of the temporal evolution of summer rainfall for the subregions found. These representative series together with appropriate climate forcings will be employed in the construction of seasonal rainfall statistical forecasts, in which the objective is to estimate the mean precipitation expected for each of the homogenous subregions of the Pampa Region.

Hereinafter, the subregions defined by the clusters will be referred to simply as *Regions* (1 to 6).

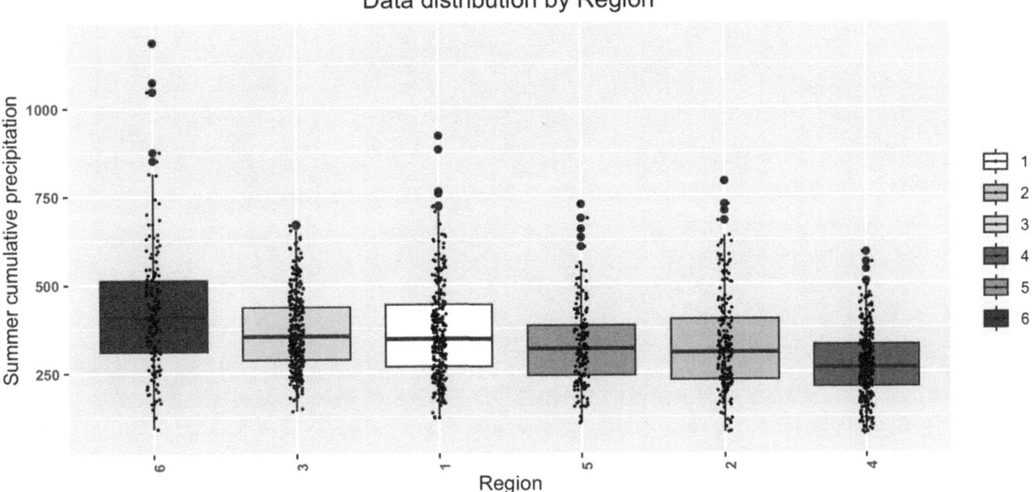

FIGURE 4.7 Boxplots of Summer cumulative precipitation observations for the six clusters derived from the K-means method.

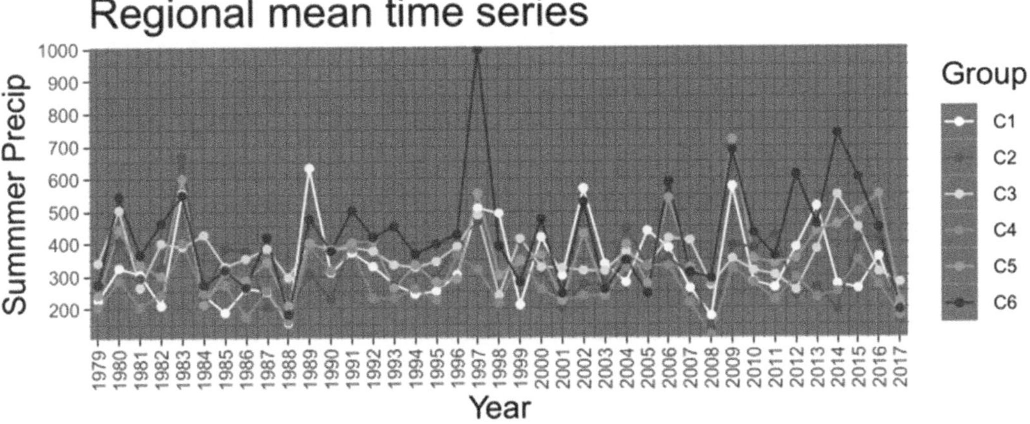

FIGURE 4.8 Regional mean time series of summer precipitation for each group.

4.3.1.2 Global Forcings Analysis

As it was explained in Section 4.2, correlation maps were calculated between the seasonal DJF precipitation observations time series and the large-scale variables from the reanalyses: geopotential height (hgt) at 1000, 500, and 200 hPa, zonal wind (u) and meridional wind (v) at 850 hPa, precipitable water from surface to 700 hPa (tcw), and sea surface temperature (sst), in the month prior to the beginning of summer (November). The linear correlations were performed on grid points for each of the seven reanalysis variables and for the mean precipitation series of each region. Therefore, there were 42 resulting correlation maps. Convex sets containing only significant correlations were delimited. Each predictor was calculated as the average of the variable over all of the grid points in the significant correlation area.

Figure 4.9 shows three of the correlation maps for Region 1 (the remaining regions' correlations are not shown). Positive and negative correlation values are denoted by a plus and a minus sign, respectively. As an example for the interpretation of the correlation fields, positive correlations in

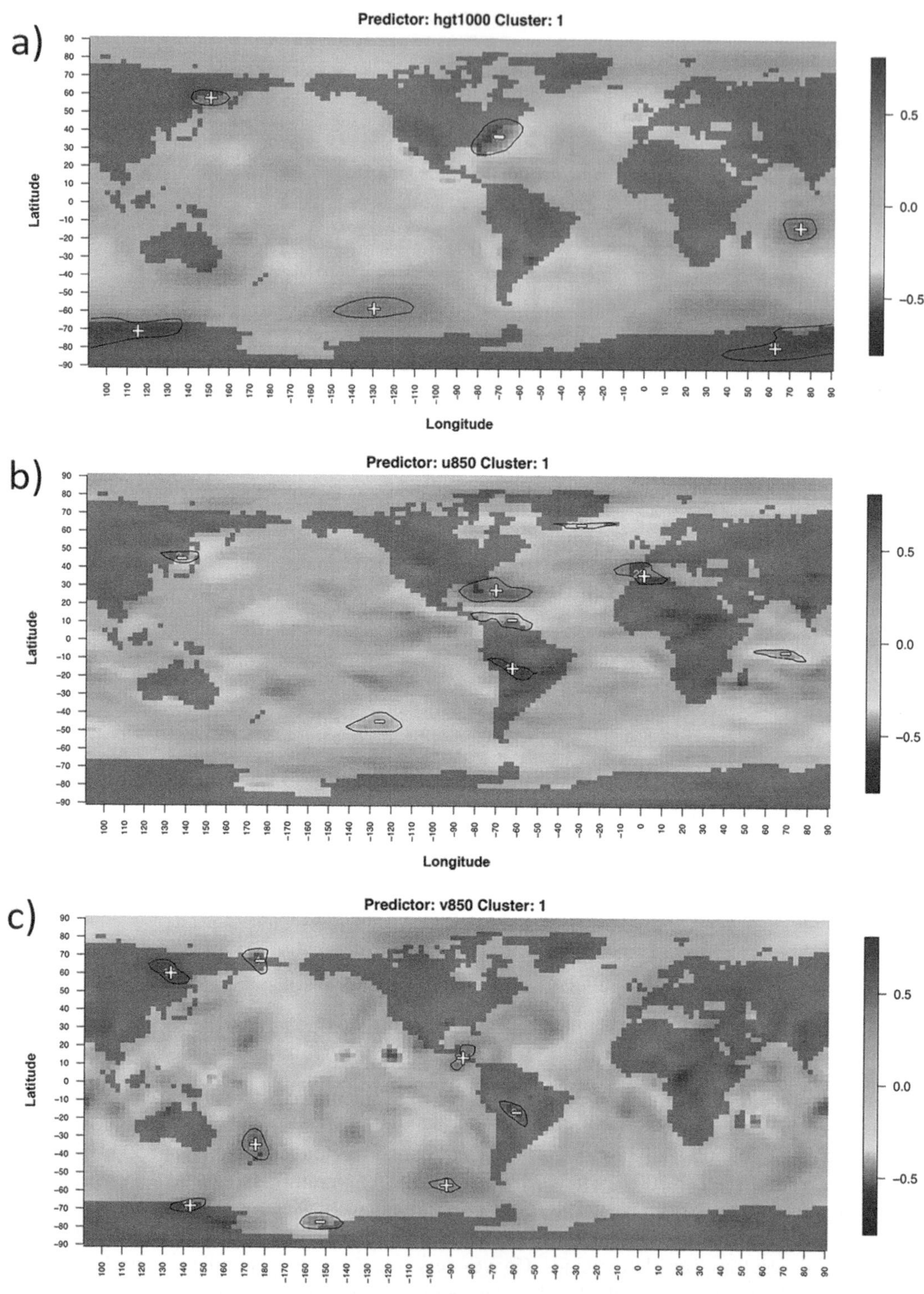

FIGURE 4.9 Distribution of the correlation coefficients between the mean Summer precipitation series of Region 1 and the predictor series of the previous month (November), for 1000 hPa geopotential height (a), 850 hPa zonal wind (b), and 850 hPa meridional wind. Black curves enclose areas with significant values. A plus (minus) sign indicates positive (negative) correlation values.

the Indian Ocean at low levels indicate that when higher(lower)-than-normal geopotential height is present in November, higher(lower)-than-normal mean summer precipitation is expected in Region 1 (Figure 4.9a). Additionally, weaker (stronger) low-level westerlies in high latitudes (statistically significant over the Pacific Ocean) in November precede more humid (drier) summers (Figure 4.9b). Finally, higher (lower) rainfall values in DJF are expected in Region 1 when stronger northerly (southerly) low-level winds in tropical and subtropical central South America are present in the previous November (Figure 4.9c).

The forcings considered in this work can be classified as follows:

- SST related: the most relevant regions are Equatorial Pacific (ENSO: El Niño–Southern Oscillation), Indian Ocean, South Atlantic Ocean, South Pacific Ocean.
- Dynamic forcings: geopotential height related (hgt100, hgt500, hgt200) and 850 hPa zonal wind (u850). The most relevant regions in this case are the Indian Ocean, South Atlantic Ocean, South Pacific Ocean.
- Precipitable water: total precipitable water (tcw) available in the column of the atmosphere over the Pampa Region.
- Low Level Jet (LLJ): meridional wind in Northern Argentina (v850).

The methods employed to select the forcings that will act as predictors in the linear multiple regression models of each region will be described in the following sections. Hereinafter, predictors for each hydrologically homogeneous region will be referred to by type of variable, associated cluster, and an identifying number. For instance, hgt500_C4_3 refers to predictor number 3 associated with cluster 4 and corresponding to 500 hPa geopotential height.

4.3.2 Pre-selection of Predictors (LASSO).

To avoid the adverse effects of collinearity problem in a linear model estimated by least squares in the case that the number of predictors is less than the number of years ($p < n$), the LASSO (Least Absolute Shrinkage and Selection Operator) regression technique was used. The LASSO methodology combines a regression model with a shrinkage process of some parameters towards the null value to select predictors, penalizing the regression coefficients. Since a predictor with a null regression coefficient does not influence the model, LASSO manages to exclude less relevant predictors. The degree of penalty is controlled by the parameter λ. When this parameter is null, the result is equivalent to that of a linear least squares model and when λ increases, the penalty also increases and more predictors are excluded. The best value of λ was sought by the cross-validation method.

To apply the LASSO method, the variables were previously standardized. Figure 4.10 shows the summary graph of applying LASSO doing "leave-one-out" type "cross-validation" (LOOCV) using data from Region 1. Lower standard deviation curves along the sequence of λ (error bars) can be seen and the cross-validation curve (MSE: mean square error) is shown as a dotted line. The upper vertical dotted lines indicate the band in which the selected predictors have a low MSE, that is, it indicates the amount of significant predictors for the regression (dotted lines indicate the λ for which MSE is minimum as well as the first value of λ whose MSE exceeds the minimum by one deviation). The top axis shows the number of nonzero coefficients in that interval.

Figure 4.11 shows the selected predictors for λ whose MSE is minimum, in the case shown it is $\lambda = 0.21$, corresponding to mid-gray predictors. Those in light-gray were considered non-significant predictors, that is, they do not provide additional information for the regression. This predictor selection method was simultaneously applied to all of the clusters, and the chosen predictors were reserved for the construction of the multiple linear regression models.

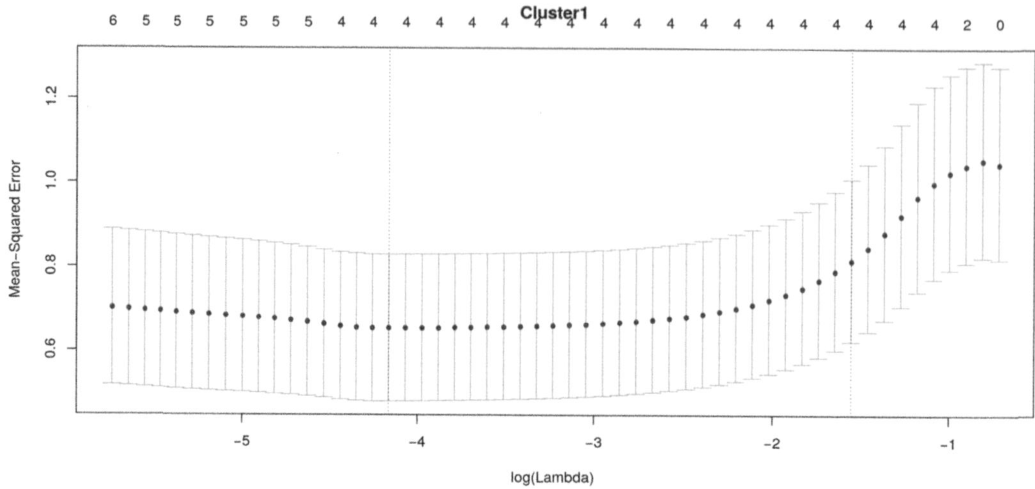

FIGURE 4.10 "Leave-one-out" cross-validation for Region 1. The cross-validation curve (dots) and lower and upper standard deviation curves (error bars) are shown. The dotted lines show both the λ with the minimum MSE (optimal value), and the lowest value of λ with an MSE greater than the minimum plus a deviation. The top axis shows the number of predictors which are significant for the regression and which are associated with each λ value.

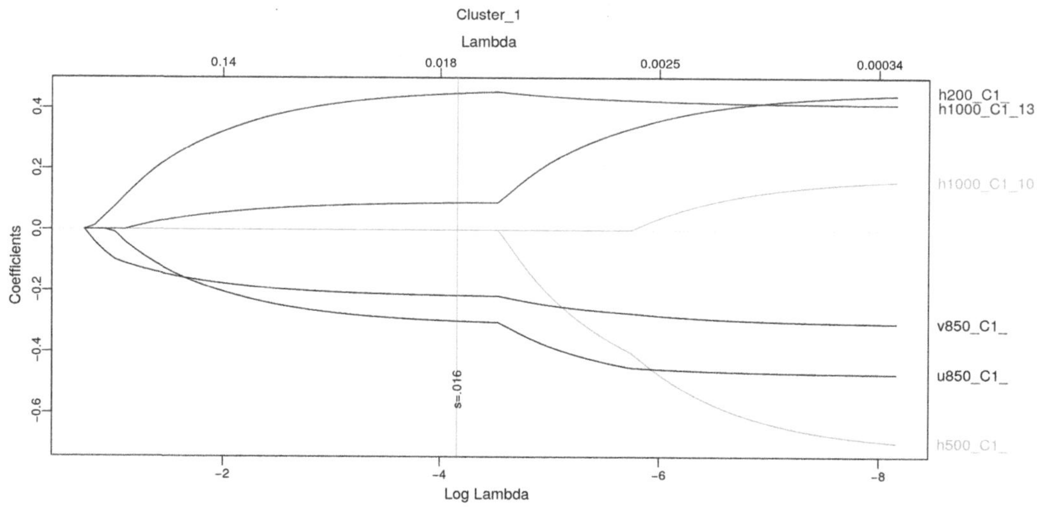

FIGURE 4.11 Coefficients and predictors selected by LASSO methodology (mid-gray curves). Light-gray curves correspond to non-significant predictors.

4.3.3 MODEL CONSTRUCTION

Predictor variables were normalized to properly apply LASSO, thus resulting in a loss of the variables' units. Therefore, after determining which predictors are more suitable for each region through LASSO, an automatic model construction method was implemented using these predictors but retaining their original unit. Stepwise regression was chosen to generate the prediction models, which sequentially adds or subtracts predictors following a pre-specified criterion. Several models were obtained by considering different sets of independent predictors. Finally, the "best" model for each region was selected by searching for the maximum adjusted coefficient of determination (hereinafter, AdjR^2), since it implies maximizing the variance of summer precipitation explained by the model.

The best models obtained for each region along with their AdjR^2 are shown in Table 4.1. After applying LASSO and stepwise regression the models for each subregion in the Pampa Region present between three and seven predictors and are composed primarily of dynamic and SST-related forcings. The predictors selected in the models present statistically significant relations with the predictand, and represent physical mechanisms that could potentially influence the summer rainfall of the Pampa Region. While predictors were defined specifically for each of the six hydrologically homogeneous regions, presenting differences in location and spatial extension of the relevant convex areas among clusters, occasionally agreement arises in the physical mechanisms behind the linear regression models.

The SST predictors are related to the modes of variability present in the tropical Pacific and Indian Oceans. Influence of ENSO on summer rainfall is present through the predictors sst_C2_5, sst_C3_21, sst_C5_8, and sst_C6_40 in the models of regions 2, 3, 5, and 6 respectively. These convex areas are part of the tongue-like pattern typical of ENSO events [29], and the sign of the correlations suggests that higher summer precipitation amounts are expected when El Niño is present in the prior November, while the opposite occurs for La Niña events. Indeed, previous authors have considered ENSO as an important remote forcing of South American precipitation by inducing anomalies in the Walker circulation cell and triggering teleconnection processes [5, 8, 10]. The Indian Ocean was also found in earlier studies to impact South American rainfall by way of its atmosphere-sea coupled modes, the Indian Ocean Dipole [13, 14] and Indian Ocean Basin-Wide Warming [30], which excite Rossby wave trains that displace from the Indian basin towards South America. Predictors that connect the Indian Ocean with summer rainfall variability in the Pampa Region are hgt1000_C1_13, hgt500_C4_4, and sst_C5_41 for the models of clusters 1, 4, and 5 respectively.

The Antarctic Oscillation is another hemispheric-scale forcing, defined by anomalies in the intensity of the subtropical and subpolar pressure systems [15]. The correlation patterns related to low-level circulation and the predictors involved in the models suggest that a negative (positive) phase of this Oscillation observed in November could lead to higher(lower)-than-normal summer precipitation for Regions 1, 2, and 4 in southern Pampa Region (due to predictor u850_C1_5, hgt1000_C2_2, hgt1000_C2_9, and hgt1000_C4_11). This result is coherent with the fact that a negative phase of the Antarctic Oscillation produces weaker westerlies in mid-latitudes, a stronger

---

TABLE 4.1

Multiple Linear Regression Models for Each Region, with the Corresponding AdjR^2 Values. Regression Coefficients for the Equations Are Indicated between Parentheses (Unit Being mm*[Unit of the Predictor]$^{-1}$).

Region	Model	AdjR^2
1	(−712.58) + (10.71)*hgt1000_C1_13 + (−15.63)*u850_C1_5 + (−21.07)*v850_C1_9	0.447
2	(19626.02) + (−70.34)*sst_C2_5 + (1.02)*hgt1000_C2_2 + (1.22)*hgt1000_C2_9 + (−1.39)*hgt200_C2_2	0.477
3	(2119.3) + (−75.27)*sst_C3_21 + (−0.86)*hgt1000_C3_5 + (−11.35)*u850_C3_11	0.566
4	(−22119.82) + (−2.87)*hgt500_C4_3 + (0.8)*hgt1000_C4_11 + (0.47)*hgt200_C4_5 + (−21.3)*u850_C4_15	0.498
5	(5202.97) + (−84.3)*sst_C5_8 + (117.58)*sst_C5_41 + (3.04)*hgt1000_C5_14 + (−0.52)*hgt200_C5_9 + (−12.39)*u850_C5_10 + (−52.64)*u850_C5_17 + (−7.21)*u850_C5_23	0.68
6	(−7648.71) + (74.67)*sst_C6_40 + (0.54)*hgt200_C6_13 + (−15.81)*u850_C6_15 + (−38.26)*u850_C6_17	0.589

meridional exchange of energy, and a greater northward incursion of precipitating systems over the South American continent, while the opposite occurs when a positive phase is present [17].

Other circulation features are discernible from the correlation patterns. The models include the influence of the upper-level southern hemisphere jets on the variability of DJF rainfall for regions 2, 4, 5, and 6 (through hgt200_C2_2, hgt200_C4_5, hgt200_C5_9, and hgt200_C6_13, respectively). Changes in the location and/or intensity of jets could potentially lead to modifications in the upper-level convergence/divergence areas, and consequently in the position of the convective areas [31]. On the other hand, models include the influence of the mid-latitude circulation through the inclusion of the predictors hgt1000_C3_5 and u850_C3_11 for Region 3; u850_C5_10, u850_C5_17, and u850_C5_23 for Region 5; and u850_C6_15 and u850_C6_17 for Region 6. Wave trains displacing over the Pacific Ocean usually relate to SST variability in low latitudes, with their associated precipitating systems interacting with the South American climate after entering the continent through its southernmost extreme [32–34].

Circulation in low latitudes is present in the models through the predictors u850_C4_15 (cluster 4) and hgt1000_C5_14 (cluster 5). These are low-level forcings situated in northeastern South America and the adjacent tropical South Atlantic Ocean, respectively. The latter suggests that weaker (stronger) equatorial lows on the Atlantic basin precede a rainier (drier) summer season in Region 5. The former predictor indicates that westerly (easterly) anomalies of 850 hPa zonal wind over the region of the upper branch of the South Atlantic High lead to higher (lower) summer precipitation values in Region 4. Indeed, Garbarini et al. [35] found that an eastward displacement of the South Atlantic High in November is followed by a rainier (drier) summer season in most subtropical Argentina. Low-level meridional wind in South America, representative of changes in the summer LLJ and moist transport towards higher latitudes, was taken as a predictor only by the model corresponding to Region 1 (v850_C1_9). For this case, stronger (weaker) northerly winds in central South America in November precede higher(lower)-than-normal summer precipitation.

4.3.4 Model Verification

In this section, the analysis for each region's model and their verification are explained. Figures 4.12 to 17 show the observations series (black line), the series derived from applying the best model (gray line), and the LOOCV-derived series (light-gray line, Y_hat). Black horizontal lines show the thresholds given by the tertiles of the observations. Using those thresholds, the observations were

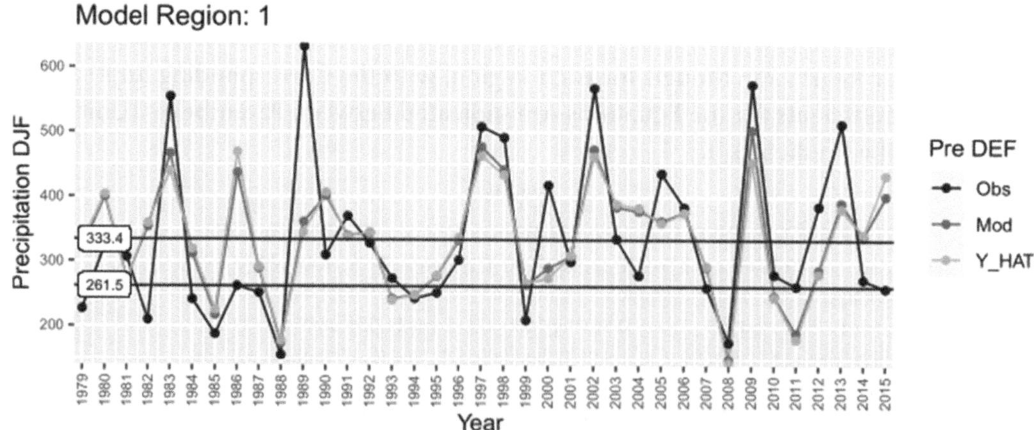

FIGURE 4.12 Series of observations (black line, Obs), series derived from applying the best model (mid-gray line, Mod) and LOOCV-derived series (light-gray line, Y_hat), for Region 1. Black horizontal lines show the thresholds given by the tertiles of the observations.

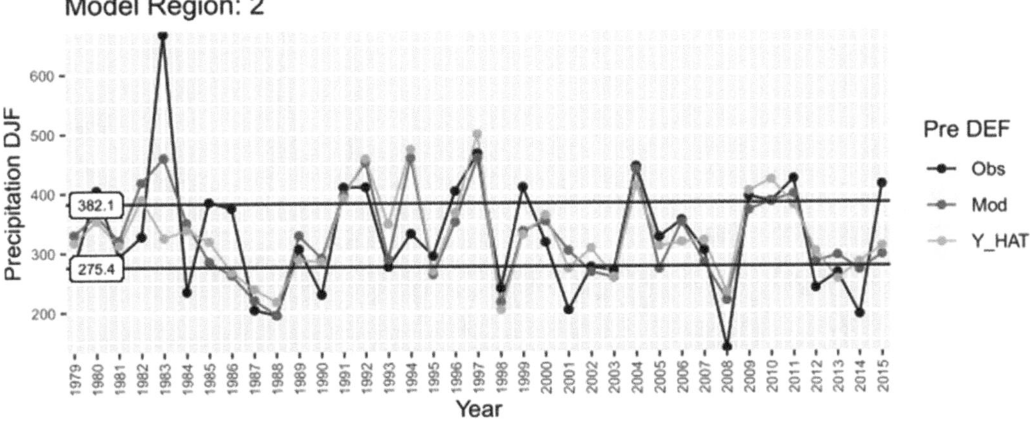

FIGURE 4.13   As in Figure 4.12 but for Region 2.

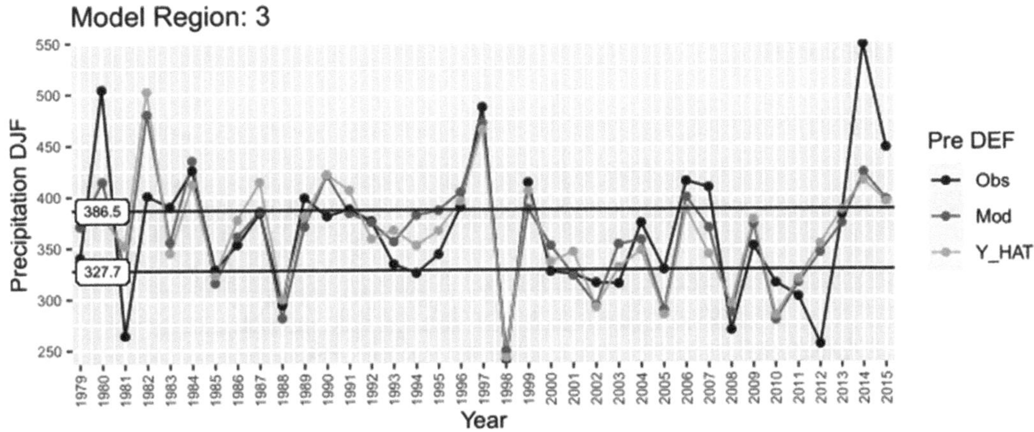

FIGURE 4.14   As in Figure 4.12 but for Region 3.

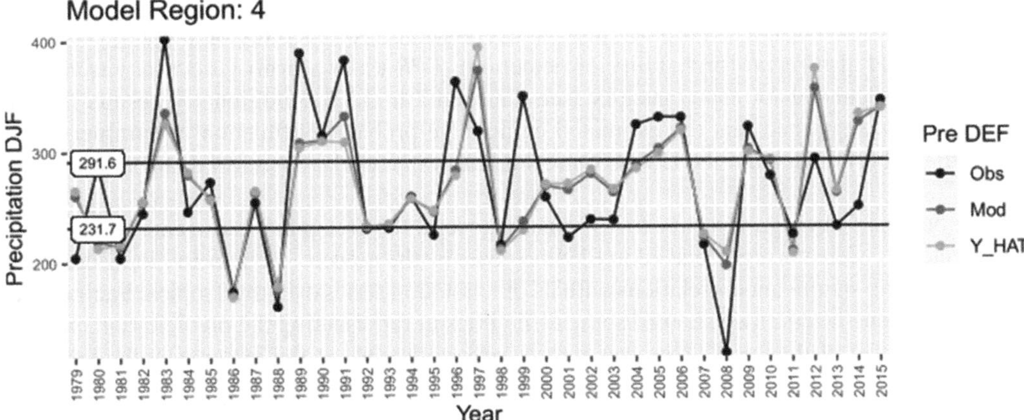

FIGURE 4.15   As in Figure 4.12 but for Region 4.

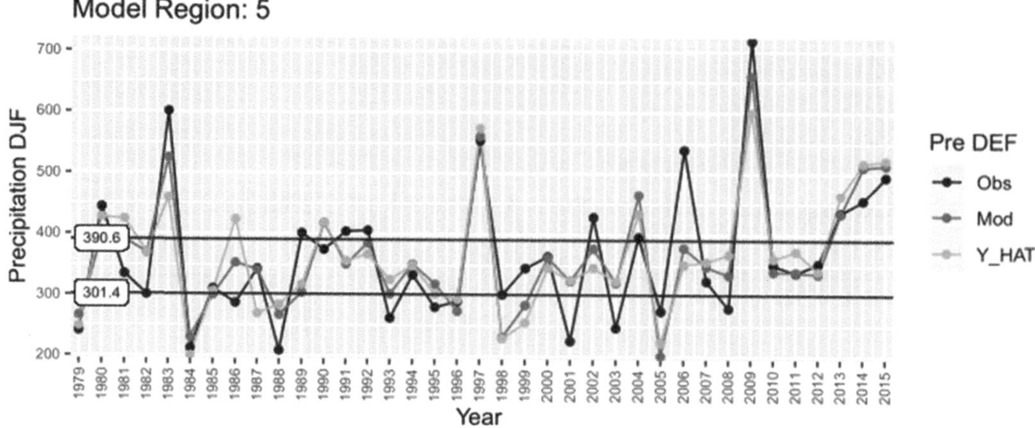

FIGURE 4.16   As in Figure 4.12 but for Region 5.

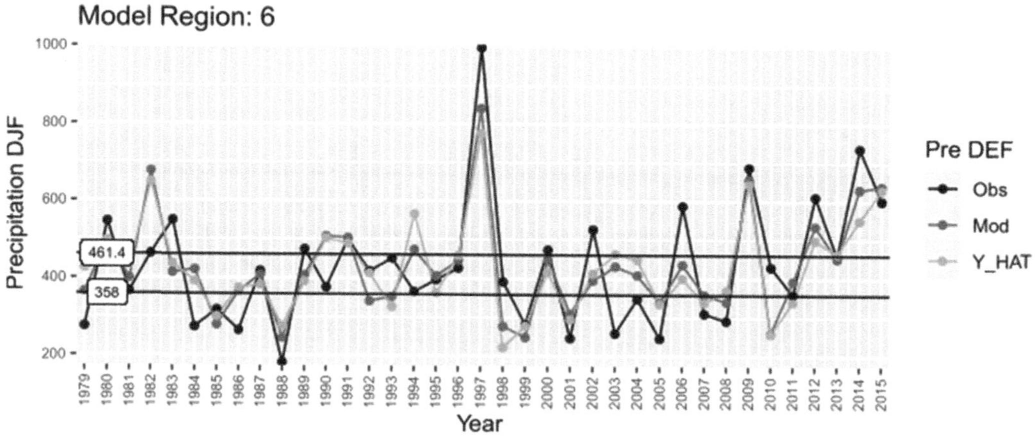

FIGURE 4.17   As in Figure 4.12 but for Region 6.

classified as "below-normal", "normal", and "above-normal". For each region a contingency table was created and also *recall* (probability of detection), *FPR* (false alarm) y *accuracy* (sensitivity) indices were calculated for each category (see Tables 4.2 to 4.13). Additionally, results from the Chi-Square test suggested that the estimated cumulative probability function of summer rainfall from both the best model and the LOOCV are similar to that of the observed distribution (p-value > 0.05).

It can be seen from Figuress. 4.12 to 4.17 that in general the precipitation extremes of the above-normal category were correctly placed in El Niño years by the models, even for Region 1 despite the SST not showing significant correlations with its precipitation series over the Tropical Pacific Ocean (not shown). Nonetheless, they showed some difficulties in representing the precipitation extremes' intensities in most of the cases. This was particularly noticeable at the extreme in the year 1983, which took place during a strong El Niño event, for most of the regions.

The equation of each model and their explained variance (AdjR^2) can be seen in Table 4.1. Further details about the models' predictors and their classification skills are detailed below, for each Region.

TABLE 4.2

Contingency Table of Modeled (MOD) versus Observed (OBS) Precipitation Values, for Region 1

MOD	OBS		
	below-normal	normal	above-normal
below-normal	4	3	0
normal	6	2	2
above-normal	2	7	11

TABLE 4.3

Accuracy, Recall, and False Positive Rate (FPR) Indices by Category, for Region 1

	Accuracy	Recall	FPR
below-normal	0.7	0.33	0.12
normal	0.51	0.17	0.32
above-normal	0.7	0.85	0.38

TABLE 4.4

As in Table 4.2 but for Region 2

MOD	OBS		
	below-normal	normal	above-normal
below-normal	7	2	0
normal	5	8	7
above-normal	0	2	6

TABLE 4.5

As in Table 4.3 but for Region 2

	Accuracy	Recall	FPR
below-normal	0.81	0.58	0.08
normal	0.57	0.67	0.48
above-normal	0.76	0.46	0.08

4.3.4.1 Region 1

The model only represented 45% of the observations' variability, as it was shown by the AdjR^2 value. The predictors which were retained by the model as precipitation forcings were 1000 hPa geopotential height (hgt1000) over the Indian Ocean, west winds (u850) in the South Pacific Ocean, and meridional wind (v850) over the LLJ area in South America (See Figure 4.9). With regard to the model's classification skill, it had a fairly good accuracy index (0.70; Table 4.3) for below-normal and above-normal categories, but it was lower for the normal category (0.51). It had the best

TABLE 4.6

As in Table 4.2 but for Region 3

		OBS		
		below-normal	normal	above-normal
MOD	below-normal	6	2	0
	normal	6	7	5
	above-normal	0	3	8

TABLE 4.7

As in Table 4.3 but for Region 3

	Accuracy	Recall	FPR
below-normal	0.78	0.5	0.08
normal	0.57	0.58	0.44
above-normal	0.78	0.62	0.12

TABLE 4.8

As in Table 4.2 but for Region 4

		OBS		
		below-normal	normal	above-normal
MOD	below-normal	7	1	1
	normal	5	9	2
	above-normal	0	2	10

TABLE 4.9

As in Table 4.3 but for Region 4

	Accuracy	Recall	FPR
below-normal	0.81	0.58	0.08
normal	0.73	0.75	0.28
above-normal	0.86	0.77	0.08

performance for the above-normal category, considering both accuracy and recall (0.85) indices, although FPR was not similarly good (0.38).

4.3.4.2 Region 2

The model could represent only 48% of the observations' variability. The predictors which were part of it were: SST in the Western Pacific Ocean, 1000 hPa geopotential height over the South Pacific and South Atlantic Oceans, and 200 hPa geopotential height over the Indian Ocean. The model's classification skill is quite better for below-normal and above-normal categories than for the normal category, according to the accuracy index (values of 0.81, 0.76, and 0.57 respectively;

TABLE 4.10

As in Table 4.2 but for Region 5

		OBS		
		below-normal	normal	above-normal
MOD	below-normal	5	3	0
	normal	6	7	5
	above-normal	1	2	8

TABLE 4.11

As in Table 4.3 but for Region 5

	Accuracy	Recall	FPR
below-normal	0.73	0.42	0.12
normal	0.57	0.58	0.44
above-normal	0.78	0.62	0.12

TABLE 4.12

As in Table 4.2 but for Region 6

		OBS		
		below-normal	normal	above-normal
MOD	below-normal	6	4	0
	normal	5	5	5
	above-normal	1	3	8

TABLE 4.13

As in Table 4.3 but for Region 6

	Accuracy	Recall	FPR
below-normal	0.73	0.5	0.16
normal	0.54	0.42	0.4
above-normal	0.76	0.62	0.17

Table 4.5). The recall index of the below-normal and normal categories showed that the model was able to detect more than a half of the possible cases within those categories (0.58 and 0.67 respectively) and slightly less than a half for the above-normal category (0.46). The highest False Positive Rate was for the normal category and it reached a value of 48%.

4.3.4.3   Region 3

The model was able to represent 57% of the observations' variability. The predictors selected to build this model were SST in the Central Pacific Ocean, 1000 hPa geopotential height in the Indian Ocean, and zonal wind in the South Atlantic Ocean. This region is particular due to its orography, conformed by several highland areas. With regard to the model's classification skill, the accuracy index is relatively high for below-normal and above-normal categories (0.78 for both; Table 4.7), but it is lower for the normal category (0.57). The recall indices show that the model classifies half or more of the events correctly in all the categories (values of 0.50, 0.58, and 0.62). The highest False Positive Rate was for the normal category and it had a value of 44%.

4.3.4.4   Region 4

The model could represent 50% of the observations' variability. The predictors retained by it were: 500 hPa geopotential height over the Indian Ocean, the same variable but in 1000 hPa over the South Atlantic Ocean and in 200 hPa over the Central Pacific Ocean, and the zonal wind in South America, where the subtropical high enters the continent. This region is located in the south of Buenos Aires Province and it is the area with the lowest rainfall of the study region. The model's classification skill was relatively good according to the accuracy indices, especially for below-normal and above-normal events (0.81, 0.73, and 0.86; Table 4.9). The recall indices showed that the events were correctly classified for more than 50% of the cases, for all the categories (values of 0.58, 0.75, and 0.77). The highest FPR was for the normal category (0.28).

4.3.4.5   Region 5

The model was able to represent 68% of the observations' variability. The predictors employed by its model were: SST in the Indian Ocean, 1000 hPa geopotential height over the Atlantic Ocean north of the Subtropical High, 200 hPa geopotential height over the South Atlantic Ocean and the wind in the westerlies belt over both the South Atlantic and South Pacific oceans. This region is the "core" of the study region. The observed accuracy indices' values were fairly good for the below-normal and above-normal categories (0.73 and 0.78, respectively; Table 4.11) but it was lower for the normal category (0.57). It can be seen in the correspondent series that the model could represent the extreme values correctly (Figure 4.16). This also agrees with the AdjR^2 value. According to the recall indices, the model classified the events in categories correctly in 42 to 62% of the cases, depending on the category (values of 0.42, 0.58, and 0.62). The highest False Positive Rate was for the normal category and it reached a value of 44%.

4.3.4.6   Region 6

The model could represent 59% of the observations' variability. The predictors which were retained by the model were SST in the Equatorial and South regions of the Pacific Ocean, 200 hPa geopotential height over the Western Pacific Ocean, and the zonal wind over the southern parts of the Indian and Atlantic Oceans. This region is located in the North of the Pampa Region and it is the area with the highest rainfall of the study region. With regard to the model's classification skill, the accuracy index is fairly good for below-normal and above-normal categories, but lower for the normal category (0.73, 0.54, and 0.76, respectively; Table 4.13). The recall indices showed that the model classified the events correctly in 42 to 62% of the cases, depending on the category (values of 0.50, 0.42, and 0.62). The highest False Positive Rate was for the normal category and it was as high as 40%.

4.4   CONCLUSIONS

An exploratory analysis of weather station observations was performed over the Pampa Region with the aim of separating them by regions, discriminating the observed effects of the different areal precipitation regimes. Different techniques of hierarchical and non-hierarchical clustering were evaluated to cluster the non-uniformly distributed stations together. From this analysis, it was found that the best clustering technique over the Pampa Region is K-means, which is a non-hierarchical method. On the other hand, good results were also observed using a hierarchical method based on variance minimization within a group which employs Euclidean distance. To validate the applied clustering techniques, internal validation techniques such as "silhouette" were applied, as well as verification techniques for the optimal number of groups. The resulting clusters were coherent considering the mean climatological precipitation variability in the study region. A total of six clusters of weather stations were obtained. In addition, a verification through analysis of variance (ANOVA) was also carried out to assess whether the groups were significantly different from each other. The results of this analysis were positive. As a result, it was determined that the selected groups were consistent in the study region, and mean regional series of precipitation were calculated for each of them.

Correlation maps were made with the values of lagged correlation between the average series of each cluster of weather stations and the climate forcings of the previous month (November). The regions of significant correlation values were found over some areas which are interesting from a point of view based on the climate forcings. Mean areal series were calculated over those significant areas, considering only the interior points of the set delimited by the isolines of the significance threshold (0.35 in absolute value), instead of using rectangular areas as the usual technique. For each Region, 15 to 40 predictors were obtained.

The LASSO technique was applied to filter the less significant predictors and to avoid the collinearity problem among them. A model of multiple linear regression was created with the resulting predictors, for each one of the six subregions of this study, using the "Leave-one-out-cross-validation" technique. The main forcings of summer precipitation involved in the models seem to be related to the influence of sea surface temperature variability in tropical Pacific and Indian Oceans, mid-latitude circulation characteristics (wave train activity and intensity of mid-latitude pressure systems and zonal wind), and low-latitude circulation near the South American continent (northerly low-level flow and moisture transport and variability of the South Atlantic High).

The obtained models could partially reproduce the precipitation variability. Most of them represented around 50% of the observations' variability, and all of them presented an adjusted coefficient of determination of less than 70%. In particular, they failed to represent the magnitude of the extreme values in most of the cases. Taking the difficulties that the models usually have in predicting exact values into account, precipitation events were classified into three different categories based on the tertiles of the mean observations series (below-normal, normal, and above-normal), for each region. The models' classification skill was better for the below-normal and above-normal categories than for the normal category.

ACKNOWLEDGMENTS

Precipitation data was provided by the Argentinean National Meteorological Service (SMN) and reanalysis data was provided by NOAA/ESRL Physical Sciences Division, Boulder Colorado from their website: http://www.cdc.noaa.gov. This research was supported by 2020–2022 UBACyT 20020190100090BA and 2018–2020 UBACyT 20620170100012BA projects.

REFERENCES

1. Liebmann, B., Vera, C., Carvalho, L., Camilloni, I., Hoerling, M., Allured, D., Barros, V., Báez, J., and Bidegain, M. (2004). An observed trend in central South American precipitation. *Journal of Climate*, 17(22), 4357–4367.
2. Barros, V., Doyle, M., and Camilloni, I. (2008). Precipitation trends in southeastern South America: Relationship with ENSO phases and the low-level circulation. *Theoretical and Applied Climatology*, 93, 19–33.
3. Saurral, R., Camilloni, I., and Barros, V. (2016). Low-frequency variability and trends in centennial precipitation stations in southern South America. *International Journal of Climatology*. Wiley Online Library. https://doi.org/10.1002/joc.4810.
4. González, M., Dominguez, D., and Nuñez, M. (2012). Long term and interannual rainfall variability in argentinean Chaco Plain region. In Martín, O., and Roberts, T. (Eds.), *Rainfall: Behavior, Forecasting and Distribution*. Nova Science Publishers, 68–89. ISBN: 978-1-62081-5519.

5. Ropelewski, C., and Halpert, M. (1987). Global and regional scale precipitation patterns associated with El Niño. *Monthly Weather Review*, 115(8), 1606–1626. https://doi.org/10.1175/1520 -0493(1987)115<1606:GARSPP>2.0.CO;2.

6. Kiladis, G. N., and Diaz, H. F. (1989). Global climatic anomalies associated with extremes in the Southern Oscillation. *Journal of Climate*, 2, 1069–1090.

7. Compagnucci, R., and Vargas, W. (1998). Inter-annual variability of the Cuyo river streamflow in the Argentinean andean mountains and ENSO events. *International Journal of Climatology*, 18, 1593–1609.

8. Grimm, A., Barros, V., and Doyle, M. (2000). Climate variability in Southern South America associated with El Niño and La Niña events. *Journal of Climate*, 13, 35–58.

9. Vera, C., Silvestri, G., Barros, V., and Carril, A. (2004). Differences in El Niño response in Southern Hemisphere. *Journal of Climate*, 17(9), 1741–1753.

10. Barreiro, M. (2010). Influence of ENSO and the south Atlantic Ocean on climate predictability over Southeastern South America. *Climate Dynamics*, 35, 1493–1508.

11. Garbarini, E., Skansi, M., Gonzalez, M. H., and Rolla, A. (2016). ENSO influence over precipitation in Argentina. In *Advances in Environmental Research*, Chapter 7, Volume 52, Ed. Justin A. Daniels. NOVA Publisher, New York, NY.

12. Ashok, K., Behera, S., Rao, S., Weng, H., and Yamagata, T. (2007). El Nino Modoki and its possible teleconnection. *Journal of Geophysical Research*, 112, C11007. https://doi.org/10.1029/2006JC003798.

13. Saji, N. H., Goswami, B. N., Vinayachandran, P. N., and Yamagata, T. (1999). A dipole mode in the tropical Indian Ocean. *Nature*, 401(23), 360–363.

14. Chan, S. C., Behera, S. K., and Yamagata, T. (2008). Indian Ocean dipole influence on South American rainfall. *Geophysical Research Letters*, 35, L14S12. https://doi.org/10.1029/2008GL034204.

15. Thompson, D. W., and Wallace, J. M. (2000). Annular modes in the extratropical circulation. Part I: month-to-month variability. *Journal of Climate*, 13, 1000–1016.

16. Nan, S., and Li, J. (2003). The relationship between summer precipitation in the Yangtse River Valley and the previous Southern Hemisphere Annular Mode. *Geophysical Research Letters*, 30(24), 2266.

17. Reboita, M. S., Ambrizzi, T., and Da Rocha, R. P. (2009). Relationship between the Southern Annular Mode and Southern Hemisphere Atmospheric Systems. *Revista Brasileira de Meteorología*, 24(1), 48–55.

18. González, M. H., and Vera, C. S. (2010). On the interannual wintertime rainfall variability in the Southern Andes. *International Journal of Climatology*, 30, 643–657. https://doi.org/10.1002/joc.1910.

19. González, M. H., Skansi, M. M., and Losano, F. (2010). A statistical study of seasonal winter rainfall prediction in the Comahue region (Argentina). *Atmosfera*, 23(3), 277–294.

20. González, M. H., and Dominguez, D. (2012). Statistical prediction of wet and dry periods in the Comahue Region (Argentina). *Atmospheric and Climate Sciences*, 2(1), 23–31. https://doi.org/10.4236/ acs.2012.21004.

21. González, M. H. (2015). Statistical seasonal rainfall forecast in the Neuquén River Basin (Comahue Region, Argentina). *Climate*, 3, 349–364. https://doi.org/10.3390/cli3020349.

22. Nnamchi, H. C., Li, J. P., and Anyadike, R. N. (2011). Does a dipole mode really exist in the South Atlantic Ocean? *Journal of Geophysical Research*, 116. https://doi.org/10.1029/2010JD015579.

23. Tan, P. N., Steinbach, M., and Kumar, V. (2006). *Introduction to Data Mining*. Pearson; Addison Wesley, Boston. ISBN-13: 9780321321367.

24. MacQueen, J. (1967). Some methods for classification and analysis of multivariate observations. *Proc. Fifth Berkeley Symp. on Math. Statist. and Prob.*, Vol. 1. University of California Press, 281–297.

25. Steinhaus, H. (1956). Sur la division des corps matériels en parties. *Bulletin de l'Académie Polonaise des Sciences*, Classe III, IV(12), 801–804.

26. Kalnay, E., Kanamitsu, M., Kistler, R., Collins, W., Deaven, D., Gandin, L., Iredell, M., Saha, S., White, G., Woollen, J., Zhu, I., Chelliah, M., Ebisuzaki, W., Higgings, W., Janowiak, J., Mo, K. C., Ropelewski, C., Wang, J., Leetmaa, A., Reynolds, R., Jenne, R., and Joseph, D. (1996). The NCEP/NCAR reanalysis 40 years- project. *Bulletin of the American Meteorological Society*, 77, 437–471.

27. Tibshirani, R. (1996). Regression shrinkage and selection via the lasso. *Journal of the Royal Statistical Society. Series B (Methodological)*, 58(1), 267–288.

28. Stone, M. (1974). Cross-validatory choice and assessment of statistical predictions. *Journal of the Royal Statistical Society*, 36(2), 111–147.

29. Rasmusson, E. M., and Carpenter, T. H. (1982). Variations in tropical sea surface temperature and surface wind fields associated with the southern Oscillation/El Niño. *Monthly Weather Review*, 110(5), 354–384.

30. Taschetto, A. S., and Ambrizzi, T. (2012). Can Indian Ocean SST anomalies influence South American rainfall? *Climate Dynamics*, 38, 1615–1628.

31. Carlson, T. N. (1998). *Mid-latitude Weather Systems*. American Meteorological Society, Boston, 507p.

32. Kidson, J. W. (1999). Principal modes of southern hemisphere low frequency variability obtained from NCEP-NCAR Reanalyses. *Journal of Climate*, 12, 2808–2830. https://doi.org/10.1175/1520 -0442(1999)012<2808:PMOSHL>2.0.CO;2.

33. Mo, K. C. (2000). Relationships between low frequency variability in the Southern Hemisphere and sea surface temperature anomalies. *Journal of Climate*, 13, 3599–3610. https://doi.org/10.1175/1520 -0442(2000)013<3599:RBLFVI>2.0.CO;2.

34. Paegle, J., and Mo, K. C. (2002). Linkages between summer rainfall variability over south america and sea surface temperature anomalies. *Journal of Climate*, 15, 1389–1407. https://doi.org/10.1175/1520 -0442(2002)015<1389:LBSRVO>2.0.CO;2.

35. Garbarini, E. M., González, M. H., and Rolla, A. L. (2019). The influence of Atlantic High on seasonal rainfall in Argentina. *International Journal of Climatology*, 39, 4688–4702. https://doi.org/10.1002/joc .6098.

Part II

Africa

5 Climate Change Impact on Sediment Yield
Case Studies in Africa

Faiza Hallouz and Mohamed Meddi

5.1 INTRODUCTION

Soil is a non-renewable natural resource dynamic that directly or indirectly affects our daily activities, and plays a vital role in maintaining life on the planet. Water movement and quality, land use, and vegetation productivity all have relationships to the soil (Briak, 2017).

In fact, soil has always been confronted with erosion, which poses a major problem in the preservation of our environment (Pimentel et al., 1995; Portenga and Bierman, 2011; Wu and Chen, 2012).

It was not until the 1950s, after the Madison Congress of the International Soil Science Association, that American methods of measuring runoff and erosion on small plots spread to French and English-speaking Africa, then Latin America, and more recently, to Asia and Europe (Roose, 1994).

The first explorers who landed in tropical Africa, seeing the abundance of vegetation under the forest, thought that the soils there must be particularly rich (Roose and De Noni, 2004). But they quickly disillusioned, because once the land was cleared, crop yields fell within a few years. In his book "Africa, a Dying Land" (translated from the French title – *Afrique terre qui meurt*), Harroy (1944) had already described the two causes of the rapid degradation of fertility in African soils: the rapid mineralization of humus in hot and humid climates and the erosion of bare soils subjected to torrential rains (Figures 5.1 and 5.2).

Until 2018, no less than 57,199 large dams (Figure 5.3) have been built to meet water and energy needs for a storage capacity of approximately 14,000 km^3 (ICOLD, 2018). Nearly half of the world's rivers have at least one large dam (ICOLD, 2018).

Climate change has been observed in recent decades and changes have been projected for decades to come (GIEC, 2007). An increase in global temperature is expected to increase evapotranspiration and cause precipitation changes, which will significantly affect the hydrological regimes of many river systems (Lu, 2005). Numerous studies have shown that climate change could have a significant impact on discharge (Nijssen et al., 2001; Menzel and Burger, 2002), soil erosion rate (Michael et al., 2005; O'Neal et al., 2005), and sediment flow (Zhu et al., 2008). Phan et al., (2011) reported annual variations of 1 to 3%, 3.9 to 11.4%, and 1.1 to 5.3% in average annual variations, wet and dry season flows of 1.2 to 4.7%, respectively, from 3.6 to 15.3% and 1.3 to 7.7% of average annual yields, wet season and dry season sediments respectively for the Song Cau watershed in northern Vietnam due to changes in temperature under climate change scenarios (Shrestha et al., 2013).

5.2 EROSION

5.2.1 Definition

Erosion comes from the Latin "erodere" meaning "gnaw" so erosion is a process whereby soil particles are removed from their environment, transported by a transport agent (water, air), and deposited (Figure 5.4) in another environment (Ammari, 2012). This process can be written as:

DOI: 10.1201/9781003473398-7

FIGURE 5.1   Soil erosion in Namibia (https://images.pexels.com/photos/3714904/pexels-photo-3714904 .jpeg?auto=compress&cs=tinysrgb&w=600).

FIGURE 5.2   Soil degradation in Morocco (Shot Hallouz, Bouregreg basin, Morocco, September 2011).

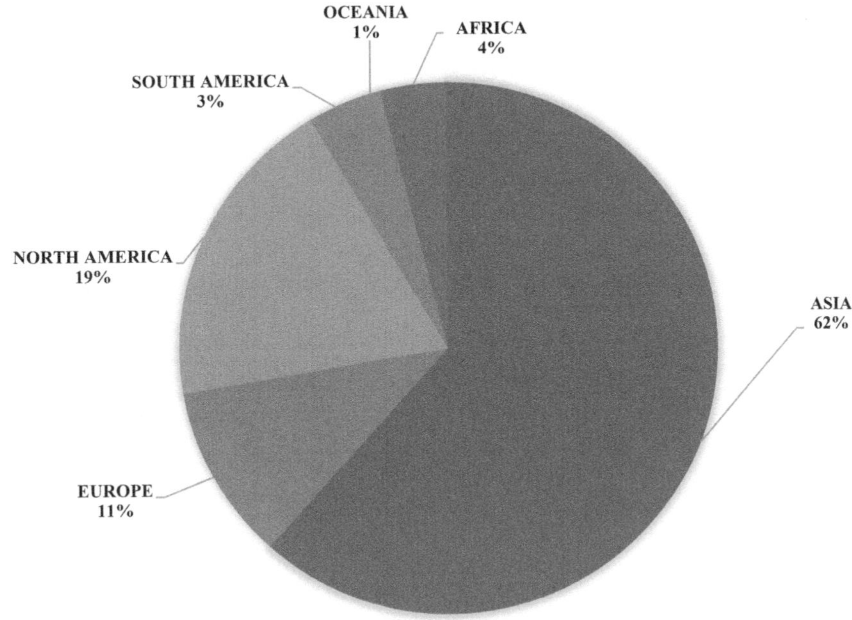

FIGURE 5.3 Distribution of large dams by continent in 2018 (ICOLD, 2018).

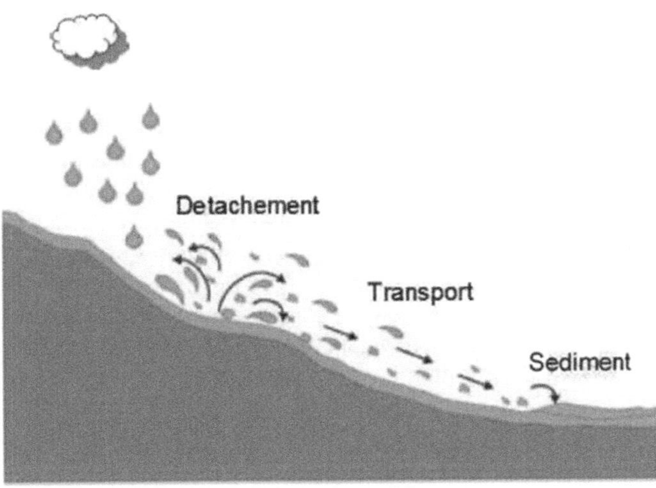

FIGURE 5.4 Water Erosion Process (Cheggour, 2008).

EROSION = DETACHEMENT → TRANSPORT → SEDIMENTATION

5.2.2 Etosion Hydrique

Water erosion could be defined as the process by which soil particles are removed by the water in their environment, transported by it, and deposited in another medium (Lagacé, 1980).

In Africa, an estimated 500 million hectares have been degraded since the 1950s (Figure 5.5), covering 65% of the region's agricultural land (Mokhtari, 2017). According to FAO studies (1990),

FIGURE 5.5　Water erosion gully process in Senegal (AI-CD website: http://aicd-africa.org/archives/818?ln =fr).

the situation continues to deteriorate: are affected by water erosion: in Morocco 40%, in Tunisia 45% of the country's area is threatened by erosion (Chevalier et al., 1995, Boussema, 1996), and in Algeria, 45% of the Tellian areas (Figure 5.3), or 12 million hectares (Chebbani et al., 1999).

Erosion is a widespread phenomenon in the various Maghreb countries, which poses a serious threat to the potential for water and soil. The intensity of the phenomenon is mainly evident during floods. Thus, the 3-day high-level floods of March 1978 drained 30×10^6 tons of sediment into the Algerian region for coastal watersheds (Demmak, 1982) (Figure 5.6). Indeed, this phenomenon continues to take on considerable proportions, especially on slopes because of the torrential nature of the rains, the high vulnerability of the land (soft rocks, fragile soils, steep slopes, and often degraded vegetation cover), overgrazing, and the adverse impact of human activities: deforestation, fires, poor conduct of agricultural works, chaotic urban planning, exploitation of quarries, etc. In addition, the region has the highest values on the planet.

5.3　SEDIMENTATION PHENOMENON

The sediment cycle begins with the erosion process, where particles or fragments are removed from rock materials by erosion agents. River sediments are referred to as water as the main agent (G.C., 2016).

The cycle ends with an accumulation and a deposit. When the energy is no longer sufficient to move the sediment, it comes to a standstill. Basins or accumulation areas include newly deposited materials from a floodplain, islands, banks of a channel, and deltas. The accumulation, even if it is considerable, is not necessarily apparent; this is the case of the one that occurs on river and lake beds. To understand the aquatic ecosystem, it is essential to know the dynamics of sediments (G.C., 2016).

The northern Maghreb ranks among the countries with the highest rate of sediment transported suspended in watersheds per unit of surface area, with an average estimated at $500\,t\,km^{-2}\,y^{-1}$ (Probst

FIGURE 5.6 Stages of degradation and transport of the surface of the ground under the action of the Rains (Shot Hallouz, Wadi Cheliff, flood of 25 January, 2017).

and Amiotte Suchet, 1992). The rate of soil-specific erosion in Maghreb countries ranges from 1000 to 5000 t km^{-2} y^{-1} (Walling, 1984). However, several authors have shown that these values can be exceeded locally. There are 7200 t km^{-2} y^{-1} for the Agrioun wadi supplying the Ighil Emda dam in Algeria (Probst et al., 1992) and the 5900 t km^{-2} y^{-1} recorded at the Nekor watershed that feeds the Mohamed El Khettabi dam in Morocco (Lahlou, 1988). In addition to these strong soil-specific erosion values, other authors have published estimates below this range. Bouanani (2004) calculated for wadi Sikkak (Algeria) 170 t km^{-2} y^{-1}, Hallouz et al. (2018b), estimated the specific erosion rate in the Mina Wadi Basin (Algeria) (Figure 5.7) and Bergaoui et al., at 864 t km^{-2} y^{-1}. (1998) estimated specific soil erosion in central Tunisia at 336 t km^{-2} y^{-1}.

5.4 SEDIMENTATION IN AFRICA

5.4.1 ALGERIA

In 1890, there were nine dams in Algeria, with a capacity of 61 Mm^3 and a mud volume of 2.7 Mm^3. In 1957, Algeria's dams with a capacity of 900 Mm^3 had accumulated nearly 200 Mm^3 of mud (Diab Djeffal, 2013). This capacity reached 1 billion m^3 in 1962, and from that date the number of dams increased considerably. The 1980s saw siltation rates of around 20 million tons per year, while the 1990s already reached 35 million tons per year, the 2000s crescendoed and climbed to the value of 45 million tons annually. All experts agree that the coming years will be very difficult for arid and semi-arid areas (Demmak, 1982; Serbah, 2011).

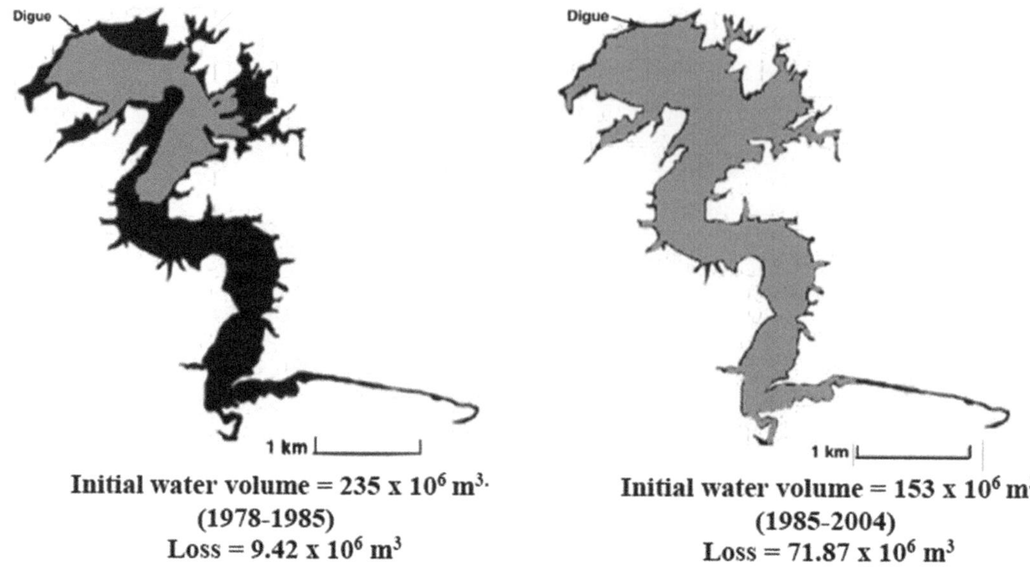

Initial water volume = 235 x 10⁶ m³·
(1978-1985)
Loss = 9.42 x 10⁶ m³

Initial water volume = 153 x 10⁶ m³
(1985-2004)
Loss = 71.87 x 10⁶ m³

FIGURE 5.7   Evolution of the basin morphology and the surface in free water of the Sidi M'Hamed Benaouda Dam (Wadi Mina) (Hallouz et al., 2018b).

TABLE 5.1

Characteristics of the Main Dams Built on the Main Wadis in Algeria

Wadi	Dam	Silting (10^6 m³·y⁻¹)	Initial capacity (10^6m³)	Date commissioned	Year: Capacity 2 (10^6m³)
Cheliff	Boughezoul	0,660	55	1934	2004: 24
	Ghrib	3.200	280	1930	2004: 115,30
	Wadi Fodda	2.246	228	1932	2004: 102.85
	Bekhadda	1.32	56	1936	2004: 44.80
Isser	Beni Amrane	0,44	16	1988	2004: 3.2
Soummam	Ain Zada	2,5	125	1985	2004: 120
Agrioun	Ighil Emda	1.83	154	1953	2004: 90
Seybouse	Bouhamdane	0.08	220	1986	2004: 184
Bou Alalah	Chaffia	0,124	171	1965	2004: 159

Source: According to Quinquis, 2017.

According to Table 5.1, the large dams, the oldest of which is the Ghrib Dam, have an estimated total sedimentary hold of 12 Mm³ y⁻¹, or more than 19 million tons y⁻¹ (Boun Heng, 2013). Based on bathymetric surveys conducted between 2005 and 2006 by the National Agency for Dams and Transfers (ANBT), the average siltation per structure would therefore be 0.79 Mm³ y⁻¹. The Ghrib Dam (Figure 5.8) and the Wadi Foda Dam on the Cheliff wadi are both the oldest and have the most important specific siltation.

5.4.2 MOROCCO

Morocco's natural water resources are among the lowest in the world. Indeed, the average efficient rain is estimated at about 25 billion m³ per year, with only 21 billion m³ of water available. It should also be noted that more than half of the water resources are concentrated in the northern basins (Madani, 2018).

FIGURE 5.8 Ghrib Dam (ANBT, 2020).

Indeed, water infrastructure records a volume loss of around 0.5% of the total storage capacity of 17.5 billion m³, or about 75 Mm³ y⁻¹. The capacity loss of large dams now exceeds 820 Mm³ (Madani, 2018).

Table 5.2 represents the rate of sedimentation of Morocco's most silted dams. The four dams Mohamed V, Mansour Eddahabi, El Massira, and Ben El Ouidane receive a total of 25 Mm³ of mud annually (Aljihad and Ouchen, 2011). Siltation varies greatly from one structure to another, ranging from 0.2 Mm³ per year for the Nakhla dam, to 11.2 Mm³ per year for the Mohamed V dam (Figure 5.9), the dam having the highest siltation among the sites selected in this table (Quinquis, 2017).

5.4.3 Tunisia

In Tunisia, water is strategically important for economic and social development due to its scarcity. Annual averages of storage capacity losses at Tunisian dams could reach 530 Mm³, or 37% of the initial capacity of all operating deductions (Ben Mammou and Louati, 2007). Tunisia had 26 dams in operation in 2003 (Abid, 2003), 13 dams observed, with an initial capacity of about 1430 Mm³, lost a total of about 250 Mm³ of their storage capacity, or about 17.7% (Abid, 1998). At the end of 2017, Tunisia has 37 dams with a total containment capacity of 2285 Mm³ (MARHP, 2017).

The hydraulic situation of the dams indicates for the date of 31 August 2017, a total useful capacity of 2078 Mm³ (for the 30 dams mentioned), while the initial capacity of these dams was 2787 Mm³ (MARHP, 2017) (Figure 5.10).

According to Table 5.3, the El Kebir Reservoir on El Kebir Wadi has the lowest siltation, with 0.2 Mm³ of sediment retained per year. Although located near the El Kebir Dam, the Sidi El Barrak Dam, built on El Zouara Wadi, holds 7.8 Mm³ per year (Quinquis, 2017).

According to the study by Quinquis (2017), the Maghrebian wadis have an average specific sediment productivity of around 1030 t km⁻² y⁻¹, well above the world average estimated at 152 t km⁻² y⁻¹ by Milliman and Meade, (1983) and twice as high as the values estimated by Probst and Amiotte Suchet (1992) which was 397 t km⁻² y⁻¹ for Mediterranean wadi and 504 t km⁻² y⁻¹ for those from across the Maghreb.

According to Table 5.4 and over the study period (1932–2007), it is the Moroccan dam Mohamed V that has retained, by far, the most sediment. This is due to its enormous specific siltation combined with a relatively long operating time (43 years). The two Algerian dams that follow, Ghrig and Wadi Foda (Cheliff), which have been in operation for 80 and 78 years, do not retain half of the sediments of the Mohamed V Dam. Finally, it is the Tunisian dam Sidi Salem that comes in third place after only 5 years of operation. We can therefore expect to see the impact of these four works on the coastlines concerned (Quinquis, 2017).

TABLE 5.2

Morocco Dam Siltation Rates in 2003 and 2004, Categorized by Watersheds

Basin	Dam	Dam capacity (Mm³)	Silting 2003 (Mm³·y⁻¹)	Silting 2004 (Mm³·y⁻¹)
Loukkos Tangé et	Nakhla	4.9	0.2	0.2
Côtiers	Smir	41.9	0.2	0.3
Méditerranés	M B A . Al Khattabi	26.6	1	1.1
	Joumoua	6.5	…	0.5
	09-avr	300	0.5	2
	Ibn Battouta	33.9	0.6	0.6
	Wadi El Makhazine	724	3.2	4.1
Moulouya	Mohamed V	383	11.6	11.2
	Hassan II	125	5.3	5.3
	Injil	12.5	…	0.5
Sebou	Al Wahda	3712.1	11	3.6
	Idriss 1er	1156.8	1.9	2.6
	Allal El Fassi	69.6	3.6	1.7
	El Kansera	230.5	1.2	1.8
	Bab Louta	35.5	…	0.3
	Sidi Chahed	169	0.3	0.3
	Sahal	62	…	0.5
	Bouhouda	55	…	0.3
	Asfalou	316	…	0.5
Bou Regreg	Sidi M. B. Abdellah	441.5	2.5	2.5
	El Mellah	5	0.3	0.3
Oum Er Rbia	Al Massira	2744	3.1	3.1
	Bin El Ouidane	1253.4	4.5	5.2
	Hassan I	245	2.5	2.5
	Moulay Youssef	161	2.5	2.5
Tensift	Lalla Takerkoust	60.6	0.2	0.7
Souss - Massa	Youssef Ben Tachfine	301.8	0.8	0.8
	Abdelmoumen	201.1	0.1	1.1
	Aoulouz	108.2	1.2	1.2
Draa, Ziz	Mansour Eddahbi	445.3	4.8	5
	Hassan Addakhil	326.8	1.2	1.2
Autres petits barrages			…	6.5
Total			**64.3**	**70**

Source: Belaabed, 2012.

5.4.4 MALI

Thus, in South Mali, for example, Diallo et al. (2004) showed that the erosion of the slopes of the Djitiko Basin (104 km²) estimated from erosion plots, soil maps, and land use is 20 times greater than the solid suspended transport observed at the outlet of the basin to the Niger River. On the other hand, in mountains where the slope of emissaries is strong, as in the Mediterranean area, the erosive energy of runoff is stronger than that of rains. Land losses on the fields can be modest, 0.1 to 15 t ha⁻¹ y⁻¹ (Heusch, 1970; Arabi and Roose, 1989; Roose et al., 1993; Laouina, 1992), while solid transport by gullies and wads exceeds 100 to 300 t ha⁻¹ y⁻¹ (Olivry and Hoorelbeck, 1990; Roose et al., 2000).

FIGURE 5.9 Mohamed V Dam in Morocco (AgriMaroc, 2023).

FIGURE 5.10 Sidi Salem Dam (Tunisia) (https://www.pexels.com/fr-fr/chercher/barrage%20de%20sidi%20salem/).

TABLE 5.3

Characteristics of Tunisian Dams Studied

Wadi	Dam	Silting (10^6 m³·y⁻¹)	Bet in service	Amount of sediment (10^6 t·y⁻¹)	Period (year)	Deficit Sedimentary outlet (10^6 t)
El Kebir	El Kebir	0,2	2002	0,32	2	0,64
Zouara	Sidi El Barrak	7,8	1999	12,40	5	62
Medjerda	Sidi Salem	4,7	1981	7,47	23	171,88
	M'Cherga	0,39	1972	0,62	32	19,84

Source: According to Quinquis, 2017.

TABLE 5.4

Sedimentary Retention Carried Out by Dams over the Period 1932–2007

Dam / Wadi	Amount of sediment (10^6 t·y⁻¹)	Operating period (year)	Period retention (10^6 Tons)
Nakhla / MARTIL	0.32	49	15.58
Ali Thlet (Ali Thelat) / LAOU	0.48	56	26.71
Abdelkarim El Khattabi / NEKOR	1.59	29	46.11
Mohamed V / MOULOUYA	18.44	43	793.09
M. Hommadi / MOULOUYA	1.59	53	84.27
Hassan II / ZA (affulent MOULOUYA)	2.26	2	4.52
Arabat / MOULOUYA	0.06	13	0.83
Enjil / MOULOUYA	0.38	13	4.96
Total Morocco	**25.12**		**976.07**
Boughezoul / CHELIFF	0.91	76	68.88
Ghrib / CHELIFF	3.96	80	316.73
Wadi Foda / CHELIFF	3.58	78	279.05
Bakhada / CHELIFF	0.13	74	9.41
Beni Amrane / ISSER	0.70	22	15.39
Taksebt / SEBAOU	0.43	3	1.29
Ain Zada / SOUMMAM	3.98	25	99.38
Kherrata / AGRIOUN	2.34	65	151.92
Eghilmda / AGRIOUN	2.91	56	162.94
Beni Harroun / EL KEBIR-RHUMEL	9.54	3	28.62
Bouhamdane / SEYBOUSE	0.13	24	3.05
Chaffia / BOU ALALAH	0.19	45	8.59
Mexanna / BOU ALALAH	0.25	18	4.58
Bouhaloufa / BOU ALALAH	0.25	62	15.77
Total Algeria	**29.29**		**1165.60**
El Kebir / EL KEBIR	0.32	2	0.64
Sidi El Barrak / EL ZOUARA	12.40	5	62.01
Sidi Salem / MEDJERDA	7.47	29	216.72
M'Chergua / MEDJERDA	0.62	38	23.56
Total Tunisia	**20.81**		**302.93**

Source: Quinquis, 2017.

In the case of young, deep-valley mountains, the larger the basin, the more concentrated and rapid the runoff, the stronger the peak flows, and the more runoff assaults the bottom and banks of the wads by causing landslides in the low terraces (Roose et al., 2000).

5.4.5 Egypt

The Nile Delta is located in a semi-arid climatic zone, but its source is in an equatorial region in Tanzania (Fanchette, 2006). Although it is the longest river in the world (6825 km), the Nile has a low annual flow (84 billion m^3 of water per year), similar to that of the Rhine (Waterbury 1979). The sedimentation power of the Nile is very low: in Aswan, the Nile carries only 110 million tons of alluvials, much of which is held behind the dam (Figure 5.11; Fanchette, 2006).

In Ethiopia, the rates of soil erosion are alarmingly high and sedimentation in reservoirs, lakes, and rivers is a serious problem (Haregeweyn et al., 2006; Tamene et al., 2006). Many reservoirs which have been established for hydroelectric power, urban water supply, and irrigation accumulate large amounts of sediment, resulting in a shortage of water supply for these functions, a decline in reservoir water storage capacity, and high costs to remove sediment from reservoirs. Some of the dams in the Amhara Region of Ethiopia, such as the dams of Adrako, Borkena, and Dana have completely silted up before their design expectation period (Amare, 2005; Kebede, 2012). Other dams in this region that have been constructed over the last decades are threatened by accelerated sedimentation (Meknnen et al., 2015). Indeed, the results of the study of the assessment of the dam's sedimentation rate in northern Ethiopia carried out by Mekonnen et al. (2015) showed that DSS trapped an average of 1584 t y^{-1} of the inflow sediments and the specific yield of the catchment sediments ranged from 8.6–55 t ha^{-1} y^{-1}.

5.4.7 Congo

The Congo River (formerly Zaire) is the second largest river in the world after the Amazon by the size of its watershed (3.70×10^6 km^2) and in terms of flow (interannual average ~ 41,000 m^3 s^{-1}) (Mouzéo, 1986; Prahl et al., 1994), its source is the Katanga highlands located in the southeast of the

FIGURE 5.11   High Aswan Dam (Joellem, 2014).

Democratic Republic of Congo. It stretches for 4700 km. It is fed by numerous tributaries located on both sides of Ecuador, a characteristic that has earned it a very stable hydrological regime, the most regular in the world (Martins and Probst, 1991; Coynel et al., 2005; Laraque et al., 2013).

The Congo has an annual sedimentary discharge of 55×10^6 tons (Wetzel, 1993). However, this value depends on the period of time the measurements are made. Recently, Laraque and his collaborators (2013) calculated a lower sedimentary discharge of 33–106 tons y^{-1} that is consistent with Olivry et al. (1988) and Laraque et al. (2009). Other estimates in the literature are that of Meade (1996) and Coynel et al. (2005) who estimate the sedimentary flow at 43×10^6 tonnes and 30.7×10^6 tons y^{-1}. These studies highlight the difficulty in estimating the specific flow of particles exported by this river (Stetten, 2015).

5.4.8 SOUTH AFRICA

Analysis of reservoir sedimentation rates in South Africa revealed an estimated average annual loss of 0.4% per year. Although the estimated global reservoir sedimentation rate is 0.8% per year (ICOLD, 2009), the sedimentation rate in South Africa is quite significant given the magnitude of the increase in water demand in southern Africa.

Nearly 25% of the total number of reservoirs analyzed in South Africa lost between 10% and 30% of their initial storage capacity. This requires increased attention to reservoir sedimentation problems. There have been extensive studies of sediment transport in rivers and reservoirs in South Africa over the past 50 years. Some of the studies have resulted in the development of the Southern African sediment yield map and a treaty on the transport of sediment into rivers and reservoirs by Rooseboom et al. (1992). The latter resulted in what was considered a basic manual for predicting sediment yield (Msadala and Basson, 2019).

5.4.9 WEST AFRICA

The studies carried out by Adam (1986) estimated that with the construction of the Nangbéto hydroelectric dam, 540,000 m³ of sediment is trapped each year in the reservoir, whereas before the dam was built, the Mono River brought 900,000 m³ of sediment to Nangbéto per year (an average specific erosion of 60 t km^{-2} y^{-1}), of which 100,000 m³ reached the outlet (Oyédé, 1991; Rossi and Blivi, 1995; Blivi, 2000, 2005). But the calculation of Nangbéto's theoretical transport capacity downstream yielded 1,800,000 m³ y^{-1} (or 10^6 t km^{-2} y^{-1}) (Rossi, 1996; Blivi, 2000). However, these are old and timely measures.

In fact, before the dam, 800,000 m³ of sediment was retained downstream of Nangbéto. After the construction of the dam, 540,000 m³ of sediment is trapped by the reservoir; thus 360,000 m³ of sediment is added to those remobilized in the lower Mono–Couffo valley to feed the lagoon system, an unknown quantity of which goes into the sea (Amousson, 2010).

5.4.10 CHAD

In Chad, for example, the Logone and the Chari associated with El Beid, deliver to Lake Chad 385,000,000 km³ of water and huge quantities of material in dissolved form (2,580,000 tons) and in solid form (2,860,000 tons) (Gac, 1980).

As a result, it is the reservoirs in arid or semi-arid areas that are most likely to suffer from serious siltation problems, with those in temperate areas less likely to experience these difficulties.

5.4.11 BURKINA FASO

In Burkina Faso, as elsewhere, the problem of sediment inputs into lakes is very little known: it is claimed that only very large reservoirs are likely to become silted up and, as a result, few studies have been conducted on the subject (Gresillon, 1976). The most recent work on this subject in Burkina Faso is that of Sanon (1998) on the Tamasgho Dam and Dipama (1992) on dams 1, 2, and 3 in Ouagadougou.

5.4.12 COTE D'IVOIRE

Côte d'Ivoire has an extensive hydrographic network with six large hydroelectric dams (Figure 5.12). The water reservoirs of these dams have been the subject of several studies of hydrological, physico-chemical, and biological characterization (Reizer, 1967; Kassoum, 1979; Galy-Lacaux et al., 1999; Ouattara, 2000; Yapo, 2002). Studies on the morphology of the bottom and the nature of the sediments that fill these reservoirs exist partially and are very recent (Kouassi, 2007); which does not allow the appreciation of the sedimentology evolution of the bottom of these lakes. However, these reservoirs, which are major assets to support the country's development, are increasingly confronted with the effects of hydrological variations and siltation problems (Konan et al., 2013). Thus, the exploitation of these reservoirs becomes difficult due to hydroclimatic fluctuations that affect the water resources of the Sahelian region and, on the other hand, because of the modification of the reservoirs by siltation (Konan et al., 2013).

5.4.13 SENEGAL

The dynamics of suspended sediments in the Senegal basin, and the behavior of the river particle load at the Bakel gauging station (218,000 km^2) during the period 1979–1984 (Kane and Diallo, 2005) showed that during this study period, the average annual particle load carried by the Senegal River was about 1.9 million tons (Kattan et al., 1987). Two approaches are used to estimate the different contributions to the suspended transport of river sediments. The main contribution comes

FIGURE 5.12 Ayamé Dam (Babi Inside, 2016).

from slope erosion, which provides 50–80% of total sediment transport, and the second contribution comes from channel erosion (Kattan et al., 1987).

Finally, in Africa, more specifically (Figure 5.13) North Africa, Ethiopia, and the Great African Rift (Remini et al., 2009; Shahin, 1993) are distinguished by their high filling speed. In other regions of Africa, there are few data directly related to the capacity lost by sedimentation: we must nevertheless cite the work of Adwubi et al. (2009) on four recent reservoirs in Ghana, in a locally very eroded Sudanian savannah environment, with a useful life varying from 22 to 190 years (annual losses between 0.5% and 4.5%), and the Sichingabula study (1997) on 66 small reservoirs in southern Zambia, with a lifespan of between 200 and 5100 years (losses between 0.02% and 0.5%). In order to obtain other elements of comparison, it is necessary to refer to the data from the specific degradation studies, expressed in $m^3\ km^{-2}\ y^{-1}$ or $t\ km^{-2}\ y^{-1}$: these illustrate relatively low values for sub-Saharan Africa, at least for the South Sudan domain. The map, drawn up by Jansson (1988), and listing the various specific erosion measures carried out on a global scale, thus shows a specific erosion generally less than $10\ t\ km^{-2}\ y^{-1}$ in the South Sudan area of West

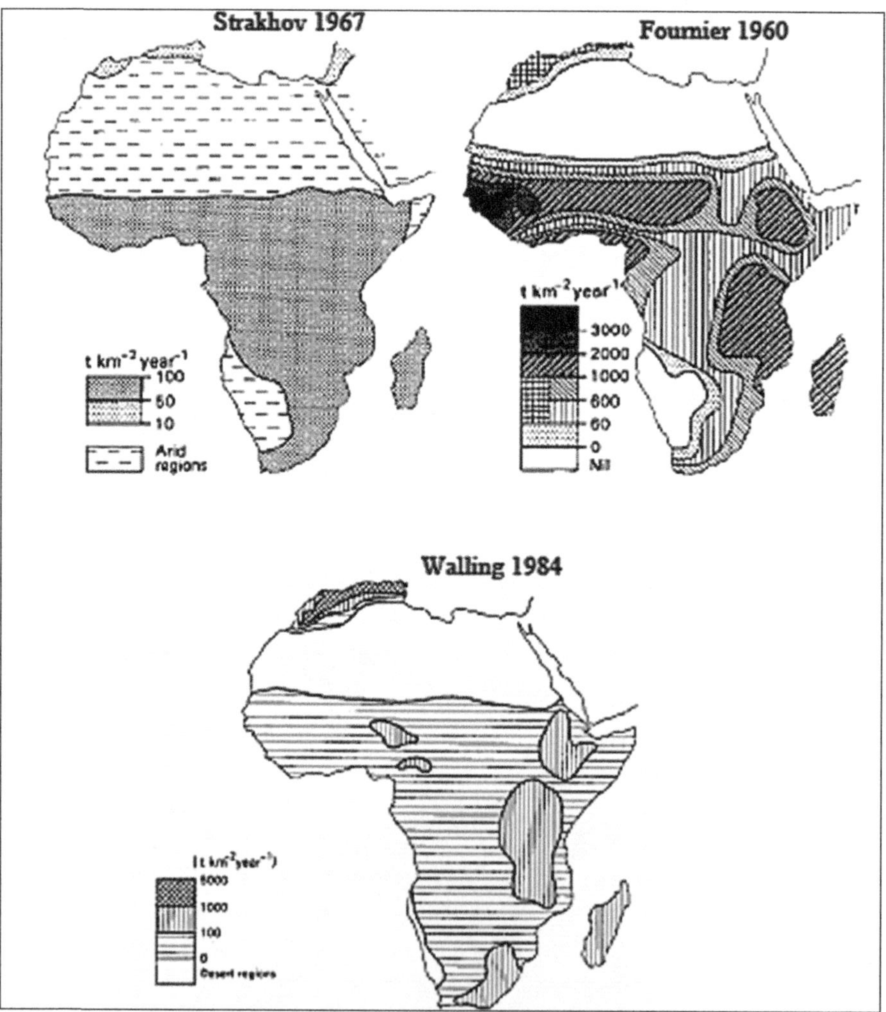

FIGURE 5.13 Different maps of suspended material transport values proposed by three authors for Africa (Liénou, 2007).

Africa. This map is confirmed by the more recent state of affairs maintained by Vanmaercke et al. (2014): in Africa, regional differences illustrate a range of values ranging from 0.2 to 15,700 t km^{-2} y^{-1} with maximums in North Africa (Atlas region) and the Great African Rift, and low values in the West and Central African watersheds (less than 60 t km^{-2} y^{-1}). A value close to 50 t km^{-2} y^{-1} is found on the scale of a large basin in Lake Volta in Ghana (Akrasi, 2005). Similar or lower values but based only on suspended loads are found downstream of the Sélingué Dam on the middle Niger (Ferry et al., 2018).

The conditional factors of reservoir sedimentation are well known: some relating to the climatic, physical, and human occupancy characteristics of the watershed, the other related to the characteristics of the reservoir itself (Liénou, 2007).

5.5   THE IMPACT OF CLIMATE CHANGE ON SEDIMENTATION

Climate change is a global phenomenon associated with the emission of greenhouse gases into the atmosphere, with the resulting effect of increasing average global temperatures. In West Africa, this overall change can lead to changes in temperatures, rainfall, rainfall patterns, and changes in the frequency and intensity of thunderstorms (IPCC, 2007). Global change occurs over long-time scales (typically for decades) and, given the high intrinsic climate variability in West Africa occurring during shorter time scales, it is very difficult to distinguish between the impacts of global change and those of natural climate variability (Kouassi et al., 2010). However, there is some evidence to suggest that global change is beginning to have a major impact on the region's climate. Thus, climate change could become the engine of the degradation of natural resources in the basin.

Therefore, adaptation to climate change is essential to the sustainable development of the basin and the use of its long-term resources (Kouassi et al., 2010).

In Algeria, the main impacts of this high erosion are the rapid drying up of reservoirs (up to 2–5% per year, Kassoul et al., 1997; Remini et al., 2009) with significant implications for water resource management in an area where sediment deposits in dam reservoirs are estimated at an average of 20 Mm^3 per year (Tidjani et al., 2000). The high efficiency of sediment transport by Algerian rivers (Colombani et al., 1984, Bourouba, 1998) has encouraged many authors to study the transfer of suspended sediments in this area (Megnounif et al., 2003; Achite, 2005; Ghenim et al., 2007; Achite, 2016; Hallouz et al., 2018a, 2018b). Among the factors favoring erosion (slope, rock nature, terrain, climate, human activities), the climate is recognized as the main factor in the semi-arid Mediterranean regions of Algeria that experience short and intense rain events, strong wind evaporation power, prolonged droughts, and freezing cycles (Touaibia, 2010; Houyou et al., 2014). In addition, Hallouz et al. (2017), in their study on the Wadi Mina Basin, found that the maximum values of the solid inputs, for all the sub-basins, are observed at the beginning of autumn and at the end of the spring, this variability is explained on the one hand by the variation of the vegetal cover (bare soils) during the year and the erosive nature of the autumn rains (high intensities) and on the other hand by the releases made by the Bakhadda Dam, so these two factors allow the first autumn floods to transport large quantities of sediment after a long dry season characterized by high temperatures and the destruction of soil aggregates. Rainfall erosivity is the potential ability of rainfall to cause soil loss (Silva, 2004). The rainfall erosivity index represents the climate influence on water-related soil erosion (Yu, 1998). Rainfall erosivity is the impact of the kinetic energy of raindrops on soil.

5.6   DISCUSSION

At present, the globe is a closed system, for the most part, and the conservation of matter implies that the destruction of continents by erosion is compensated by correlative sedimentation. In fact, the existence of this erosion–sedimentation couple has been highlighted for a very long time. The idea of this relationship is already clearly expressed, for the first time, it seems, as early as the end of the 18th century in the work of the Scottish geologist (Hutton, 1795 in Ellenberger et al., 1982).

The problem of solid transport and the extent of the siltation phenomenon in dams have attracted the interest of many researchers since the 1950s. Many explanatory models of solid transport from parameters such as liquid flow, runoff water, the area and average slope of the watershed, soil moisture, and rain have been developed by various authors for the Maghreb (Tadrist et al., 2016).

The sedimentary load of a river is sensitive to both climate change and a wide range of human activities in its watershed. These factors could influence sediment mobilization and transfer through actions such as land clearing, agricultural development, mineral extraction, urbanization and infrastructure development, dam and reservoir soil construction, and conservation and sediment control programs (Walling, 2008). Although researchers have emphasized the potential importance of climate change in increasing global soil erosion rates and possibly increasing the amount of sediment suspended in rivers, the response of suspended sediment flow in a particular location varies because it is strongly affected by the physical characteristics of the watershed and human activities (Zhang and Nearing, 2005). In any event, it is clear that there is a better understanding of the potential impact of climate change on the sedimentary load of dams. In addition, possible changes in sediment load need to be assessed to determine the sensitivity of the river system to drivers of change, to understand the implications for future reservoir development, and to assess their effects (changes) on future management strategies, as Walling (2008) pointed out.

It should be noted that in the Maghreb Region, the semi-arid climate is a major factor in the severity of the phenomenon of water erosion. The semi-arid climate is characterized by dry and humid periods as well as sudden spatial-temporal variations in precipitation amounts. Potential evapotranspiration often exceeds precipitation. This rainfall in mountainous areas causes torrential flows (Yahiaoui, 2012). The flows resulting from this torrential regime pull particles from the ground and transport them into the wadis and sediment behind the dams. Climate change in recent years has accentuated the drought periods that have affected all Mediterranean areas (Medejerab et al., 2011). Dry periods have made it more difficult to provide drinking water from reserves stored at dams.

According to a study by Bastone and Torre (2011), globe-wide climate simulations predict a very likely increase in precipitation at high latitudes and likely declines in most tropical lands as well as a trend of increasing days of intense rain in many regions, added to this, the influence of man on the climate system that is clear and increasing, with impacts observed on all continents (GIEC, 2014).

In addition, this climate change can lead to major changes in the erosivity of rainfall, which can have significant effects on the deterioration and loss of land. Indeed, on a global scale, Panagos et al. (2017) showed that the average erosivity of precipitation was estimated at 2190 MJ mm ha^{-1} h^{-1} y^{-1}, with the highest values in South America and the Caribbean, Central East Africa, and Southeast Asia. Similarly, Essays, (2018) found that high precipitation was the main factor that increased the spatial and temporal erosivity of precipitation runoff in the lower Niger basin (West Africa). In Central Asia, higher precipitation was the main cause of the increase in spatio-temporal precipitation erosivity (Duulatov et al., 2019). Also, climate change in the western Rif mountains shows a small evolution of erosivity on an annual calendar but a very strong evolution of the latter according to the seasons with a reduction of the R factor in winter and spring, and a pronounced increase in summer and autumn (Choukri et al., 2016). In Tunisia, the regime of heavy rains plays a leading role in the dynamics of gully in the semi-arid Tunisian context (Rebai et al., 2017), in particular, the so-called erosive showers or greater than 12.7 mm which occupy the largest share (Hechmi and Hammadi, 2017).

In order to be able to guide anthropogenic activities (e.g. agricultural practices) and soil protection developments, it is essential to predict the likely evolution of precipitation erosivity according to present and future climate conditions.

5.7 CONCLUSION

In this chapter, the study of the impact of climate change on the sedimentation of deductions in Africa was discussed. At first, a complete presentation of the concepts and definitions of erosion and sedimentation was given. Then, a description of the climate in Africa and its impact on sedimentation in dams reservoirs was presented, followed by the different values recorded in Africa in general and the Maghreb in particular. It is necessary at the African level to continue to work on the climate and sedimentation (siltation) of dams and to find better methods that can also explain the impact of climate change on the sedimentation of water reservoirs in hydraulic structures. However, the impacts of climate change on this phenomenon are not yet systematically recorded across the continent and there are constraints such as the lack of sufficient and appropriate data to model the risks of this impact. Currently, international efforts to develop and implement standards for quantification and monitoring sedimentation in dams, lakes, estuaries, etc., as well as systematically considering climate losses and impacts, are an important step towards assessing the risks of the impact of climate change on the sedimentation phenomenon. All of these elements eventually contribute to the construction of credible models or methods of climate impact on sedimentation at all scales, from local to global.

BIBLIOGRAPHY

Abid, M. (1998). Envasement des barrages en Tunisie. Direction Générale des Barrages et des Grands Travaux Hydrauliques (DGBGTH), ministère de l'Agriculture, Tunisie, 69p.

Abid, M. (2003). Gestion de l'envasement dans les retenues des grands barrages tunisiens. Direction Générale des Barrages et des Grands Travaux Hydrauliques (DGBGTH), ministère de l'Agriculture, Tunisie, 22p.

Achite, M., and Meddi, M. (2005). Variabilité spatio-temporelle des apports liquide et solide en zone semi-aride. Cas du bassin versant de l'oued Mina (nord-ouest algérien). *Revue des Sciences de l'Eau Journal of Water Science*, 18, 37–56. https://doi.org/10.7202/705575ar.

Achite, M., and Sylvain, O. (2016). Recent changes in climate, hydrology and sediment load in the Wadi Abd, Algeria (1970–2010). *Hydrology and Earth System Sciences*, 20(4), 1355–1372. ISSN 1027-5606.

Adwubi, A., Amegashie, B., Agyare, W., Tamene, L., Odai, S., Quansah, C., and Vlek, P. (2009). Assessing sediment inputs to small reservoirs in Upper East region, Ghana. *Lakes & Reservoirs: Research and Management*, 14, 279–287. https://doi.org/10.1111/j.1440-1770.2009.00410.x.

Adam K.S., (1986). Les impacts environnementaux du barrage de Nangbéto (Togo). Geo-Eco-Trop, 13: 103-112.

AgriMaroc. (2023). La capacité du barrage Mohamed V à Nador sera multipliée par 4. https://www.agrimaroc.ma/barrage-mohamed-nador/.

Akrasi, S. (2005). The assessment of suspended sediment inputs to Volta Lake. *Lakes & Reservoirs: Research and Management*, 10, 179–186. https://doi.org/10.1111/j.1440-1770.2005.00272.x.

Al Jihad Yassine et Ouchen Lahcen. (2011). Contribution à l'élaboration d'une approche pour la délimitation du domaine publique hydrauliques des oueds étude cas: oued Tassaoute. Mémoire de troisième cycle pour l'obtention du diplome d'ingénieur d'État de l'Institut Agronomique et Vétérinaire Hassan II à Rabat, Maroc, 200p.

Amare, A. (2005). Study of sediment yield from the watershed to Angereb reservoir. Matser's thesis, Department of Agricultural Engineering. Alemaya university, Ethiopia.

Ammari, A. (2012). Vulnérabilité à l'Envasement des Barrages (cas du bassin Hydrographique des Côtiers Algérois). Thèse de Doctorat en Hydraulique. Institut Université Mohamed Khider, Biskra, 195p.

Amoussou, E. (2010). Variabilité pluviométrique et dynamique hydro-sédimentaire du bassin-versant du complexe fluvio-lagunaire Mono-Ahémé-Couffo (Afrique de l'Ouest). Thèse de Doctorat l'Université de Bourgogne, France, 315p.

ANBT (Agence Nationale des Barrages et Transferts). (2020). Notre technique Barrage Ghrib, document interne.

Arabi, M., and Roose, E. (1989). Influence du système de production et du sol sur l'érosion et le ruissellement en nappe en milieu montagnard méditerranéen (station de Ouzera). Bulletin Réseau Érosion N°, IRD, Montpellier (France).

Babi Inside. (2016). Ayamé, ses barrages et son château, 18 avril 2016. https://www.babiinside.com/ayameses -barrages-et-son-chateau.

Bastone, V., et De la Torre, Y. (2011). Etude préliminaire de l'impact du changement climatique sur les risques naturels à la Réunion. 135p.

Belaabed, H. (2012). Apport de la télédétection spatiale dans l'évaluation du taux d'envasement des retenues des barrages cas de la retenue du barrage El Kansera. Mémoire de troisième cycle pour l'obtention du diplôme d'Ingénieur en Topographie de l'Institut Agronomique et Vétérinaire Hassan II à Rabat, Maroc, 104p.

Benchetrit, M. (1972). L'érosion actuelle et ses conséquences sur l'aménagement en Algérie. Édit. Presses Universitaires de France, Paris, 216p.

Ben Mammou, A., et Louati, M. H. (2007). Évolution temporelle de l'envasement des retenues de barrages de Tunisie. *Revue des sciences de l'eau / Journal of Water Science*, 20(2), 201–210. http://id.erudit.org/iderudit/015813ar. https://doi.org/10.7202/015813ar.

Bergaoui, M., Camus, H., and Nouvelot, J. F. (1998). Essai de modélisation du transport solide sur les micros-bassins versants de Tebaga (Tunisie centrale). *Sécheresse*, 9(1), 51–57.

Blivi, A. B. (2000). Effet du barrage de Nangbéto sur l'évolution du trait de côte: une analyse prévisionnelle sédimentologique. *Journal de la Recherche Scientifique; Univ. Bénin (Togo)*, 4(1), 29–41.

Blivi, A. (2005). Érosion côtière dans le Golfe de Guinée en Afrique de l'Ouest: Exemple du Togo. Communication en Nairobi du 17 au 29 octobre 2005, REDDA/ RIOD/ UNCCD, 9p.

Bouanani, A. (2004). Hydrologie, transport solide et modélisation. Étude de quelques sous-bassins de la Tafna. Thèse de doctorat: Université de Tlemcen (Algérie).

Boun Heng, M. (2013). La sédimentation dans les lacs de barrage à Java, Indonésie processus, rythmes et impacts. Thèse de Doctorat de l'université Paris 1 Panthéon, Géographie, 301p.

Bourouba, M. (1998). Contribution à l'étude de l'érosion et des transports solides de l'Oued Medjerda supérieur (Algérie orientale). *Bulletin Réseau Érosion*, 18, 76–97. (In French).

Boussema, H.-R. (1996). Système d'information pour la conservation et la gestion des ressources naturelles. Colloque international sur le rôle des technologies de télécommunications et de l'information en matière de protection de l'environnement, Tunis, 112–116.

Briak, H. (2017). Estimation de l'Érosion des Sols et Modélisation Hydrologique Spatialisée: Cas d'un Bassin Versant Tangérois au Nord du Maroc (Thèse de doctorat). Université Abdelmalek Essaâdi, Faculté des Sciences et Techniques de Tanger, Maroc, 210.

Chebbani, R., Djilli, K., and Roose, E. (1999). Étude des risques d'érosion dans le bassin versant de l'Isser, Algérie. *Bulletin Réseau Érosion*, 19, 85–95.

Cheggour, A. (2008). Mesures de l'érosion hydrique à différentes échelles spatiales dans un bassin versant montagneux semi-aride et spatialisation par des S.I.G.: Application au bassin versant de la Rhéraya, Haut Atlas, Maroc. Thèse de Doctorat de l' Université Cadi Ayyad, Maroc, 231p.

Chevalier, J.-J., Pouliot, J., Thomson, K., et Boussema, M.-R. (1995). Systèmes d'aide à la planification pour la conservation des eaux et des sols (Tunisie). Systèmes d'information géographique utilisant les données de télédétection. Actes du colloque scientifique international, Hammamet, Tunisie.

Choukri, F., Chikhaoui, M., et Naimi, M. (2016). Impact Du Changement Climatique Sur L'évolution De L'érosivité des Pluies Dans Le Rif Occidental (Nord Du Maroc). *European Scientific Journal*, 12(32). ISSN: 1857 – 7881 (Print) e - ISSN 1857- 7431.

Colombani, J., Olivry, J. C., and Kallel, R. (1984). Phénomènes exceptionnels d'érosion et de transport solide en Afrique aride et semi-aride. In *Challenges in African Hydrology and Water, Ressources, Proceedings of the Harare Symposium*, Harare, 23–27 July 1984; IAHS Publication, Oxfordshire, Volume 144, 295–300. (In French).

Coynel, A., Seyler, P., Etcheber, H., Meybeck, M., and Orange, D. (2005). Spatial and seasonal dynamics of total suspended sediment and organic carbon species in the Congo River. *Global Biogeochemical Cycles*, 19, 4.

Demmak, A. (1982). Contribution à L'étude de L'érosion et des Transports Solides en Algérie Septentrionale. Ph.D. Thesis, Université de Pierre et Marie Curie, Paris. (In French).

Diab Djefal, I. (2013). L'Envasement dans Les Barrages de l'Algérie. Silting in the Dams of Algeria. *Proceeding du Séminaire International sur l'Hydrogéologie et l'Environnement SIHE*, Ouargla, 415–418.

Diallo, D., Barthès, B., Orange, D., and Roose, E. (2004). Stabilité des agrégats et des morts comparées aux risques de ruissellement et d'érosion en nappe mesurés sur parcelles en zone soudanienne du Mali. *Sécheresse*, 15, 57–64.

Dipama, J. M. (1992). La sédimentation des barrages n'l, no2 et no3 et ses impacts socio-économiques. Mémoire de Maitrise de Géographie. FLASHS. Ouagadougou, 97p.

Duulatov, E., Chen, X., Amanambu, A. C., Ochege, F. U., Orozbaev, R., Issanova, G., and Omurakunova, G. (2019). Projected rainfall erosivity over central Asia based on CMIP5 climate models. *Water*, 11, 897.

Essays, UK. (November 2018). Spatio-temporal variation in rainfall-runoff erosivity due to climate change in the lower Niger Basin, West Africa. https://ukdiss.com/examples/rainfall-runoff-erosivity-climate -change.php?vref=1.

Fanchette, S. (2006). Le Delta du Nil: Enjeux et limites du contrôle territorial par l'état. La Découverte | « Hérodote » 2006/2 no. 121 | pages 165 à 189. ISSN 0338-487X ISBN 2707148954.

FAO. (1990). Conservation des ressources naturelles en zones arides et semi-arides. *Cahiers FAO: Conservation des sols*, 3, 135p.

Ferry, L., Mietton, M., Fujiki, K., Laval, M., Coulibaly, N., Braquet, N., and Martin, D. (2018). Faiblesse de la sédimentation dans le barrage-réservoir de Sélingué (Mali-Guinée). Témoin de la stabilité des savanes sud-soudaniennes à l'échelle d'un grand bassin versant très peu anthropisé. G-Eau Working Papers No.5a. Montpellier. http://www.g-eau.net/.

Gac, J. Y. (1980). Géochimie du bassin du lac Tchad. Bilan de l'altération, de l'érosion et de la sédimentation. Trav. Doc. O.R.S.T.O.M., 251p.

Galy-Lacaux, C., Delmas, R., Kouadio, G., Richard, S., and Gosse, P. (1999). Long-terme greenhouse gas emissions from hydroelectric reservoirs in tropical forest regions. *Global Biogeochemical Cycles*, 13, 503–517.

G.C. (Gouvernement Canada). (2016). Pollution de l'eau: érosion et sedimentation. https://www.canada.ca/fr /environnement-changementclimatique/services/eau-apercu/pollution-causes-effects/erosion sedimen-tation.html#sec1.

Ghenim, A., Terfous, A., and Seddini, A. (2007). Étude du transport solide en suspension dans les régions semi-arides méditerranéennes: Cas du bassin versant de l'Oued Sebdou (Nord-Ouest Algérien). *Sécheresse*, 18, 39–44. (In French).

GIEC. (2007). Groupe d'experts intergouvernemental sur l'évolution du climat (GIEC). Rapport 2001 et 2007 sur l'évolution du climat.

GIEC. (2014). Changements climatiques 2014: conséquences, adaptation et vulnérabilité. https://www.notre -planete.info/actualites/3986-rapport-GIEC-consequences-changement-climatique.

Gresillon, J. M. (1976). Petits barrages en terre en Afrique occidentale. EIER, Ouagadougou, 120p.

Grimm, M., Jones, R., and Montanarella, L. (2002). *Soil Erosion Risk in Europe*. Office for Official Publications of the European Communities, Luxembourg.

Hallouz, F., Meddi, M., and Mahé, G. (2017). Régimes des matières en suspension dans le bassin versant de l'oued Mina sur l'oued Cheliff (Nord-Ouest algérien). *La Houille Blanche*, 4, 61–71. https://doi.org/10 .1051/lhb/2017034. (In French).

Hallouz, F., Meddi, M., Mahé, G., Alirahmani, S., and Keddar, A. (2018a). Modeling of discharge and sedi-ment transport through the SWAT model in the basin of Harraza (North west of Algeria). *Water Science*, 32, 79–88.

Hallouz, F., Meddi, M., Mahé, G., Toumi, S., and Rahmani, S. E. A. (2018b). Erosion, suspended sediment transport and sedimentation on the Wadi Mina at the Sidi M'Hamed Ben Aouda Dam, Algeria. *Water*, 10(7), 895. https://doi.org/10.3390/w10070895.

Haregeweyn, N., Poesen, J., Neyssen, J., Govers, G., Verstraeten, G., de Vente, J., Deckers, J., Moeyersons, J., and Haile, M. (2008). Sediment yield variability in Northen Ethiopia: A quantative analysis of its controlling factors. *Cantena*, 75, 65–76.

Harroy, J. P. (1944). Afrique terre qui meurt. La dégradation des sols africains sous l'influence de la colonisa-tion. Éditions Marcel Hayez, Bruxelles, 557p.

Hechmi, B., et Habaieb, H. (2017). Etude des intensités instantanées en 15' et 30', répercussions sur l'érosivité des pluies et les pertes en terres et mesures d'adaptation en milieu semi-aride Tunisien. *Journal International Sciences et Technique de l'Eau et de l'Environnement*, 2(3). ISSN (electronic): 1737-9350; ISSN (printed): 1737-6688.

Heusch, B. (1970). L'érosion du Pré-Rif. Une étude quantitative. *Annales de la Recherche Forestière du Maroc*, 12, 9–176.

Houyou, Z., Bielders, C. L., Benhorma, H. A., Dellal, A., and Boutemdjet, A. (2014). Evidence of strong land degradation by wind erosion as a result of rainfed cropping in the Algerian steppe: A case study at Laghouat. *Land Degradation & Development*, 27, 1788–1796. https://doi.org/10.1002/ldr.2295.

http://aicd-africa.org/archives/818?ln=fr (AI-CD website).

Hutton, J. (1795). *Theory of the Earth, with Proofs and Illustrations*, 2 vols. Printed for Messrs Cadell, London and Edinburgh, Junior and Davies; and William Creech, 1795. Voir Ellenberger F., Marcel Bertrand et "l'orogenèse programmée", Geol. Rundschau, Bd 71, 2, 1982, 463–474.

ICOL. (2018). Distribution des grands barrages par continent en 2018. https://www.icoldcigb.org/article/FR/registre_des_barrages/synthese_generale/nombre-de-barrages-par-pays-membre.

ICOLD. (2009). (International Commission on Large Dams). Sedimentation and sustainable development of dams in river systems. ICOLD Bulletin, Paris, France: ICOLD.

Jansson, M. (1988). A global survey of sediment yield. *Geografiska Annaler. Series A, Physical Geography*, 70, 81–98. https://doi.org/10.2307/521127.

Joellem2. (2014). Naissance et enjeux du Haut barrage d'Assouan. Posted by Joellem2 in barrage d'Assouan. Hydrodiplo du Nil A nos projets ~ protéger la vie. https://eauxdunil.wordpress.com/2014/09/20/naissance-du-grand-barrage-dassouan/.

Kane, H., and Diallo, A. (2005). Etude portant sur l'évaluation de l'état de l'environnement des ressources naturelles et des ressources en eau dans la partie guinéenne du bassin du fleuve Sénégal, en se servant du système d'indicateurs de l'Observatoire de l'environnement de l'OMVS, 154p.

Kattan, Z., Gac, J. Y., and Probst, J. L. (1987). Suspended sediment load and mechanical erosion in the Senegal Basin – Estimation of the surface runoff concentration and relative contributions of channel and slope erosion. *Journal of Hydrology*, 92(1–2), 59–76. https://doi.org/10.1016/0022-1694(87)90089-8.

Kassoul, M., Abdelgader, A., and Belorgey, M. (1997). Caractérisation de la sédimentation des barrages en Algérie. *Revue des sciences de l'eau / Journal of Water Science*, 10(3), 339–358. https://doi.org/10.7202/705283ar.

Kassoum, T. (1979). Caractéristiques limnologiques du lac de Kossou (Côte d'Ivoire). *Annales de l'Université d'Abidjan: Écologie. Série E*, Tome XII, 30–69.

Kebede, W. (2012). Watershed Manual: An opyion to sustain Dam and Reservoir Functions in Ethiopia. Journal of Environmental Sciences and Technology 5, 262-273.

Konan, K. S., Kouassi, K. L., Konan, K. F., Kouame, K. I., Aka, K., and Gnakri, D. (2013). Solid load estimating and hydrochemical characterization of the lake of the hydroelectric dam of Ayamé 1 (Ivory Coast). *Bulletin de l'Institut Scientifique, Rabat, Section Sciences de la Terre*, (35), 17–25.

Kouassi, K.L., Wognin, A.V.I., Gnagne, T., N'go, Y.A., Courivaud, J., Kassy, P., Deme, M. and Aka, K. (2007) Sand Characterization and Morphology of the Lake Bottom of the Taabo Hydroelectric Dam (Ivory Coast). The Science of Nature, 4, 93-103.

Kouassi, A. M., Kouamé, K. F., Koffi, Y. B., Dje, K. B., Paturel, J. E., and Oulare, S. (2010). Analyse de la variabilité climatique et de ses influences sur les régimes pluviométriques saisonniers en Afrique de l'Ouest: cas du bassin versant du N'zi (Bandama) en Côte d'Ivoire. *Cybergeo: European Journal of Geography* [Online]. Environment, Nature, Landscape, document 513. http://journals.openedition.org/cybergeo/23388. https://doi.org/https://doi.org/10.4000/cybergeo.23388.

Lagacé, R. (1980). L'équation universelle de pertes de sols: un outil. Dans Lagacé, R. (Éd.), *Érosion et conservation des sols, 8ème Colloque de génie rural*. Département de génie rural, Université Laval, Québec, 37–60.

Lahlou, A. (1988). Étude actualisée de l'envasement des barrages au Maroc. *Revue Des Sciences De L'Eau*, 6(3), 337–356.

Laouina, A. (1992). Recherches actuelles sur l'érosion au Maroc. *Bull Réseau Érosion*, 12, 292–299.

Laraque, A., Bricquet, J.-P., Pandi, A., and Olivry, J.-C. (2009). A review of material transport by the Congo River and its tributaries. *Hydrological Processes*, 23, 3216–3224. https://doi.org/10.1002/hyp.7395.

Laraque, A., Castellanos, B., Steiger, J., Lopez, J. L., Pandi, A., Rodriguez, M., Rosales, J., Adele, G., Perez, J., and Lagane, C. (2013). A comparison of the suspended and dissolved matter dynamics of two large intertropical rivers draining into the Atlantic Ocean: The Congo and the Orinoco. *Hydrological Processes*, 27, 2153–2170.

Liénou, G. (2007). Impacts de la variabilité climatique sur les ressources en eau et les transports de matières en suspension de quelques bassins versants représentatifs au Cameroun. Thèse de Doctorat de l'Université de YAOUNDE I, Cameroun, 486p.

Lu, X. (2005). Spatial variability and temporal change of water dis- charge and sediment flux in the lower Jinsha tributary: Impact of environmental changes. *River Research and Applications*, 21, 229–243.

Madani, A. (2018). Quantification des MES au Bassin versant du Bouregreg, méthode MUSLE mémoire de fin d'étude université Mohamed V, faculté des sciences Rabat.

MARHP. (2017). Rapport National du secteur de l'eau- 2017. Ministère de l'Agriculture, des Ressources Hydrauliques et de la Pêche. Bureau de la Planification et des Equilibres Hydrauliques Bureau de la Planification et des Equilibres Hydrauliques, Republique Tunisienne, 153p.

Meade, R. H. (1996). River-sediment inputs to major deltas. In Milliman, J. D., and Haq, B. U. (Eds.), *Sea-level Rise and Coastal Subsidence.* Kluwer Academic Publisher, Dordrecht, 63–85.

Medejerab, A., and Henia, L. (2011). Variations spatiotemporelles de la sécheresse climatique en Algérie nord-occidentale. *Courrier du Savoir*, 11, 71–79.

Megnounif, A., Terfous, A., and Bouanani, A. (2003). Production et transport des matières solides en suspension dans le bassin versant de la Haute-Tafna (Nord-Ouest algérien). *The Revue des Sciences de l'Eau / Journal of Water Science*, 16, 369–380. https://doi.org/10.7202/705513ar.

Mekonnen, M., Keesstra, S. D., Baartman, J. E., Ritsema, C. J., and Melesse, A. M. (2015). Evaluating sediment storage dams: Structural off-site sediment trapping measures in Northwest Etgiopia. *Cuadernos de Investigación Geográfica*, N-41(1), 7–22. ISSN 0211-6820. https://doi.org/ 10.18172/cig.2643. Universidad de La Rioja.

Menzel, L., and Burger, G. (2002). Climate change scenarios and runoff response in the Mulde catchment (Southern Elbe, Germany). *Journal of Hydrology*, 267, 53–64.

Merzouk, A., Rayan, J., and Kacemi, M. (1994). A perspective on soil erosion in Morroco's dry land semi-arid zone. Actes du colloque international des Sciences du Sol: #Sciences du sol au développement#, Rabat, Maroc, 6-I Avril 1993, 12p.

Michael, A., Schmidt, J., Enke, W., Deutschlander, T., and Maltiz, G. (2005). Impact of expected increase in precipitation intensities on soil loss results of comapritive model simulations. *Catena*, 61, 155–164.

Milliman, J. D., and Meade, R. H. (1983). World-wide delivery of river sediments to the oceans. *Geology*, 91, 1–21.

Mokhtari, El. Hadj. (2017). Impact de l'érosion hydrique sur l'envasement du barrage Ghrib. Thèse de Doctorat. Université Hassiba Ben Bouali, Chlef. Algérie, 272p.

Mouzeo, K. (1986). Transport particulaire actuel du fleuve Congo et de quelques affluents enregistrement quaternaire dans l'éventail détritique profond (sédimentologie, minéralogie et géochimie). Thèse de doctorat, Université de Perpignan, 262pp.

Msadala, V. C., and Basson, G. R. (2019). Revised regional sediment yield prediction methodology for ungauged catchments in South Africa. *Journal of the South African Institution of Civil Engineering*, 59(2), 28–36. ISSN 1021-2019. Paper 1326.

Nijssen, B., O'Donnell, G., Hamlet, A., and Lettenmaier, D. (2001). Hydrologic sensitivity of global rivers to climate change. *Climate Change*, 50, 143–175.

Olivry, J. C., Bricquet, J. P., Thiébaux, J. P., and Sigha-Nkamdjou, L. (1988). Transport de matière sur les grands fleuves des régions intertropicales: les premiers résultats des mesures de flux particulaires sur le bassin du fleuve Congo. Sediment Budgets, *IAHS Publ.*, 174, 509–521.

Olivry, J. C., et Hoorelbeck, J. (1990). Erodibilité des Terres Noires de la vallée du Buëch. *Cahiers ORSTOM, Pédologie*, XXV(I–2), 97–112.

O'Neal, M. R., Nearing, M. A., Vining, R. C., Southworth, J., and Pfeifer, R. A. (2005). Climate change impacts on soil erosion in Mid- west United States with changes in crop management. *Catena*, 61, 165–184.

Ouattara, A. (2000). Premières données systématiques et écologiques du phytoplankton du lac d'Ayamé (Côte d'Ivoire). Thèse de Doctorat Unique, Université Katholieke de Louvain, Belgique, 207pp.

Oyédé, L. M. (1991). Dynamique sédimentaire actuelle et messages enregistrés dans les séquences quarternaires et néogènes du domaine margino littoral du Bénin (l'Afrique de l'Ouest). Thèse présentée pour l'obtention du doctorat en géologie sédimentaire, nouveau régime. Université de Bourgogne, Paris, 302p.

Panagos, P., Borrelli, P., Meusburger, K., Yu, B., Klik A., Lim, K.J., Yang, J.E, Ni, J., Miao, C., Chattopadhyay, N., Sadeghi, S.H., Hazbavi, Z., Zabihi, M., Larionov, G.A., Krasnov, S.F., Garobets, A., Levi, Y., Erpul, G., Birkel, C., Hoyos, N., Naipal, V., Oliveira, P.T.S., Bonilla, C.A., Meddi, M., Nel, W., Dashti, H., Boni, M., Diodato, N., Van Oost, K., Nearing, M.A., Ballabio, C. (2017). Global rainfall erosivity assessment based on high-temporal resolution rainfall records. Scientific reports, 7(1), 4175.

Phan, D. B., Wu, C. C., and Hsieh, S. C. (2011). Impact of climate change on stream discharge and sediment yield in Northern Viet Nam. *Water Resources*, 38, 827–836.

Pimentel, D., Harvey, C., Resosudarmo, P., Sinclair, K., Kurz, D., McNair, M., Crist, S., Shpritz, L., Fitton, L., Saffouri, R., Blair, R. (1995). Environmental and economic costs of soil erosion and conservation benefits. *Science*, 267, 1117–1123.

Portenga, E. W., and Bierman, P. R. (2011). Understanding earth's eroding surface with 10Be. *GSA Today*, 21, 4–10.

Prahl, F. G., Ertel, J. R., Goñi, M. A., Sparrow, M. A., and Eversmeyer, B. (1994). Terrestrial organic carbon contributions to sediments on the Washington margin. *Geochimica and Cosmochimica Acta*, 58, 3035–3048.

Probst, J. L., Nkounkou, R. R., Krempp, G., Bricquet, J. P., Thiebaux, J. P., and Olivry, J. C. (1992). Dissolved major elements exported by the Congo and the Ubangi rivers during the period 1987–1989. *Journal of Hydrology*, 135, 237–257.

Quinquis, M. (2017). Relations entre bassins versants et cellules sédimentaires littorales: les exemples du Maroc, de l'Algérie et de la Tunisie. Thèse présentée pour obtenir le grade universitaire de Docteur de l'université d'Aix-Marseille, France, 327p.

Rebai, H., Raclot, D., et Ben Ouezdou, H. (2017). Impact du régime des pluies sur la dynamique érosive d'un bassin versant vertique et gypseux: cas de Fidh Ali (Kairouan). *Journal International Sciences et Technique de l'Eau et de l'Environnement*. ISSN (electronic): 1737-9350; ISSN (printed): 1737-6688; Volume 2 - Numéro 3 - Juin 2017.

Reizer, C. (1967). Aménagement piscicole du lac artificiel d'Ayamé. Centre technique Forestier Tropical. 108pp.

Remini, B., Leduc, C., and Hallouche, W. (2009). Evolution des grands barrages en régions arides: quelques exemples algériens. *Sécheresse*, 20(1), 96–103.

Roose, E. (1980). Dynamique actuelle de sols ferralitiques et ferrugineux tropicaux d'Afrique occidentale. Étude expérimentale des transferts hydrologiques et biologiques de matières sous végétations naturelles ou cultivées. Thèse Doct. En Sciences, Université d'Orléans, 587p. In Travaux et Documents de l'ORSTOM, Paris, n° 130, 569p.

Roose É. (1994) - Introduction à la gestion conservatoire de l'eau, de la biomasse et de la fertilité des sols (GCES). FAO Soils Bulletin, vol. 70, Rome, 420 p.

Roose, E. (1987). Évolution des stratégies de lutte antiérosive. Nouvelle démarche proposée en Algérie: la GCES. *Bulletin Réseau Érosion*, 7, 91–96.

Roose, E., and De Noni, G. (2004). Recherches sur l'érosion hydrique en Afrique: revue et perspectives. *Sécheresse*, 15(1). https://horizon.documentation.ird.fr/exl-doc/pleins_textes/divers10-02/010033576.pdf.

Roose, E., Arabi, M., Brahamia, K., Chebbani, R., Mazour, M., et Morsli, B. (1993). Érosion en nappe et ruissellement en montagne algérienne. *Cahiers ORSTOM: Série pédologie*, 28, 289–308.

Roose, E., Barthes, B., and Prat, C. (2000). Agrégation du sol, ruissellement et érosion à l'échelle parcellaire dans trois régions intertropicales (Bénin, Cameroun, Mexique). *Bulletin Réseau Érosion*, (20), 373–387.

Roose, E., Chebbani, R., and Bourougaa, L. (2000). Ravinement en Algérie. Typologie, facteurs de contrôle, quantification et réhabilitation. *Sécheresse*, 11(4), 317–326.

Rooseboom, A., Verster, E., Zietsman, H. L., and Lotriet, H. H. (1992). The development of the new sediment yield map of South Africa. WRC Report No. 297/2/92. Water Research Commission, Pretoria.

Rossi, G. (1996). L'impact des barrages de la vallée du Mono (Togo-Benin). *La gestion de l'incertitude. Géomorphologie: Relief, processus, environnement*, 2(2), 55–68. http://www.persee.fr.

Rossi, G., et Blivi, A. B. (1995). Les conséquences des aménagements hydrauliques de la vallée du Mono (Togo-Bénin). *S'aura-t-on gérer l'avenir ? Cahiers d'Outre-Mer*, 48(192), 435–452.

Sanon, O. I. (1998). Contribution à l'étude de l'envasement des retenues d'eau en milieu Tropical: exemple du barrage de Tamasgho dans la province du Sanmatenga (BF). Mémoire de Géographie, Université, FLASHS, Ouaga, 137p.

Serbah, B. (2011). Etude et valorisation des sédiments de dragage du barrage bakhadda Tiaret. Mémoire de Magister, Université de Tlemcen.

Shahin, M. (1993). An overview of reservoir sedimentation in some African river basins. In *Proceedings of the Yokohama Symposium*. Sediment problems: Strategies for monitoring, prediction and control, IAHS Publications, Yokohama, 93–100.

Shrestha, B., Babel, M. S., Maskey, S., van Griensven, A., Uhlenbrook, S., Green, A., and Akkharath, I. (2013). Impact of climate change on sediment yield in the Mekong River basin: A case study of the Nam Ou basin, Lao PDR. *Hydrology and Earth System Sciences*, 17, 1–20. www.hydrol-earth-syst-sci.net/17/1/2013/. https://doi.org/10.5194/hess-17-1-2013.

Sichingabula, H. (1997). Problems of sedimentation in small dams in Zambia. In *Proceedings of the Rabat Symposium. Human Impact on Erosion and Sedimentation*, IAHS Publications, Rabat, 251–259.

Silva, A. M. (2004). Rainfall erosivity map for Brazil. *Catena*, 57, 251–259.

Smiri, A. (1987). Type of erosion and quantification of soil losses. Université internationale de Casablanca. http://www.ma.auf.org/erosion/chapitre1/Chap1-sommaire.html.

Stetten, E. (2015). Origine, distribution et réactivité de la matière organique associée aux lobes terminaux du système turbiditique du Congo. Thèse de Doctorat de l'Université Pierre et Marie Curie, Spécialité Biogéochimie marine. École doctorale Géosciences, Ressources Naturelles et Environnement (ED 398), soutenue publiquement le 24 novembre 2015, 276p.

Tadrist, N., Debauche, O., Remini, B., Dimitri Xanthoulis, D., and Degré, A. (2016). Impact de l'érosion sur l'envasement des barrages, la recharge des nappes phréatiques côtières et les intrusions marines dans la zone semi-aride méditerranéenne: cas du barrage de Boukourdane (Algérie). *Biotechnology, Agronomy, Society and Environment*, 20(4), 453–454.

Tamene, I., Park, S., Dikau, R., and Vlek, P. (2006). Reservoir siltation in the semi-arid highlands of northern Ethiopcatchment area relationship and a semi-quantitative approach for predicting sediment yield. *Earth Surface Processes and Landforms*, 31(11), 1364–1383.

Tidjani, A. E., Yebdri, D., and Cherif, E. A. (2000). *Ampleur de L'envasement Dans les Barrages Algériens*. Documents Techniques en Hydrologie; UNESCO, Paris, 2000; Volume 29, pp. 121–128. (In French).

Touaibia, B. (2010). Problématique de l'érosion et du transport solide en Algérie septentrionale. *Secheresse*, 21(1), 1–6.

Vanmaercke, M., Poesen, J., Broeckx, J., and Nyssen, J. (2014). Sediment yield in Africa. *Earth-Science Reviews*, 136, 350–368. https://doi.org/10.1016/j.earscirev.2014.06.004.

Walling, D. E. (1984). The sediment yields of African rivers. *I.A.H.S. Publ., Harare Symp.*, 144, 265–283.

Walling, D. E. (1984). The changing sediment loads of the world's rivers. Sediment Dynamics in Changing Environments (Proceedings of a symposium held 323 in Christchurch, New Zealand, December 2008). *IAHS Publ.*, 325, 2008.

Waterbury J. (1979). Hydro-politics of the Nile Valley, Syracuse, Syracuse University Press, 1979, p. 78.

Wetzel, A. (1993). The transfer of river load to deep-sea fans: A quantitative approach. *American Association of Petroleum Geologists Bulletin*, 77, 1679–1692.

Wu, Y., and Chen, J. (2012). Modeling of soil erosion and sediment transport in the East River Basin in southern China. *Science of the Total Environment*, 44, 159–168.

Yahiaoui, B. (2012). Inondations Torrentielles Cartographie des Zones Vulnérables en Algérie du Nord (Cas de l'oued Mekerra, Wilaya de Sidi Bel Abbès). Thèse de doctorat, École Nationale Polytechnique.

Yapo, O. B. (2002). Évaluation de l'état trophique du lac de Buyo (Côte d'Ivoire). Thèse de Doctorat Unique, Université d'Abobo-Adjamé, Côte d'Ivoire, 296pp.

Yu, B. (1998). Rainfall erosivity and its estimation for Australia's tropics. *Australian Journal of Soil Research*, 36, 143–166.

Zhang, X.C., Nearing M.A. (2005). Impact of climate change on soil erosion, runoff, and wheat productivity in central Oklahoma, CATENA, Volume 61, Issues 2–3, 2005, Pages 185-195, ISSN 0341-8162, https://doi.org/10.1016/j.catena.2005.03.009.

Zhu, Y.-M., Lu, X. X., and Zhou, Y. (2008). Sediment flux sensitivity to climate change: A case study in the Longchuanjiang catchment of the upper Yangtze River, China. *Global and Planetary Change*, 60, 429–442.

6 Basin Management Case Studies in North Africa

Youssef Brouziyne, Fouad Moudden,
Lahcen Benaabidate, Lhoussaine Bouchaou,
Younes Hamed, Oumaima Attar, Mohammed El Hafyani,
Abdelhadi El ouali, Ali Essahlaoui, and Abdelghani Chehbouni

6.1 INTRODUCTION

Being threatened by the risk of water scarcity, the North African region is faced with increasing water demand and competition between the different uses linked to economic development and population growth. This situation, which is worsening due to the effect of climate variability and change, in addition to the consequences of poor management of available water resources (Brouziyne et al., 2020). Bordering the Mediterranean and the most arid desert in the world, a significant part of the territories of North African countries are subject to the Mediterranean climate (Nouaceur, 2017). Already marked by their strong temporal and spatial variability, precipitation in this area is largely impacted by ongoing climate change. In addition, population growth, accelerated urbanization, and economic development have considerably increased water needs in all sectors.

North Africa constitutes a climatic transition zone between the temperate zone and the arid subtropical zone. The climate in this area is characterized by summer drought and, depending on the sub-regions, autumn or winter maximum rainfall (Norrant, 2004). Almost a third of the disturbances affecting the central Maghreb originate from the Atlantic. For the eastern part of the Mediterranean basin, the disturbed systems are reactivated during their passage over the Mediterranean Sea (west to north-west trajectories). The importance of this body of water on a regional scale explains the great spatial extension of the Mediterranean climatic domain. The rest of the active cells that affect this region are of local origin and are linked to thermal contrast in air masses.

The other significant parts of the North African territory are drylands (Figure 6.1). These regions have been characterized in recent decades by significant climatic variability. This is evidenced by the drop in rainfall and the rise in temperatures, with negative consequences on ecosystems and production systems, making this part of the world one of the most vulnerable areas to climate change (IPCC, 2007; Fathian et al., 2016).

According to the Millennium Ecosystem Assessment (2005), desertification is one of the greatest environmental challenges and a major obstacle to meeting the basic needs of populations in arid zones. Most often, it is associated with a number of physical and socioeconomic manifestations such as silting up, desert encroachment, soil erosion and degradation, deforestation, decline in the biological productivity of land, population growth, and inappropriate use of technology (Swift, 1996).

Per capita freshwater availability in North Africa, and more specifically in Maghreb countries, has fallen by over 60% over the past 40 years. Currently, these countries are considered to be in a

DOI: 10.1201/9781003473398-8

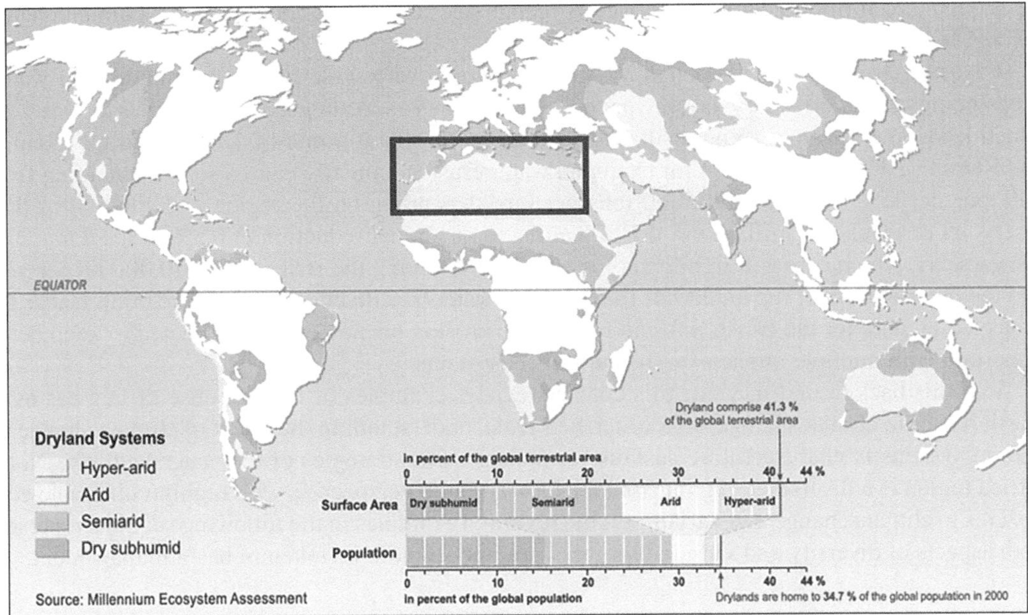

FIGURE 6.1 Drylands systems in the world with a frame on the North Africa region.

Source: Millennium Ecosystem Assessment.

water shortage situation with renewable water resources below 1000 m³/inhabitant/year. In 2014, while Morocco still had 844 m³ per capita per year, Algeria and Tunisia only benefited from 294 and 410 m³ per capita per year, respectively (FAO, 2016). These low endowments take on a misleading meaning in that they refer to renewable resources which, however, cannot all be mobilized. The mobilized water resources are generally 20 to 30% lower compared to renewable resources in this part of the world (Nouaceur, 2017). This observation clearly illustrates the future challenges of the Maghreb countries in terms of water security.

Regarding groundwater resources, waters from shallow groundwater are generally renewable and easy to extract and, therefore, are considered to be conventional. However, when groundwater is at great depths such as in the Sahara, it is less accessible and often non-renewable. As a result, they are part of unconventional resources and their pumping and treatment require specific expensive techniques because they are highly mineralized (South of Algeria, Lybia, Egypt).

Identifying the impact of climate change on the natural environments and renewable resources of the local ecosystems is a major challenge for decision-makers in North Africa for the 21st century. According to the results of the different climate forecasting models, most of the North African region is expected to experience an average temperature rise of 3 to 4°C by 2100, a decrease in precipitation, and an increase in extreme events (IPCC, 2013). The work of Giannakopoulos et al. (2005) leads to the same result with a reduction in rainfall in the southern Mediterranean of the order of 20% under a scenario of +2°C global warming. Giorgi et al. (2006) also estimated that a significant decrease in Mediterranean precipitation is expected in relation to an intensification of the anticyclonic circulation. Gael (2007) showed that the regions of the MENA region (Middle East and North Africa) will be severely affected by climate change, since most of the countries located in this zone (except Iran and Iraq) have reached the water stress threshold (1700 m³ of renewable water per capita per year). Algeria and Tunisia, which are part of the MENA zone, have reached, according to this study, a state of absolute water scarcity. According to many studies, the combined effect of climate change and human impact in the Mediterranean area would result in water scarcity for around 290 million people. This area is recognized today as a "hot spot" for climate change (Giorgi et al., 2006).

In North Africa, the phenomenon of water erosion is very widespread, the majority of watersheds being characterized by severe specific degradations exceeding 2000 tons per km² per year, which leads to an average annual siltation of dam reservoirs at a rate of 125 million m³ (Remini and Remini, 2003). In Morocco, for example, water erosion annually causes soil loss ranging from 500 tons per km² to more than 5000 tons per km² depending on the region and siltation of dam reservoirs of around 75 million m³, this is leading to an annual reduction of 0.5% of the dams storage capacity, which causes a significant loss of water allowing the irrigation of 10,000 ha per year and the deterioration of the quality of the drinking water mobilized (HCEFLCD, 2008). Under the projected climate for the North African region, the erosion phenomenon is expected to keep being important with multiple impacts on the natural ecosystems.

With this background in view, this chapter reviews examples of the response of key basins in North Africa to climate change. At a watershed scale, understanding the effect of climate change on natural systems is vital for future sustainable management strategies of the watershed. The North Africa region is a diversified region from a hydro-climatic perspective with highly vulnerable ecosystems to climate change and variability; the reviewed examples in the following sections represent both aspects of diversity and vulnerability that are important to be taken in basin management.

6.2 CLIMATE CHANGE AND WATER QUALITY AND QUANTITY

6.2.1 Gafssa Basin

In southwestern Tunisia, where industrial activities are held around mining (especially phosphates), the water resources are under the effects of climatic and anthropogenic effects. Climate-related factors include irregularity of the rainwater cycle, desertification, increase of evapo-transpiration, and landslides; while human activity–related factors include urbanization, intense agricultural activities, industrial activities, poor management of water resources (storage and use and re-use), water resources over-exploitation, poor distribution, and pricing between the different socio-economic sectors in this part of the country.

The study area, Gafsa Basin, is located in the southwestern part of Tunisia. Overall, the Tunisian climate is a typical Mediterranean climate with some Atlantic oscillations, with dry to desertic summers and rainy autumns and springs (Hamed et al., 2012, 2013, 2014).

The effects of climate variability are, furthermore, evaluated by the frequency of extreme events, especially during the recharge periods. Indeed, droughts and floods are affecting the surface water quality and quantity in direct ways (interference with water availability and management decisions), and in indirect ways (exacerbation of quality issues, impact on groundwater pumping rate, flood damage, sanitary reasons, etc.). Being amplified by the anthropogenic activities, the availability of surface water in the study basin, moreover, affected by the high pollution risks related mainly to the industrial waste of the phosphate mining basin (Gafsa, Sfax, and Gabes Regions) and the agricultural runoff (Gafsa, Tozeur, and Kebili Regions). Indeed, the increasing expansion of economic activities is producing huge volumes of nitrate-contaminated runoff water and (or) highly polluted industrial wastes released to the transboundary areas of the Chott Djerid hydrographic network (Tuniso–Algerian Basin) and wetlands on the surface. These quantities reduce the surface water supply in terms of sufficient quantity and adequate quality (concentrations of dissolved substances, enhancing dissolution of the leached lands, etc.). (Hamed et al., 2018).

In southwestern Tunisia, the demand for water provokes frequent conflicts of use between the agro-industrial operators (phosphates) and large urban agglomerations, and between agriculture and tourism sectors. The climate variability expressed by the decrease of precipitations amount and the variability of their intensity and frequency will undoubtedly influence the recharge amount and the groundwater processes. However, the different responses to these "stressors" are related to the different characteristics of hydrogeological system components, namely the confined aquifer, generally ranging from poorly to non-renewable (transboundary aquifer between southwestern Tunisia, Algeria, and Libya). The unconfined aquifer is highly influenced by inter-seasonal climate variability, with low recharge amount and generally fed by flows from deep aquifers and coastal aquifers embodying fresh rainfed water and vulnerable to saline intrusion (Sea–Chott–Sebkha) (Hamed et al., 2018). According to recent studies, the total dissolved solids (TDS) of groundwater in some boreholes in the Gafsa Region (M'Dilla Region), is almost three times that of seawater (90 g/l). These strongly saline waters are confined and not renewed. Hence, future orientations should be towards the desalinization of seawater and the geo-valorization of wastewater in this part of the country.

6.2.2 Souss–Massa Basin

Despite its geographical situation on the Atlantic coast of Morocco (Figure 6.2), the Souss–Massa Basin is one of the most arid areas of the country; it experiences highly variable rainfall and recurrent droughts. The limited water resources are threatened by increasing demands and accelerated quality degradation. Most climate models that simulate global and regional mean precipitation show consistent results of significant unoptimistic climate change in this part of the kingdom (IPCC, 2013). According to Seif-Ennasr et al. (2017), temperature values are expected to increase up to two degrees during the period 2030–2049 and up to four degrees during the period 2090–2100 under the scenario RCP8.5 compared to the baseline (1986–2005). Precipitation will also decrease, especially in high-altitude areas, and the extreme change (under the RCP8.5) will result in –53% precipitation by the end of the 21st century. Clearly, these changes are expected to have several impacts, especially on agricultural demand, dam storage, and renewable-recharge groundwater. Consequently, projections for future renewable water resources in the Souss–Massa Basin are bleak, and climate change coupled with increasing water demands are likely to amplify the water crisis in the area (Hssaisoune et al., 2020). The major economic assets of the Souss–Massa area are agriculture, tourism, and fishing. The first two sectors depend on water availability and quality, and therefore

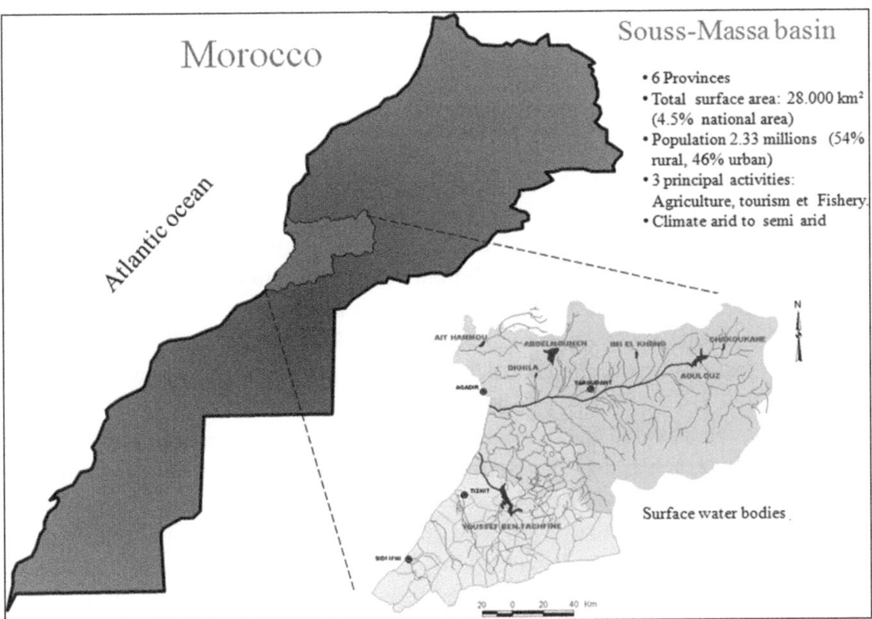

FIGURE 6.2 Souss–Massa Basin characteristics.

possible decline in water availability and quality might have severe consequences for the fast-growing economy of southwestern Morocco.

Water resources in Souss–Massa are over-exploited resulting in dramatic declines of groundwater levels, forcing well owners to drill deeper wells. This situation is leading to the drying up of springs and surface waters, marine intrusion of seawater into coastal aquifers, and enhancing desertification phenomena in some areas in the southern part of the area.

Current overexploitation phenomena of groundwater in the Souss–Massa Basin have limited the operation of many agricultural activities, and several farms in the central Souss Valley had to cease operations completely due to the non-availability of the water resources. More than 25,000 wells and boreholes were drilled across the aquifer. Meanwhile, unfortunately, there is no consistent monitoring of the water quality, nor real-time measurement of water quantity wasted (spillage, loss from the system) in the basin. Water management is therefore very limited and can act only after the crisis of deterioration occurs and groundwater is becoming too saline for agricultural operation.

The unconfined alluvial aquifer of Souss–Massa is the main water resource in this part of Morocco. The development of agricultural activities during the last four decades has led to a decrease of the water table by 0.5 to 2.5 m.y^{-1}; in addition to an increase of salinity up to levels of 7 g.L^{-1} and high amount of nitrates in large sections of this basin. The groundwater (GW) level declines coupled with high salinity make this aquifer very vulnerable, especially in areas stressed by intensive pumping. The hydrological simulation results indicated an annual flow from the deeper Turonian aquifer of 9.5 million m^3. The main recharge mechanism is infiltration through wadi beds and return flow for irrigation in a few areas. The main recharge area is from the High-Atlas Mountains on the right side of Oued Souss and the Anti-Atlas Mountains on the left side (Bouchaou et al., 2008). The water balance analysis indicates that the aquifer does not have enough time to replenish water abstractions hence it has been overexploited. Dynamic modeling was used to calculate water residence time which identified younger ages in the middle and upper plains. In absolute values, the ages estimated in the model are younger than those estimated with isotopic tools (Bouchaou et al., 2008).

From a quality point of view, the main origins of high salinity and nitrate rates in water resources of the Souss–Massa Basin are the water and rock interaction in geological formation containing Halite and Gypsum; in addition to marine intrusion in the coastal zone and anthropogenic activities (wastewater and fertilizers) are impacted mainly by salinity and nitrates (Malki et al., 2016).

6.3 CLIMATE CHANGE AND EROSION

6.3.1 Bouregreg Basin

With a total drainage area of 3690 km^2, the Bouregreg watershed is located in the northwest of Morocco and is considered one of the most strategic watersheds in the country. It is bounded to the north by the Sidi Mohamed Ben Abdellah Dam, to the south by the limits of Oum Rabia River catchment, to the west by the Grou River catchment bounds, and to the east by the Beht River catchment bounds. Laying on heterogeneous landscapes, its altitudes range from 60 m in the downstream plain part to 1627 m in the upstream mountainous area. This geographic situation bestows to the sub-basin a bioclimatic stage extending from the semi-arid to the sub-humid with an average annual rainfall of 600 mm.

The watershed can be divided into five main sub-basins, which are from the SE to the NW: the Agunnour River, Ksiksou River, Middle Bouregreg, Tbahart River, and Low Bouregreg.

To map and assess the water-induced soil erosion rates within the sub-basin, several models are available and range from empirical (ex: USLE, RUSLE, and MUSLE) to physical models (e.g., CREAMS, WEPP). In this case, the Revised Universal Soil Loss Equation (RUSLE) (Wischmeier and Smith, 1978) was adopted which is considered the most commonly used model worldwide. It consists of the multiplication of the main erosive process drivers and is expressed as follows:

$$A = R * K * L S * C * P \qquad\qquad (6.1)$$

Where: A is the potential long-term average annual soil loss (t per ha per year), R is the rainfall and runoff erosivity index (MJ mm ha^{-1}•h^{-1}•yr^{-1}), K is the soil erodibility factor (t h MJ^{-1}•mm^{-1}), LS is the slope length and steepness factor; C is: the cover management factor, and P is the conservation practices factor.

Calculation of these factors from both long time-series data (1994 to 2014) and different available thematic maps (pedological, geological, topographic, land use, land cover) made it possible to generate their spatial distribution throughout the whole basin.

Conservation practices' factor (P) is assigned the value 1 as only a few insignificant Soil and Water Conservation (SWC) measures were found across the whole study area. LS map is derived from the Digital Elevation Model maps. To obtain the C factor maps that reflect the land cover evolution from 1994 to 2014, we used two Landsat 8 Oli images acquired from www.earthexplorer.usgs.gov respectively in May 2000 and June 2014 with a resolution of 30 m. After performing the required corrections (geometric, radiometric, and atmospheric), we proceeded to the supervised classification using the Maximum Likelihood method to differentiate prevailing land use/land cover types across the sub-basin. Classification results were compared to those of 20 field surveys' geolocated samples. The Normalized Difference Vegetation Index (NDVI) maps derived from the satellite images perfectly reflect the C factor distribution within the study area. The obtained C factor distribution maps had been separately multiplied with those of R, K, and LS factors in order to generate the average annual soil loss map for both years 1994 and 2014.

The diachronic analysis of soil erosion rates between 1994 and 2014 revealed that a shift towards more severe erosion forms has occurred throughout the study area. Classes of slight (0 to 5 t per ha per y) and moderate (05 to 10 t per ha per y) erosion severity decreased in importance by −67.78 and −46.44 respectively. In the meantime, classes of severe (20 to 30 t per ha per y) to very severe (>30 t per ha per y) erosional process were likely to score a positive strong variation with +156.80 and +170.58 respectively.

Moreover, the watershed has witnessed an overall increase in annual average soil loss rates of 54.55%. It evolved from 10.64 (t per ha per yr) in 1994 with a total loss of 4,208,875 tons to 16.44 (t per ha per yr) in 2014 with a total loss of 6,504,824 tons (Table 6.1).

The highest annual average soil loss rates in 2014 are those of the Ksiksou River, Tabahhart River, and Low Bouregreg sub-basins with respectively 19.19, 19.02, and 17.01 (t per ha per yr). The high rate in Ksiksou Rive is due to obvious climate change impacts. The latter led the local population to cultivate fields with steep slopes (>30%) besides exerting a high deforestation pressure making soils less protected from rainfall splash effect. As an agro-sylvo-pastoral area, the Tabahhart River sub-basin is under high overgrazing pressure. Higher temperatures trigger a lower vegetative cover in pasturelands resulting in a higher herd density per unit area. Meanwhile, the lower Bouregreg is witnessing a drop-down in agricultural yields due to a lack of rainfall records allowing realizing former performances. This results in a massive wave of available area conversion into croplands with intensive agricultural production systems.

Nevertheless, the most significant changes in the 1994–2014 time period were recorded in the Low Bouregreg, Aguennour River, and Ksiksou River sub-basins with respectively +117.52%, +58.87%, and +51.58% in terms of annual average soil loss rate. In fact, the intensive cropping system acts jointly with extreme weather events (intensive rainfalls) to make the plain part of the watershed downstream more vulnerable to flood hazards. Furthermore, in rugged areas with high altitudes (up to 1600 m) such as the Aguennour River and Ksiksou River sub-basins, weather pattern shifts are behind the vegetative cover degradation making soil particles more easily detachable and transportable. Such intensive land erosion exposes downstream water bodies (Sidi Mohamed Benabdellah Dam) to experience unprecedented siltation rates (Gourfi, 2020).

TABLE 6.1
Soil Loss Rate Variation 1994–2014

		1994		2014		
Sub-basin	Area (ha)	Average soil loss (t per ha per yr)	Total loss (t per yr)	Average soil loss (t per ha per yr)	Total loss (t per yr)	Average soil loss variation (%)
Ksiksou river	110,700	12.66	1,401,462	19.19	2,124,333	51.58
Tabahart river	94,850	14.18	1,344,973	19.02	1,804,047	34.13
Low Bouregreg	70,830	7.82	553,891	17.01	1,204,818	117.52
Middle Bouregreg	44,040	8.80	387,552	12.35	543,894	40.34
Aguennour river	75,180	6.93	520,997	11.01	827,732	58.87
Watershed	395,600	10.64	4,208,875	16.44	6,504,824	54.55

6.4 CONSIDERATIONS FOR BASIN MANAGEMENT IN NORTH AFRICA

Water and other resource management within the watershed territory has become a major concern. Thus, the floods and the water shortage episodes in North African countries were always occasions to revive the debate on climate change and its possible consequences, but also on the responsibilities of the decision-makers in the maintenance of waterways. However, while concerns about flood and drought risks have been around for a long time, water pollution is a more recent concern. Indeed, the consequences of the evolution of economic activities in these countries since their independence (in the middle of the last century) started to be exacerbated by most investigations with dramatic implications in some parts of this region. One of the common features of North African countries is that they are emerging economies with sustained growth rates; the expansion of diverse economic growth will definitely have effects on water and the other natural resources within the watershed unit. It is highly recommended to take into consideration the local ecosystem characteristics of the watershed while planning for the (economic) development in this region.

Moreover, small-scale agriculture is a major component in North African societies with an important position in social architecture and with significant contribution to the local economies. Most of the activities of this social category are around pasture, forestry, and subsistence farming. For any successful basin-based management strategy, the involvement of this component of the society is of key importance. Despite being vulnerable to many factors (climate, market, finance, etc.), small-scale agriculture in North Africa has shown great ingenuity and resilience over the different crises and represents an adapted ally in sustainable watershed management strategies in this region.

The regulatory aspects can play an important role in the basin's management in North Africa. Morocco offers an advanced example of proactivity when it comes to watershed management. Since the 1960s, the state has adopted the policy of building dams in the major watersheds for drinking water and for large irrigation perimeters. Meanwhile, a very developed set of laws and decrees have been promulgated to manage water within the watershed unit. The state has also established public institutions, namely basin agencies, to be in charge of the monitoring, planning, and supply of water within their respective territories; thus, there are nine basin agencies in Morocco acting as decentralized organizations between the different water users in their respective regions.

6.5 CONCLUSIONS

The North African region is faced with serious challenges related to climate change impacts on natural resources and their potential consequences on the development track adopted by the region's nations. North Africa is part of the Mediterranean region, one of the global "hot spots" for climate change, which makes this part of the world particularly vulnerable to the impacts of future rising temperatures and decreasing average precipitation, interannual variability, and extreme weather events.

The over-exploitation of freshwater resources, water quality degradation, and erosion are the key common challenges of river basins in northern Africa. In addition to the anthropogenic contribution to these issues, climate change is interfering (directly or indirectly) to make the situation even more complicated for all stakeholders involved in either the cycle of natural resources management or community prosperity planning.

River basin development, as an approach to conserve natural resources and to improve livelihoods in the region, is already emerging as an important planning tool for sustainable development and for tackling climate change in most of North African countries. A holistic paradigm is still needed to monitor and evaluate the natural processes within the river basin unit with regard to climate change impacts; and together with the incorporation of the region (social, bio-physical, historic) specificities and the involvement of local populations, they can increase the resilience of North African river basins to climate change.

REFERENCES

Bouchaou, L., Michelot, J. L., Vengosh, A., Hsissou, Y., Qurtobi, M., Gaye, C. B., Bullen, T. D., and Zuppi, G. M. (2008). Application of multiple isotopic and geochemical tracers for investigation of recharge, salinization, and residence time of water in the Souss-Massa aquifer, Southwest of Morocco. *Journal of Hydrology*, 352(3–4), 267–287.

Brouziyne, Y., Benaabidate, L., Abouabdillah, A., Bouabid, R., and Chehbouni, A. (2020). Modeling hydrologic processes and potential responses to climate change in an agro-silvo-pastoral watershed in the Mediterranean area. *Proceedings of IAHS*, 383, 151–158.

Da Silva, M. R., Santos, C. A. G., Moreira, M., Corte-Real, J., Silva, V. C. L., and Medeiros, I. C. (2015). Rainfall and river flow trends using Mann–Kendall and Sen's slope estimator statistical tests in the Cobres River basin. *Natural Hazards*, 77(2), 1205–1221. https://doi.org/10.1007/s11069-015-1644-7.

FAO. (2016). Base de Données Principale d'AQUASTAT, Organisation des Nations Unies pour l'alimentation et l'agriculture (FAO). Site web consulté le [31/12/2020 18:12].

Fathian, F., Dehghan, Z., and Eslamian, S. (2016). Evaluating the impact of changes in land cover and climate variability on streamflow trends (Case study: Eastern subbasins of Lake Urmia, Iran). *International Journal of Hydrology Science and Technology*, 6(1), 1–26.

Gael, G. (2007). Climate change adaptation in the water sector in the Middle East and North Africa, Technical Note prepared by METAP (Mediterranean Environmental Technical Assistance Program) under the EC funded SMAP III project "Promoting awareness and enabling a policy framework for environment and development integration in the Mediterranean with a focus on Integrated coastal Zone Management", 25p.

Giannakopoulos, M., Bindi, M., Moriondo, M., and Tin, T. (2005). Climate change impacts in the Mediterranean resulting from a 2°C global temperature rise, A report for WWF. http://www.fao.org/fileadmin/user_upload/rome2007/docs/Climate_Change_Adaptation_Water_Sect or_NENA.pdf.

Giorgi, F. (2006). Climate change hot-spots. *Geophysical Research Letters*, 33(8), L08707.

Gourfi, A., Daoudi, L., and de Vente, J. (2020). A new simple approach to assess sediment yield in a large scale known by a great landscape diversity: Example of Morocco. *Journal of African Earth Sciences*, 168, 103871.

Hamed, Y., Ahmadi, R., Hadji, R., Mokadem, N., Ben Dhia, H., and Wassim, A. (2012). Groundwater evolution of the continental intercalaire aquifer of southern Tunisia and a part of southern Algeria: Use of geochemical and isotopic indicators. *Desalination and Water Treatment Journal*. https://doi.org/10.1080/19443994.2013.806221.

Hamed, Y., Al-Gamal, S. A., Ali, W., Nahid, A., and Ben Dhia, H. B. (2013). Palaeoenvironments of the continental intercalaire fossil from the Late Cretaceous (Barremian-Albian) in North Africa: A case study of southern Tunisia. *Arabian Journal of Geosciences.* https://doi.org/10.1007/s12517-012-0804-2.

Hamed, Y., Ahmadi, R., Demdoum, A., Bouri, S., Gargouri, I., Ben Dhia, H., Al-Gamal, S. A., Laouar, R., and Choura, A. (2014). Use of geochemical, isotopic, and age tracer data to develop modelsof groundwater flow: A case study of Gafsa mining basin-Southern Tunisia. *Journal of African Earth Sciences*, 100, 418–436.

Hamed, Y., Hadji, R., Redhaounia, B., Zighmi, K., Bâali, F., and El Gayar, A. (2018). Climate impact on surface and groundwater in North Africa-A global synthesis of findings and recommendations. *Euro-Mediterranean Journal for Environmental Integration.* https://doi.org/10.1007/s41207-018-0067-8.

HCEFLCD. (2008). Rapports annuels des incendies de forêts et bases de données du Service de la Protection des Forêts, Bilans internes, Haut-Commissariat aux Eaux et Forêts et à la Lutte contre la Désertification, Rabat-Chellah, Maroc.

Hssaisoune, M., Bouchaou, L., Sifeddine, A., Bouimetarhan, I., and Chehbouni, A. (2020). Moroccan groundwater resources and evolution with global climate changes. *Geosciences*, 10(2), 81. https://doi.org/10.3390/geosciences10020081.

IPCC. (2007). Climate change, 2007: Impacts, adaptation and vulnerability. Working Group II Contribution to the Fourth Assessment Report (Summary for Policymakers and Technical Summary), 93p.

IPCC. (2013). Changements climatiques en 2013, Les éléments scientifiques, résumé à l'intention des décideurs, service d'appui technique du groupe de travail I GTI.

Malki, M., Bouchaou, L., Hirich, A., Ait Brahim, Y., and Choukr-Allah, R. (2016). Impact of agricultural practices on groundwater quality in intensive irrigated area of Chtouka-Massa, Morocco. *Science of the Total Environment*, 574, 760–770.

Millennium Ecosystem Assessment (MEA). (2005). *Ecosystems and Human Well-being: Desertification Synthesis.* Island Press, Washington, DC, 36p.

Norrant, C. (2004). Tendances pluviométriques indicatrices d'un changement climatique dans le bassin méditerranéen de 1950 à 2000. Étude diagnostique, Thèse, Université Aix Marseille I, Université de Provence, France, 266p.

Nouaceur, Z. (2017). *Eau et climat en Afrique du Nord et au Moyen Orient.* Editions Transversal, 11–25. 978–606-605-166-8.

Remini, W. and Remini, B. 2003. LA SEDIMENTATION DANS LES BARRAGES DE L'AFRIQUE DU NORD, Larhyss Journal, ISSN 1112-3680 (n 02), 45–54.

Seif-Ennasr, M., Hirich, A., El Morjani, Z. E. A., Choukr-Allah, R., Zaaboul, R., Nrhira, A., Malki, M., Bouchaou, L., and Beraaouz, E. H. (2017). Assessment of global change impacts on groundwater resources in Souss-Massa basin. In Osman Abdalla, Anvar Kacimov, Mingjie Chen, Ali Al-Maktoumi, Talal Al-Hosni, and Ian Clark (Eds.), *Water Resources in Arid Areas: The Way Forward.* Springer, Berlin/Heidelberg, 115–140.

Swift, J. (1996). Desertification: Narratives, winners and losers. In Leach, M., and Mearns, R. (Eds.), *In the Lie of the Land Challenging Received Wisdom on the African Environment.* The International African Institute, in association with James Currey Ltd, London, 73–90.

Wischmeier, W. H., and Smith, D. D. (1978). Predicting rainfall erosion losses. Agriculture Handbook No. 537, U.S. Dept. of Agr., Science and Education Administration.

7 Global Climate Change and Free Market Economy

Double Exposure Influence on Co-operatives in Tanzania: Hopes and Fears

Luka Sabas Njau and Neema Kumburu

7.1 INTRODUCTION

The world is changing due to the change in climate. Climate change translates into a change in our way of life. As countries change to become more industrialized and increase the production of goods, such industries emit greenhouse gases, which are the main contributors to climate change. It is widely considered that climate change is mainly due to various human activities such as agriculture, industrial production, consumer practices, and transportation. The resultant effect is global warming due to the accumulation of greenhouse gases leading to increased flood frequencies, extreme temperatures, drought conditions, unpredictable rain patterns, and rising sea levels among others (Zamani Nuri et al., 2013). Studies have projected that the frequency and intensity of extreme weather events will increase, causing severe socio-economic consequences, particularly to marginalized groups in society. Poor communities especially in developing countries face a disproportionate impact from climate change due to their limited resource capabilities and their high dependence on climate-sensitive livelihood activities (URT, 2012). Studies have indicated that climate change and economic globalization may mutually affect a region, or a sub-group leading to the so-called double exposure (Grineski et al., 2001) effect.

7.2 DOUBLE EXPOSURE CONCEPT: CLIMATE CHANGE AND GLOBALIZATION FORCES

Global climate change and globalization forces are real challenges today. The forces are serious obstacles to most developing countries' communities which largely depend on smallholder farming activities as their main source of livelihood and/or survival. The interplay between climate change and globalization forces and their resultant negative effects especially on poor communities has resulted in these forces being labeled as double exposure effects. Double exposure is a concept proposed and coined by O'Brien and Leichenko (2000), and Leichenko and O'Brien (2008), to emphasize and demonstrate how processes of globalization and global environmental change redefine risk and encourage new, interrelated responses to social and ecological transitions. It provides a generalized approach to understanding the interactions between global environmental and economic changes, directing attention to the ways that the two interacting processes spread risk, and vulnerability over both time and space. The concept refers to the fact that certain regions, sectors, ecosystems, and social groups will be confronted both by the impacts of climate change and by the consequences of globalization. It is from this concept and recognition that the joint forces of climate change and economic globalization are expected to exert pressure on smallholder farmers in Tanzania and co-operative societies in particular leading to multiple or double exposure effects. Such double exposure effects are considered to result in both hopes and fears that should be taken into account by co-operatives as they strive to overcome such forces.

For many years ago, co-operatives have been considered the only true member-based organizations established, owned, and governed by members, with their services used by the members who establish them. Such qualities have granted co-operatives their own unique values, principles, and definitions. Several definitions of co-operatives by different scholars exist. Nevertheless, the most commonly applied and internationally recognized definition of a co-operative is the one formulated by the International Co-operative Alliance (ICA), a non-governmental co-operative federation representing co-operatives and the co-operative movement worldwide. The ICA defines a co-operative as "an autonomous association of persons united voluntarily to meet their common economic, social, and cultural needs and aspirations through a jointly-owned and democratically-controlled enterprise based on the values of self-help, self-responsibility, democracy, equality, equity and solidarity" (ICA, 1995; ICA, 2017). This is a unique organization that has managed to survive many years of ups and downs and hence an appropriate tool for addressing diverse global challenges including climate change and free market forces.

7.3 CO-OPERATIVES AND THEIR POTENTIAL TO ADDRESS DIVERSE GLOBAL CHALLENGES

Over the decades, co-operatives have been considered organizations with the potential to foster socio-economic development and pull communities out of poverty (FAO, 2012; Vicari and De Muro, 2012). This postulation is supported by the unique characteristics and qualities inherent in co-operative societies over other organizations. Karki (2006) identified three distinguishing features of co-operatives: first, those who own the co-operative are those who use the co-operative; second, those who use the co-operative govern/control the co-operative and lastly, those who use the co-operative are the primary benefactors of the co-operative. The primary objective of co-operatives is to increase the economic well-being of the producer/owner of the co-operative (Barton et al., 1993). Increased economies of scale translate into lower per-unit costs during the production and processing of raw produce and the distribution of products to retailers and end-users (Hill and Jones, 2007). Co-operative societies, therefore, offer numerous opportunities to members in terms of reduced production cost, agricultural input accessibility, product value additions, marketing information, increased bargaining power, collective marketing, promotion of intensive farming, advocating agroforestry practices, stimulation of demand for the outputs and loans among many other benefits.

Globally, over 1 billion people are directly benefiting from co-operatives as members or clients (UNDESA, 2014). Besides that, employment has been one of the most important contributions made by co-operatives throughout the world (CICOPA, 2017). ILO (2022) and UNDESA (2014) indicated that the co-operative sector was one of the major job providers, employing at least 100 million people worldwide. Based on data from 156 countries, the updated estimate in 2017 shows that, employment in or within the scope of co-operatives concerns at least 279.4 million people across the globe, equivalent to 9.5% of the world's employed population (CIPOPA, 2017). Moreover, the co-operative sector has more than 800 million individual members across the globe (ILO, 2022). It has been estimated that the livelihoods of nearly half of the world's population are made secure by co-operative enterprises (ILO, 2012, 2022; Wanyama, 2014). Consequently, Africa is home to 70 million co-operative members.

Majurin (2012) indicated that approximately 7% of the African population is affiliated with the co-operative movement. In Tanzania alone, about 1.6 million people are co-operative members and about 8 million people benefit directly from them in terms of service access, employment, etc. (Sumelius et al., 2013). Despite these interesting and encouraging facts, the performance of the co-operative sector in Tanzania is likely to be compromised by the evident effects of globalization and climate variability and change. Agriculture and particularly smallholder farming activities remain the sole and priority food security determinant for the majority of Tanzanians. Nevertheless, despite such importance, smallholder farmers' activities are highly vulnerable to the potential impacts of climate variability and change. The sector alone contributes approximately 3.7% of greenhouse gas emissions (GHGs) globally besides being a key driver in deforestation activities, which also adds a further 7–14% of greenhouse gas emissions on a global scale (FAO, 2013).

Stern (2010) established the situation to be exacerbated by the existence of weak and fragile smallholder farmers' support systems in many of the developing countries. Nonetheless, if facilitated, the agricultural sector, co-operative sector in particular, contains the potential to contribute to climate change mitigation and increasing resilience through adaptation (Nyang'au et al., 2021). For instance, co-operative societies have been contributing to carbon emissions reduction and enhancing carbon sinks by getting involved in land use and land-use changes. Marketing of agricultural products, supply of agricultural inputs, provision of extension services, promotion of intensive farming, and advocating agroforestry practices have a significant contribution to climate change mitigation. Nevertheless, for co-operatives to be able to effectively and sufficiently undertake such noble functions they must be enabled to successfully address the climate change and free market forces currently facing them.

7.4 THE GENESIS OF CO-OPERATIVES IN TANZANIA AND THEIR INTERPLAY WITH TRADE LIBERALIZATION FORCES

In Tanzania, the co-operative idea dates back to the colonial era in 1925 with the formation of the Kilimanjaro Native Planters Association (KNPA), the first-ever indigenous association of African coffee farmers. This was later transformed and registered as the Kilimanjaro Native Co-operative Union (KNCU) in 1933 through the 1932 Co-operative Ordinance. The social and economic unrest wrought by white settlers and Asians forced small-scale peasants to establish co-operatives for social and economic emancipation. The co-operatives were established as a remedy to challenge white settlers and Asian middlemen who in most cases underpriced the commodities in the market and sold exorbitantly for super profit. Sometimes such middlemen bought farmers' crops on credit with delayed payment and at times without paying them at all. It is estimated, for instance, that 40% of cotton bought by Asian businessmen during the 1930s and 1940s was not paid for (URT, 1997).

7.4.1 THE POST-INDEPENDENCE CO-OPERATIVES IN TANZANIA

To date, the co-operative movement in Tanzania has gone through three eras, each with its unique characteristics. First, the colonial era witnessed a dynamic and independent association of predominantly smallholder cash crop farmers mainly in coffee, cotton, and tobacco. These, among other things, built the early capital/assets base in the form of buildings, machinery, and factories. The examples include the Kilimanjaro Native Co-operative Union (hotels, offices, schools, etc), Bahaya Co-operative Union (hotels, schools, and office buildings), Rungwe African Co-operative Union (hotels, offices, etc), Victoria Federation of Co-operative Unions (offices, buildings, hotels, ginneries, etc), among others. The post-independence era, i.e., the period after Tanganyika's independence in 1961 witnessed the growth of co-operatives particularly in terms of numbers, even in the non-cash crop-growing regions of the country. The third era in the development of co-operatives in Tanzania is from the period around the late 1980s' with the emergence of the liberalization of the economy and therefore the co-operative sub-sector. This sub-section will focus on the post-independence eras in Tanzania.

7.4.2 The Central Planning Era or State-controlled Co-operatives

The central planning era in Tanzania refers to the period between 1961–1982 characterized by independence attainment and *ujamaa* (brotherhood) and villagization policies. After Tanganyika's independence in 1961, on account of the successful co-operative movement inherited from the colonial epoch in some pockets of the country, the government took the lead in co-operative development. The state actively promoted the establishment of co-operatives. In 1963, the government promulgated the Co-operative Societies Ordinance of 1963 that aimed at introducing co-operatives throughout the country through the institutionalization of a unified co-operative model as a means to challenge the exploitation of rural class (Bergeron, 2015). After a special presidential commission of inquiry in 1966, 16 co-operative unions and hundreds of primary co-operatives were taken over by the state (Develtere et al., 2008). Until that time, the co-operative societies' performance was so impressive and encouraging. In 1967, the marketing co-operative movement in Tanzania had grown to the tune of being labeled the co-operative giant of Africa, taking third place in the world after Denmark and Israel. During this time, co-operatives were the only organizations outside the government that managed to bring to the village, the largest sums of money used to purchase the crops even in the most remote areas of the country (Chambo, 2018).

It is worth noting that, Tanganyika declared independence in 1961 under the leadership and socialist vision of Mwalimu Julius Kambarage Nyerere, the founder and first president of free Tanganyika. Tanganyika later united with Zanzibar in 1964 to form a union country/government named the United Republic of Tanzania. The basic tenets of Mwalimu Nyerere's socialism vision were cemented in the Arusha Declaration of 1967, which called for the elimination of privilege and exploitation, and improvement in the quality of life of the peasant population (Saul, 2002). The most dramatic step to accomplish these goals was villagization, also known as *ujamaa,* through the grouping of scattered populations into co-operatively based villages (Saul, 2002). The policy aimed at ensuring a unified egalitarian society with a centralized political system at its core. Similarly, the policy re-organized co-operative movement to allow co-operatives to participate in the government's socialist projects and national development objectives (Mhando, 2014). Despite its emphasis on equality, co-operation, and democracy, the era following the Arusha Declaration was characterized by heavy government intervention, land nationalization, and strict agricultural marketing control (Lyimo, 2012). During this period, all major means of production, e.g., land, infrastructural capital, banks, machinery, factories, natural capital, and others were all owned by the state. Thus, co-operatives were viewed by the government as vehicles for achieving agro-socialist transformation through the co-operative principles of collective ownership, management of the means of production, and allocation of resources rather than as grassroots member-driven organizations (Lyimo, 2012). The co-operatives that attempted to operate as member-driven and autonomous entities were suppressed while branded by politicians as capitalist organizations. The term *mawakala wa mabeberu* literally meaning "capitalist agents or organizations" was famous during this time just like it is famous today to some politicians branding those whom they perceive to be against their political agenda or providing alternative solutions or strategies that challenge them.

7.4.3 The Free Market Economy or Liberalization Era

The co-operatives movement in Tanzania entered into the free market economy or liberalization era in the period around the 1980s through the 2000s. The year 1982 deserves a special mention in co-operatives history in Tanzania because it was the period when the co-operatives were reinstated by the government after their dissolution in 1976. During and after this period, however, reinstated co-operatives continued to suffer from strict government control, placed under the patronage of the ruling party, i.e., Chama Cha Mapinduzi (CCM). The co-operative leaders were elected and made accountable to the ruling party rather than addressing the needs of its membership to the extent that the co-operative movement itself became weak, undemocratic, and non-transparent with respect to marketing activities (Maghimbi, 1992; Banturaki, 2000). While this was the situation in co-operatives, the general national economy was in shamble as well. Prior to the liberalization era, the country was facing both internal and external challenges that necessitated the government to re-think its economic strategies. The local internal forces that include economic mismanagement, underperformance, corruption, incompetence, and non-accountability; and foreign ones that included the 1978/79 war with neighboring Uganda, the rapid rise of oil prices, and declining agricultural commodity prices necessitated the government re-thinking of its economic strategies to rescue the economy from a total collapse.

In an effort to rescue the economic situation that was otherwise worsening, the locally improvised rescue package, the so-called Economic Recovery Programmes (ERPS) I and II were introduced. Concurrently, the externally imposed Structural Adjustment Programme (SAPS) by the International Monetary Fund (IFM) and World Bank were put into operation. The ERPS emphasized more production for more exports largely from agriculture while SAPs strongly advocated for the free market, deregulation, decontrols, devaluation of local currency against major currencies, and privatization of government-owned businesses/parastatals. Similarly, the government was forced to cease the provision of agricultural subsidies which had before sustained the smallholder farmers and monopolistic co-operatives.

7.5 TANZANIAN CO-OPERATIVES UNDER FREE MARKET ECONOMY AND CLIMATE CHANGE EXPOSURE

The trade liberalization era marked the tremendous decline and/or withdrawal of government control of co-operative societies in Tanzania. The aim here was to enable co-operatives to operate viable, self-reliant, democratically, and professionally managed member-oriented co-operative societies. It is worth noting that, while the government's withdrawal from direct control of co-operatives was possibly a good intention, most of such co-operatives were already facing climate change forces exposing them to the resultant negative effects. This means that during this era co-operative societies were left in a situation where they had to face both the liberalization and climate change forces with little or no government support. This is because the role of the state remained limited to a few aspects such as regulation and promotion of co-operatives as opposed to direct interventions in its operations. This regulation means that the government remains responsible for formulating and overseeing necessary laws and regulations regarding the formation, registration, operation, and deregistration of co-operative societies. It also offers legal advice to co-operative societies. Similarly, it is responsible for formulating the necessary co-operative policies/policy and standards to ensure the smooth operation of the co-operative societies. On the promotion side, the government is responsible for ensuring a facilitative environment for the smooth operation of co-operatives in terms of providing education and training to members, board members/leaders, co-operative staff, and the wider community.

It is worth noting that, while direct state control in terms of co-operative operation, governance, financing, subsidies, etc., has ceased and/or tremendously declined in the liberalization era, the government has continued to play a substantial role in regulating and promoting the co-operative societies. Given the delicate nature of most co-operative societies, the government's extended hand in terms of regulation and promotion is necessary to enable such co-operatives to face and overcome the challenges and embrace the opportunities offered by trade liberalization and/or climate change such as carbon trading and others. Nevertheless, its execution needs to be keenly undertaken to avoid any sort of mechanisms that are likely to suppress the co-operative principles and values. The type and quality of co-operative education and training offered also need to be re-assessed as there have been concerns that the education and training offered are more theoretical than the real-life situations in the co-operative field.

7.6 FREE MARKET ECONOMY, CLIMATE CHANGE, AND TANZANIA'S CO-OPERATIVES READINESS TO THRIVE AND COMPETE

The implementation of the free market economy in Tanzania came at a time when the co-operative societies were still under strict government control and at the government's mercy. Thus it came at a time when co-operative societies were neither prepared to compete (Bibby, 2006) nor ready to fully overcome the climate change negative effects. The free market economy, for instance, was accompanied by the removal of agricultural subsidies which were in those times sustained by the monopolistic co-operative societies and of course smallholder farmers. Under a free market economy, private enterprises such as private individuals, non-co-operative related companies, etc., were allowed to do business in various sectors including agricultural commodities. This includes procuring and marketing crops, agricultural inputs, equipment, etc. Nevertheless, the interest of most of these private operators was more on profit and/or wealth-making at the expense of the environment and smallholder farmers. This implies that issues relating to crop productivity (land use and land use changes consideration), environmental and/or biodiversity protection, affordable agro-inputs supply, and related activities received no or little consideration from such private operators. With the door wide opened, many co-operative societies found themselves "physically lacking/unfit" and thus unprepared to get into the trade liberalization race in terms of resources, e.g., finance/capital, skilled personnel, physical facilities and infrastructures, technologies, etc. While this was the case, it should be borne in mind that in addition to the liberalization forces, the same co-operatives were already prone to the global climate change effects resulting in unpredictable climate-related weather events and disasters. UNEP (2007) established that such major disasters include famines, loss of traditional lifestyles, drought, excessive rainfall leading to floods, increased spread of pests and diseases, and biodiversity losses, among others.

It is worth noting that, developing countries have very different individual circumstances and the specific impacts of climate change on a country depend on the climate it experiences as well as its geographical, social, cultural, economic, and political situations. As a result, countries require a diversity of adaptation measures very much depending on individual circumstances. However, there are cross-cutting issues which apply across countries and regions (UNFCCC, 2007). Given the co-operatives uniqueness and far-reaching coverage both in urban and remote rural areas they stand a better chance/position in addressing the climate change and globalization effects as compared to other organizations.

7.7 THE HOPES AND/OR OPPORTUNITIES OFFERED BY TRADE LIBERALIZATION AND CLIMATE CHANGE TO TANZANIAN CO-OPERATIVES

7.7.1 Hopes Offered by Trade Liberalization to Co-operatives

In Tanzania, the economic liberalization era refers to the period that followed the implementation of SAPs beginning in the 1980s through the 2000s. There are differing views from various co-operative stakeholders and practitioners regarding the influence of trade liberalization on co-operative societies in Tanzania. There are those who are of the view that trade liberalization has made co-operative societies stronger and more resilient and those that believe that it has intensified the weakening of the already weak co-operative societies. A study by Wanyama (2006) on the impact of liberalization on the co-operative movement in Kenya established that there are variations with regard to the performance of co-operatives in the country since the liberalization of the sector. Whereas the performance of agricultural co-operatives has generally been on the decline in Kenya, that of non-agricultural co-operatives, especially savings and credit co-operative societies (SACCOS) has been on the rise.

In analyzing the Tanzanian scenario, some brief descriptions of the hopes and/or opportunities offered by trade liberalization to co-operative societies are analyzed. Subsequently, the fears emanating from the same are described as well. The discussion will begin with the hopes and/or opportunities offered to agricultural-based co-operatives, particularly the agricultural marketing co-operative societies (AMCOS), and later on a glimpse into non-agricultural based co-operatives particularly the savings and credit co-operative societies (SACCOS) will be narrated. The two categories of co-operative societies will represent others because, first they are the main categories of co-operatives in Tanzania, and second, trade liberalization impacted almost all other co-operatives in the country in nearly similar manner.

Gibbon (2001) indicated that Agricultural Marketing Co-operatives (AMCOS) have had a monopoly role in marketing key cash crops since Tanzanian independence in 1961 until 1995/1996 when the monopsony ended with the liberalization of the market. Basically, the primary co-operative societies were the sole buyers of crops from smallholder farmers. Since liberalization in 1995, co-operatives have had to compete with private dealers in this function (Msonganzila, 2013). Trade liberalization allowed private buyers to buy crops directly from farmers, and they quickly took a large portion of the market share, because they were able to offer farmers a better service or more attractive prices (Gibbon, 2001). Thus one of the opportunities offered to co-operative societies under this arrangement is that, as private buyers offered competitive prices to farmers, co-operative societies were forced to look for the means to attract and retain their members by offering at least reasonable prices close to those offered by private buyers.

This in turn enabled co-operative society members to enjoy a bit of competitive prices to their crops. Nevertheless, not all co-operative societies were able to compete with the private buyers in terms of crop prices. In addition to that, Wanyama (2006) established that trade liberalization in Kenya triggered a structural transformation of co-operative movement that had seen the fading away of inefficient federative and apex co-operative organizations as co-operative societies seek better services. This was the case also in Tanzania where some primary co-operative societies specifically in the coffee sub-sector in the Kilimanjaro region did exactly the same. With the doors open for trade liberalization and the realization that primary co-operative societies could reap more benefits if they organized their own marketing channels within and outside Tanzania and hence more income, they decided to do away with the union systems/bureaucracies that existed in the famous union organization, i.e., the Kilimanjaro Native Co-operative Union (KNCU). To them, the apex was considered to be inefficient and costly to maintain. Thus, some of the primary co-operative societies decided to withdraw their membership from KNCU to form their own intermediary registered as the Kilimanjaro New Co-operative Initiative-Joint Venture Enterprises Limited (KNCI-JV Ltd) initially consisting of 32 primary co-operative societies commonly referred to as G32.

The G32 were/are the third window allowing farmers to sell their produce/coffee at premium prices. This is to say that; trade liberalization has enabled some primary co-operative societies to enjoy price hikes in the crops they sell. Wanyama (2006) attested that co-operatives that had adapted to the new liberalization environment have come out stronger than they were before the liberalization of the sector. Apart from best prices, trade liberalization has also opened up avenues for some co-operative societies to engage in fair trade arrangements whereby at their own consents, they entered contracts with coffee buyers in various markets of the world to produce and supply organic produce, particularly, coffee. Under this arrangement, such primary co-operative societies have found themselves earning more income without necessarily being subjected to costly and demanding second-tier, e.g., co-operative union bureaucracies.

The other hope offered by trade liberalization to co-operative societies in Tanzania is that, while the withdrawal by some primary co-operative societies from the union organizations, e.g., KNCU may be detrimental to such union organizations in terms of declining membership, bargaining power, and income, it has offered the same apex organizations an opportunity to keenly re-assess their relevance for their existence. This is in tandem with re-assessing the services they are offering to the co-operative societies by making them more competitive and user-friendly so that they may be able to retain the remaining members and/or reclaim those who have withdrawn their membership. This implies that, as trade liberalization has provided co-operative societies with more options to do their business, the same options demand co-operative organizations' structure to be innovatively maintained to attract and retain members.

In addition, trade liberalization has provided the co-operative societies with more options and opportunities to diversify their activities unlike in the past when they concentrated on only limited service provisions such as agricultural produce marketing. Trade liberalization has enabled some co-operatives to think out of the box and conduct numerous productive activities within and beyond their lines of operations. For instance, some primary co-operative societies in Tanzania that were primarily agricultural marketing based, e.g., Mamsera AMCOS have diversified their activities to offer extra services such as brick making and selling, and selling of other building materials (i.e., operating a hardware business) and other related services. Others have diversified their activities by offering mobile phone money services such as M-Pesa, Halopesa, Tigo Pesa, and Airtel Money operating as banking services provision agencies, etc. All these are the positive results of trade liberalization to co-operative societies that have forced them to re-think alternative financing mechanisms after government support ceased. There are concerns, however, that when co-operative societies operate beyond their lines of establishment such as by diversifying away from their core activities, they may be breaking and/or going against their founding purpose. This concern, however, can be accommodated within the co-operative societies' laws and regulations. Furthermore, trade liberalization has also enabled some co-operative societies to re-think their operational relationships in the context of establishing some symbiotic (mutualism) relationships among themselves in the so-called integrated co-operatives model (ICM).

The ICM is the product of the co-operative practitioners rethinking the way to complement one another in terms of services they offer after ceasing the government and donor supports that were earlier extended to co-operative societies before the liberalization era. Under the ICM, some AMCOS have been offering services to SACCOS in terms of storage facilities (warehousing) while such SACCOS have been extending loans to the AMCOS members to enable the acquisition of agro-inputs, land preparation, packaging, and transporting of harvested crops to the warehouses. The SACCOS have also been extending loans to AMCOS members while their crops in store are awaiting good prices, especially after harvesting seasons when in most cases crops fetch low prices. An example of ICM arrangement in Tanzania is found in the Mbinga District in the Ruvuma region between Kimuli AMCOS and Muungano SACCOS.

The other thing to note in Tanzania is that, with many co-operative societies hit hard by trade liberalization forces, the government of Tanzania has taken numerous steps including establishing, elevating, and strengthening some organizations specifically mandated at overseeing the co-operatives growth and development. The organizations include the Tanzania Co-operative Development Commission (TCDC), the Moshi Co-operative University (MoCU), and the Co-operatives Audit and Supervision Audit (COASCO), to name a few. The establishment of TCDC in 2013 and the elevation of the former Co-operative College of Moshi into a university status (MoCU) in 2014 has brought new impetus into the co-operative sector in Tanzania in terms of expectations to supply well-trained researchers, think tanks, and skilled co-operative practitioners that are likely to be productive change agents.

While MoCU is accredited with producing a skilled co-operatives workforce at various cadres ranging from certificate to post-graduate levels, TCDC is more focused on making use of such expertise to advance co-operatives development in the country. COASCO on the other hand undertakes co-operative societies auditing and supervision roles. It is expected that with such organizations' technical and professional blend in place, co-operative societies are likely to benefit from them in various dimensions.

There are also those who believe that, in the current globalization era, there are minimal chances for co-operative societies especially in poor countries such as Tanzania to be able to overcome the liberalized capitalist marketing challenges and emerge as winners. This is because, there are fears that when co-operatives decide to fiercely engage in the liberalization marketing opportunities struggle, they have in most cases been on the losing side. Co-operative societies and other stakeholders including the government therefore should strive to reduce and/or eliminate such fears especially those which are taking too long to be addressed such as too long and endless litigations that have resulted in members losing hope in co-operatives. Numerous efforts may be undertaken by the government to address this including establishing co-operatives legal clinic at MoCU to make use of the available co-operatives legal gurus to deal with co-operatives cases.

This can be done at a low cost because such legal experts may be working to assist co-operative societies in Tanzania as part of their duties or else they may do it at a low and/or subsidized cost as compared to other legal service providers who are in most cases expensive for most of the co-operatives to afford. The government may also decide to establish a special legal window that specifically focuses on co-operative matters only similar to the famous corruption litigation unit famously known in Swahili as the *mahakama ya mafisadi* which was established some few years back during the late President John Magufuli's era. Nevertheless, the operation and intended outputs from such litigation unit especially on grand corruption matters have remained questionable. Despite numerous co-operative societies' hope to excel in the trade liberalization context, it is worth noting that, unfortunately, the recorded hopes are not evenly distributed among all co-operative societies in Tanzania. The recorded achievements are only context-specific, occurring in a small segment or portion of the many dormant and/or almost dead co-operative societies in Tanzania. Efforts therefore should be taken to ensure sharing and learning from the best practices occurring in some of such few co-operatives to the many co-operatives that are still struggling, idle, or even almost dead. If those few organizations have emerged to be winners in the liberalization era, there is no doubt that, by learning from them, those others can do the same and/or even better.

7.7.2 Hopes Offered by Climate Change to Co-operatives in Tanzania

Studies have indicated that the winners from climate change will include the middle and high-latitude regions, whereas losers will include marginal lands in Africa and countries with low-lying coastal zones (McCarthy et al., 2001). Despite this general observation, it should be borne in mind that some climate change effects are context-specific, and thus some regions, individuals, or organizations will experience its effects differently. This means that some climate change adaptation and mitigation strategies are likely to be context-specific as well. It is from this recognition that co-operative societies in Tanzania and elsewhere in developing countries are considered the key smallholder farmers' organizations better suited to addressing the climate change challenges than other farmer organizations. One of the areas where co-operative societies in Tanzania may benefit from climate change is carbon trading. A study by Bamanyisa (2019) in Tanzania established that co-operative societies fit the requirements of the carbon market by bundling or stacking carbon credits into bigger volumes eligible for trading. Therefore, the co-operative marketing approach through stacking carbon credits makes smallholder farmers eligible for carbon projects and therefore can earn revenue from both carbon credits and agricultural produce. This implies smallholder farmers organized in co-operatives stand a chance to engage in carbon trading as compared to individual farmers who are in most cases ineligible for the same due to their relatively small land sizes and hence smaller carbon credits.

The other aspect is that it is widely acknowledged that climate change will result in the loss of some habitats and biodiversities, and increased crop pests and diseases among the other adverse effects. Given this scenario, co-operative societies stand a better chance of organizing smallholder farmers towards appropriate adaptation and mitigation strategies such as climate change awareness campaigns, and cultivation of climate sensitive and/or alternative crops, among others. This is possible because of their wide coverage, values, and principles inherent in co-operative societies as compared to the other farmer organizations. Moreover, co-operative societies have been efficient tools in representing smallholder farmers in various national and international agendas. In Tanzania for example, co-operatives have been important tools for influencing national policies including important political agendas during and after elections. Given this necessity, co-operatives are and remain important tools for representing smallholder farmers in Tanzania and elsewhere at various national and international climate change fora aiming at strategizing on best approaches to deal with the climate change challenges.

7.8 TRADE LIBERALIZATION AND CLIMATE CHANGE FEARS TO TANZANIAN CO-OPERATIVES

7.8.1 Trade Liberalization Fears on Co-operatives Discourse

Many co-operative stakeholders acknowledge that trade liberalization had come at a time when many co-operative societies were not ready to get into the competitive race. What is not said in such an argument is, however, how long co-operative societies could have waited to become ready to enter into trade liberalization. The answer may not be easy because trade liberalization came to Tanzania at a time when most co-operative societies were very weak: lacking and/ or with inadequate working capital, declining crop prices coupled with disintegrating international markets, poor co-operatives governance, embezzlement, and a general declining of members trust and in a time when co-operatives prospects for success were enormously trending down. It is worth noting also that, significant trade liberalization forces were externally imposed on co-operatives giving them little or no room to negotiate its arrival. Given such circumstances fears were inevitable. Wanyama (2006) indicated that the liberalization of the economy saw many observers and analysts fear that the monopolistic tendencies that had been embedded in the co-operative movement over the years would not enable the co-operatives to withstand the competitive market.

This was therefore translated into the so-called lack of readiness by co-operative societies to compete. However, these fears were not common to all of the co-operatives as some were able to disapprove and overcome the fears in the liberalization era. Apart from the lack of co-operative societies' readiness to compete, while trade liberalization through private buyers challenged agricultural marketing co-operative societies by offering higher prices than those offered by co-operatives, most co-operatives found themselves unable to compete by offering similar and/or more attractive prices. Msonganzila (2013) found that, at that time, AMCOS were not prepared to overcome price challenges and therefore were not able to offer competitive prices or keenly priced farm inputs to their members. As a result, some co-operative societies' members started selling their cash crops to private buyers. This in turn reduced the quantities of crops sold through co-operatives. Since then, private buyers continued to offer unchecked competition to co-operative societies because since they are profit-oriented, their main interest has been on cash crop buying rather than overseeing the production, quality assurance, and marketing of such crops which were the sole responsibility of co-operatives before trade liberalization.

This implies that private buyers have partly contributed to the declining co-operative societies and smallholder farmers' income and crop productivity per unit area since little attention was given to productivity. It is also worth noting that, despite the fact that private buyers offered higher prices than co-operative societies, such prices were frequently fluctuating usually set high at the beginning of the crop seasons and low thereafter. Likewise, the prices offered were just slightly higher (between TZS 20–100 per kilogram) than those offered by co-operatives and usually set soon after co-operative unions announce their prices so that they may indulge co-operative societies members to sell to them (Msonganzila, 2013). By law, the state must protect its citizens and economy, the question as to why the co-operatives be the first to indicate crop prices signifies government weakness in the areas of price control and regulation.

For all of the practical purposes, trade liberalization simply means survival of the fittest. Thus, the advent of trade liberalization in Tanzania witnessed many weak co-operative societies and organizations at various levels collapsing and/or totally fading away from the business. The country has witnessed the collapse of the once labeled co-operative giants such as the famous Rift Valley Co-operative Union (RIVACU), the Nyanza Co-operative Union (NCU), the Karagwe Co-operative Union (KCU), the KNCU, and others disintegrating. It has also witnessed the inability of many co-operative societies to manage and revive the former co-operative societies' projects such as tractor projects, ginneries, factories, transport, estates, agro-inputs supply, and others.

Apart from the collapse of the co-operative giants, trade liberalization has also led to the fall of some key financial institutions that were predominantly co-operative-based. Such institutions including co-operative banks, e.g., the famous Co-operative Bank of Tanganyika (see sub-section 7.1.3.1.1), which died because partly because they were dependent on the government for subsidies to operate. The removal of subsidies as one of SAPs conditions left such institutions incapable of attending co-operative societies. Others including the Co-operative and Rural Development Bank were privatized so that it can be rescued from collapsing and/or dying. Few co-operative banks that were established by co-operative members as a means to overcome the liberalization challenges had/have continuously remained weak in terms of capital formation to the extent of being deregistered by the Bank of Tanzania (BoT). Likewise, the banks have been suffering from a lack of carrier of professional bankers, boards, and management. They have/had also been unable to extend their services beyond the main office, e.g., the former Kagera Farmers' Co-operative Bank Ltd had no branch apart from its headquarters.

The same applies to Kilimanjaro Co-operative Bank Ltd (KCBL). Recently, however, in the year 2018, Kagera Farmers' Co-operative Bank Limited was deregistered by the BoT while Kilimanjaro Co-operative Bank Ltd (KCBL Ltd) managed to survive the deregistration hammer after a stiff toiling by co-operative stakeholders to raise the required minimum capital. Since then, the co-operative sector has mainly been depending on private financial providers including private banks to finance some of its operations. There are concerns, however, that private financial providers, e.g., banks, credit companies, and others usually charge extremely high-interest rates such that they are in most cases not in favor of borrowers including co-operative societies.

7.8.2 CLIMATE CHANGE FEARS ON CO-OPERATIVES DISCOURSE

There is no single nation that can claim to be entirely safe from climate change effects. Thus, all nations may lose in absolute terms but the nations that lose less would be the relative winners. Co-operative societies particularly agricultural-based co-operatives as with other smallholder farmers in Tanzania and elsewhere in developing countries are all subject to the fears and effects of climate change. The co-operative societies in Tanzania therefore are all subjected to the realities and fears of increased drought conditions and unpredictable rain patterns. This by itself is likely to affect the co-operatives members' income negatively and hence their inability to fully participate in their core co-operative activities such as farming activities, savings, and credit financial services, etc. Moreover, the same co-operatives are faced with fears of increased climate-related disasters such as flood incidences, pests and disease outbreaks, extreme temperatures, rising sea levels, declining membership, etc. It is worth noting that most co-operative societies lack resources to address climate change effects and are therefore unable to quickly adapt and implement available mitigation strategies. There is an obvious fear that some co-operative societies will shrink and/or collapse as a result of their failure to implement appropriate strategies to deal with climate change challenges. Co-operative societies therefore are likely to register less revenues, membership, and other co-operative benefits as a result of climate change effects. This is because some of the negative effects of climate change such as loss of biodiversity, increased pests and diseases, drought, and the others which are likely to face poor countries, especially smallholder farmers, will equally affect the co-operative societies as they are all lacking sufficient coping and mitigation strategies.

7.9 OVERCOMING CO-OPERATIVE THREATS AND CHALLENGES UNDER THE FREE MARKET ECONOMY AND CLIMATE CHANGE

The co-operative societies' threats and challenges regarding trade liberalization and climate change are real and one can hardly claim their inexistence. Such fears and challenges have been destabilizing co-operative societies operations and their impact on co-operatives is likely to persist. It is therefore upon the co-operative societies' members, leaders, and all other key stakeholders such as the government of Tanzania, TFC, SCCULT, ICA, and the private sector, to mention but a few, to all work together to overcome such realities. This is because, as most of the fears and challenges facing co-operatives are global, it is nearly impossible to effectively address them in isolation. Nevertheless, the measures that need to be taken may vary depending on the context, type, and nature of co-operative society. This is to say, any efforts/measures to address the co-operative fears and/or challenges in Tanzania should be customized according to the context, type, and nature of the co-operative society at hand. Numerous factors including the co-operatives core mandates, type of products or services offered, prevailing markets and marketing information access, services or products, consumer preferences, leadership skills and techniques, etc., may all contribute to the success or failure of co-operative societies. It is worth noting that, as far as most co-operative societies in Tanzania, particularly AMCOS, have continued to rely on the external markets for their produce by largely producing what they do not consume, under such circumstances the liberalization fears and challenges identified above are likely to continue haunting them.

Experience has shown that a reliance on external markets for raw cash crops has been so precarious and unpredictable as market holders have continued to switch prices on and off resulting in price fluctuations which in most cases are not in favor of co-operative societies and smallholder farmers. In this section, therefore, a few key remedies for redressing the fears and challenges facing co-operative societies in Tanzania are suggested. They include co-operatives investment into member training and empowerment, embracing co-operatives entrepreneurship and innovation, and re-thinking co-operative societies' products and/or services to customers and markets.

7.9.1 THE CO-OPERATIVES MEMBER TRAINING AND EMPOWERMENT

Co-operatives member training and empowerment is a key component in building co-operative societies members with the capability to take charge of their own fate. This is to say that, co-operative societies' members are responsible for positioning themselves in the driver's seat and taking full charge in terms of co-operative operations. Taking full charge in the driver's seat implies that the driver should be well-versed with the measures necessary to ensure the co-operative truck remains focused. In other words, it is the driver's role to always bear in mind that the road to co-operatives development is never smooth and thus the only way to remain focused is to learn and understand the rough parts of the road. This will be possible only when the driver is capacitated with important training, that is, in terms of appropriate skills, attitudes, and knowledge. It is widely acknowledged that the key tool for fighting the main development enemies in Tanzania, which are poverty, diseases, ignorance, and corruption, is education. Thus, co-operative education is crucial in enabling co-operative societies' members to acquire the necessary skills and knowledge for analyzing their own situation including trade liberalization and climate change challenges and obstacles, and finding their own solutions to overcome them. It is also crucial to enable co-operative members, leaders, and staff to better understand their responsibilities and structures that support them as well as exert proper control on their performance in a participatory and democratic way. Furthermore, training and education are important in enabling the co-operative members as owners of their organizations to oversee and supervise their leaders, e.g., board members, committee members, and staff to manage their co-operative societies professionally, profitably, and competitively. Equally, education and training are necessary to enable co-operative sustainability under liberalized competitive markets coupled with climate change challenges circumstances.

This is because, as a result of globalization that has made the world a small village, and climate change that has changed the way of our lives, education, and training are important since it facilitates co-operative societies make use of science and technologies in almost all aspects of its operational nodes such as production, marketing information, e.g., during market research using information science and technologies (ICTs) such as the Internet, online marketing, inputs procurement, etc. URT (2005) emphasized that co-operative societies members' education and training are very crucial because they increase the bargaining and negotiation capacity of co-operatives with regard to finance and credit, transport, and processing alternatives. It also enhances knowledge translation and transparency regarding co-operative business information, production, quality, and price information and develops the capacity to generate sustainable business solutions through increased member participation and decision-making. All of these are necessary qualities that can enable co-operative societies to face and overcome the trade liberalization and climate change hurdles. It is therefore upon co-operative societies and all other co-operative stakeholders to work on them and genuinely invest in enabling co-operative members' education and training. There is a saying that "the roots of education are bitter but its fruits are sweet" and another that says "no pain, no gain".

The two statements remind us that any investment in education of whatever kind must be enduring, and in that case, costful. Therefore, if we really want to see co-operative societies' growth and development in Tanzania, investing in co-operative societies' member education and training is inevitable. To make this practical, prioritization in the allocation of resources such as financial, human, technological, and physical resources for co-operative societies' member education is necessary. While there is a say that "charity begins at home"; co-operative societies in Tanzania should make sure they set aside necessary resources including some financing as per the legal requirements which demand each co-operative society to allocate some funds for education and training from their annual income (URT, 2013). This is because experience has shown that many co-operative societies have not been allocating such funds for various reasons including inadequate funds or revenues.

It is also worth noting that, co-operative education and training should not only focus on conventional training such as members' rights and responsibilities, but rather it should be research-based. The problem with the current education and training provided to co-operative societies in Tanzania is that it is rarely backed up with comprehensive research. As a result, in most cases, the types of training that have been offered are top-down, mainly flowing from the so-called co-operative experts such as university trainers, co-operative officers, community development officers, and other providers to the co-operative members, leaders, and staff. Therefore, co-operative societies and other stakeholders should work to collaboratively research and invest in translating the research outputs into practice so that they can benefit them. It is important to notice that, in researching and establishing co-operative societies' education and training needs, co-operative societies should be part and parcel of the process and they should always be enabled to take full charge of all matters concerning them. Any support in any form should facilitate or complement co-operative societies' initiatives and not otherwise.

7.9.2 The Co-operatives Innovation and Entrepreneurship

Co-operatives innovation and entrepreneurship are considered to be important factors towards co-operatives growth and development in today's globalized and climate change–influenced world. Thus, the environments in which current co-operative societies in Tanzania and elsewhere operate highly demand them to be innovative and entrepreneurial. Throughout emerging and advancing economies innovation and/or entrepreneurship is recognized as a major source of improved productivity, competitiveness, and growth (OECD, 2009). Innovation and/or entrepreneurship is considered a tool for making co-operatives economically viable and service-oriented since innovative and/or entrepreneurial practices make use of new methods, ideas, or skills to ensure the production of quality consumer goods, joint use of capital, joint use of facilities, and carefully planned and well-executed marketing. The trade liberalization has witnessed the removal of a monopoly status and greater freedom for the members of co-operatives that have in turn set into motion large-scale re-organization among co-operatives and support organizations. This repositioning is more critical for co-operative organizational structures, which need to address new member needs rather than their predominant orientation towards the provision of services along the marketing chain of the crops (Sizya, 2001).

Besides structural re-organization of the co-operative movement, liberalization, and the climate change influences have also triggered the diversification of some co-operative ventures to enable them to survive market and climate change–related forces while offering competitive services to the members. The loss of monopoly status, coupled with the business-oriented demands of the market, is increasingly seeing co-operatives redesign and innovate their activities competitively (Wambua, 2014). As a result, some co-operative societies in Tanzania have managed to make use of innovative and entrepreneurial techniques such as diversification of ventures to spread the risks and/or accrue benefits across diverse areas of operations unlike in the past when diversification was limited. Other co-operative societies have undergone innovative transformation to form structures that accommodate diverse co-operative societies members' needs such as integrated co-operative model as well as venturing into many other activities such as alternative crop production, climate-sensitive organic farming, and mobile money banking services. Wambua (2014) indicated that, apart from deepening financial resources, especially in terms of access to credit to individuals and firms, some co-operative societies have been innovative in terms of technology use by expanding and providing more services to its members, such as mobile money services which are safe and easy to access, offering new payment products as well as access to information (Wambua, 2014).

The success recorded from such efforts provides hope that co-operative societies' growth and development can be attained through embracing innovation and/or entrepreneurship as a key competitive element. Thus, various co-operative fears and challenges pertaining to the shift from monopolistic productivity to liberalized productivity, co-operatives financing, climate change, technologies, and others can all find a solution upon embracing co-operatives innovation and entrepreneurship. This is possible by co-operative societies and other supporting organizations thinking out of the box by operating beyond the traditional co-operative societies' boundaries by daring to mobilize and establish new co-operative ventures, enabling product or service diversifications, etc.

7.9.3 RE-THINKING CO-OPERATIVES PRODUCTS AND/OR SERVICES CUSTOMERS AND MARKETS

Since prior to Tanganyika's independence to date, co-operative societies, particularly AMCOS, have been dealing with the production and marketing of cash crops. Most of such cash crops have been mainly exported as raw materials with very minimal or no value addition. It has been also a long-time practice and tradition, either by default or by design that, most crops particularly coffee, cotton, tea, cashew nuts, and tobacco which are the major crops marketed through AMCOS, are largely not consumed by either the farmers who produce them or by other Tanzanians. This calls for new thinking outside of the box as most co-operatives have been largely depending on external markets, mainly European and Asian countries as their main markets for their produce. Unfortunately, for many years, prices of such crops have been fluctuating tremendously to the detriment of the producers, co-operative societies inclusive. As a result, co-operative societies have found themselves accepting prices and income far below the resources and/or inputs invested. The advent of trade liberalization and climate change influence in Tanzania has not reversed this situation and as such it is likely to exacerbate it. This therefore has remained one of the co-operative societies' fears and challenges. This is because, with trade liberalization and climate change influence, co-operative societies' capacity in terms of crops value addition, processing, packaging, venturing into new and sustainable markets, etc. was expected to advance, but to the contrary, little has been achieved in that direction. To address such fears and challenges, co-operative societies and all other key stakeholders particularly the government of Tanzania should work to mobilize and organize a pool of local cash crop markets and/or consumers. The stakeholders should work to mobilize the ever-growing population of youth and other categories of Tanzanians to realize the health and economic benefits of consuming our own crops such as cashew nuts, coffee, cotton oil, cotton clothes, tea, etc.

The stakeholders should establish special programs in public and private institutions such as primary schools, secondary schools, colleges, universities, workplaces, etc., and through mass and social media enumerating the benefits of such products. Such programs may include some special offers and promotion strategies on product testing such as coffee and its products/types, free cashew nuts testing and supply, etc. By so doing Tanzanians, especially the young generation will find themselves acclimatized with their own products while building interest in consumption whereby with time the number of buyers will grow and by so doing decline external market reliance which is in most cases unpredictable. The program may extend beyond Tanzanian boundaries to take advantage of the growing population in the existing regional blocks such as the East African Community (EAC) and the other potential regional blocks' co-operation. Vivid examples of the same practice in other parts of the world do exist. Ethiopia for instance, is the largest producer of coffee in Africa with significant quantities of its coffee being domestically consumed. India is the largest producer of tea in the world. It also consumes significant quantities of its tea domestically.

Nevertheless, the proposed program should go hand in hand with governments' determination and commitment to investing in value-addition industries so that co-operative societies can take advantage of such industries. This is to say that, the problem of co-operative societies' regular crop price fluctuation may be effectively addressed by co-operative societies re-thinking of mobilizing and organizing Tanzanians to understand the necessity and benefits inherent within their own cash crops and/or other goods and services. Given their unique features and founding principles and values, co-operatives are capable organizations at ensuring and maintaining members' business competitive edge even in highly challenging (i.e., globalized and climate change influenced) situations. Co-operatives can be competitive even in highly competitive environments due to the fact that they have a greater potential to mobilize resources through co-operative networks, credit unions, multi-stakeholder co-operatives, and facilitative legal environments. Morris (2009) asserted that "competitive markets and co-operative organizational forms are all quite compatible; ceteris paribus, the case can be made that a world without co-operatives is, at a minimum, one that is poorer". It is from this argument that we can say co-operative societies have the capacity to invest in different situations. For that matter, the government in collaboration with other stakeholders including the private sector should encourage more investment in the cash crops value chain.

Instead of allowing the export of raw cash crops, investors should be encouraged to invest in processing industries which will create more employment for Tanzanians as well as increase the pool of industrial goods produced in Tanzania. Given that there are different forms of partnerships, foreign investors or local private investors can enter into partnerships with the co-operative societies in building and operating processing industries. Moreover, the government of Tanzania can also enter into a public–private partnership with the co-operative societies as one of the means to strengthen the co-operative societies' competitive capacity in the climate change strained situations and free market economy. While the above aspects need to be genuinely addressed by the key co-operative stakeholders such as the central government, MoCU, TCDC, COASCO, and others, the same should be in a position to re-assess themselves as to whether they do/conduct annual or periodic evaluation on their role in supporting co-operatives. They should also ask themselves what are tangible innovations, research outputs, or entrepreneurial achievements they have managed to design and disseminate to co-operatives since trade liberalization.

7.10 CONCLUSIONS

The climate change and free-market economy forces are real and there to stay with co-operative societies and all other organizations in Tanzania and elsewhere. Since it is a reality that we cannot avoid, the best option is for co-operative societies and all other stakeholders to design and organize the best and most beneficial alternatives to overcome them. There is no single remedy to all fears/ threats and challenges of climate change and trade liberalization that is suitable to all co-operative societies in Tanzania since some of such challenges may be context-oriented. Thus, it is upon the co-operative societies in Tanzania and all other stakeholders to think and design the best options that can work and become productive based on the context they are operating. It is time now for co-operative societies and the other stakeholders and all co-operatives well-wishers to act appropriately and promptly. The appropriate time is now, let us all act; let us all play our party.

REFERENCES

Banturaki, J. (2000). *Co-operatives and Poverty Alleviation.* TEMA Publishers, Dar es Salaam, 180pp.

Barton, D. G., Schroeder, T. C., and Featherstone, A. M. (1993). Evaluating the feasibility of local cooperative consolidations: A case study. *Agribusiness,* 9(3), 281–294.

Bergeron, J. (2015). Exploring the dimensions of women empowerment in agricultural co-operatives in Kilimanjaro, Tanzania: The case of Mamsera AMCOS. Master of Arts Degree Dissertation in Public Policy and Administration, Carleton University, Ottawa, 188pp.

Bibby, A. (2006). *World of Work Magazine,* No. 58, December 2006. The International Labour Office, Geneva, 8–11.

Chambo, S. (2018). Importance of autonomous cooperatives. Habari Leo [http//www.habarileo.co.tz] site visited on 29 June 2018.

CICOPA. (2017). Cooperatives and employment: Second global report. ILO Office, Geneva, 60.

Develtere, P., Ignace, P., and Wanyama, F. (Eds.). (2008). Cooperating out of poverty: The renaissance of the African cooperative movement. ILO, World Bank Institute, 57.

FAO. (2012). Towards self-sustaining and market-oriented producer organizations: Issue Brief Series, Rome.

FAO. (2013). Climate-smart agriculture sourcebook. Food and Agriculture Organization of the United Nations Report, Rome.

Grineski, S., Collins, T., McDonald, Y., Aldouri, R., Aboargob, F., Eldeb, A., Aguilar, M., and Gibbon, P. (2001). Cooperative cotton marketing, liberalization and civil society in Tanzania. *Journal of Agrarian Change,* 1(3), 389–439.

Hill, C., and Jones, G. (2007). *Strategic Management Theory: A Integrated Strategic Approach.* Seventh Edition. Houghton Mifflin Company, New York, 202pp.

ICA. (2017). Co-ops for 2030: A movement achieving sustainable development for all. https://www.ica.coop/en/media/library/research-and-reviews/coops-2030-movement-achieving-sustainable-development-all.

ILO. (2012). The informal economy and decent work: A policy resource guide-supporting transition to formality. Geneva: ILO Office. 507pp.

ILO. (2022). Guide to Recommendation 189: Job creation in small and medium sized enterprises recommendation. Geneva: ILO Office. 42pp.

International Co-operative Alliance. (1995). Macpherson, Ian. *Co-operative Principles for the 21st Century.* ICA, Brussels.

Karki, B. B. (2006). Strategic planning in co-operative sector: A study of dairy enterprises. *Journal of Nepalese Business Studies,* 2(1), 72–80.

Leichenko, R., and O'Brien, K. (2008). *Environmental Change and Globalization: Double Exposures.* Online edition. Oxford Academic, New York. https://doi.org/10.1093/acprof:oso/9780195177329.001.0001. Accessed 19 January 2024.

Lyimo, F. (2012). *Rural Co-operation in the Cooperative Movement in Tanzania.* Mkuki and Nyota Publishers, Dar es Salaam.

Maghimbi, S. (1992). Abolition of peasant's cooperatives and crisis in rural economy in Tanzania. In Fosters, G., and Maghimbi, S. (Eds.), *Tanzania Peasantry: Economy in Crisis.* Grower Publishing Company Ltd, Avebury, 81–100.

Majurin, E. (2012). *How Women Fair in East African Cooperatives: The Case of Kenya, Tanzania and Uganda.* International Labour Office, Dar es Salaam.

McCarthy, J. J., Canziani, O. F., Leary, N. A., Dokken, D. J., and White, K. S. (2001). *Climate Change 2001: Impacts, Adaptation, and Vulnerability.* Cambridge University Press, Cambridge. https://doi.org/10.1002/joc.775.

Mhando, D. (2014). Conflict as motivation for change: The case of coffee farmers' co-operatives in Moshi, Tanzania. *African Study Monographs,* 50, 137–154.

Morris, A. (2009). Co-operatives, history and theories of cooperatives. School of Economics and Finance, Victoria University of Wellington, Wellington, 14pp.

Msonganzila, M. (2013). Gender, cooperative organization and participatory intervention in rural Tanzania. Published PhD Thesis, Wageningen University, The Netherlands, 224.

Nyang'au, J. O., Mohamed, J. H., Mango, N., Makate, C., and Wangeci, A. N. (2021). Smallholder farmers' perception of climate change and adoption of climate smart agriculture practices in Masaba South Sub-county, Kisii, Kenya. *Heliyon*, 7(4), e06789. https://doi.org/10.1016/j.heliyon.2021.e06789.

O'Brien, K., and Leichenko, R. (2000). Double exposure: Assessing the impacts of climate change within the context of economic globalization. *Global Environmental Change*, 10(3), 221–232.

OECD (Organization for Economic Co-operation & Development). (2009). Growing prosperity. Agriculture, economic renewal & development. Draft Document from the Experts Meeting "Innovating Out of Poverty," OECD, April 6–7, Paris.

Saul, J. (2002). Poverty alleviation and the revolutionary-socialist imperative: Learning from Nyerere's Tanzania. *Journal of Global Policy Analysis*, 57(2), 193–207.

Sizya, M. J. (2001). The role co-operatives play in poverty reduction in Tanzania. Paper Presented at the United Nations in observance of the International Day for the Eradication of Poverty on 17 October 2001, 14pp.

Stern, N. (2010). Climate: What you need to know. *New York Review of Books*, Times Books, 57(11), 253pp.

Sumelius, J. (2013). Co-operatives as tool for poverty reduction and promoting business in Tanzania. University of Helsinki, Helsinki.

UNDESA. (2014). Measuring the size and scope of the cooperative –Results of the 2014 global census on co-operatives. United Nations, 8pp.

UNEP. (2007). Climate change: Processes, characteristics and threats. UNEP/Grid-Arendal maps and graphics library, United Nation.

UNFCCC. (2007). Climate change: Impact, vulnerabilities and adaptations in developing countries. Climate Change Secretariat, Bonn, 64pp.

United Republic of Tanzania, URT. (1997). Cultural policy: Policy statement of 1997. Dar es Salaam.

United Republic of Tanzania, URT. (2012). National baseline survey report for micro, small, and medium enterprises in Tanzania. Ministry of Trade and Industry, Dar es Salaam.

URT. (2005). National strategy for growth and poverty reduction. Ministry of Planning, Dar es Salaam.

Vicari, S., and De Muro, P. (2012). The co-operative as an institution for human development. Working Paper No. 156, Roma Tre University, Italy.

Wambua, J. M. (2014). Factors influencing dairy productivity in Machakos County: A case of Wamunyu dairy Farmers Co-operative Society. Master of Arts in Project Planning and Management, University of Nairobi, Nairobi. http://hdl.handle.net/11295/74285.

Wanyama, F. O. (2006). Interfacing the state and the voluntary sector for african development: Lessons from Kenya. *Africa Development*, 31(4), 73–103.

Wanyama, F. O. (2014). Co-operatives and the sustainable development goals-A contribution to the post 2015 development debate. ILO Office, Geneva.

World Employment and Social Outlook: Trends 2022. (2022). International Labour Office, Geneva.

Zamani Nuri, A., Farzaneh, M. R., Fakhri, M., Dokoohaki, H., Eslamian, S., and Khordadi, M. J. (2013). Assessment of future climate classification on Urmia Lake basin under effect of climate change. *International Journal of Hydrology Science and Technology*, 3(2), 128–140.

8 Assessing Environmental Causes and Disaster Risk Management of Erosion in Agulu–Nanka Community

Victor Otti, Saeid Eslamian, and Mir Bintul Huda

8.1 INTRODUCTION

Gully erosion is defined as a relatively deep vertical-walled channel recently formed within a valley where no well-defined channel previously existed, Akpokodje et al. (2010). Active erosion occurs when the erosion is actively moving up in the landscape by head cut mitigation. Gully erosion is a process whereby the surface layer of the soil is detached and carried by agents of denudation and a cover in the soil is exposed leaving a topographic roughness on the resulting landscapes. Gully erosion occurs when soil is eaten away by natural agencies and is simply the removal of soil, including plant nutrients, from the land surface by various agents of denudation (Ofomata, 2000). Erosion could be a smoothing or leveling process with soil or rock particles being carried, rolled, or washed away by the force of gravity, altering the Earth's landscape and constituting a global environmental problems.

Onyeagocha (1986) described erosion as an accelerated phenomenon which resulted from the movement of soil by water, wind, or other agents and the deposition of such detached soil elsewhere. On the other hand, Onwuka (2008) studied erosion as inclusive of soil forming as well as soil eroding process which maintains soil in favorable balance/deterioration and loss of soil because of man's activities. Egboka (1993) asserted that the state of Anambra is the most erosion-devastated state in Nigeria and adjudged as the most erosion-pruned and erosion-devastated landscape in the world.

Anambra state suffers from a large amount of erosion sites and is threatened by a physical environment that seems ready to slip into nothingness, Nwajide and Reijers (1986).

Erosion ogre is a progressive albatross in the southeast, which some geologists attributed to the unsavory development of war activities, such as the indiscriminate digging of trenches by soldiers as well as the detonation of explosives and bombs in the southeast in the Biafan war, where war actively raged, Igbokwe et al. (2008).

Erosion has caused horrendous and incalculable havoc in the southeast of Nigeria, although various efforts made by the communities affected to check the menace amounted to naught in the face of a rampaging landslide. Some efforts made by the state government to contain this menace through practical measures and enlightenment campaigns on the indigenous population in erosion-ravaged localities yielded nothing (Obiekezie et al., 2002). Therefore, erosion is one of the surface sculptures of the Earth's landscape and constitutes one of the global environmental problems. More so, gully erosion is the most impressive and striking ogre and most serious mechanism of land degradation in the southeastern part of Nigeria and one of the major global environmental problems the country is facing.

Gully erosion within settlements requires attention because of the threat it poses to buildings and other structures that endanger human lives (Barrocu and Eslamian, 2022). Gully erosion in

DOI: 10.1201/9781003473398-10

Agulu–Nanka has posed numerous threats to the inhabitants of the area and has caused many residential buildings and worship centers to collapse, destroying road networks, and other infrastructure and degrading land for commercial and agricultural purposes, Igbozurike (1989).

Ezechi and Okagbue (1989) stated Agulu–Nanka communities are of the opinion that the gully erosion caused impairment of relationships when people left their homes and became refugees in the neighboring towns, which could lead to poor health conditions and an increase in crime rate. A total of 787 houses and 325 hectares of farmland belonging to the 567 affected households were gulped by the landslip because of gully erosion in the area as observed by Igbokwe et al. (2008).

The Agulu–Nanka erosion complex has posed a big challenge to foresting, companies, the local population, and the state in general, in terms of deforestation. The problem has defied all efforts to control it as the erosion is spreading such as wildfire to other neighborhoods such as Umuchima–Ekwulobia, Oko, and others. The landslide has become more potent and active, gulping everything in its course. People are retreating, abandoning homes livestock, economic trees, and farm lands. The devastating conditions have led to untold socio-economic problems.

Apart from the destruction of houses, farmlands, and economic trees, erosion has caused the siltation of rivers, and streams and the consequent loss of biodiversity and water supplies for domestic purposes, Albert et al. (2000).

There is also the destruction of aquatic life in the affected Odor River and in the Nanka and Agulu Lakes. The gullies are a nuisance, interrupting communication as well as despoiling farmlands. More so, there is the impoverishment of soil over wider areas resulting from deforestation and its replacement by open-to-tussocky grassland. It also led to the problem of finding alternative plots for those whose houses and farmlands had disappeared in the landslide and those whose lands were put out of cultivation as a result of conservation regulation passed by the local authority. If compensation and resettlement are to be made to the displace the source of such compensation presents a problem.

There is also a problem of population swelling in the neighborhood towns and villages because those lands have been swallowed by erosion, some trooping to other areas in search of new homes and means of sustaining life. The erosion has further hindered the infrastructural development of the area.

Erosion and flood, especially in both urban and rural areas have fast become the most reoccurring disaster in many communities in Anambra state. Many communities' infrastructure and farmland have been destroyed by these hazards.

Gully erosion is a single major process responsible for the loss of a vast amount of soil in the state as seen in the study of Akpokodje et al. (2010).

Nnodu (2005) observed that one kilometer of gully would produce 10,000 cubic meters of settlement per km^2 of land. He further emphasized that such happens for a gully aged 100 years the mean annual rate of erosion would be 1.5 tonnes per hectare per year.

Obidimma and Oluruntemi were of the view that Anambra state is besieged by serious environmental degradation, resulting in gully erosion due to very high intensive rainfall resulting in heavy runoff and soil loss. The problem has adversely affected agricultural productivity thus casting doubt on food scarcity in the zone. The ecological and social settings in the zone are often distorted sometimes leading to losses in human and material capital. Most often high torrential rainfall in the southeastern states of Nigeria creates an environment for catastrophic soil erosion in the region Ndubude (1982). Surely erosion is the greatest threat to the environmental setting of southeastern Nigeria and is gradually and constantly dissecting the landscape of the Agulu–Nanka area.

The aim is to access and determine the causes and effects of erosion in the study area as given below:

- Determine the surface run-off occurrences whenever there is excess water on the stop that cannot be absorbed into the soil.
- Ascertain infiltration due to soil compaction and increase in run-off.
- Determine the effect of run-off from an agricultural level that is greatness during the rainy seasons.

- Create awareness of exposure to erodible soil that has poor structure and lower organic matter.
- Encourage the protection of soil from raindrop impact by covering the soil with vegetation.
- Avoid the consideration of small fields into large ones which often result in large slop lengths of scouring.
- Avoid the effects of tillage operations that could cause potential soil erosion.

8.2   STUDY AREA

Agulu–Nanka lies between 7° and 7°30E and 6° and 5.30°N of the equator. The erosion complex itself lies about 95 km southwest of Enugu and about 52 km east of the Niger River (Figure 8.1). Although the site lies within the moist rainfall belt of southern Nigeria, the vegetation is predominantly

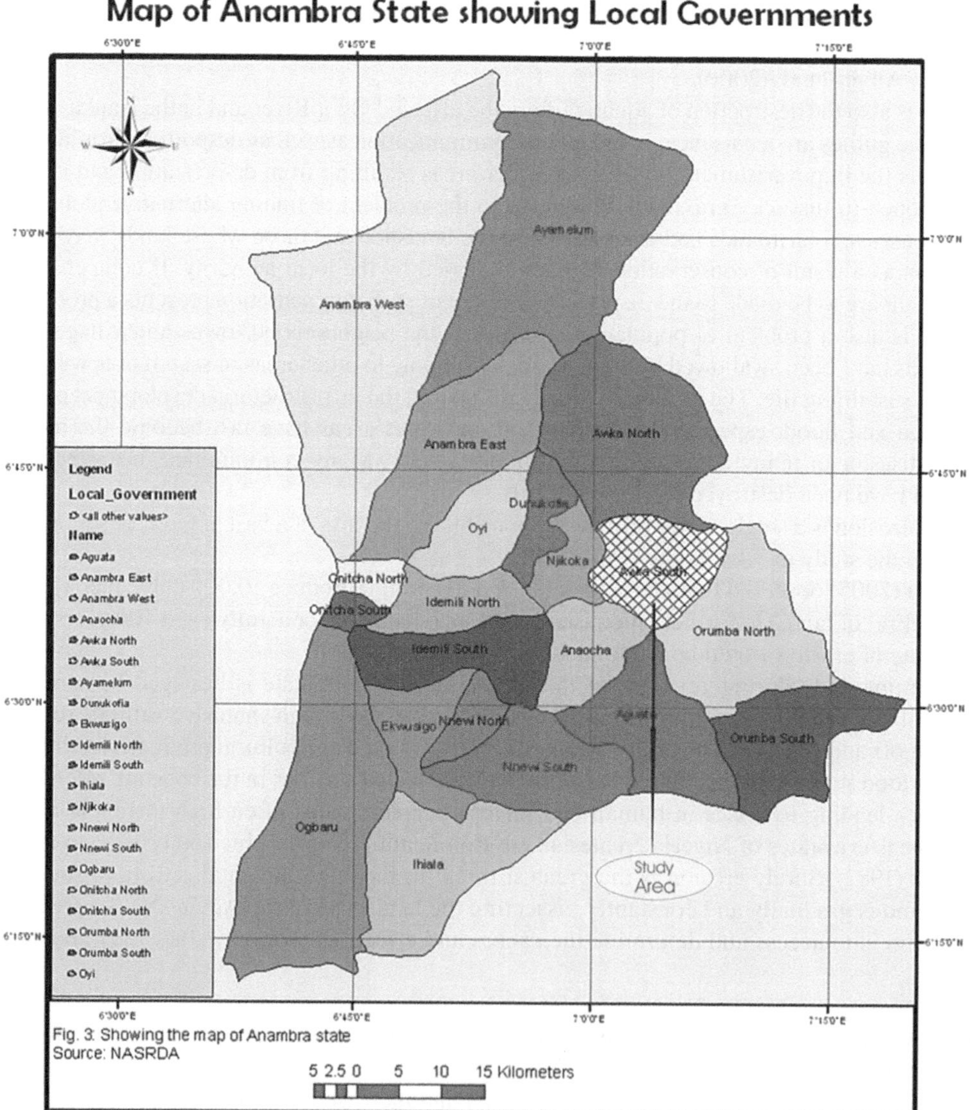

FIGURE 8.1   The study area.

derived from savanna and dominated by a continuous occurrence of oil palm trees (Okafor, 1986). The rainy season is between April and October. The soil of Agulu–Nanka consists of acid sand, laterite, and alluvial. The people of Agulu–Nanka area are occupationally farmers, businessmen, and traders. The agricultural crops cultivated include cassava, yam, maize, banana, plantain, and vegetables.

8.3 CAUSES OF GULLY EROSION

Gully erosion is caused by several factors working simultaneously or individually to detach, transport, and deposit soil particles in a different place other than where they were formed Nwajide and Hague (1979). Moreover, gully erosion can be caused in a number of ways having different mechanism modes and conditions of formation, some of which are directly related to the underlying geology on the surface geology and soil cover, Ezechi and Okagbue (1989) summarized the type of gully erosion with respect to their modes and conditions of formation, and common advance mechanism. The resultant effects of these phenomena are deep cuttings and ravines which dissect the entire Agulu–Nanka area. It is a well-established fact that a number of environmental factors as well as pedagogical parameters influence the extent of soil, erosion wherever it occurs globally (Eze, 2000). The study indicated that the nature of the underlying bed has a bearing on the initiation and propagation of gullies. Observations have also shown that gully erosion in Agulu–Nanka is more predominant in sedimentary terrains and perhaps in the basement/sediment contact areas like Agulu-Nanka.

This accounts for why its occurrence is more skewed Aguata–Orumba local government areas of Anambra state. The factors that cause erosion in Anambra state are guided by human factors known as anthropogenic factors. Actually, man has helped in reshaping and preserving the Earth's surface yet man has also helped in causing instability of equilibrium in the natural ecology and hence the rapid spread of environmental problems such as soil erosion.

Anthropogenic factors are technical factors comprising mainly land use and tillage methods, the choice and distribution of culture, and the nature of agro-technology. Vegetation clearance intensive harvesting and overgrazing leaving the soil bare are among the factors that encourage soil erosion. Soil compaction is caused by heavy-duty machinery which reduces the infiltration capacity of the soil thus promoting excessive water runoff and soil erosion rainfall and soil factors are the main agents that determine the extent of soil erosion hazard. The soil factors represented the soil erodibility which is also a product of geology and soil characteristics.

Anthropogenic influences are comprised of:

- Misuse of land.
- Poor farming systems contributed to the collapse of soil structure and encouraged accelerated runoff and soil loss due to erosion.
- Uncontrollable grazing caused by the nomads has resulted in deforestation of the land.
- Indiscriminate foot parts created on the landscape have helped incipient channels on the landscape to form the channels eventually metamorphose into gullies especially when they are not checked at inception.
- Road construction including uncontrolled infrastructural developments have contributed significantly to gully development. Some road networks under construction have been abandoned in the areas due to gully formation.
- Improper channeling of runoff water.
- Dumping of refuge on the waterways.
- Poor construction is evident by the absence of anticipated runoff in the designed roads.
- Roads were constructed without drainages to channel the runoff into the nearby stream or river.
- Haphazard erection of buildings on steep terrains and waterways.

In Anambra state most of the gullies took advantage of the loosely consolidated and friable rocks such as Ajalli sandstones. The cause of gully erosion with respect to the geology settings as suggested by the earlier studies was numerous (Ofomata, 2000). Moreover, some of the identified natural causes include:

- Tectonism and uplift
- Geotechnical properties of soil and others

Some of the geological, hydrogeological, geotechnical, and hydrogeo-chemical characteristics of the area and human activity have contributed to the gully development and growth. Major aquifers and aquitards from multiaquifer systems and heavy rainfall caused the rise in the water table. The increase in hydraulic head produced rapid flow rates that enhanced the gully process. Expansion and contraction of clayey soil and shales in the rainy and dry seasons respectively led to slumping and landslides.

8.4　IMPACT ASSESSMENT OF AGULU–NANKA EROSION

Agulu–Nanka gully erosion in contemporary is put as the greatest scourge ravaging the two communities. This gully erosion and landslides have terminal and cancerous ecological diseases that destroy within minutes or at most hours land formed over thousands of years. According to Egboka (2007), the problems that result from gully erosion in Nanka–Agulu were many and varied too. They included human, material, political, psychological, sociological, economic, and spiritual all rolled into one.

8.4.1　Physiological Impact Assessment

The physiological effect of erosion activities is the most immediate reaction to environmental stress. Some community members were induced to hypertension and high blood pressure even leading to loss of life. Some were full of anxiety, anguish and stress, despair, anger, and a life full of despondency.

The effect of erosion menace on the lives of people led to the loss of lives some either fell into the gullies and sustained various degrees of injury or died. Some people were reported drowned in some of the gully sites. In the past few years, some adults and children have been reported to lose their lives in a single event of gulling activities in both Nanka and Agulu towns in Orumba and Aguata local government areas. Also, thousands of people have been displaced and evacuated from their homes following the gully incident. The erosion in Agulu–Nanka and Oko communities in Anambra state has created deep gullies and wide craters, threatening to sweep away the homes of about 1825 families as the channels are continuously expanding at an alarming rate. This has caused the people, to live with palpable fear and psychological trauma whenever the rainy season commences thus justifying that the activities affect the people mentally. This erosion ogre brought the problem of overpopulation of the neighborhood towns and villages because their land has been swallowed by erosion and they trooped to other areas in search of new homes and greener pastures for sustainability of life.

8.4.2　Socio-cultural Impact Assessment

This impact caused the isolation of villages and towns. Gully erosion in these communities resulted in the separation of adjacent villages and towns which involved the collapse of box culverts linking them together, therefore the lives of the populace were severely affected as there was a scarcity of amenities. This had negative impacts on these communities since the facilities such as schools, hospitals, water supplies, and road networks shared by the affected neighboring communities have become inaccessible. Transportation of farm produce was also affected and this also led to a loss of agricultural products, especially the perishable ones. Traders who used to go to these communities for their trade were also cut off from their normal day-to-day business. Moreover, their dead bodies or corpses are being carried to faraway places for burial and even erosion has unearthed their sleeping heroes or ancestors from resting places (graves). Also, the destruction of social infrastructure caused impairment of relationships among families, churches, and schools.

Obviously, erosion affected the social standard of the communities, resulting in the increase of social vices that impacted adversely on some socio-cultural values of the people. Among these are youth restiveness, due to the non-availability of employment, which would have been available from farming, local industries, and small businesses. These also increased the level of poverty in these communities. Some important cultural ceremonies were denied or disrupted by the people because families were displaced and settled elsewhere.

8.4.3 ENVIRONMENTAL IMPACT ASSESSMENT

Erosion activities led to the destruction of aquatic life in the Odor River in the Nanka and Agulu Lake, despoiling farmland. It overburdened farmland resulting from deforestation and its replacement by open tussocky grassland. The aftermath of erosion activities in these communities brought the destruction of agricultural land, loss of ancestral homes and properties, and devastation of social networks. Agulu–Nanka erosion activities are a threat to vegetation, it has resulted in the loss of vegetation and wild animal species, as its continuous expansion encroached into these communities that were hitherto forest, leading to the falling of trees and exposure of more surface areas to gully activities. The erosion phenomenon if continues and remains unchecked might ultimately lead to climate changes locally.

The environmental impact affected most properties in this area. Several properties whose value could not be quantified accurately were destroyed and others were under threat by the menace especially houses and other properties located on the flood plain about 13 houses were lost in a single event of gully erosion at Agulu–Nanka areas in Anambra state alone as a result of erosion activities on separate note. The federal government discovered 21 gully sites in Agulu–Nanka, Ekwulobia, Oko, Awka, Nnewi, and others, all in Anambra state.

Apart from the untimely evacuation from these such as pipelines, utility cables roads, and houses also suffer from these hazardous events.

8.4.4 ECONOMIC IMPACT ASSESSMENT

Erosion displaced hundreds of families and rendered them homeless, and some lost their livestock and farmland. Money was lost due to the economic setback of these deserted (unsettled) families. The few people who were staying could not make any significant increase in their income.

Surely the movement of people away from their communities led to the loss of business, farmland, loss of farm produce, loss of properties, an increase in the price of goods and services, unemployment, and incidences of impoverishment and loss of local industries. Consequently, the loss of vast areas of farmland due to erosion menace has led to a drastic decrease in agricultural productivity and ultimately food shortage that could lead to famine. This gully erosion in these communities has given rise to infertile and barren farmland that needed to be reclaimed. This usually brought untold hardship to the inhabitants, if the land is still inhabitable but has been several affected. According to the Anambra State Ministry of Environment, the state has lost over 30% of its land, and over 40% of the total area of land and homes is being threatened by the ugly incident.

8.4.5 HEALTH IMPACT ASSESSMENT

An increase in soil erosion, an increase in deforestation, and a reduction of oxygen supply by the trees as well as a lower amount of carbon dioxide being removed from the atmosphere could lead to ill-health.

Erosion deposits from farmland to the Odor River in the Nanka and Agulu Lakes have a lot of nitrate and phosphate due to the application of fertilizer on the crops. This singular act creates soil salinity down to the river and lake. The deposits also created eutrophication in Agulu Lake due to nitrate and phosphate. This eutrophication created room for mosquito breeds and caused malaria disease.

These deposits sometime in 1997 were directly linked to the incidence of river blindness in Nanka. The blackfly and mosquitos transmitted river blinding and malaria into the people through bite. Also, tapeworm from cattle, a parasitic disease invaded the Odor River in Nanka due to erosion sediment deposited in the river. The disease infected the people with rainfall urination and abdominal pains. The impact on health led to a loss of many hours that would have been invested in labor as a result of being infected by these diseases even could lead to loss of lives.

8.4.6 Legal Impact Assessment

Anambra state has enacted a law against individuals or companies channeling rainwater from their compounds to the roads. Every individual or company must have a big sealed ditch or sump to reduce the cause of erosion through the reduction of flood velocity, which could gradually destroy the roads (Alaghmand et al., 2012). The implementation of this law has helped in the reduction of erosion menace in the state. Agulu and Nanka indigenous have benefited from the law as the stakeholders (communities) are empowered to participate in the regulating and monitoring.

8.5 CONCLUSIONS

Gully erosion constitutes the major ecological problem in Anambra state and requires adequate scientific and proper technical competence in the prevention and control of this menace. Effective control of gully erosion is not possible unless and until the principles and mechanism underlying its behavior and distribution over time, and space are fully understood, according to Eze (2000) in his strategies for enhancing the adoption of soil conservation technology. However, incorrect information, incomplete data, or wrong concepts in the application of the method of erosion control have aggravated the gully erosion problem.

Anambra state has suffered heavily from the havoc of gully erosion and its causes include natural and anthropogenic sources. The impacts included loss of human and animal lives, loss of properties, and land resources. Some of the solutions to erosion menace proffered by the experts included improved farming techniques, cultural methods of gully control, and the enactment of laws against any activities that favor gully growth.

However, poor or lack of implementation has hindered the complete evaluation of the proposal solution. Some poor quality of work led to greater erosion, as in the case of road construction probably due to poor supervision, poor funding, and corruption.

8.6 RECOMMENDATIONS

Anambra state, local government authority, and stakeholders in environmental management should sensitize the people on the cause, impacts, and problem of erosion menace. The prevention of the processes and mechanisms that result in, or advance, gully erosion should be of paramount importance to all stakeholders in environmental management in the state organic carbon, chemical properties textural characteristics of moisture context of the soil which is the most useful factor in control of erosion must be considered in the detailed survey (Osadebe and Enuvie, 2008), in the bid to better design preventive measure.

The preventive measures, used to curb the ugly menace of gully erosion are as follows:

- Avoid poor farming techniques, avoiding farm technique which is a contributing factor to the growth of gully erosion.

- Avoid refuse dumping along the river course, which impedes the flow of water loading especially during heavy rainfall.
- Maintain a good cultural method or vegetation technique of erosion in control.
- Planting cash crops such as plantain banana and some grass species on the flood plain has been found to be effective and suitable, especially for slope stability and erosion control.
- Adequate awareness and enlightenment of the effects of human activities on both flood plains and river channels to avoid misuse of the area.
- Efforts should be made by the state government to enact a law against locating engineering structures on waterways.
- Anambra state should take the gully erosion menace as a matter of urgency to yield to addressing issues relating to erosion especially gully erosion at an early stage so as to avoid loss of people's lives and properties.

REFERENCES

Akpokodje E.G., Tse A.C. and Ekeocha N (2010). Gully Erosion Geohazards in Southeastern Nigeria and management implications. Scientia Africana, 9(1), pp.20–36.

Alaghmand, S., Bin Abdullah, R., Abustan, I. and S. Eslamian, 2012, Comparison between capabilities of HEC-RAS and MIKE11 hydraulic models in river flood risk modeling (a case study of Sungai Kayu Ara River basin, Malaysia), International Journal of Hydrological Science and Technology, Vol. 2, No. 3, 270–291.

Albet A.A., Samson A.A. Peter O.O. and Olufunmilayo A.O. (2000). An Assessment of Socio Economic Impacts of Soil Erosion in South Eastern Nigeria, Shaping the change, XXIIFIG Congress Munich Germany, Germany, pp.12.

Barrocu, G. and Eslamian, S., (2022). River Flood Erosion and Land Development and Management, Ch. 20 in Flood Handbook, Vol. 3: Flood Impact and Management, Ed. by Eslamian, S., and Eslamian, F., Taylor and Francis, CRC Group, USA

Egboka, B.C.E. (1993). The Raging War.A publication of Anambra State Government of Nigeria, Awka.Ministry of Environment (2006).Landforms in Anambra State.Vol. 17 Anambra State, Nigeria.

Egboka, B.C.E., Orji, A.E., and Nwankwoala, H.O. (2019). Gully Erosion and landslides in southeastern Nigeria: causes, consequences and control measures. *Global Journal of Engineering Science* Vol.2 No 4, 11p. https://doi.org/10.33552/GJES.2019.02.000541

Eze, A.N. (2000). Strategies for Enhancing the Adoption of Soil Conservation Technologies in School Farms in Anambra State. Environmental Review Vol.3 No. 2., 412–419.

Ezechi, J. I. and Okagbue, C.A. (1989). Genetic Classification of Cullies in Eastern Nigeria and Its implication on control measure. Journal of African Earth Science, 8, pp 716.

Igbokwe, J.I. Akinyede, J.O. Dang, B. Alaga, T. Ono, M.N. Nnodu, V.C. and Anike, L.O. (2008).Mapping and Monitoring of the Impact of Gully Erosion in South Eastern Nigeria with Satellite Remote Sensing and Geographic information system.The International Archives of the Photogrammetry. Remote Sensory and Spatial Information Sciences Vol. XXXVII Part B8 Beiging, China.

Igbozurike, U. M. (1989). Erosion in the Agulu-Nanka Eastern Region of Nigeria.Suggestion on Control Measures Paper. Presented at the Conference of the Soil Science Society of Nigeria, Nsukka, Nigeria

Ndubude, R.K. (1982). Soil Erosion in Anambra State. A Document Prepared, Guideline of UNESCO Visiting Erosion Sites in Anambra State, France, Unpublished.

Nnodu, V.C. (2005). Environmental Renewal for Sustainable Development of Anambra State". A Special Paper Presented at 3 Day Stakeholders Workshop on Environment By Ministry of Environment, Awka, Nigeria, pp.15–17.

Nwajide, C.A. and Hogue (1979). Gullying Processes in South-Eastern Nigeria: The Field Journal, Vol. 44 No.2, pp 65–74.

Nwajide, C.S. and Reijers, T.J.A. (1986). Geology of the Southern Anambra Basin, In: Reijers T.J.A (Ed) Selected Chapters on Geology. SPDC, Warri, Nigeria, pp 133–148.

Obidimma, C.E. and Olurunfemi, A. (2011). Resolving the Gully Erosion Problem in South-Eastern Nigeria Innovation Through Public Awareness and Community-Based Approach, Journal of Soil Science and Environmental Management, 2(10), 286–291.

Obiekezie, S.O. Ono, M.N. and Igbokwe, J.I. (2002). Monitoring and Auditing of Impacts of Environmental Degradation. Environmental Review Vol. 3, No. 2, 309–321.

Ofomata, G.E.K. (2000). Classification of Soil. Erosion with specific reference to Anambra State Environmental review, Vol. 3, No. 2, 252–255.

Okafor, J.C. (1986). Soil Erosion in Anambra State Paper Presented at the Geological Science Seminar at Anambra State University of Technology Enugu, Nigeria.

Onwuka, S.U. (2008). Urban Erosion Problem In Nigeria, Rex Charles and Patrick Ltd., Enugu, Nigeria, pp.139–142,

Onyeagocha, S.C. (1986). The role of forestry in Erosion Control and Watershed management in Nigeria. A Report Working Document No.4 of August 1980, Lagos, Nigeria.

Osadebe, C. C. and Enuvie, G. (2008). Factor Analysis of Soil Variability in Gully Erosion Area of Southeastern Nigeria: Case Study of Agulu-Nanka, Oko Area. Scientia Africana Vol.7, No.2, pp. 45.

9 Land Cover/Use Changes and Urban Flood Incidence in the Osun River Basin, Southwest Nigeria

Adebayo Oluwole Eludoyin

9.1 INTRODUCTION

Changes in land cover/use have become of scientific importance because they are indicators of anthropogenic and/or natural impairment of natural ecosystems (Fathian et al., 2015; Lamine et al., 2018; Wang et al., 2022). Studies have expressed land cover in the form of the physical and biological composition of the outer lithosphere, including rocks, water bodies, vegetation types, and other forms of biodiversity (both fauna and flora) whereas land use describes the functionalities attached to the biodiversity by humans, including farming, built-up and recreation, among others (Eludoyin et al., 2019). Land use activities may also include mining, industrial activities, intensive agriculture, and construction as well as other activities that are capable of altering the physical, chemical, and biological nature of the earth's surface, including the earth's compositions of soil, water and vegetation diversities (Geist, 2017).

Land cover/use change often proves consequential to the natural habitats of animals and the entire ecosystem at varying (local, regional, national, and global) scales (Rimal et al., 2019). Factors, including population increases, population pressure, urbanization, as well as developments in transportation, climate change, and extreme events and natural disasters are among those which have raised concerns for research on the impact of land cover/use across localities and continents (Rimba et al., 2021; de Jong et al., 2021). Brown et al. (2018) conceptualized the relationships between land use and land cover change in terms of a complex scenario that includes the climatic, economic, and societal drivers of land use changes, human activities, such as agriculture and urbanization, as well as their links with the hydrosphere (Figure 9.1).

In many sub-Saharan Africa and South East Asian countries, a large proportion of economic products are focused on land resources, and the majority of those who work in the extractive and farming industries are of low socio-economic status. Friedman's core-periphery concept, which tends to suggest an imbalance in regional development between privileged urban and unprivileged but naturally rich (in natural resources) rural communities appears to be more suitable in the explanation of the situation where natural resources become degraded and development becomes polarized (Parnreiter, 2021). Damages to natural resource bases have become the significant focus of global interests for many years, having been characterized by protests, and social and civil unrest (Le Billon, 2013; Buur et al., 2019). Studies have listed experiences in Angola, Iraq, and the Niger Delta in Nigeria (see Le Billon, 2013; Omodanisi et al., 2014) as examples of countries where a generous endowment of natural resources that should have favored rapid economic and social development have proven to be a curse rather than a blessing due to exploitations by a repressive government and rebel groups, at huge cost to local populations. Many local populations are subjected to impoverished environmental, social, and economic conditions that include environmental

DOI: 10.1201/9781003473398-11

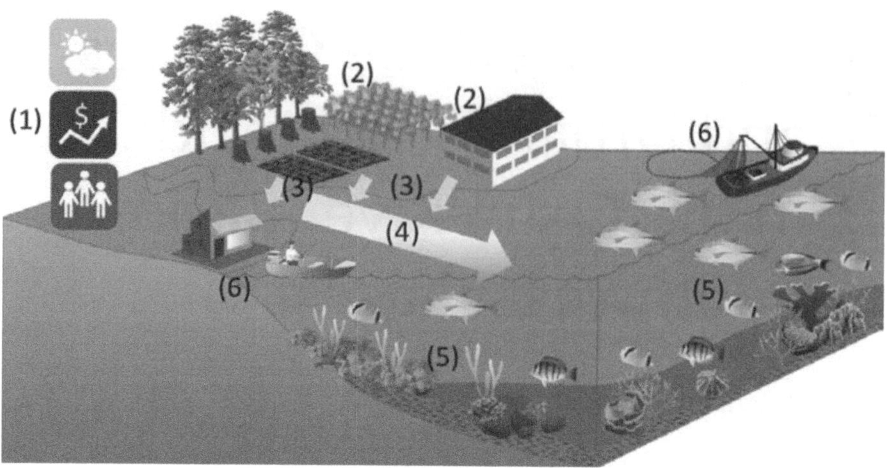

FIGURE 9.1 A conceptual perspective of the interaction between land cover/use changes and the other components of the environment. (1) Climatic, economic, and societal factors as drivers, (2). human activities, including forestry, agriculture, and urbanization; and consequences, such as (3) runoffs, (4) threats of freshwater pollution (5) loss of water-based organisms, and (6) implications to socio-economic and ecological values of ecosystem services (modified from Brown et al., 2018).

pollution, degradation, flooding, and even high mortality rates in extreme cases (Eludoyin et al., 2017, Omodanisi et al., 2014, 2015; Adeniji et al., 2022).

One of the main consequences of ineffective or poor land management strategies is interference with the river basins, whose effects include flooding. Flooding is induced by climatic elements, especially rainfall, and can be exacerbated by a number of natural and anthropogenic factors (Figure 9.2). Floods refer to overland flows, which may have occurred either as infiltration-excess, saturation-excess, or a more direct preferential (of a specific source such as dam failure, or effect of a sudden climatic extreme event, including hurricane, cyclones, etc) (Nanni et al., 2021; Petrucci, 2022; Ologunorisa et al., 2022a). They may also be distinguished based on the location of incidence, into the coastal, river, or fluvial and urban or pluvial flooding; urban floods characterize many towns and cities with impervious landscapes (Cea and Costabile, 2022; Kumari et al., 2022). The occurrence and intensity of floods have been noted to be exacerbated by global warming, climate change (Wasko, 2021; Nanditha and Mishra, 2022), land uses, misuse of the physical environment (Kuller et al., 2021; Mu et al., 2021), urbanization (Mu et al., 2021) deforestation, dam failures, poor waste management, and lack of flood control measures as well as combined influence of natural and socio-cultural factors that may be linked with modification of river flow, runoff enhancement, among others (Pielke, 2000; Hooijer et al., 2004; Ologunorisa et al., 2022b). Figure 9.2 is descriptive of the occurrence of urban floods, which studies have shown to be influenced by land cover/ use change. According to Pérez Molina (2019), flood patterns are vulnerable to the influence of land cover patterns, especially urban impervious surface increases, and population and built-up growth. The relationship between climate and landscape is also dependent on rainfall amount and physical characteristics of the natural landscape – this also determines the rate of infiltration and consequently flood vulnerability in an area. Flood occurrence poses a major risk to floodplains and riverside populations. Flooding is often exacerbated by anthropogenic activities, including built-up and urban activities that are capable of distorting river flow, including refuse dumps, poorly planned intensive agriculture, and other activities that are capable of upsetting the water movement and hydrological balance (Fletcher, 2021). Floods have also been linked with major disasters, globally; with a significantly high rate of mortality and displacements (Askew, 1999; UN Waters, 2011; Bashir et al., 2012; Ologunorisa et al., 2022; Mir et al 2022).

Improvements in computer knowledge, particularly in Remote Sensing (RS) and Geographic Information Systems (GIS) have been very useful in flood management (Opolot, 2013). With the use

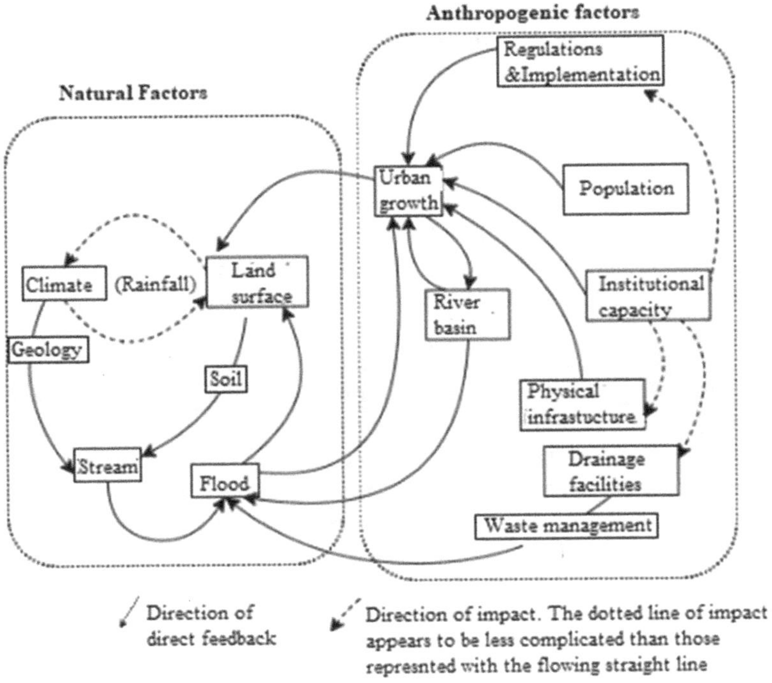

FIGURE 9.2 Interactive factors that influence flood occurrence as well as the interaction between them. Modified from Pérez Molina (2019).

of RS, GIS, and modeling, floods can be predicted and the areas prone to flooding can be mapped out. This information is crucial to both the policymakers and the public, most importantly those residing in the affected areas, so that early warnings can be provided, evacuation exercises carried out, and general preparedness undertaken. Remote sensing refers to the techniques that are used to analyze objects which are at a distance, i.e., acquiring information about an object without being in close contact with the object. Objects or phenomena of interest are studied and analyzed based on their unique electromagnetic wave characteristics so as to acquire information about the features and state of the object. Reflection and radiation of electromagnetic waves are the most frequently used methods because different objects reflect and radiate electromagnetic waves differently. The geographic information system is a system in which spatial and non-spatial data can be displayed and referenced using computers.

9.2 CASE STUDY APPLICATION OF REMOTE SENSING AND GEOGRAPHICAL INFORMATION SYSTEMS

Studies have shown the impact of spatial technology in the decision-making process, globally, and developing countries in African and Asian regions have also improved their capacities for smarter, technology-based decision-making processes in recent decades. In the area of environmental management, remote sensing, and geographic information technology have been proven useful for mapping flood hazard and risk areas, especially with the concept of an analytical hierarchy process-based multi-criteria decision analysis (Danumah et al., 2019). Studies (e.g., Youssef and Hegab, 2019) also argued that a multi-criteria analysis is suitable for solving complex problems with multiple variables and alternatives, and a high degree of uncertainty due to data challenges. The analytic hierarchy process compares relatively fairer than some other weighing approaches to multi-criteria analysis (including data envelopment analysis, gray target model, principal component analysis, fuzzy comprehensive evaluation model, set pair analysis, and variable sets method (Hu et al., 2017), hence it is been considered more frequently for adoption (see Alfa et al., 2018; Mahmoud and Gan, 2018; Rincón et al., 2018; Hoque et al., 2019; Youssef and Hegab, 2019). Consequently, the AHP-based multi-criteria analysis in the Geographic Information System was adopted for the case study presented in the chapter. The advantages of the AHP include improved decisions by coupled qualitative and quantitative factors for ranking and evaluating alternative scenarios (Youssef and Hegab, 2019) and better decision alternative that suits different options and criteria (Rincón et al., 2018). The AHP requires weight assignment as values of relative importance to variables. Relative importance usually ranges from "1" to "9"; "1" represents equal importance and "9" represents the weight of very importance (Komi et al., 2016; Rincón et al., 2018). Specifically, weight attached to a variable or factor can be obtained using one or more of the following criteria (e.g., Costache et al., 2019; Mishra and Sinha, 2020):

i. A pairwise comparison matrix based on Saaty's (1980) scale. The scale uses a sequence of absolute numbers from *1* to *9* or a different rating scale subject to the experience, intuition, or perception/knowledge of users (Saaty and Vargas, 2001).
ii. Use of (estimated) eigenvalue to determine the relative weight of each decision indicator as well as sub-indicators. This may be achieved with the adoption of factor analysis or by multiplying all the elements in each row in the matrix and then taking its nth root for each element using (Equation 9.1).

$$\text{(Estimated) Eigen Value} = \sqrt[N]{a_a * a_b * a_c * a_d * ... a_N} \qquad (9.1)$$

Where a_a, a_b, a_c, a_d are the values of the row elements and N is the number of the row elements

iii. Estimation of relative importance weight: This provides an opportunity for ranking the variables, and it is estimated using (Equation 9.2).

$$\text{Relative importance weight (RIW)} = \frac{\sqrt[N]{a_a * a_b * a_c * a_d * ... a_N}}{EE_1 + EE_2 + EE_3 + EE_4 + ... EE_N} \qquad (9.2)$$

Where EE_1, EE_2, EE_3, EE_4, ... EE_N are the estimated Eigen Value of each element.

iv. Computation of Consistency Ratio (CR) to prove the quality of the ranked comparison as elucidated in Mishra and Sinha (2020). The judgment or preference is consistent only if the CR is less than 0.10. Saaty's Consistency Ratio CR (Equation 9.3a–b) is used to check the pair-wise comparison (Saaty, 1980).

$$CR = \frac{CI}{RI} \qquad (9.3a)$$

$$CI = \frac{\lambda \max - n}{n - 1} \qquad (9.3b)$$

Where:
CI = Consistency Index which reflects consistency,
RI = Random Inconsistency Index; dependent on the sample size,
n = Number of criteria,
λ_{max} = Average of the value of the consistency vector (calculated factor weight).

The acceptable judgment range is from CR 0–1; 0 being most preferred/consistent (Saaty, 2008). Values outside 0–0.1 would require the assignment of the criterion weight again; CR $\geq$ 0.1 signifies inconsistent judgment (Mishra and Sinha, 2020).

Studies that have applied AHP in flood risk assessment include Alfa et al., (2018) on flood causative factors for flood hazard map development, and Rincón et al., (2018) which used the approach to develop flood risk maps along River Don in Toronto, Canada, as well as Radwan et al., 2018) on Riyadh city in Saudi Arabia. In all cases, weights of the various indicators of flood risk were defined and quantified using the approach and indicators, such as the digital elevation model, the population data, the streams, land use, and soil type layers were integrated for robust multicriteria flood risk map, which proved more useful than complex hydraulic and hydrological models in the generation of flood hazard maps. The multicriteria approach was coupled in GIS for the dynamically integrated maps.

The AHP is nonetheless still evolving in improved precision of visualized features since the method still requires a lot of improvement (Mallick et al., 2015; Kayastha et al., 2013). The axiomatic foundation, priorities, the 1–9 measurement scale, and the rank reversal problem are also being increasingly debated (Millet and Saaty, 2000), indicating that the approach is significantly evolving and has improved over time. For instance, Buckle (1985) and Hsieh et al. (2004), among others have extended Saaty's AHP to the case where the evaluators are allowed to employ fuzzy ratios in place of exact ratios to handle the difficulty for people to assign exact ratios when comparing two criteria and derive the fuzzy weights of criteria by geometric mean method (see also Wang et al., 2009; Gorsevski and Jankowski, 2010).

9.3 CASE STUDY: OSOGBO, OSUN STATE, NIGERIA

A case study analysis was undertaken in the Osun River basin in southwestern Nigeria. The basin is situated in the southern part of Osogbo, the administrative capital of Osun state. Osogbo became the state capital when Osun state was created in 1991 (Egunjobi, 1995). The state capital exists between 7.751°N–7.794°N and 4.48°E–4.55°E. Figure 9.3 shows a rough demarcation of the river basin within the Osogbo township. The basin area and the perimeter are 152.73 km² and 49.47 km², respectively. The basin is characterized by urban development and it occupies one of the urbanized areas of the state capital, Osogbo, the administrative capital of Osun state, Nigeria. Osogbo area grew from the status of a provincial headquarters in the western region to that of an administrative capital city in 1991 and has thus witnessed significant levels of urban growth and population increase through the years (Egunjobi, 1995). From a cultural center whose outward growth from the Oja-Oba (King's market) and palace (Mabogunje, 1965; Ojo, 1996) that is similar to the concentric urban setting described by Burgess (2004) as the emergence of population and built-up growth around spatially different nuclei similar to the one described by the multiple-nuclei model. Osogbo has grown over the years, and in some areas, different tribes and ethnic populations have emerged in a similar way that was described in Hoyt's sectoral model of urban development. Since it became an administrative capital city and seat of the Osun State Government in 1991, Osogbo has been transformed as home to many private and public institutions and has thus received an influx of migrants probably for economic reasons. The increase in population and urbanization has been linked with infrastructure development that has transformed into "leapfrog development" and "urban sprawl" (Aguda and Adegboyega, 2013; Adedotun, 2015). According to Hovinen (1977), leapfrog development or scattered development often results in continuous built-up, low-density development of many suburbs, and may be a causal factor of many urban sprawls. The population of the Osogbo local government area within which the Osun River Basin is largely situated in Osogbo grew from 155,507 (NPC, 2006) in 2006 to an estimated 217,544 in 2020 according to Geo-Referenced Infrastructure and Demographic Data for Development (GRID³) Nigeria Project (www.grid3.gov.ng), marking an annual growth rate of 3.99%.

Consequently, urban growth has increased at different rates across Osogbo, and one of the areas that have attracted such growth is around the river Osun in the southern part of the town. The river Osun basin with its hitherto well-developed riparian vegetation was preserved for farming, especially market gardening and cultural activities. Whereas the farming activities appeared to have given way to built-up areas, the cultural activities have reduced to a relatively protected section of the basin, now referred to as Osun Groove.

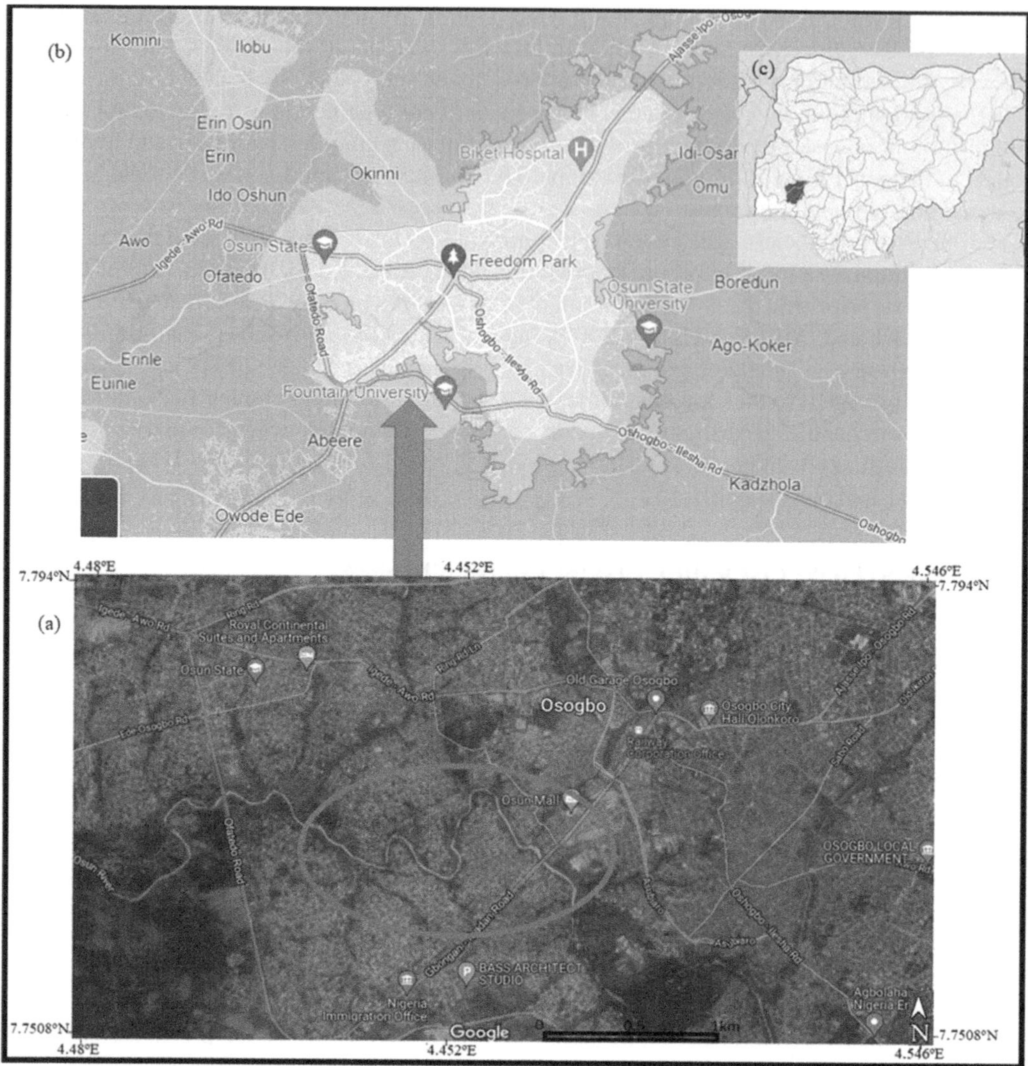

FIGURE 9.3 Topographical delineation (in red circle) and land use map of Osogbo in Osun State.

9.3.1 DATA

Data used were the shapefiles of Osogbo township as well as the freely available Landsat multi-spectral (MSS) image of the study area in 1986, Landsat Thematic Mapper™ of 1996, Landsat Enhanced Thematic Mapper (ETM) of 2006, and Landsat Operational Land Imager (OLI) of 2020 were obtained and investigated (Figure 9.4). While the shapefiles were obtained from the office of the Physical Planning Department of the Osun State Capital Territory Development Authority in Osogbo. The Landsat images were downloaded from the archive of the United States Geological Survey (USGS). The main characteristics of the Landsat imageries used are reported in Table 9.1. The Landsat images were selected for the study because of their temporal range and cost advantages over other available remotely sensed images. The Landsat data were the known image data that spanned between 1986 and 2020 and their free accessibility as well as medium spatial resolution made them a better candidate for monitor projects in developing countries including Nigeria (Eludoyin and Iyanda, 2018). In addition, the 12.5 m Alos Palsar Digital Elevation Model, which was sourced from the archive of the Alaska Satellite Facilities (search.asf.alaska.gov) was used to

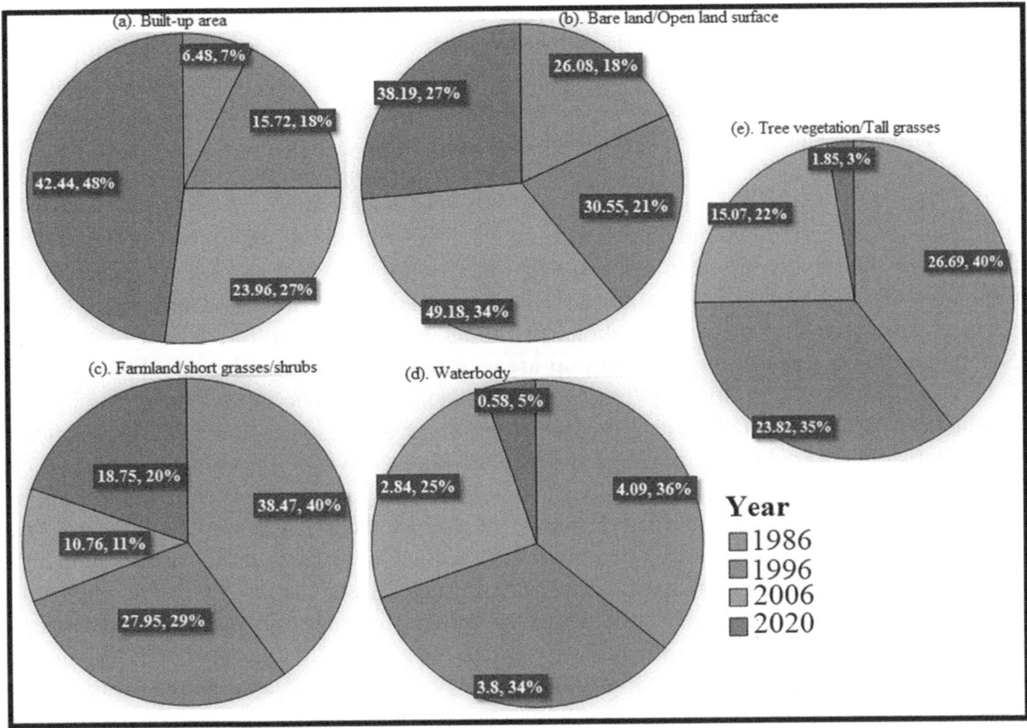

FIGURE 9.4 Distribution of different land use/cover (area in km², percentage of total area) in the different Landsat images of the Osun River basin in Osogbo.

TABLE 9.1

Characteristics of Landsat images used for the study (Path/Row = 190/55)

Sensor	Spatial Resolution	Spectral Resolution	Date of Acquisition
Multispectral scanner (MSS)	30 m	5 Bands	17-12-1986
Thematic Mapper (TM)	30 m	7	16-11-1996
Enhanced Thematic Mapper (TM)	30 m	8	12-01-2010
Operation Land Imager/Thermal Infrared Sensor (OLI/TIRS)	30 m	11	14-02-2020

Source: United States Geological Surveys (USGS). https://earthexplorer.usgs.gov/

delineate the drainage characteristics of the Osun drainage basin, and drainage variables, including elevation, slope, nearness to drainage and drainage density were derived using appropriate methods.

9.3.2 Data Analysis

The software used for the study is Quantum GIS (also known as QGIS) and it was preferred because it is open access and for novel contribution to knowledge because QGIS has enjoyed a rather minimal use by researchers for LULC classification by the time of this research. First, the Landsat images were corrected for geometric and radiometric errors in the QGIS software environment.

This means that the images were first georeferenced using the previously acquired coordinates of established landmarks on the field as described by Eludoyin and Iyanda (2018) following a guide from the QGIS manual (Lillesand et al., 2008; Congedo, 2017). Also, the process of radiometric enhancement includes pan-sharpening and other image rectifying processes as described in the Semi-automatic Classification Plugin (SCP) documentation (Congedo, 2017). Using the Semi-automatic Classification Plugin (SCP) in QGIS, the pre-processing tab was opened and Landsat was selected. The directory containing Landsat bands and the corresponding metadata was specified. DOS1-atmospheric correction and pansharpening were selected and the result was saved as a band set containing the pre-processed imageries. From the pre-processed imageries, a false color band combination that displays the near-infrared (NIR) and parts of the visible (red and green) bands of the images was derived to enhance visual interpretation, with the assumption that the red coloration indicates vegetation, cyan represents built-up areas, dirty ash represents bare surface, and black depicts water bodies as described by Eludoyin and Iyanda (2018). The images were classified into different land uses as adapted from a modified version of Anderson et al.'s (1976) classification schema (Table 9.2). Based on the model, a training set of pixels was selected for a supervised classification using a Maximum Likelihood (ML) classification algorithm for five land (built-up, bare surface, farmland, vegetation, waterbody) classes, after which the LULC classification was derived.

Consequently, the results of the supervised classification were assessed for accuracy using an error or confusion matrix (Jensen and Cowen, 1999; Congalton and Green, 2008). The error matrix is useful to assess the accuracy of land cover classification, in order to identify and measure map errors. Accuracy assessment under the post-processing tab of the SCP dock was selected, and classified data and ground-truth data which serve as a reference were added. The result returns the error matrix, the overall accuracy, the producer's accuracy, the user's accuracy, and the Kappa coefficient (Table 9.3) which was exported as *.xlsx* format for better visualization (Equations 9.5–9.7).

$$\text{User's accuracy} = \frac{\text{number of correct object of a specific class}}{\text{the total number of objects assigned to that class}} \tag{9.5}$$

$$\text{Producer's accuracy} = \frac{\text{number of correct objects of a specific class}}{\text{actual number of reference data object for that class}} \tag{9.6}$$

$$\text{Kappa Coefficient} = \frac{\text{observed accuracy} - \text{channce agreement}}{1 - \text{chance agreement}} \tag{9.7}$$

A flood risk zone was delineated and a flood risk map was generated using distance-to-drainage, elevation, slope, land cover type, flow accumulation, topographic wetness index (TWI), drainage density, soil type, geology, and normalized difference water index (NDWI) as criteria, based on information from previous studies (e.g., Ouma and Tateishi, 2014; Das, 2019). To achieve the AHP

TABLE 9.2

Classification Schema of Landuse/cover

Code	Landuse/cover	Description
1	Bare surface	Lands devoid of vegetation; lands with exposed soil
2	Built-up	Lands used for residential, commercial, industrial, etc. purposes
3	Farmland/Agricultural Land	Lands used for farming (plantation, cropland)
4	Vegetation/Natural Forest	Lands covered with natural vegetation
5	Waterbody	Reservoirs, rivers, streams, lakes

Source: Anderson *et al.* (1976); Eludoyin and Iyanda (2018).

TABLE 9.3
Classification Accuracy Assessment Report

Class	1986		1996		2006		2020	
	PA	UA	PA	UA	PA	UA	PA	UA
Built-Up	96.18	99.21	99.25	100	97.97	100.00	96.42	100.00
Bare Surface	98.31	92.06	94.83	98.66	97.23	96.84	98.68	90.89
Farmland	94.44	100.00	99.26	99.01	91.94	82.61	89.17	91.45
Vegetation	100	92.42	100.00	76.25	93.10	91.53	80.82	100.00
Waterbody	100	100.00	95.24	95.24	88.23	78.95	92.85	100.00
Overall Accuracy	96.86		98.00		98.00		94.95	
Kappa Coefficient	0.9578		0.9716		0.9492		0.9256	

analysis, each thematic layer was reclassified using the *reclassify* tool in ArcMap software, and different values were assigned to different classes within each layer with a rating scale of *1 to 9* depending on how they affect the vulnerability of humans, flow of surface water and water permeability (Mwangi, 2016). Classes within each dataset that aid flood formation the most were assigned value *9*, while classes which do not encourage flood occurrence at all were assigned value *1;* intermediate classes were ranked 2–8. The ranked parameters were thus assigned a weight (percentage influence) derived using the fuzzy AHP method. Flood risk zones were generated using the weighted overlay analysis in ArcGIS (version 6.4).

9.3.3 RESULTS

9.3.3.1 Changes in Land Use/Cover over Osun River Basin

Analysis of the multi-date (1986, 1999, 2006, and 2020) Landsat images show comparative changes in the classified land use/cover of the Osun basin area over the selected study period. As of 1986, the built-up area was 6.48 km² which accounted for 6.4% of the entire area, while the bare land surface/open land space was 26.08 km² or 26.6% of the total area. Areas such as farmland or short vegetation (including shrubs) occupied 34.5 km² or 37.9% while tree and tall grasses vegetation was 26.7 km² or 26.2%. Waterbodies were 4.1 km², occupying 4.02%, the smallest proportion (when compared with other land use/cover) of the 101.8 km² total Osun basin area in Osogbo. Table 9.4 also shows that while built-up areas and bare land/open land surfaces were on the increase within the study area, areas covered by waterbodies, tree vegetation/tall grasses have declined.

Areas occupied by farmland/short grasses/shrubs first declined between 1986 and 2006 but have recently (2006–2020) experienced a growth of 7.8% of what existed between 1996 and 2006. Obviously, the area has witnessed increases in a number of populations through immigration, and accompanying accommodation and facilities, following the status of Osogbo as an administrative state capital. The increase in population and built-up areas would understandably be linked with the reduction in open land surface, tree vegetation, and waterbodies, especially as there is evidence of construction in areas that hitherto were riparian vegetation zones (Figure 9.4).

9.3.3.2 Flood Risk Mapping of the River Basin

Results of the multi-criteria analysis used to determine the vulnerability of the study area to flood risks identified five major risk zones based on the level of vulnerability. Table 9.5 provides the results of the list of the average and range (as appropriate for proximity to the Osun River or major tributaries, drainage density, elevation, slope, dominant land use/cover distance, topographic wetness index (TWI), normalized difference water index (NDWI), flow accumulation geology, and

TABLE 9.4

Land Use/Cover Change (Area in km², and % Change in Parenthesis) between 1986 and 2020 in the Osun River Basin Area of Osogbo

Land use/cover type	Area covered in 1986 (km²)	Change from 1986 to 1996	Change from 1996 to 2006	Change from 2006 to 2020
			%	
Built-up	6.5	9.2 (9.1)	8.2 (8.1)	18.48 (18.2)
Bare land/open land surface	26.18	4.5 (4.4)	18.6 (18.3)	11.0 (10.8)
Farmland/Short grasses/shrubs	38.5	−10.5 (−10.3)	−17.2 (−16.9)	8.0 (7.8)
Tree vegetation/tall grasses	26.7	−2.9 (−2.8)	−8.75 (−8.6)	−13.2 (−13.0)
Waterbody	4.1	−0.3 (−0.3)	−1.0 (−0.9)	−2.3 (−2.2)

soil class/group) of the criteria used. The characteristics of each of the zones are described in the following sections.

9.3.3.2.1 Zone 1: The Very High Flood-risk Zone

This is the most vulnerable region to flood incidence in the study area (Figure 9.5). The area covers about 4.5 km². The zone is characterized by 0 –728.5 km around the Osun River or its tributaries, drainage density in the zone is 0.5–5.2, and elevation above the mean sea level varied between 307 m and 357 m with 0–8.5° slope. The Normalized Difference Water Index varied from −0.6 to 0.4 while the Topographic Wetness Index in this zone was 4.6–19.7 in the region; flow accumulation varied from zero to 439,577. Major land uses in the zone are built-up areas, bareland/open land surface, and farmland/short grasses/shrubs. Tree vegetation/tall grasses are essentially lacking in most parts of the region. The geological underlain is essentially blended gneiss and schist and the soil composition is fluvisol and lixisols. The geological underlain and soils as well as land use/cover are common to zones 2 and 3 which experienced comparatively high and moderate risks.

9.3.3.2.2 Zone 2: The High Flood-risk Zone

This zone is highly vulnerable to flooding (Figure 9.5). The area covers around 22.4 km². The zone is characterized by 0–924.1 km around the Osun River or its tributaries, drainage density in the zone is 0.5–5.19, and elevation above the mean sea level varied between 305 m and 369 m with 0–14.7° slope. The Normalized Difference Water Index varied from −0.75 to 0.89 while the Topographic Wetness Index in this zone was 4.14–19.07 in the region; flow accumulation varied from zero to 439,673. Major land uses in the zone are built-up areas, bareland/open land surface, and farmland/short grasses/shrubs. Tree vegetation/tall grasses are essentially lacking in most parts of the region. The geological underlain is essentially banded gneiss and schist and the soil composition is fluvisol and lixisols. The geological underlain and soils as well as land use/cover are common to zones 2 and 3 which experienced comparatively high and moderate risks.

9.3.3.2.3 Zone 3: The Moderate Flood-risk Zone

This risk of flooding in this zone is moderate except when triggered by excess anthropogenic and natural factors such as rainfall and blocked drainages. The area covers around 39.6 km². The zone is characterized by 0–1573.0 km around the Osun River or its tributaries, drainage density in the zone is 0.5–3.8, and elevation above the mean sea level varied between 307 m and 384 m with 0–18.9° slope. The Normalized Difference Water Index varied from −0.78 to 0.76 while the Topographic Wetness Index in this zone was 3.74–15.68 in the region; flow accumulation varied from zero to 36,315. Major land uses in the zone are built-up areas, bareland/open land surface, and farmland/

TABLE 9.5

Identified Attributes of the Different Classes of Flood Risk Zone in the Osun River Basin in Osogbo, Osun State

Classified vulnerable zone	Dominant Land use	Proximity to Osun River/ tributary (km)	Drainage Density	Elevation (m)	Slope (degree)	NDWI (No unit)	TWI (No unit)	Flow Acc.	Geology	Soil
Very High	• Built-up,	0–728.5	0.5–5.2	307–357	0–8.5	–0.6–0.4	4.6–19.7	0–439,577	Banded gneiss, Schist	Fluvisols, Lixisols
High	• Bare land/open surface	0–924.07	0–5.19	305–369	0–14.70	–0.75–0.89	4.14 – 19.07	0–439,673		
Moderate	• Farmland/short grasses/ shrubs	0–1573.03	0–3.83	307–384	0–18.88	–0.78 –0.76	3.74 – 15.68	0–36315		
Low	• Tree vegetation/tall grasses	12.6–1941.7	0–2.18	309–405	0–18.66	–0.79–0.43	4.14–19.07	0–4737		
Very Low	• Built-up, • Bare land/open surface • Farmland/short grasses/ shrubs	328.3–1945.4	0–0.71	326–418	0–23.21	–0.70–0.01	4.60–19.13	0–956		Lixisols

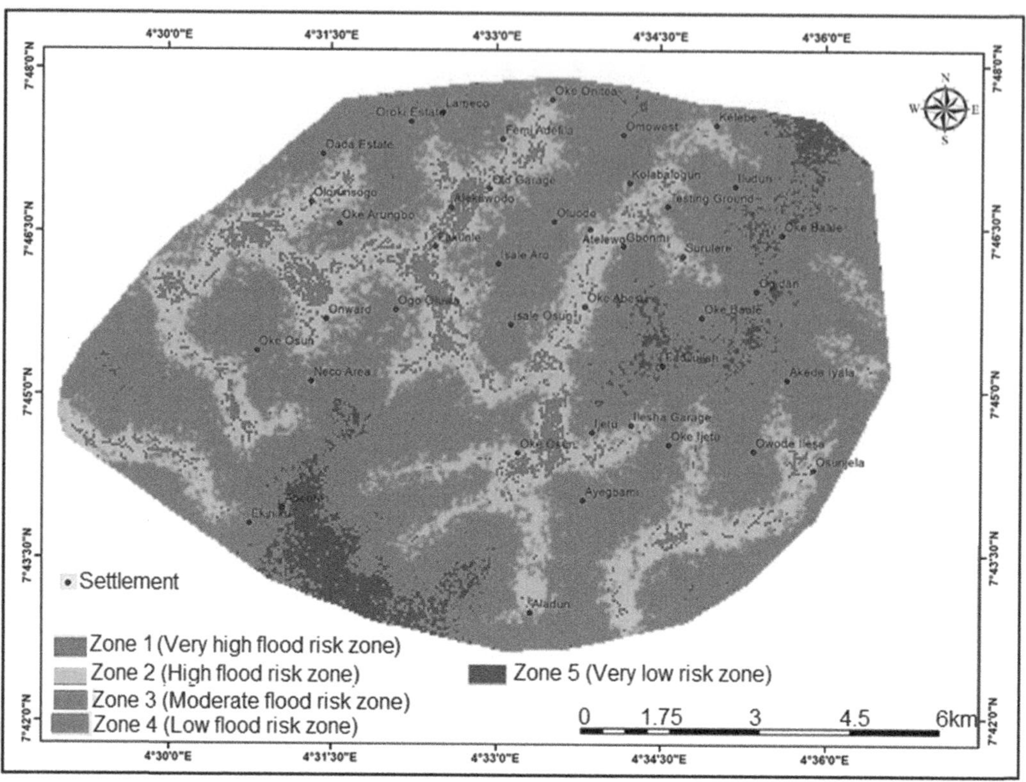

FIGURE 9.5 Classified flood risk zones based on selected multi-criteria weights.

short grasses/shrubs. Tree vegetation/tall grasses are essentially lacking in most parts of the region. The geological underlain is essentially banded gneiss and schist and the soil composition is fluvisol and lixisols. The geological underlain and soils as well as land use/cover are common to zones 2 and 3 which experienced comparatively high and moderate risks.

9.3.3.2.4   *Zones 4 and 5: The Very Low and Low Flood-risk Zone*

These zones are the least vulnerable to flooding incidence in the study area. The area covers about 33.5 km^2. The zone is characterized by 12.6–1945.4 km around the Osun River or its tributaries, drainage density in the zone is 0–0.71 and elevation above the mean sea level varied between 309 m and 418 m with 0–23.2° slope. The Normalized Difference Water Index varied from −0.79 to 0.01 while the Topographic Wetness Index in this zone was 4.14–19.13 in the region; flow accumulation varied from zero to 956. Major land uses in the zone are built-up area, bareland/open land surface, tree vegetation/tall grasses, and farmland/short grasses/shrubs. The geological underlain is essentially banded gneiss and schist and the soil composition is mainly lixisols.

9.4   DISCUSSION

The study reports the findings from an empirical study on the impact of land use/cover on an urban river basin in southwestern Nigeria. Major findings include the fact that vegetation became replaced with built-up and bare surfaces over the years, indicating urban growth, and this may have replaced a significant proportion of areas that were hitherto green. In a similar study, Salami and Akinyede (2006) argued that land area in savanna and forested part of southwestern Nigeria had between 1986 and 2014 been replaced by built-up areas, due to urbanization, intensive agriculture, construction activities, infrastructure, and some previously built-up area became deserted due to crises and environmental degradation (especially mining sites). Built-up and abandoned open areas have been areas with a tendency for increased temperature as explained by the concept of urban heat island (see Akinbode et al., 2004; Guyekye, 2013), flooding (Akinwumiju, 2016), biodiversity loss (Rain et al., 2011), and loss of water sources (Onanuga et al., 2022). Studies (Mengistu and Salami, 2007; Bennet and Saunders, 2010) argued that poorly planned or monitored urban growth would enhance chances for the creation of shanty towns or slums, the effect of which can be threatening to environmental cleanliness desired in many urban areas. In the study area, built-up in the river basin is inadequately monitored and/or regulated. This is probably because the implementation of environmental regulatory rules is often impaired by lack of or inadequate punitive measures, among others. According to Eludoyin (2023), the study area, being a state capital and core of the development in the state, has become a central place with significant influence for in-migration, and capacity for the increase in population and growth that exceeds the capacity of the existing resources.

The decrease in waterbody observed over the study period suggests that development has encroached into streams and waterways. This may be caused by a number of factors such as haphazard planning of the city, failure of the residents to follow laid-out plans when building which leads to encroachment into streams and access roads, and the cheap cost of riparian land which encourages its purchase and consequent use for siting of permanent structures. Cultivated lands probably decreased due to urban growth and increased land demand for constructions, especially with a change in status from a previously (although relatively) unpopular status to the status of an administrative capital city. There is strong evidence of outmigration from congested built-up areas to the newly developing areas – which were previously largely unoccupied, and this generally threatens the land ecosystem and could change the ecosystem, in a drastic way if not adequately planned.

The Fuzzy AHP was used to derive the weights assigned to each thematic layer and to generate the flood risk map of the study area after which the weighted overlay tool in ArcGIS was used to assign ranks and weights to the thematic layers, produced five flood risk zones, namely the very high-risk zone, the high-risk zone, the moderate zone, the low-risk zone, and the very low-risk zone. The very low-risk zones and low-risk zones are the locations where the probability of a flood occurring is close to zero while the moderate-risk zones are locations where the chances of the flood occurring are also quite low but can be triggered by extreme anthropogenic activities such as the release of water from the dam during the peak of a rainy season (Njoku et al., 2018). The high and very high-risk zones are areas where flood occurrences can be easily triggered by any rainfall event especially if it lasts for a sustained period. The flood risk map shows that the moderate-risk zone at 39.60% has a higher percentage, followed by the low-risk zone (28.18%). high-risk zone accounts for 22.39% of the study area while very low and very high account for 5.37% and 4.46% respectively. A couple of the vulnerable communities have been reported by Adegoke and Sajobi (2015) and Adegboyega et. al. (2018) to be areas with high flood risk potentials. The communities were inundated and the residents suffered significant losses in 2016.

The regions which are highly vulnerable to flooding generally show a combined characteristic of built-up areas with low elevation, lower degree of slope, and high proximity to the drainage network as also observed by Mmom and Ayakpo (2014), Cao et al. (2016) and Das (2019). Built-up areas are usually paved and concretized which makes them impermeable and greatly prevents water infiltration. Impermeable land covers typically decrease water hold-up time which leads to a decrease in lag time and an increase in the peak discharge which enhances flood occurrence (Krouska and Parcharidis, 2014). Low slope angles and elevation also increase the vulnerability of an area to flooding (Korah and Lopez, 2015). People who live in these areas should be aware of the potential hazards caused by flash flooding and pay attention to weather forecasts for heavy rain and evacuate in advance. In conclusion, urban growth in the Osogbo River basin, Osogbo, has been significant over the years and this has had a corresponding effect on the incidence of flooding. It is recommended that sustainable waste management and urban drainage cleaning practices in the study area, and similar urban environments be encouraged. Planning's set-back rules, which disallow buildings to be too close to the river channels should be strictly enforced. The need to improve data collection and availability at river basins/catchments at a national and regional level should be emphasized. This will encourage more model-based decision-support research into the area.

REFERENCES

Adedotun, S. B. (2015). A study of urban transportation system in Osogbo, Osun State, Nigeria. *European Journal of Sustainable Development*, 4(3), 93–101.

Adegboyega, S. A., Onuoha, O. C., Adesuji, K. A., Olajuyigbe, A. E., Olufemi, A. A., and Ibitoye, M. O. (2018). An integrated approach to modelling of flood hazards in the rapidly growing city of Osogbo, Osun State, Nigeria. *American Journal of Space Science*, 4, 1–15.

Adegoke, O., and Sajobi, A. (2015). Climate change impact on infrastructure in Osogbo Metropolis, South-West Nigeria C.W. *Journal of Emerging Trends in Engineering and Applied Sciences*, 6(3), 156–165.

Adeniji, O. A., Adejoba, A. L., Irunokhai, E. A., Osaguona, P. O., Alaye, S. A., and Ojo, B. S. (2022). Assessment of deforestation in Mashegu Local Government Area of Niger State, Nigeria: Causes, effects, and reduction strategies. *Agro-Science*, 21(2), 95–99.

Aguda, A. S., and Adegboyega, S. A. (2013). Evaluation of spatio-temporal dynamics of urban sprawl in Osogbo, Nigeria using satellite imagery and GIS techniques. *International Journal of Multidisciplinary and Current Research*, 1, 60–73.

Ahmed, M. Y., and Mahmoud, A. H. (2019). Flood-hazard assessment modeling using multicriteria analysis and GIS: A case study—Ras Gharib Area, Egypt, Chapter 10. In Pourghasemi, H. R., and Gokceoglu, C. (Eds.), *Spatial Modeling in GIS and R for Earth and Environmental Sciences*. Elsevier, Amsterdam.

Akinwumiju, A. S. (2016). Morphometric and land use analysis: Implication on flood hazards in Ilesa and Osogbo Metropolis, Osun State, Nigeria. *Ethiopian Journal of Environmental Studies and Management*, 10(2), 229–240.

Alfa, M. I., Ajibike, M. A., and Daffi, R. E. (2018). Application of analytic hierarchy process and geographic information system techniques in flood risk assessment: A case of Ofu river catchment in Nigeria. *Journal of Degraded and Mining Lands Management*, 5(4), 1363.

Anderson, J. R., Hardy, E. E., Roach, J. T., and Witmer, R. E. (1976). A land use and land cover classification system for use with remote sensor data. Geological Survey Professional Paper No. 964, U.S Government Printing Office, Washington, DC, 28.

Askew, A. J. (1999). Water in the international decade for natural disaster reduction. Destructive water: Water-caused natural disaster, their abatement and control. IAHS Publication, 239.

Bashir, O., Adedeji, O. H., and Oladosu, J. O. (2012). Floods of fury in Nigerian cities. *Journal of Sustainable Development*, 5(7), 69.

Bennet, A. F., and Saunders, D. (2010). Habitat fragmentation and landscape. *Change Conservation Biology for All*, 5, 88–106.

Brown, G., Sanders, S., and Reed, P. (2018). Using public participatory mapping to inform general land use planning and zoning. *Landscape and Urban Planning*, 177, 64–74.

Buckley, J. J. (1985). Fuzzy hierarchical analysis. *Fuzzy Sets and Systems*, 17(3), 233–247.

Burgess, R. (2004). Good neighbour, bad neighbour, no neighbour: The debate on urban fragmentation. *Paper Presented to the 8th International Conference of the ALFA-IBIS Network on Globalisation and Large Urban Projects in 25 Cities Department of Architecture, Shanghai Jiao Tong University Shanghai China*, Nov. 23rd–24th 2004.

Buur, L., Pedersen, R. H., Nystrand, M. J., and Macuane, J. J. (2019). Understanding the three key relationships in natural resource investments in Africa: An analytical framework. *The Extractive Industries and Society*, 6(4), 1195–1204.

Cao, C., Xu, P., Wang, Y., Chen, Y., Zheng, L., and Niu, C. (2016). Flash flood hazard susceptibility mapping using frequency ratio and statistical index methods in coalmine subsidence areas. *Sustainability*, 8, 948–960.

Cea, L., and Costabile, P. (2022). Flood risk in urban areas: Modelling, management and adaptation to climate change. A review. *Hydrology*, 9(3), 50.

Congalton, R. G., and Green, K. (2008). *Assessing the Accuracy of Remotely Sensed Data Principles and Practices*. 2nd Edition. CRC Press, Taylor and Francis Group, Boca Raton, FL.

Congedo, L. (2017). Semi-automatic classification plugin documentation. Semi-automatic classification plugin: A Python tool for the download and processing of remote sensing images in QGIS. *Journal of Open Source Software*, 6(64), 3172. https://doi.org/10.21105/joss.03172

Costache, R., Pham, Q. B., Sharifi, E., Linh, N. T. T., Abba, S. I., Vojtek, M., ... Khoi, D. N. (2019). Flash-flood susceptibility assessment using multi-criteria decision making and machine learning supported by remote sensing and GIS techniques. *Remote Sensing*, 12(1), 106.

Danumah, J. H., Odai, S. N., Saley, B. M., Szarzynski, J., Thiel, M., Kwaku, A., and Akpa, L. Y. (2019). Flood risk assessment and mapping in Abidjan district using multi-criteria analysis (AHP) model and geoinformation techniques, (cote d'ivoire). *Geoenvironmental Disasters*, 3(1), 1–13.

Das, S. (2019). Geospatial mapping of flood susceptibility and hydro-geomorphic response to the floods in Ulhas Basin, India. *Remote Sensing Applications: Society and Environment*, 14, 60–74.

de Jong, L., De Bruin, S., Knoop, J., and van Vliet, J. (2021). Understanding land-use change conflict: A systematic review of case studies. *Journal of Land Use Science*, 16(3), 223–239.

Eludoyin, A. O. (2023). Domestic water quality and the discriminatory influence of socio-economic stratification on accessibility to safe water in a part of Osun State, Nigeria. *Aegean Journal of Geography*, 32(1), 19–31. Ege University, Izmir, Turkey (In press).

Eludoyin, A. O., and Iyanda, O. O. (2018). Land cover change and forest management strategies in Ife nature reserve, Nigeria. *GeoJournal*, 84(6), 1531–1548.

Eludoyin, A. O., Ojo, A. T., Ojo, T. O., and Awotoye, O. O. (2017). Effects of artisanal gold mining activities on soil properties in a part of southwestern Nigeria. *Cogent Environmental Science*, 3(1), 1305650.

Eludoyin, A. O., Omotoso, I., Eludoyin, O. M., and Popoola, K. S. (2019). Remote sensing technology for evaluation of variations in land surface temperature, and case study analysis from southwest Nigeria. In Koutsopoulos, K., González, R. M., and Donert, K. (Eds.), *Geospatial Challenges in the 21st Century, Key Challenges in Geography*. Springer Nature, Switzerland AG, 154–170. ISSN: 2522-8420.

Fathian, F., Prasad, A. D., Dehghan, Z., and Eslamian, S. (2015). Influence of land use/land cover change on land surface temperature using RS and GIS techniques. *International Journal of Hydrology Science and Technology*, 5(3), 195–207.

Fletcher, M. (2021). Flood management. In Ferrier, Robert, and Jenkins, Alan (Eds.), *Handbook of Catchment Management 2ème*. John Wiley & Sons, Ltd., New Jersey, 205–244.

Geist, H. (2017). *The Causes and Progression of Desertification*. Routledge, London.

Guyekye, A. K. (2013). Environmental change and flooding in Accra, Ghana. *Sacha Journal of Environmental Studies*, 3(1), 65–80.

Hooijer, A., Klijn, F., Pedroli, B., and Van, O. A. (2004). Towards sustainable flood risk management in the Rhine and Meuse river basins: Synopsis of the findings of IRMA-SPONGE. *River Research and Applications*, 20, 343–357.

Hoque, M. A. A., Tasfia, S., Ahmed, N., and Pradhan, B. (2019). Assessing spatial flood vulnerability at Kalapara Upazila in Bangladesh using an analytic hierarchy process. *Sensors*, 19(6), 1302.

Hovinen, G. R. (1977). Leapfrog developments in Lancaster County: A study of residents' perceptions and attitudes. *The Professional Geographer*, 29(2), 194–199.

Hsieh, T. Y., Lu, S. T., and Tzeng, G. H. (2004). Fuzzy MCDM approach for planning and design tenders' selection in public office buildings. *International Journal of Project Management*, 22(7), 573–584.

Hu, S., Brixner, D., Maniadakis, N., Kaló, Z., Shen, J., and Wijaya, K. (2017). Applying multi-criteria decision analysis (MCDA) simple scoring as an evidence-based HTA methodology for evaluating off-patent pharmaceuticals (OPPs) in emerging markets. *Value in Health Regional Issues*, 13, 1–6.

Jensen, J. R., and Cowen, D. C. (1999). Remote sensing of urban/suburban infrastructure and socio-economic attributes. *Photogrammetric Engineering and Remote Sensing*, 65, 611–622.

Kayastha, P., Dhital, M. R., and De Smedt, F. (2013). Application of the analytical hierarchy process (AHP) for landslide susceptibility mapping: A case study from the Tinau watershed, west Nepal. *Computers and Geosciences*, 52, 398–408.

Komi, K., Amisigo, B. A., and Diekkrüger, B. (2016). Integrated flood risk assessment of rural communities in the Oti River Basin, West Africa. *Hydrology*, 3(4), 42.

Korah, P. I., and Lopez, F. M. J. (2015). Mapping flood vulnerable areas in Quetzaltenango, Guatemala using GIS. *Journal of Environment and Earth Science*, 5(6), 132–145.

Krouska, Z., and Parcharidis, I. (2014). Hazard maps for flash-floods in the Thriassion Plain. The 10th International Congress of Hellenic Geographical Society, Greece.

Kuller, M., Schoenholzer, K., and Lienert, J. (2021). Creating effective flood warnings: A framework from a critical review. *Journal of Hydrology*, 602, 126708.

Kumari, R., Kamal, V., Mukherjee, S., and Eslamian, S. (2022). Catchment morphometric characteristics impact on floods management, Ch. 8. In Eslamian, S., and Eslamian, F. (Eds.), *Flood Handbook, Vol. 3: Flood Impact and Management*. Taylor and Francis, CRC Group, USA.

Lamine, S., Petropoulos, G. P., Singh, S. K., Szabó, S., Bachari, N. E. I., Srivastava, P. K., and Suman, S. (2018). Quantifying land use/land cover spatio-temporal landscape pattern dynamics from Hyperion using SVMs classifier and FRAGSTATS®. *Geocarto International*, 33(8), 862–878.

Le Billon, P. (2013). *Fuelling War: Natural Resources and Armed Conflicts*. Routledge, 128p. ISBN 9781315019529.

Lillesand, T. M., Kiefer, R. W., and Chipman, J. W. (2008). *Remote Sensing and Image Interpretation*. 6th Edition. John Wiley and Sons, New York.

Mabogunje, A. (1965). Urbanization in Nigeria: A constraint on economic development. *Economic Development and Cultural Change*, 13(4), 413–438.

Mahmoud, S. H., and Gan, T. Y. (2018). Multi-criteria approach to develop flood susceptibility maps in arid regions of Middle East. *Journal of Cleaner Production*, 196, 216–229.

Mallick, J., Khan, R. A., Ahmed, M., Alqadhi, A. S., Alsubih, M., Falqi, I., and Hasan, M. A. (2015). Modeling groundwater potential zone in a semi-arid region of Aseer using Fuzzy-AHP and geoinformation techniques. *Water*, 11(12), 26–56.

Mengistu, D. A., and Salami, A. T. (2007). Application of remote sensing and GIS in land use/land cover mapping and change detection in a part of South Western Nigeria. *African Journal of Environmental Science and Technology*, 1(5), 99–109.

Millet, I., and Saaty, T. L. (2000). On the relativity of relative measures–accommodating both rank preservation and rank reversals in the AHP. *European Journal of Operational Research*, 121(1), 205–212.

Mir, B. H., Rather, N. A., and Eslamian, S. (2022). *Social Aspects of Flooding*, Flood Handbook, Volume 1, 1st Edition, Ch. 6. CRC Press, Taylor & Francis Group, Boca Raton. https://doi.org/10.1201/9781003262640-8 ISBN: 9781138584938

Mishra, K., and Sinha, R. (2020). Flood risk assessment in the Kosi megafan using multi-criteria decision analysis: A hydro-geomorphic approach. *Geomorphology*, 350, 106861.

Mmom, P. C., and Ayakpo, A. (2014). Spatial analysis of flood vulnerability levels in Sagbama local government area using Geographic Information Systems (GIS). *International Journal of Research in Environmental Studies*, 1, 1–8.

Mu, D., Luo, P., Lyu, J., Zhou, M., Huo, A., Duan, W., ... Zhao, X. (2021). Impact of temporal rainfall patterns on flash floods in Hue City, Vietnam. *Journal of Flood Risk Management*, 14(1), e12668.

Mwangi, M. P. (2016). Department of environmental studies and community development, Kenyatta University. The Role of Land Use and Land Cover Changes and GIS in Flood Risk Mapping in Kilifi County, Kenya. Department of Environmental Studies and Community, Doctoral dissertation, Kenyatta University, Kenya.

Nanditha, J. S., and Mishra, V. (2022). Multiday precipitation is a prominent driver of floods in Indian river Basins. *Water Resources Research*, 58(7), e2022WR032723.

Nanni, P., Peres, D. J., Musumeci, R. E., and Cancelliere, A. (2021). Worry about climate change and urban flooding risk preparedness in Southern Italy: A survey in the Simeto River Valley (Sicily, Italy). *Resources*, 10(3), 25.

National Population Commission (NPC). (2006). Nigeria national census: Population distribution by sex, state, LGAs and senatorial district: 2006 census priority tables (Vol. 3).

Njoku, C. G., Effiong, J., and Uzozie, A. C. (2018). A GIS multi-criteria evaluation for flood risk vulnerability mapping of Ikom local government area, cross river state. *Journal of Geography, Environment and Earth Science*, 15(2), 1–17.

Ojo, G. A. (1996). *Yoruba Culture*. University of Ife and London University Press, Nigeria and London, 131–157.

Ologunorisa, T. E., Eludoyin, A. O., and Lateef, B. (2022a). An evaluation of flood fatalities in Nigeria. *Weather, Climate, and Society*, 14(3), 709–720.

Ologunorisa, T. E., Obioma, O., and Eludoyin, A. O. (2022b). Urban flood event and associated damage in the Benue valley, Nigeria. *Natural Hazards*, 111(1), 261–282.

Omodanisi, E. O., Eludoyin, A. O., and Salami, A. T. (2014). A multi-perspective view of the effects of a pipeline explosion in Nigeria. *International Journal of Disaster Risk Reduction*, 7, 68–77.

Omodanisi, E. O., Eludoyin, A. O., and Salami, A. T. (2015). Ecological effects and perceptions of victims of pipeline explosion in a developing country. *International Journal of Environmental Science and Technology*, 12, 1635–1646.

Onanuga, M. Y., Eludoyin, A. O., and Ofoezie, I. E. (2022). Urbanization and its effects on land and water resources in Ijebuland, southwestern Nigeria. *Environment, Development and Sustainability*, 24(1), 592–616.

Opolot, E. (2013). Application of remote sensing and geographical information systems in flood management: A review. *Research Journal of Applied Sciences*, 6(10), 1884–1894.

Ouma, Y. O., and Tateishi, R. (2014). Urban flood vulnerability and risk mapping using integrated multi-parametric AHP and GIS: Methodological overview and case study assessment. *Water*, 6, 1515–1545.

Parnreiter, C. (2021). Global cities, centripetal wealth transfer and uneven development. In Neal, Zachary P., and Rozenblat, CÃ©line (Eds.), *Handbook of Cities and Networks*, Edward Elgar Publishing, Cheltenham, 618–632.

Pece, V. G., and Jankowski, P. (2010). An optimized solution of multi-criteria evaluation analysis of landslide susceptibility using fuzzy sets and Kalman filter. *Computers & Geosciences*, 36(8), 1005–1020.

Perez Molina, E. (2019). Spatial planning, growth, and flooding: Contrasting urban processes in Kigali and Kampala, ITC dissertation number 369. Printed by: ITC Printing Department, Enschede, The Netherlands. https://doi.org/10.3990/1.9789036548830.

Petrucci, O. (2022). Factors leading to the occurrence of flood fatalities: A systematic review of research papers published between 2010 and 2020. *Natural Hazards and Earth System Sciences*, 22(1), 71–83.

Pielke, R. A. (2000). Flood impacts on society, damaging floods as a framework for assessment. In Parker, D. J. (Ed.), *Floods*. Routledge Hazards and Disasters Series. Routledge, London, 133–155.

Radwan, F., Alazba, A. A., and Mossad, A. (2018). Flood risk assessment and mapping using AHP in arid and semiarid regions. *Acta Geophysica*, 67(1), 215–229.

Rain, D., Engstrom, R., Ludlow, C., and Antos, S. (2011). Accra Ghana: A city vulnerable to flooding and drought-induced migration. *Global Report on Human Settlement*, 2011, 3–9.

Rimal, B., Sharma, R., Kunwar, R., Keshtkar, H., Stork, N. E., Rijal, S., and Baral, H. (2019). Effects of land use and land cover change on ecosystem services in the Koshi River Basin, Eastern Nepal. *Ecosystem Services*, 38, 100963.

Rimba, A. B., Mohan, G., Chapagain, S. K., Arumansawang, A., Payus, C., Fukushi, K., and Avtar, R. (2021). Impact of population growth and land use and land cover (LULC) changes on water quality in tourism-dependent economies using a geographically weighted regression approach. *Environmental Science and Pollution Research*, 28(20), 25920–25938.

Rincón, D., Khan, U. T., and Armenakis, C. (2018). Flood risk mapping using GIS and multi-criteria analysis: A greater Toronto Area case study. *Geosciences*, 8(8), 275.

Saaty, T. L. (1980). *The Analytical Hierarchy Process: Planning, Priority Setting, Resource Allocation*. McGraw-Hill, New York.

Saaty, T. L. (2008). Decision making with the analytic hierarchy process. *International Journal of Services Sciences*, 1, 83. https://doi.org/10.1504/IJSSCI.2008.017590.

Saaty, T. L., and Vargas, L. G. (2001). *How to Make a Decision in Models, Methods, Concepts and Applications of the Analytic Hierarchy Process*. Springer, 1–25.

Salami, A. T., and Akinyede, J. (2006). Space technology for monitoring and managing forest in Nigeria. International Symposium on Space and Forests, United Nations Committee on Peaceful Uses of Outer Space (UNOOSA), Vienna.

United Nations. (2011). The human right to water and sanitation: Media brief. UN-Water Decade Programme on Advocacy and Communication and Water Supply and Sanitation Collaborative Council.

Wang, W. D., Xie, C. M., and Du, X. G. (2009). Landslides susceptibility mapping based on geographical information system, GuiZhou, south-west China. *Environmental Geology*, 58(1), 33–43.

Wang, Z. J., Liu, S. J., Li, J. H., Pan, C., Wu, J. L., Ran, J., and Su, Y. (2022). Remarkable improvement of ecosystem service values promoted by land use/land cover changes on the Yungui Plateau of China during 2001–2020. *Ecological Indicators*, 142, 109303.

Wasko, C., Nathan, R., Stein, L., and O'Shea, D. (2021). Evidence of shorter more extreme rainfalls and increased flood variability under climate change. *Journal of Hydrology*, 603, 126994.

10 Sediment Transport Index and Stream Power Index Using GIS
A Case Study in the Bou Saâda Wadi–Sub-basin, Algeria

Zohra Abdelkrim, Bachir Faid, and Saeid Eslamian

10.1 INTRODUCTION

Sediment damage is a problem that our society suffers from, this phenomenon poses a specific danger to the health, environment, and the economic system, thus, negatively affecting water supplies; water is an important element of the economic and social development of people.

The importance of dams and reservoirs is represented in power generation, flood protection, agricultural water supply, and drinking water. All over the world, dams have been exposed to the phenomenon of sedimentation including the United States, Russia, Taiwan, China, Iran, Sudan, and North Africa (Amamra et al., 2018).

"Around 40,000 large reservoirs suffer from sedimentation and it is estimated that between 0.5% and 1% of the total storage capacity is lost per year," (Merina et al., 2016).

Erosion and desertification are two of the most important causes of sediment. The United Nations Environmental Program reported that soil erosion and degradation were the main causes of low crop productivity (United Nations Environmental Program, 1991; Lim et al., 2005).

Sediment is deposited as a result of soil erosion due to rain, runoff, melting snow, and human activities such as high population density (rapid population growth) and deforestation.

This drifting soil is hydraulically transported, and becomes deposited (Merina et al., 2016), eroded soils are targeted rivers, streams, and the catchment areas of the reservoirs. These sediments accumulate in the reservoir, causing the water storage capacity to reduce, a study (Kanito et al., 2023) confirms that reservoirs lose 2% of their designed storage volume each year due to sedimentation. Additionally, turbine blades are affected by sand and silt that are carried by water becoming corroded and cracked, this significantly reduces their efficiency in power generation (Ikwap and Babirye, 2018).

Erosion/landslides result from the evolution of the geomorphic terrain and are a natural denudation process (Vijith and Dodge-Wan, 2019), soil erosion is a three-stage process: detachment, transport, and deposition (Jahun et al., 2015). Where detachment is the detachment of soil particles. One result of raindrop impact was the detachment of sediment from the soil surface (Merritt et al., 2003). The study says water is responsible for 80% of soil erosion worldwide, while transport is the movement of the detached material from the accumulated flow; and deposition occurs when the transport forces are depleted (Sotiri, 2020).

Topographic characteristics, land use, soil texture, location of sediment sources, and catchment area are considered factors that depend on it to know the sediment patterns and transport rates (Kanito et al., 2023). So these are the most important factors that should be considered.

DOI: 10.1201/9781003473398-12

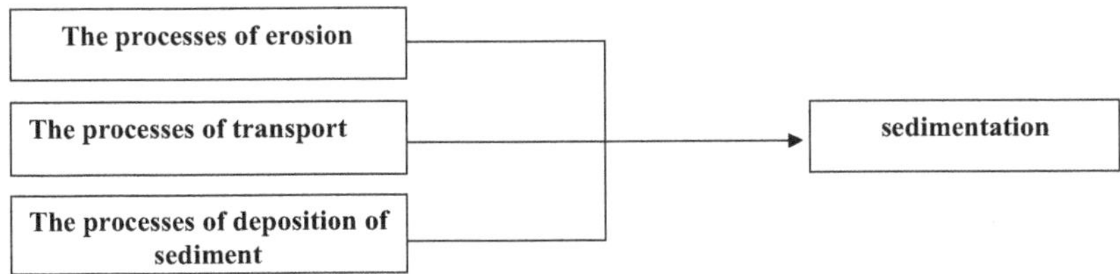

According to (Ikwap and Babirye, 2018), the rate of erosion depends on the following factors:

- Climate such as precipitation and runoff
- Geotechnics such as geology and soils
- Topography such as slope, catchment orientation, drainage basin area, and drainage density
- Vegetation cover, land use, and human impact

Among the disadvantages of soil erosion is; sediment deposition, presence of large quantities of sediment in the drainage system, hydropower generation problems, sediment may cause the riverbed levels to rise and lead to an increased risk of floods. Accordingly, sediment deposition has a negative impact on socio-economic life and the environment.

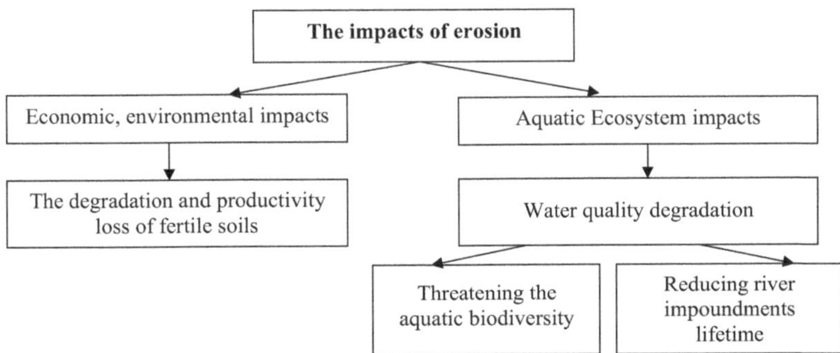

Six million hectares of land have been exposed to erosion and 120 million tons of the sediment have been transported through the water annually. In dams, water loss annually was estimated at 20 million m^3 due to silt (Dehni et al., 2015).

In Algeria, there are approximately 113 dams, including 52 large dams with a total storage capacity of about 5.2 million m^3, the dams receive approximately 32 m^3 of silt annually, heavy rains, and also caused floods during October 2001 in Bab el-Oued, Algiers. It lasted for two days, and resulted in as much as 80 million tons of sediment (Remini and Hallouche, 2004).

The liquid and solid yields of all wadis of the Hodna Basin, are received by the dry salt lake "Chott El Hodna" (1150 km²), the study shows that about 403 million m^3 of water and 11 million tons of sediment are lost annually in the Hodna watershed (Hasbaia, 2012), which poses problems for inundation and sediment deposition.

The main goal that needs to be achieved with sediment transport is to ensure the restoration of rivers and the design of hydraulic structures (Tayfur, 2023; Rather et al., 2020a; 2021). Knowledge of information regarding sediment transport is very necessary to maintain a fluvial environment for the river (Harun et al., 2023; Rather et al., 2020b).

There exist several models for use in simulating sediment transport, which differ in terms of their complexity, and the data required for model use (Merritt et al., 2003).

The best practice for estimating the sediment transport is the relationship proposed by Moore and Wilson (1992) called the sediment transport index (STI), which is based on unit stream power theory, a non-linear function between a specific discharge and slope, a dimensionless index; as for the stream power index (SPI), it is used to describe the ability to transfer sediment in channel streams and helps in demarcating areas that have the potential to be saturated with water.

Sediment indicators (STI, SPI) are used in scientific research to the erosion hazard mapping and areas associated with flash floods. These hydrological indices were used for mapping landslide susceptibility in the Himalayas of Nepal (Bannari et al., 2017).

The progress achieved in the area of science and technology offers us the possibility of finding the sediment yield using remote sensing (RS) techniques and Geographic Information Systems (GIS) (Merina et al., 2016). The main advantage of GIS is that the results would be presented quickly and efficiently (Abdelkrim and Nouibat, 2022).

This study reviewed the processing of sediment indicators (STI, SPI) in the Bou Saâda Wadi–Sub-basin, using the digital elevation model DEM and GIS tools, it is essential to know their importance in the decision-making process. This indicator is increasingly used in the field of erosion mapping, Experimental soil erosion models (RUSLE) have underestimated the evaluation of these indicators (Dehni et al., 2015).

10.2 SITE OF STUDY AND METHODOLOGY

10.2.1 SITE OF STUDY

The Hodna Basin is located at the south of the Mediterranean coast, and has a drainage area of 26,000 km². According to the Algerian Agency of Water Resources (ANRH), the Hodna Basin can be divided into 24 sub-basins (Hasbaia et al., 2012).

The Bou Saâda-Sub Basin is located in southwestern Hodna Basin in the highlands and has a drainage area of 1008 km² comprising of built-up areas, vegetation (e.g., forest, grass), and open land (e.g., mountains, soil) (Abdelkrim et al., 2023) (Figure 10.1).

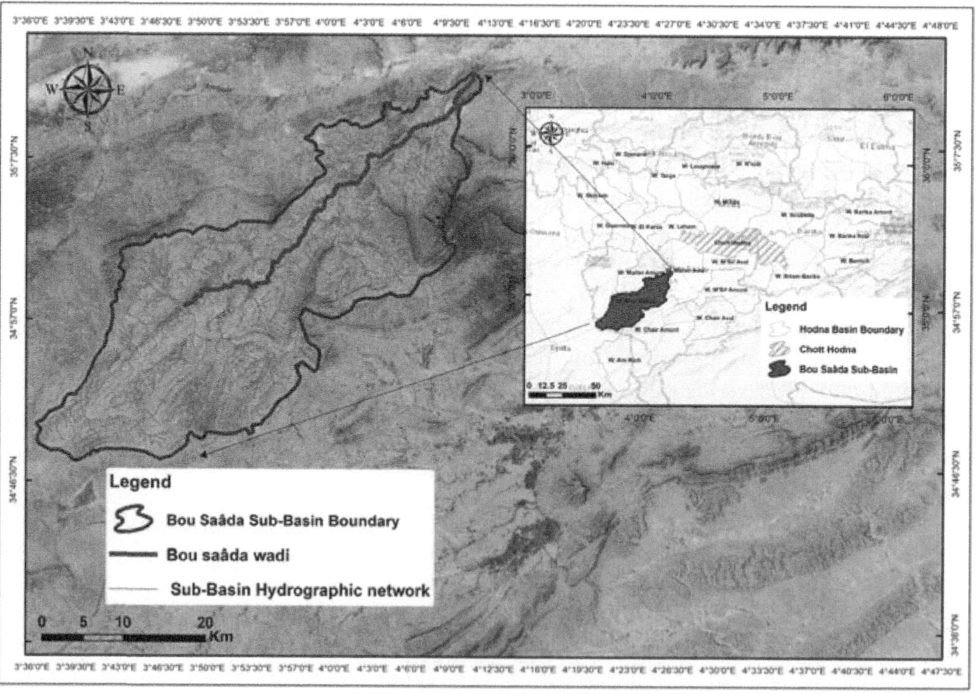

FIGURE 10.1   The geographical location of Bou Saâda Wadi–Sub-basin.

The Bou Saâda Wadi–Sub-basin is bordered to the north by the upstream Maitar Sub-Basin and to the south by the upstream Chair Sub-Basin that is characterized by the height of a mountain in the northeastern part of the basin. The study area suffers from natural disasters and human-caused disasters including the growing occupation of floodplains, increased runoff, and silted-up drainage, resulting in the occurrence of flash floods and erosion, and sediment transport and accumulation.

The Bou Saâda Wadi–Sub-basin has a climate of semi-aridness and is characterized by irregular rains which are associated with strong fluctuations. Heavy rains often cause erosion in the rainy months when the rainfall is highest the erosivity is also high, rainfall displaces soil particles on sloped surfaces and is transported into downslope in the form of sediments.

10.2.2 Methodology

Due to increased developments in computing power, the exploration of catchment erosion and sediment has become a necessity. Quantitative analysis of the digital elevation model (DEM) by Geographic Information Systems (GIS) provides various types of information about landscapes, digital elevation models (DEM) are most used for land surface characterization (Chowdhury, 2023) (Figure 10.4).

The research methodology is to use Geographic Information Systems (GIS) for studying the sediment indicators, including the sediment transportation index (STI) and stream power index (SPI) in the Bou Saâda Wadi–Sub-basin. The STI and SPI are among the most important indicators for the study of sediment, accumulation, and erosion. These indicators are derived from the digital elevation model DEM by using GIS.

10.2.2.1 Definition of STI

The sediment transportation index (STI) provides information on sediment transport capacity and accumulation and their spatial distribution (Chowdhury, 2023) (Figure 10.2).

$$STI = (As / 22.13)0.6 \times \sin(\beta / 0.0896)1.3 \tag{10.1}$$

Where As is the flow accumulation and β "is the basin slope.

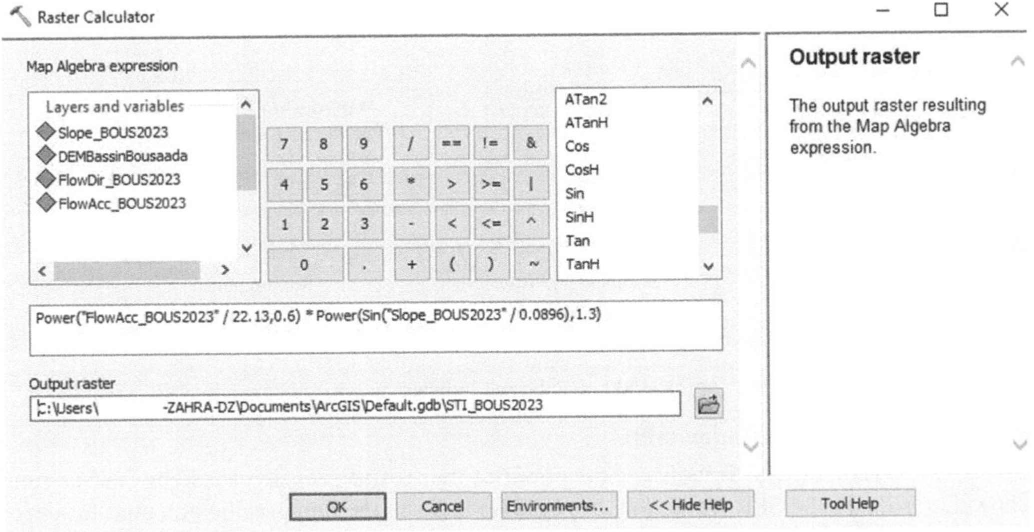

FIGURE 10.2   The calculation process of STI, in the raster calculator of ArcMap software.

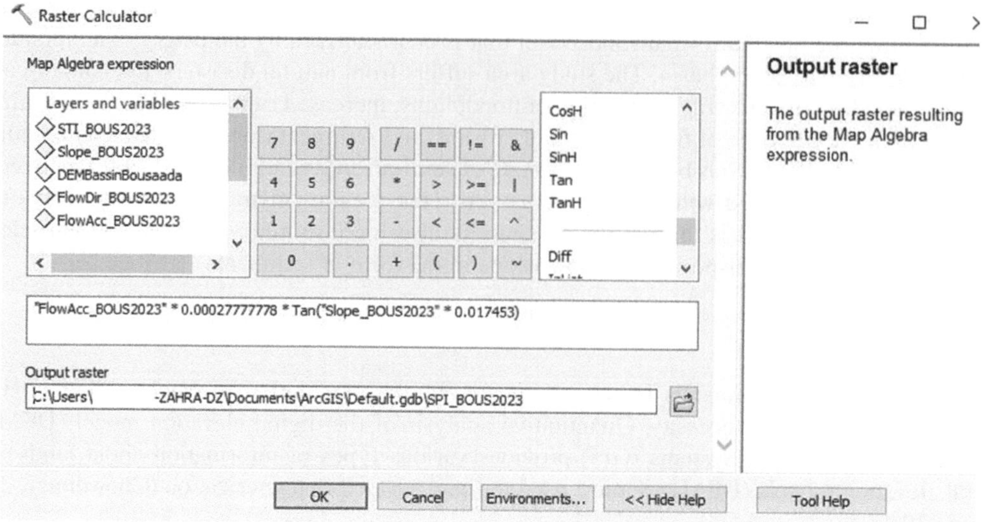

FIGURE 10.3 The calculation process of SPI, in the raster calculator of ArcMap software.

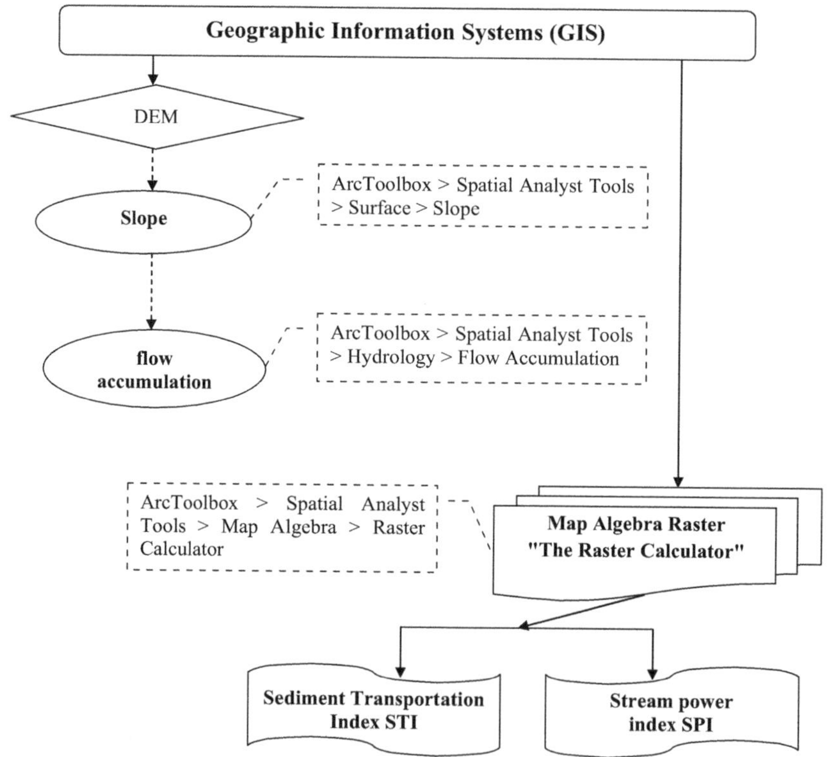

FIGURE 10.4 A methodology for the study of sediment indicators.

10.2.2.2 Stream Power Index (SPI)

The stream power index (SPI) indicates the power of the erosive process caused by surface runoff (Chowdhury, 2023), the SPI values (low or negative) indicate the topographic potential for deposition, while (positive) values indicate erosive areas (Figure 10.3).

$$SPI = (As * \tan \beta)$$

(10.2)

Where *As* is the basin flow accumulation and β is the basin slope.

Sediment indices are calculated based upon the processing of the DEM in GIS using a map algebra raster, "The Raster Calculator" in the ArcMap software. The slope layer and flow accumulation are created by using the spatial analyst toolbox in ArcGIS.

10.3 RESULTS AND DISCUSSION

The results of indices of the sediment using GIS in regard to calculating the STI and SPI in the Bou Saâda Wadi–Sub-basin are as follows:

10.3.1 SLOPE

The steeper slopes make areas prone to severe erosion, in comparison to low slopes.

The value of the slope gradient varies from 0 to 32.05. This has been reclassified into five classes, i.e., Class 1 (0–2%), Class 2 (2–3.39%), Class 3 (3.39–6%) Class 4 (6–9.30%), and Class 5 (9.30–32.05%). The study area shows a steep slope in the northern portion whereas in the southern portion, the study area is showing low to very low slopes (Figure 10.5).

10.3.2 SEDIMENT TRANSPORTATION INDEX (STI)

The STI index map is reclassified into four classes (Table 10.1).

The highest values (62.77–296.44) are corresponded to the highest levels of slopes, indicating a significant degree of soil erosion, and consequently, a significant degree of sediment transportation.

The lowest values (5.81–23.25) correspond to the lowest levels of slopes, indicating the low degree of soil erosion, and consequently, the accumulation of the sediment. Figure 10.6 shows the STI map and illustrates the spatial distribution of the sediment transport capacity and accumulation.

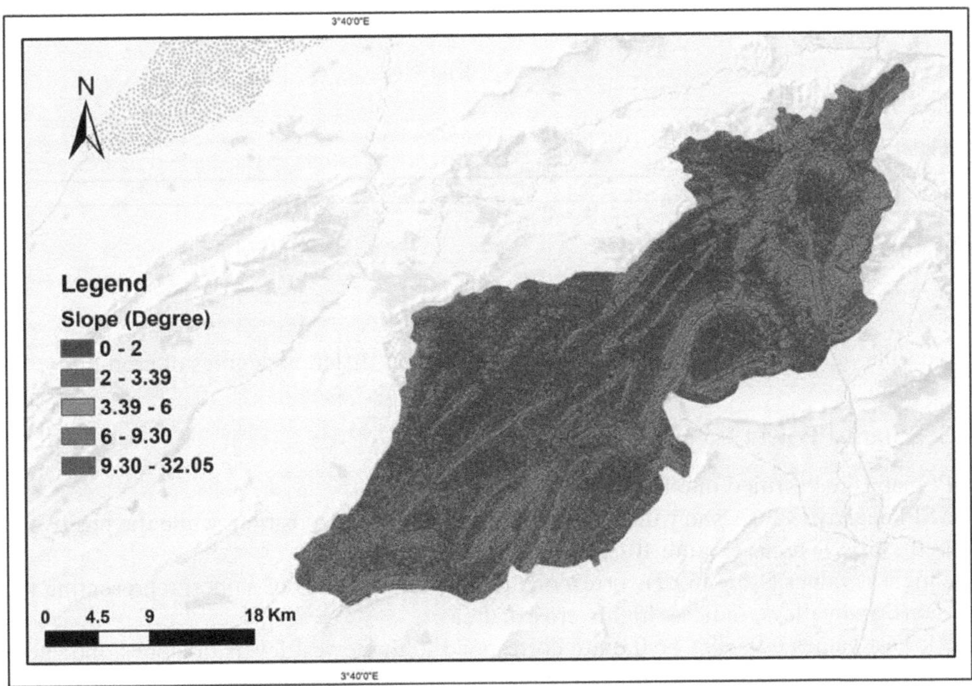

FIGURE 10.5   Slope map of the surface Bou Saâda Wadi–Sub-basin.

TABLE 10.1

Reclassification of the STI Map

The Classes	Sediment Transportation Index
Class 1	0–5.812666993
Class 2	5.812666994–23.25066797
Class 3	23.25066798–62.77680352
Class 4	62.77680353–296.4460166

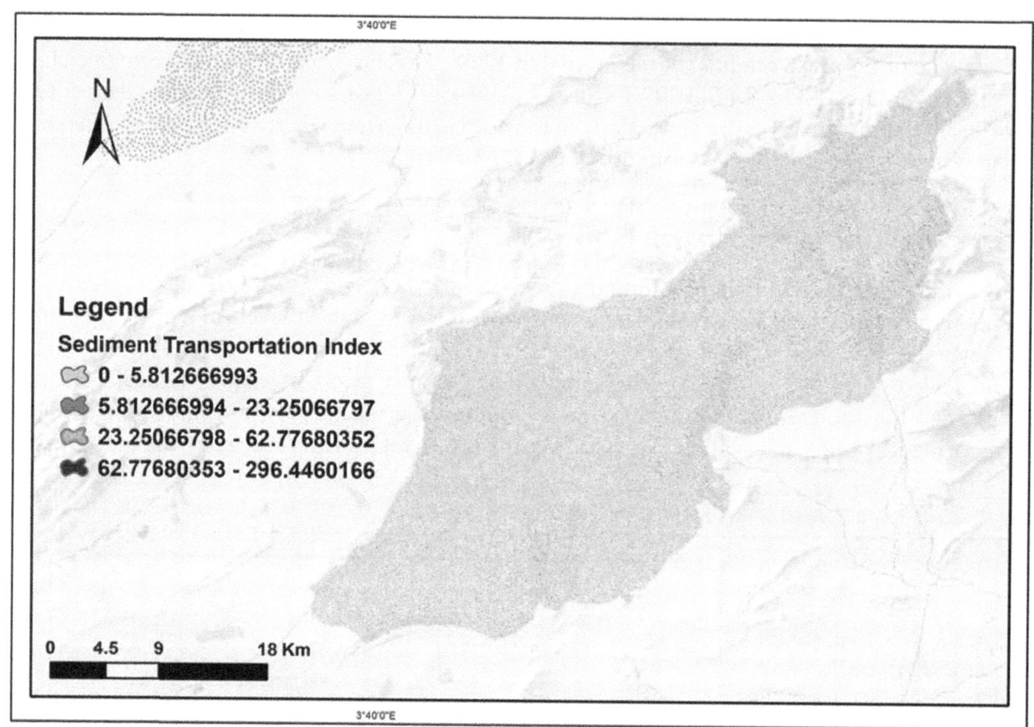

FIGURE 10.6   The STI map.

The results suggest that Bou Saâda Wadi–Sub-basin has different degrees of erodibility.

10.3.3 STREAM POWER INDEX (SPI)

The SPI map is reclassified into four classes (Table 10.2).

The SPI negative values show the topographic potential for deposition, while the positive values indicate the erosive areas (Figure 10.7).

The highest values (1.01–16.87) correspond to the highest levels of slopes, representing that the depressions, and valleys, indicate highly erosive areas.

The lowest values (−95.49– −6.03) are correspond to the lowest levels of slopes, indicating the sediment deposition and accumulation increases.

TABLE 10.2

Reclassification of the SPI Map

The Classes	Stream Power Index
Class 1	−95.49840042– −42.175665
Class 2	−42.17566499– −6.039596361
Class 3	−6.03959636–1.011343861
Class 4	1.011343862–16.87595936

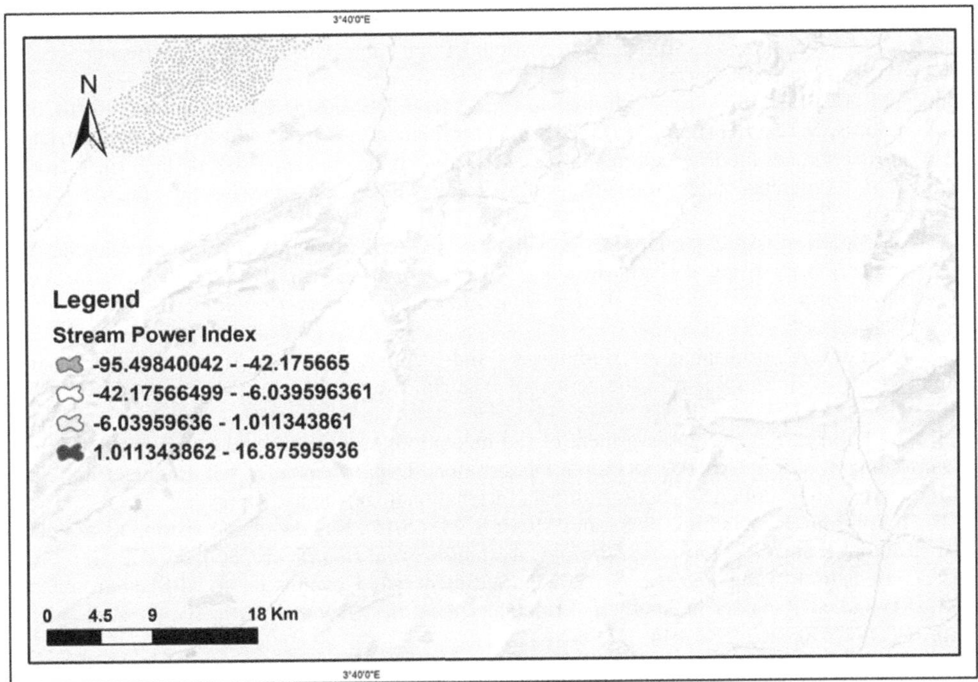

FIGURE 10.7   The SPI map.

10.4   CONCLUSIONS

The objective of this chapter is the study of sediment indicators in the Bou Saâda Wadi–Sub-basin, Algeria including the sediment transportation index (STI) and stream power index (SPI); the extraction stages of sediment indicators by processing digital elevation model (DEM) in Geographic Information Systems (GIS) were discussed.

The output map shows the different levels of erosion the in Bou Saâda Wadi–Sub-basin. It is worth recalling that the soil erosion and land degradation in the study area are caused due to the terrain morphology, this in addition to high-intensity rainfall would contribute to the erosion.

The GIS software has been used as a suitable input for the delineation of the sediment transport capacity, accumulation, and soil erosion represented by the STI and SPI maps.

The obtained results show that GIS has an important role in the study of sediment indicators in the Bou Saâda Wadi–Sub-basin. The GIS software tools have thus become an essential step towards crisis response, e.g., erosion and desertification are two of the most important causes of sediment.

REFERENCES

Abdelkrim, Z., and Nouibat, B. (2022). Assessing flood exposure in informal neighbourhoods: A case study of Bou Saâda, Algeria. *International Journal of Hydrology Science and Technology*, 13(1), 74–91.

Abdelkrim, Z., Nouibat, B., and Eslamian, S. (2023). Hydrological-Hydraulic modeling of floodplain inundation: A case study in Bou Saâda Wadi—Subbasin_Algeria, in Ch. 14. In Eslamian, S., and Eslamian, F. (Eds.), *Handbook of Hydroinformatics: Volume 3: Water Data Management Best Practices*. Elsevier, 219–232.

Amamra, A., Khanchoul, K., Eslamian, S., and Hadj Zobir, S. (2018). Suspended sediment estimation using regression and artificial neural network models: Kebir watershed, northeast of Algeria, North Africa. *International Journal of Hydrology Science and Technology*, 8(4), 352–371.

Bannari, A., Ghadeer, A., El-Battay, A., Hameed, N. A., and Rouai, M. (2017). Detection of areas associated with flash floods and erosion caused by rainfall storm using topographic attributes, hydrologic indices, and GIS. In Pirasteh, S., and Li, J. (Eds.), *Global Changes and Natural Disaster Management: Geoinformation Technologies*. Springer International Publishing, Cham, 155–174.https://doi.org/10.1007/978-3-319-51844-2_13

Chowdhury, M. S. (2023). Modelling hydrological factors from DEM using GIS. *MethodsX*, 10, 102062.

Dehni, A., Lounis, M., and Hassani, M. I. (2015). Géotraitement des indices hydro-morphométriques pour l'automatisation des modèles sédimentaires et érosifs (Application sur le BV de la Tafna – Nord-Ouest Algérien). Conférence internationale sur l'hydrologie des grands bassins fluviaux de l'AfriqueAt: Hammamet, Tunisie.

Harun, M. A., Ghani, A. A., Eslamian, S., and Chang, C. K. (2023). Sediment transport with soft computing application for tropical rivers, in Ch. 26. In Eslamian, S., and Eslamian, F. (Eds.), *Handbook of Hydroinformatics: Volume 3: Water Data Management Best Practices*. Elsevier, Amsterdam, 379–394.

Hasbaia, M., Seddi, A., Bournane, A., Hedjazi, A., and Paquier, A. (2012). Study of the water and sediment yields of Hodna Basin in the centre of Algeria, examination of their Impacts. ICSE6, Paris, 103–110.

Ikwap, F., and Babirye, S. (2018). Assessment of sedimentation level using remote sensing and GIS in Kiira Dam, Jinja District, Uganda. Doctoral Dissertation, Department of Civil Engineering, School of Engineering and Applied Sciences, Kampala International University, Kampala.

Jahun, B. G., Ibrahim, R., Dlamini, N. S., and Musa, S. M. (2015). Review of soil erosion assessment using RUSLE model and GIS. *Journal of Biology, Agriculture and Healthcare*, 5(9), 36–47.

Kanito, D., Bedadi, B., and Feyissa, S. (2023). Sediment yield estimation in GIS environment using RUSLE and SDR model in Southern Ethiopia. *Geomatics, Natural Hazards and Risk*, 14(1). DOI: 10.1080/19475705.2023.2167614.

Lim, K. J., Sagong, M., Engel, B. A., Tang, Z., Choi, J., and Kim, K. S. (2005). GIS-based sediment assessment tool. *CATENA*, 64(1), 61–80.

Merina, R. N., Sashikkumar, M. C., Rizvana, N., and Adlin, R. (2016). Sedimentation study in a reservoir using remote sensing technique. *Applied Ecology and Environmental Research*, 14(4), 296–304.

Merritt, W. S., Letcher, R. A., and Jakeman, A. J. (2003). A review of erosion and sediment transport models. *Environmental Modelling & Software*, 18(8–9), 761–799.

Moore, I. D., and Wilson, J. P. (1992). Length-slope factors for the Revised Universal Soil Loss Equation: Simplified method of estimation. *Journal of Soil and Water Conservation*, 47(5), 423–428.

Rather, N.A., Lone, M.A., Dar, A.Q., and Mir, B.H. (2021). Determination of hydraulic conductivity for granular filters based on constriction size and shape parameters. Water Supply, IWA Publication, Vol. 21, No. 8, 4121. https://doi.org/10.2166/ws.2021.167

Rather, N.A., Lone, M.A., Dar, A.Q., and Mir, B.H. (2020a). The influence of particle shape and gradation parameters on the permeability of filters media. *International Journal of Hydrology Science and Technology, Inderscience Publication*, 10(3), 267–286.

Rather, N.A., Lone, M.A., Dar, A.Q., Eslamian, S., Mir, B.H., and Dar, B.A. (2020b). Design criteria of protective filters based on particle shape and gradation parameters. *International Journal of Hydrology Science and Technology, Inderscience Publication*, 10(1), 102–122.

Remini, B., and Hallouche, W. (2004). La sédimentation dans les barrages algériens. *La Houille Blanche*, 90(1), 60–64.

Sotiri, K. (2020). Integrated sediment yield and stock assessment for the Passaúna Reservoir, Brazil. Doctoral dissertation, Dissertation, Karlsruhe, Karlsruher Institut für Technologie (KIT).

Tayfur, G. (2023). Developments in sediment transport modeling in alluvial channels. In Eslamian, S., and Eslamian, F. (Eds.), *Handbook of Hydroinformatics, Volume II: Water Data Management Best Practices*. Elsevier, Amsterdam, 257–264.

United Nation Environmental Program. (1991). Status of desertification and implementation of the UN plan of action to combat desertification. UNEP, Nairobi.

Vijith, H., and Dodge-Wan, D. (2019). Modelling terrain erosion susceptibility of logged and regenerated forested region in northern Borneo through the Analytical Hierarchy Process (AHP) and GIS techniques. *Geoenvironmental Disasters*, 6(1), 1–18.

11 Flash Food Hazard in Arid Regions
A review and Case Study

Ismail Abd-Elaty, Amir S. Ibrahim, Mahmoud M. Afify, and Abdelaziz El Shinawi

11.1 INTRODUCTION

Hydro-hazards are exciting events related to the movement and circulation of water (i.e., droughts and floods) with a combination of physical processes that occur across multiple spatial and temporal scales. Hydrological hazard (HH) principles are related to environmental hazards such as flash floods, especially in urban areas (El-Shinawi and Naymushina, 2015). Also, HH and their impacts are associated with climate variability, demographic trends, land-cover change, and other causative factors and could be exasperated by global climate change (CC) (Trajkovic et al., 2016 and Abu Salem et al., 2022). Droughts are affecting the flow to the groundwater and increasing aquifer abstraction rates for providing freshwater resources, the withdrawals of groundwater tripled in the last 50 years with unequal volumes in arid and semi-arid zones. It will certainly cause instabilities, where sustainable management on the basis of conservation to adapt the new climatic changes by sea level rise (SLR), the salinity of groundwater in coastal regions and inversion are observed in the saltwater intrusion (Abdel-Aty et al., 2024; Shinawi et al., 2020, 2022).

SLR is a natural phenomenon and its acceleration that is currently observed, such as the thermal expansion of the oceans and seas and the melting of glaciers and ice caps including the Greenland and Antarctic ice sheets. According to the Intergovernmental Panel on Climate Change (IPCC, 2013), in the year 2100 about 95% of the coastal areas in the world will be considerably affected by SLR, hence increasing the risk of inundation in internal land and saltwater intrusion (SWI) in coastal aquifers (Agren and Svensson, 2007; Abdel-Aty et al., 2019a; 2023).

Saltwater intrusion (SWI) degrades groundwater quality due to an increase in the aquifer salinity to levels exceeding acceptable drinking water and irrigation standards in many coastal areas and is accelerated by freshwater boundary changes due to over-pumping and the impact of CC by decreasing aquifer recharge due to the droughts and SLR (Abdel-Aty et al., 2021a). Coastal regions are influenced by natural and artificial changes, the changing of the coastal aquifers boundary conditions including SLR, decreasing inland recharge, and overexploitation of production wells' water resources (Abdel-Aty et al., 2022a).

Land subsidence is a common environmental consequence of over-pumping groundwater due to groundwater resource usage for domestic and industrial purposes. Commonly, it leads to reduction, unceasing distortions of the land surface, and, consequently, to economic and social consequences (Attwa and El Shinawi, 2014). Simple aquifer system compaction according to its drainage is observed in many urban areas in the world (Guzy and Malinowska, 2020; El Shinawi et al., 2022).

Water balance is the cyclical movement of water between the atmosphere and the ground surface, considering precipitation, evaporation, and runoff (Whittow, 1984) (Figure 11.1). *Water balance* equation presents elementary parts of water balance: precipitation, evapotranspiration, and runoff, expressed as the volume of water (Sposób, 2011).

DOI: 10.1201/9781003473398-13

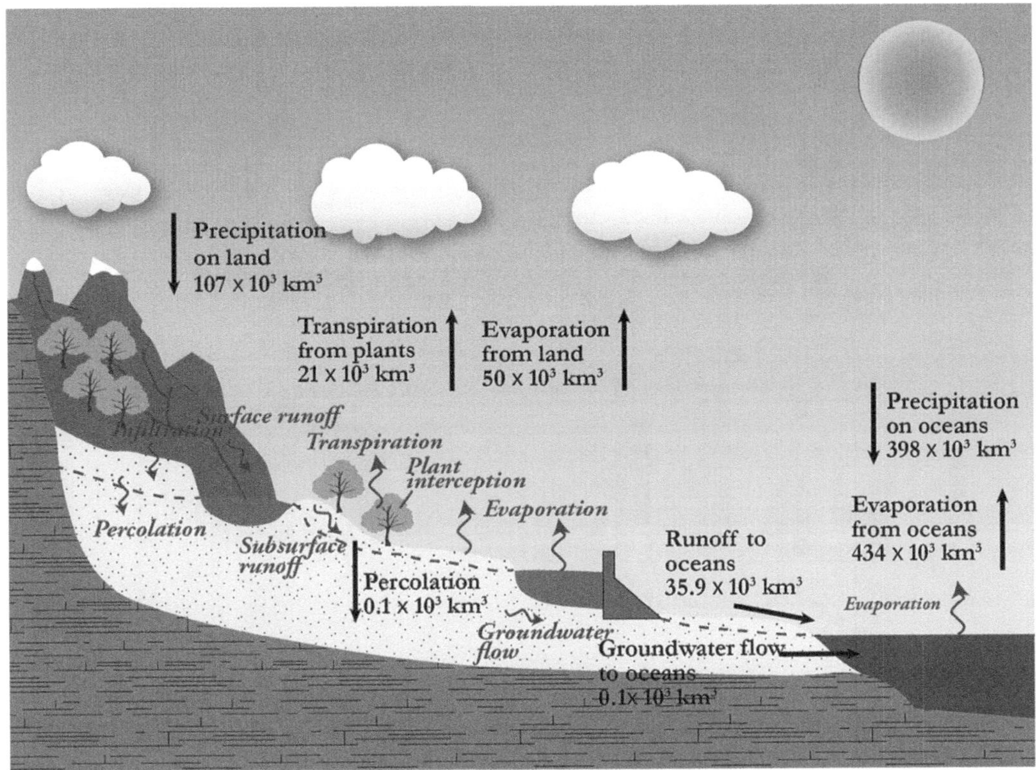

FIGURE 11.1   Hydrological cycle estimates the main fluxes in km³/year (Mascaro, G., 2018; from http://scied.ucar.edu/longcontent/water-cycle).

11.2   CLIMATE CHANGE AND HYDROLOGICAL SYSTEM

In many countries, the concerns of the manufacturing sector are concerning global warming and the possible impacts of current CC, particularly those associated with changes in the average surface temperature of the Earth (Zamani Nuri et al., 2013). Additionally, CC affects global temperature and precipitation patterns (see Figure 11.2). These effects, in turn, influence the intensity and, in some cases, the frequency of extreme environmental events, such as forest fires, hurricanes, heat waves, floods, droughts, storms, and saltwater intrusion (SWI) (Abd-Elaty and Straface, 2022).

The increase in greenhouse gases in the atmosphere will continue leading to global warming and an intensification of the hydrological cycle (HC), making studies of hydrological extremes more complex and challenging (Trajkovic et al., 2016).

11.2.1   IMPACT OF CLIMATE CHANGE ON PRECIPITATION

The Intergovernmental Panel on Climate Change (IPCC, 2001) predicted that the global average water vapor concentration and precipitation will increase in the 21st century. In the second half of the 21st century, winter precipitation is likely to increase in northern mid-to-high latitudes and in Antarctica, while there are both regional increases and decreases over land areas at low latitudes. Tabari (2020) based on the ensemble median of the Coupled Model Intercomparison Project Phase 5 (CMIP5) (Taylor et al., 2012) global climate models (GCMs), water-limited regions are mainly located in North Africa and the Middle East (MENA), and Australia, while water-abundant regions are located in the mid-latitudes and the tropics (Figure 11.2a). About 72% of global land is likely to undergo aridification in the future, with substantial aridification (aridity increase of > 30%) in MENA, south Europe, South Africa, and Australia leading to a shift in climate regimes (Figure 11.2b).

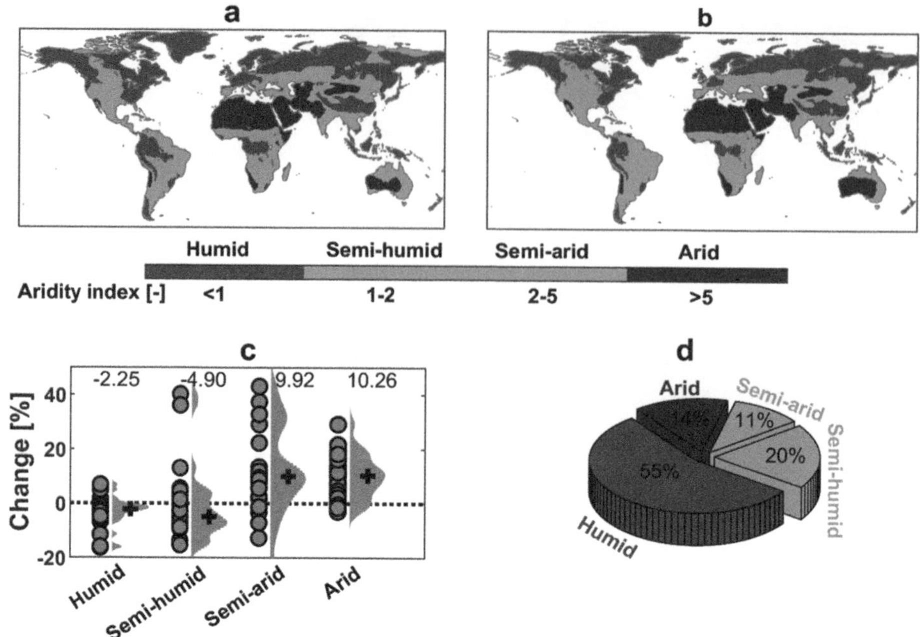

FIGURE 11.2 Aridity index and its expected future changes for (a) historical (1971–2000) and (b) future (2070–2099) climates. (c) projected change in area coverage of each climate regime for the period 2070–2099 and (d) ensemble median projected area coverage of each of the five climate regimes in percent of total terrestrial land area for 2070–2099 (URL https://www.mathworks.com/products/mapping.html); Tabari, 2020).

Globally, arid and semi-arid regions would expand by 10.3% and 9.9%, respectively, while humid and semi-humid regions would decline by 2.3% and 4.9%, respectively (Figure 11.2c). This makes the area coverage of humid, semi-humid, semi-arid, and arid climates at the end of the 21st century equal to 55%, 20%, 11%, and 14% of the total terrestrial land area, respectively (Figure 11.2d). Sarkar and Maity (2021) showed that the shift in climate regimes around the 1970s caused an overall improvement in extreme precipitation across the world with a specific spatial distribution pattern.

11.2.2 Impact of Climate Change on Sea Level Rise

The tide gauges and satellite altimeters are used for SLR measurements. Tide gauge stations from around the world have measured the daily high and low tides for more than a century, using a variety of manual and automatic sensors. Since the early 1990s, sea levels have been measured from space using radar altimeters, which determine the height of the sea surface by measuring the return speed and intensity of a radar pulse directed at the ocean (Figure 11.3) (IPCC, 2007; UNDRR; 2017).

Estimating SLR due to thermal expansion, scientists measure sea surface temperature by using moored and drifting buoys, satellites, and water samples collected by ships. As global temperatures continue to warm, further SLR is inevitable and depends mostly on future rate of greenhouse gas emissions. In the 2022 report, with the lowest possible greenhouse gas emissions and warming (1.5°C), global mean SLR of at least 0.3 m while very high rates of emissions that trigger rapid ice sheet collapse, sea level could be as much as 2 m higher in 2100 than it was in 2000. Moreover, SLR is expected between 18 and 58 cm in 2100 (IPCC, 2007; UNDRR; 2017).

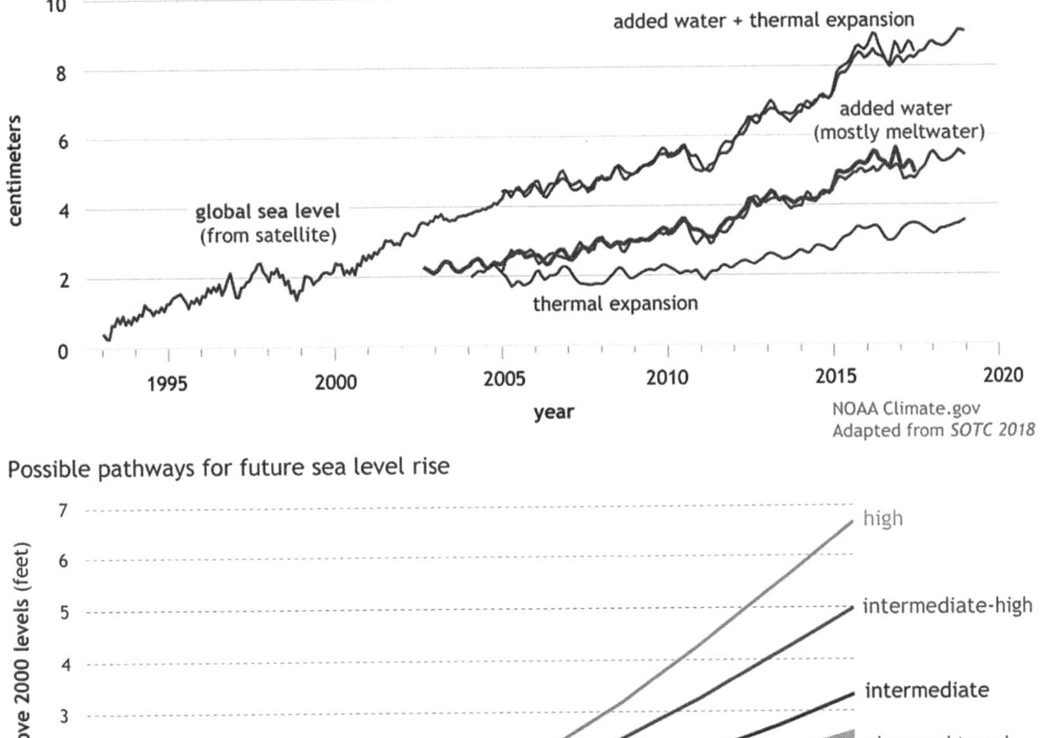

FIGURE 11.3   SLR estimation and future prediction for (a) observed sea level since the start of the satellite altimeter record in 1993, plus independent estimates of the different contributions to SLR: thermal expansion and added water, mostly due to glacier melt. Added together (Blunden et al., 2019; https://www.climate.gov/media/12868) and (b) observed sea level from 2000–2018, with future sea level through 2100 for six future pathways (Sweet et al., 2022; https://www.climate.gov/media/14136).

11.3   HAZARD AND DISASTER

A hazard is defined as a process, phenomenon, or human activity that has the potential to cause loss of life, injury, health impacts, property damage, social and economic disruption, or environmental degradation. Hazards may be natural, anthropogenic, or socio-natural in origin. The hazard classification schemes vary across different research institutions and governments, but these can be divided the following subsections. Natural hazards are further categorized into biological, environmental, technological, hydrological, geological or geophysical, hydro-meteorological, technological and hydrological hazards (UNDRR; 2017; Prasad and Francescutti, 2017).

11.3.1.1   Biological Hazards

It is of organic origin or conveyed by biological vectors, including pathogenic microorganisms, toxins, and bioactive substances. Examples are bacteria, viruses, or parasites, as well as venomous wildlife and insects, poisonous plants, and mosquitoes carrying disease-causing agents. An example of biological hazards is a room, a bar, a classroom, and how the coronavirus is spread through the air (UNDRR; 2017). Biological hazards are events that involve the rapid incidence and prevalence of vector-driven diseases, toxins, or pathogens. The emergence of severe acute respiratory syndrome in 2003 is an example of a biological hazard that resulted in disastrous effects (spread through air travel), killing over 400 people (U.N. ISDR, 2004).

11.3.1.2   Environmental Hazards

These hazards may include chemical, natural, and biological hazards. It created by environmental degradation, physical, or chemical pollution in the air, water, or soil. Moreover, many of the processes and phenomena falling into this category can be characterized as hazard and risks rather than hazards in themselves, such as soil degradation, deforestation, loss of biodiversity, salinization, and SLR. An example of an environmental hazard: SLR which it may erode the development in Africa due to the flooding of lands (UNDRR; 2017).

11.3.1.3   Geological or Geophysical Hazards

It originates from internal Earth processes. Examples are earthquakes, volcanic activity and emissions, and related geophysical processes such as mass movements, landslides, rockslides, surface collapses, and debris or mud flows. Hydro-meteorological factors are important contributors to some of these processes (UNDRR; 2017; Nabih and El Sinawi, 2020). Tsunamis are difficult to categorize: although they are triggered by undersea earthquakes and other geological events, they essentially become an oceanic process that is manifested as a coastal water-related hazard. Geological hazards involve changes due to tectonic plates or fault shifting (tsunamis result from shifting of faults) or mass movement of solid mass (U.N. ISDR, 2004).

11.3.1.4   Hydro-meteorological Hazards

The risk includes atmospheric, hydrological, or oceanographic origin. Examples of hydro-meteorological hazards are tropical cyclones also known as typhoons and hurricanes; flash floods; drought; heatwaves and cold spells; and coastal storm surges. These risk conditions are a factor in other hazards such as landslides, wildland fires, locust plagues, epidemics, and transport and dispersal of toxic substances and volcanic eruption material (Kramarenko et al., 2016; UNDRR, 2017; El Shinawi et al., 2021). An example of a hydro-meteorological hazard: CC causes hurricanes that make landfall to stay for longer and be stronger (UNDRR, 2017). Hydro-meterological hazards, the most common natural hazards, are related to the climate, oceans, or movement of wet mass such as floods, droughts, and heat waves like the 2010 Russian heat wave which resulted in a disaster that claimed 56000 lives (U.N. ISDR, 2011).

11.3.1.5   Technological Hazards

It originates from technological or industrial conditions, dangerous procedures, infrastructure failures, or specific human activities. The risk examples are including industrial pollution, nuclear radiation, toxic wastes, dam failures, transport accidents, factory explosions, fires, and chemical spills. It is a result of the impacts of a natural hazard event (UNDRR, 2017; El Shinawi et al., 2021).

The examples of technological hazards are industrial pollution, nuclear leak, or structural dam collapse, while environmental degradation hazards include events that disrupt the environment, ecosystem, or natural resources (i.e., deforestation, forest fires, changes in climate) (U.N. ISDR, 2011).

Each hazard often triggers a sub-set of hazards. For example, tropical cyclones (known as hurricanes in the Atlantic Ocean, cyclones in the Indian Ocean, and typhoons in the Northern Pacific Ocean) can cause intense winds, storm surges, and heavy precipitation, as well as trigger secondary risks, for example landslides. A series of triggering relationships can cause a domino or cascading effect, for example in the case of the tsunami–earthquake–nuclear crisis in Japan, in 2011 (U.N. ISDR, 2011; UNDRR, 2017).

11.4 HYDROLOGICAL HAZARDS

Hydrological hazards ("hydro-hazards") are defined as extreme events associated with the occurrence, movement, and distribution of water, specifically resulting in floods and droughts. Due to global CC, these hazards are expected to change in the future, with areas of the globe becoming "hotspots" for the intensification of these extremes (Beevers et al., 2022). As global surface temperatures increase, more droughts are likely to occur. As more water evaporates, increasing heat in the atmosphere and raising ocean surface temperatures, it can lead to higher wind speeds during tropical storms. Rising sea levels expose higher locations not usually subjected to the power of the sea and to the erosive forces of waves and currents (USGS; https://www.usgs.gov/faqs/how-can-climate-change-affect-natural-disasters).

11.4.1 CLIMATE CHANGE AND HYDROLOGICAL HAZARDS

Climate change (CC) is likely to increase the stress on currently stressed resources, especially in the developing world. Studies have shown that most systems are sensitive to both the magnitude and the rate of CC (e.g., Gleick, 1998; Abdel-Aty et al., 2022b). However, the vulnerability of a system to the expected change depends on economic strength and existing infrastructure as well as overall country resilience to cope with different risks. Most developing countries are generally more vulnerable and less able to adapt. On the other hand, the challenges of CC are uncertain in terms of impacts, magnitude, spatial distribution, and onset. Being recognized recently, the older strategies did not include sufficient actions to face its serious consequences which cannot be ignored (Khedr, 2017).

In order to reduce the expected impacts of CC, it is necessary to both reduce (mitigate) emissions of heat-trapping pollution and build resilience (adapt) to the impacts of CC. However, even with strong programs to reduce greenhouse gas emissions (which proved to be a very difficult process), the effects of CC will persist due to the longevity of certain greenhouse gases in the atmosphere and the absorption of heat by the oceans. Therefore, adaptation has a major role to play in reducing the impacts of CC on people, businesses, and society at large (Trajkovicet al., 2016).

HH of various types present a myriad of technical and public policy issues worldwide. It defined as extreme events associated with water occurrence, movement, and distribution, HH include droughts, flooding, and related events (e.g., landslides, river scour, and deposition). HH and their impacts are associated with climate variability, demographic trends, land cover change, and other causative factors and could be exasperated by global CC (Trajkovicet al., 2016).

The increase in greenhouse gases in the atmosphere will continue leading to global warming and an intensification of the HC, making studies of hydrological extremes more complex and challenging (Trajkovicet al., 2016).

11.4.2 TYPES OF HYDROLOGICAL HAZARDS

11.4.2.1 Drought

The existing rise in temperature and weather conditions have a major impact on rainfall. The increase in rainfall has probably resulted in flooding and/or drought. The primary cause of any drought is below-average rainfall. Drought is different from other hazards in that it develops slowly, sometimes over years, and its onset can be masked by a number of factors. Drought can be devastating: water supplies dry up, crops fail to grow, animals die, and malnutrition and ill health become widespread. An increase or decrease in the ambient temperature, in different regions, will allow the existence of new living conditions experienced by the scenarios that will be conditioned by the presence and/ or absence of the superficial or underground H_2O depending on the phreatic depth and the water retention capacity of different types of soil. Henceforth, it is recommended in the places where the ambient temperature decreases or increases, studying the availability of water resources from a point of view where the variables are related by the geographical effects allows the development of climatic conditions. The decrease in water availability results in aridity and when the pluviometric manifestation occurs in large volumes that exceed the saturation threshold of the soil, flooding occurs (Naymushina et al., 2010). At present, concepts of drought and a weather condition in the absence of rain can be tolerated by all plants. Water deficit refers to the water content of a tissue or cell in a plant below the highest water content that can resist, and is exhibited in the state of greatest hydration. The complexity in the analysis of this type of phenomena is attributed to the forms that originate them, as well as the lack of detailed quantitative information on the changes in the infiltration capacity of the different soil types and the fertile profiles of hydraulic conductivity with respect to the soil depth, moisture retention capacity, and the depth of development of plant roots, in front of CC risks, the presence of systematic sampling campaigns is demanded in order to perform numerical models of associated extreme flow rates in hydrological basins (Taiz and Zeiger, 2009).

11.4.2.2   Flash Flood

Floods and flash floods can occur anywhere after heavy rain events. All floodplains are vulnerable and heavy rain or thunderstorms can cause flash flooding in any part of the world. Flash floods can also occur after a period of dry conditions when moderate or heavy rain falls onto very dry, hard ground that the water cannot penetrate. Floods come in a number of forms, from small flash floods to sheets of water covering extensive areas of land. They can be triggered by severe thunderstorms, tropical cyclones, large low-pressure systems, monsoons, ice jams, or melting snow. In coastal areas, storm surges caused by tropical cyclones, tsunamis, or rivers swollen by exceptionally high tides can cause flooding. Dikes or flood levees can overtop causing floods when the rivers carry large amounts of snowmelt. Dam breaks or sudden regulatory operations such as the release of water for hydroelectric power generation can also cause catastrophic flooding. Floods threaten human life and property worldwide. Some 1.5 billion people were affected by floods in the last decade of the 20th century (https://public.wmo.int/en/our-mandate/focus-areas/natural-hazards-and-disaster-risk-reduction). For flash flood management actions to be effective, detailed planning is key for the participation of users such as local communities. Flash flood management touches upon several sector-specific areas, so responsibilities and roles in each stage of management should be defined within a solid legal and institutional framework. For flash floods, local communities especially have much more decisive responsibilities than those for riverine flood management (WMO, 2012). The hazard is defined as a flash flood magnitude which is defined by probability of occurrence. This information can be translated to inundation maps showing the areas at risk as well as depths of inundation and related water velocities. The exposure describes human activity and the environment in the hazard zone (who and what are endangered by flash floods). The third parameter that characterizes the general notion of risk is the vulnerability of a given region to flood damage. Presently, various definitions of this parameter are applied depending on the needs. One of these concerns is the susceptibility of a region to flood losses, which is defined via the geophysical, economic, and societal attributes of a region (WMO, 2023).

11.4.2.3   Tropical Cyclones

A tropical cyclone is a rapidly rotating storm originating over tropical oceans. It is typically around 200 to 500 km in diameter but can reach 1000 km. A tropical cyclone brings very violent winds, torrential rain, high waves, and, in some cases, very destructive storm surges and coastal flooding (WMO, 2023).

11.4.2.4   Landslides, River Scour and Deposition

Mudslides and landslides are local events and are usually unexpected. They occur when heavy rain; rapid snow or ice melt; or an overflowing crater lake loosens vulnerable parts of the landscape on steep slopes, resulting in large amounts of earth, rock, sand, or mud flowing swiftly down the slope. Hillsides or mountainsides that are bare or have had their vegetation cover degraded through clearance or by forest or brush fires may be especially at risk. They can reach speeds of over 50 km/h and can bury, crush, or carry people away, objects, and buildings. In Venezuela in 1999, after two weeks of continuous rain, landslides and mudflows slid down a mountain, destroying towns and causing an estimated 15000 fatalities (https://www.safecommunitiesportugal.com/natural-hazards-types/).

11.4.2.5 Heavy Rain and Snow, Strong Winds

Heavy rain and snow are dangerous for vulnerable communities. They can exacerbate rescue and rehabilitation activities after a major disaster, such as the earthquake in Pakistan in October 2005. They bring havoc to road and rail transportation, infrastructure, and communication networks. An accumulation of snow can cause the roofs of buildings to collapse. Strong winds are a danger for aviation, sailors, and fishermen, as well as for tall structures such as towers, masts, and cranes. Blizzards are violent storms combining below-freezing temperatures with strong winds and blowing snow. They are a danger to people and livestock. They cause airports to close and bring havoc to roads and railways (https://www.safecommunitiesportugal.com/natural-hazards-types/).

11.5 IMPACT OF HYDROLOGICAL HAZARDS

Hydrological risk are influenced by both climatic factors (e.g., temperature, rainfall patterns, and intensity) and hydrological factors (e.g., land use, soil, and bedrock). With the impact of CC on global and regional weather, where is growing evidence that there will be changes in the severity and frequency of weather and climate-related natural hazards (Collet et al., 2018; Visser-Quinn et al., 2021; Beevers et al., 2022).

11.5.1 LAND-USE OR LAND-COVER CHANGE

The environmental changes in land use and land cover (LULC) such as urbanization have an irreparable impact on urban ecosystems (Lu et al., 2019). Because LULC terms are frequently used together, they have to be defined clearly (Rujoiu-Mare and Mihai, 2016). Accordingly, the characterization of HH by studying the geomorphological features and detecting the LULC changes is critical to dealing with the negative environmental impacts of urban growth especially in desert lands. The LULC change is a major issue of concern regarding change in the global environment. Due to the rising population over the years, lots of pressure has been imposed on land resources and CC. Therefore, the expansion of cultivation, decline of grassland, and other unsustainable land uses have become major concerns, especially in semi-arid/arid regions (Lu et al., 2019).

11.5.2 CLIMATE VARIABILITY

Climate variability includes all the variations in the climate that last longer than individual weather events, whereas the term CC only refers to those variations that persist for a longer period of time, typically decades or more. Climate variability may result from natural internal processes within the climate system (internal variability) or from variations in natural or anthropogenic external forces (external variability).

11.5.3 DEMOGRAPHIC TRENDS

Demographic change trends can affect the growth rate of the economy, structural productivity, living criteria, reserves rates, consumption, and investment; and they can influence the long-run unemployment rate and equilibrium interest rate, housing market trends, and the demand for financial assets. The five main demographic segments are age, gender, occupation, cultural background, and family status. Demographic segmentation groups customers and potential customers together by focusing on certain traits such as age, gender, income, occupation, and family status.

The time it takes for water to travel through the terrestrial HC and critical zone is of great interest in Earth system sciences with broad implications for water quality and quantity. To date, most water age studies have focused on individual compartments (or sub-disciplines) of the HC such as the unsaturated or saturated zone, vegetation, atmosphere, or rivers. However, recent studies have shown that processes at interfaces between the hydrological compartments (e.g., soil–atmosphere or soil–groundwater) govern the age distribution of water fluxes between these compartments and thus can therefore greatly affect the travel times of water. The broad variation from complete to almost absent water mixing at these interfaces affects the water ages in the compartments (Sprenger et al., 2019).

This is particularly the case for the highly heterogeneous critical zone located between the top of the vegetation and the bottom of the aquifers. Here, a wide variety of studies on water ages in the critical zone were reviewed and provide (i) an overview of new perspectives and challenges in the use of hydrological tracers to study water ages, (ii) a discussion of the limiting assumptions related to our lack of process understanding and methodological transfer of water age estimations to individual disciplines or compartments, and (iii) a vision of how to enhance future interdisciplinary efforts to better understand the feedbacks between the atmosphere, vegetation, soil, groundwater, and surface water that control water ages in the critical zone (Sprenger et al., 2019).

11.6.4 Infrastructure

The relevant infrastructures include the drainage system, the hydropower generation, and the irrigation and drainage pumping stations. The drainage system protects the irrigated soils against salt accumulation and enables recycling of the irrigation water. It consists of an extensive drainage network of field drains, sub-collectors, and collectors and main drains, which either convey the drainage water or discharge it into coastal or inland lakes.

However, the effect of land subsidence on infrastructures is the compaction or compression of unconsolidated and partly consolidated sediments. Consequently, this damages the roads, railways, bridges, and pipelines, putting them at risk for structural failure. Subsidence reduces channel and canal flow capacity and freeboard and increases flooding (El-Shinawi and Naymushina, 2015). In addition, natural hazards such as collapsing roads, broken pipe utilities, damaged canals, and cracked foundations/separation. Furthermore, earth fissures can create harm to humans and animals and rapidly damage infrastructure, constantly changing elevation data. In accordance, groundwater over-pumping, degrading groundwater quality, and increased salinity (coastal and delta areas) potentially increased the groundwater contamination. Land subsidence can also lead to major economic losses such as structural damage and high maintenance costs for roads, railways, dikes, pipelines, and buildings. The total bill worldwide amounts to many billions of dollars annually. It can only rise further in the future with population growth and the intensification of economic activities in delta areas. Finally, the plan for subsidence increased freeboard requirements for infrastructure which increased planned maintenance budgets and developed requirements for well permits, groundwater modeling, and sustainability (https://www.deltares.nl/en/expertise/areas-of-expertise/land-subsidence).

11.6.5 Natural Environments

The natural environment underpins all aspects of our lives. It will be affected by CC, and yet we will be increasingly reliant on it to help us manage the impacts that a changing climate will bring. It is a complex system that is affected by a vast range of people and organizations across all sectors and a range of pressures beyond those to do with the changing climate. This is a new kind of challenge and will require new approaches from all of us (Defra, 2010).

The resources and services that we receive from our environment cannot be adequately understood in isolation from one another. We will be able to find more efficient and effective solutions to the challenges posed by a changing climate if we consider individual parts of the system, such as water, biodiversity, agriculture, and landscape as components of a complex and highly interconnected whole. We need to develop sustainable solutions that work effectively for people, the economy, and the natural environment. There is a strong case for close and collaborative working with other experts, stakeholders, and delivery partners to encourage a coherent response to the challenges ahead.

The natural environment is an integral part of the climate system, it is both affected by and affects climate globally and locally. To implement effective adaptation measures in the UK account of both needs to be taken. It is important adaptation measures take a holistic, cross-sectoral approach bringing together different sectors. Taking an integrated approach to action may result in mutual benefits by reducing emissions, CC impacts, and biodiversity loss (Sposób, 2011).

11.7 MITIGATION OF CLIMATE CHANGES

The negative impacts of hazards, particularly natural hazards, often cannot be completely avoided, but their scale or severity can be significantly mitigated through various strategies and actions. Mitigation measures include engineering techniques and hazard-resistant construction as well as improved environmental and social policies and public awareness. It should be noted that in climate change policy, the term "mitigation" is defined differently and refers to the reduction of greenhouse gas emissions that cause climate change (UNDRR 2017).

Improving our knowledge of hazards and conducting hazard assessments can help us locate and, in the case of some hazards, anticipate over different time periods when they might occur. Anticipation ranges from probabilistic analysis of long-term hazard occurrence to monthly, daily, or even hourly hazard detection and monitoring, in order to inform early warning systems (EWS). Improving our knowledge of hazards and carrying out hazard assessments can help us to locate and, in the case of certain hazards, anticipate their occurrence over different periods of time . Anticipation ranges from probabilistic analysis of long-term hazard occurrence to monthly, daily, or even hourly detection and monitoring to inform EWS (UNDRR 2017).

Warning systems must be accompanied by disaster risk reduction strategies to reduce vulnerability and improve the capacity of populations to respond and recover from a disaster. In the case of slow-onset hazards, if early indicators of a potential crisis are detected then a warning can be a key tool for building resilience, as exemplified by food security EWS. An effective EWS includes four components: (i) hazard detection, monitoring, and forecasting; (ii) analysis of the risks involved; (iii) dissemination of timely and authoritative warnings; and (iv) activation of emergency preparedness and response plans. These need to be coordinated across many agencies at the national and community levels for the system to work – failure of one component, or lack of coordination, can lead to the failure of the whole (UNDRR, 2017).

FIGURE 11.4   Study area location.

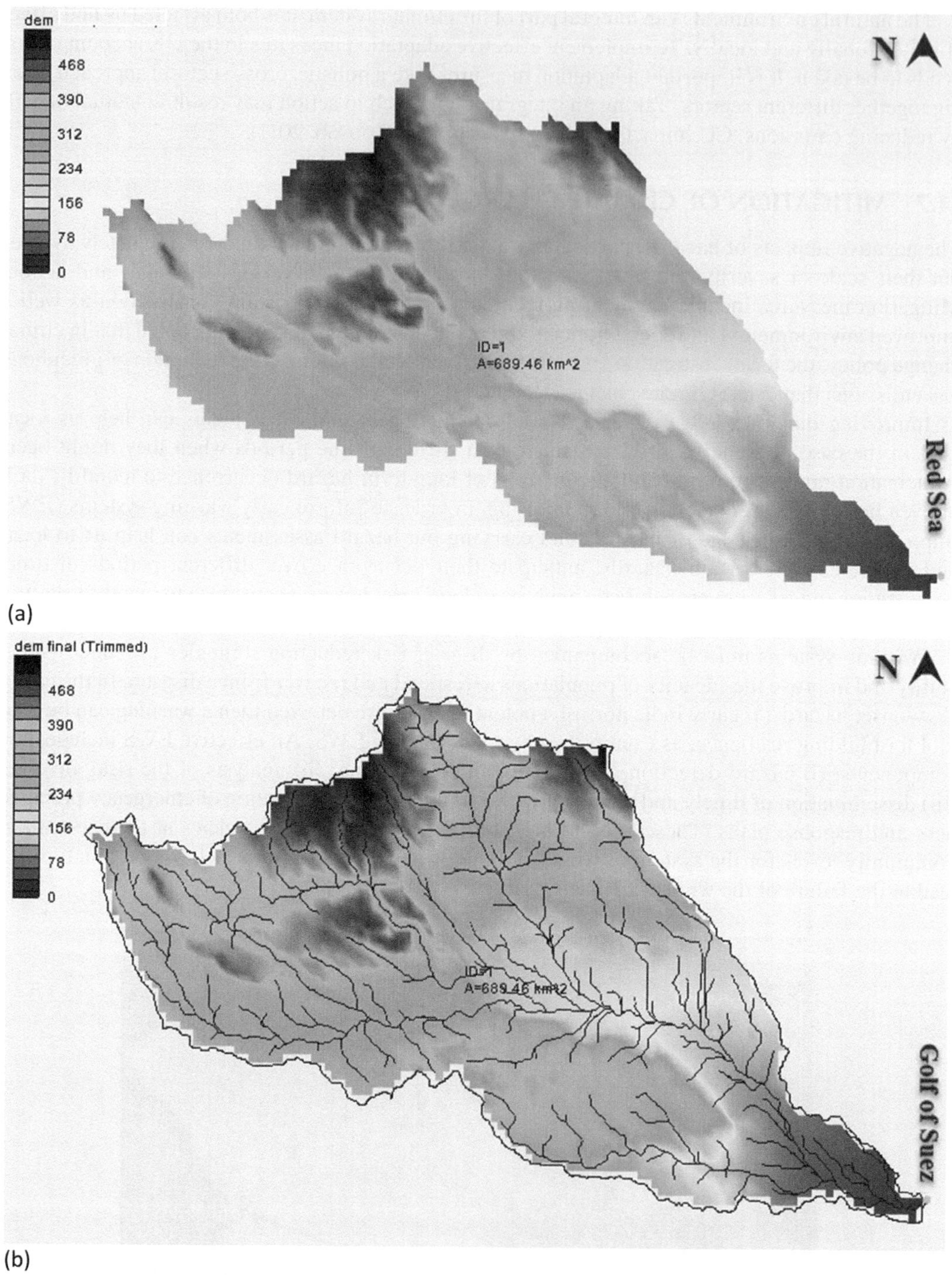

FIGURE 11.5 Watershed for (a) elevation and (b) streams.

11.8 STUDY AREA

11.8.1 Location Map and Topography

Figure 11.4 presents the location map for the study area, it is a coastal tourism area located at Sokhna, Suez,, Egypt, and it is located along the Gulf of Suez at the Red Sea coast. Digital elevation models (DEM) are used in the current study to delineate the flow paths and the stream flow using 30-m-resolution, the elevation ranges from 0 m to 468 m above mean sea level (see Figure 11.5a).

11.8.2 Watershed Modeling

The watershed modeling system (WMS) (https://www.aquaveo.com/downloads.wms) was applied in this study to delineate the stream flow volume and discharges and used in the calculation of the main flood factors (Chow et al., 1988). The watershed streams are presented in Figure 11.5 in which the flood flows from the West to the East.

11.8.3 Rainfall Analysis

The rainfall for the study area was measuredfrom 1995 to 2021, the maximum daily rainfall was 71.10 mm, which was in 2021. The watershed streams are shown in Figure 11.5b where the flow direction is from the West to the East at the Gulf of Suez. These data are used in the WMS for the estimation of the peak flooding volume and discharge.

Hydrological frequency analysis (hydrant-plus) software v2.1 was used to estimate the different recurrence intervals, the best fit was exponential distribution, and the precipitation reached 53.50 mm, 65.10 mm, and 76.80 mm at recurrence intervals of 25, 50, and 100 years.

11.8.4 Flash Flood Risk Investigation

The WMS results showed that increasing the recurrence intervals increased the flood discharge to 237.62 m^3 s^{-1}, 359.84 m^3 s^{-1}, and 495.37 m^3 s^{-1} at recurrence intervals of 25 years, 50 years, and 100 years, respectively (Table 11.1).

The flood volume reached 11003864 m^3, 16333842 m^3, and 22201092 m^3 at recurrence intervals of 25, 50, and 100 years respectively (see Figure 11.6).

TABLE 11.1

Peak Discharge and Volume for Recurrence Intervals

Recurrence intervals	25Y		50Y		100Y	
Unit	Q (m^3 s^{-1})	V(m^3)	Q (m^3 s^{-1})	V(m^3)	Q (m^3 s^{-1})	V(m^3)
Basin	237.62	11 003 864	359.84	16 333 842	495.37	22 201 092

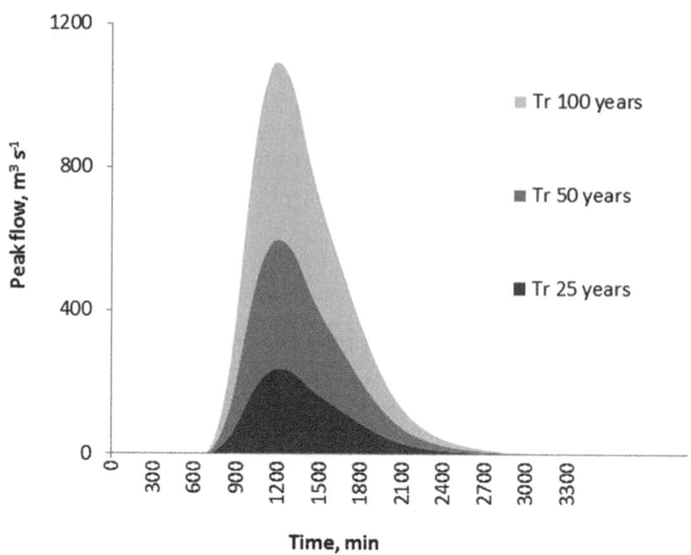

FIGURE 11.6   Runoff discharge for different recurrence interval.

11.9   CONCLUSIONS

Hydrological hazards (HHs) are critical issues for the communities planning and development; it is highly sensitive and accelerated by the influence of climate change (CC). Droughts and floods are related to the changes in hydro-hazards. The hazards vary across different research institutions and governments, but these can be divided into biological hazards, environmental hazards, geological or geophysical hazards, hydro-meteorological hazards, and technological hazards. The HH include drought, flash floods, tropical cyclones, landslides, river scour and deposition, heavy rain and snow, and strong winds.

This chapter is applied to identify the HH and CC in arid and semi-arid regions. The study was applied to a coastal area located at Sokhna, Suez, Egypt, located along the Gulf of Suez at the Red Sea coast. The WMS results showed that increasing the recurrence intervals increased the flood discharge to 237.62 m^3 s^{-1}, 359.84 m^3 s^{-1}, and 495.37 m^3 s^{-1} at recurrence intervals of 25, 50, and 100 years, respectively. The flood volume reached 11003864 m^3, 16333842 m^3, and 22201092 m^3 at recurrence intervals of 25, 50, and 100 years, respectively.

The flash flood predictions under the impact of HH and CC are useful for the decision-makers and engineers to consider in future planning and development in desert regions around the world in which the water resources are limited.

Through this chapter, it is recommended to consider CC management due to HH and mitigate the flooding impacts. This will support and meet the growing demands for freshwater resources using rainwater harvesting by aquifer recharge especially in arid and semiarid regions.

ACKNOWLEDGEMENTS

The authors thank the Department of Water and Water Structures Engineering, Faculty of Engineering, Zagazig University, Zagazig 44519, Egypt, for constant support during the study.

REFERENCES

Abd-Elaty, I., El Shinawi, A., Kuriqi, A., Puglies, L., and Zelenakova, M. (2023). Mitigation of urban water-logging from flash floods hazards in vulnerable watersheds. *Journal of Hydrology: Regional Studies*, 47, 101429. ISSN 2214-5818. https://doi.org/10.1016/j.ejrh.2023.101429.

Abd-Elaty, I., Shoshah, H., Zeleňáková, M., Kushwaha, N. L., and El-Dean, O. W. (2022b). Forecasting of flash floods peak flow for environmental hazards and water harvesting in desert area of El-Qaa Plain, Sinai. *International Journal of Environmental Research and Public Health*, 19(10), 6049. https://doi.org/10.3390/ijerph19106049.

Abd-Elaty, I., and Straface, S. (2022). Mathematical models ensuring freshwater of coastal zones in arid and semiarid regions. In Furze, J. N., Eslamian, S., Raafat, S. M., and Swing, K. (Eds.), *Earth Systems Protection and Sustainability*. Springer, Cham. https://doi.org/10.1007/978-3-030-98584-4_3.

Abd-Elaty, I., Sallam, G. A., Straface, S. and Scozzari, A. (2019). Effects of climate change on the design of subsurface drainage systems in coastal aquifers in arid/semi-arid regions: Case study of the Nile delta. *Science of the Total Environment*, 672, 283–295. https://doi.org/10.1016/j.scitotenv.2019.03.483.

Abd-Elaty, I., Kuriqi, A., Ramadan, E. M., and Ahmed, A. A. (2024). Hazards of sea level rise and dams built on the River Nile on water budget and salinity of the Nile Delta aquifer. *Journal of Hydrology: Regional Studies*, 51, 101600. ISSN 2214-5818, https://doi.org/10.1016/j.ejrh.2023.101600.

Abd-Elaty, I., Straface, S., and Kuriqi, A. (2021). Sustainable saltwater intrusion management in coastal aquifers under climatic changes for humid and hyper-arid regions. *Ecological Engineering*, 171. https://doi.org/10.1016/j.ecoleng.2021.106382.

Abd-Elaty, I., Kuriqi, A., Bhat, S. A., and Zelenakova, Z. (2022a). Sustainable management of two-directional lateral and upconing saltwater intrusion in coastline aquifers to alleviate water scarcity. *Hydrological Process Journal*, 36. https://doi.org/10.1002/hyp.14646.

Abu Salem, H. S., Gemail, K. S., Junakova, N., Ibrahim, A., and Nosair, A. M. (2022). An integrated approach for deciphering hydrogeochemical processes during seawater intrusion in coastal aquifers. *Water*, 14, 1165. https://doi.org/10.3390/w14071165.

Agren, J., and Svensson, R. (2007). *Postglacial land uplift model and system definition for the new Swedish height system RG 2000*. Lantmäteriet, Gävle.

Attwa, M., and El Shinawi, A. (2014). *Geoelectrical and geotechnical investigations at tenth of Ramadan city, Egypt – A structure-based (SB) model application*. Near Surface-Geoscience, 20th European Meeting of Environmental and Engineering Geophysics.

Beevers, L., Popescu, I., Pregnolato, M., and Liu, Y. (2022). Identifying hotspots of hydro-hazards under global change: A worldwide review. *Frontiers in Water*. https://doi.org/10.3389/frwa.2022.879536.

Blunden, J., and Arndt, D. S. (Eds.). (2019). State of the climate in 2018. *Bulletin of the American Meteorological Society*, 100(9), Si–S305. https://doi.org/10.1175/2019BAMSStateoftheClimate.1.

Chow, V. T., Maidment, D. R., and Mays, L. W. (1988). *Applied Hydrology*. McGraw-Hill Book Company, New York.

Collet, L., Harrigan, S., Prudhomme, C., Formetta, G., and Beevers, L. (2018). Future hot-spots for hydro-hazards in Great Britain: A probabilistic assessment. *Hydrology and Earth System Sciences*, 22, 5387–5401. https://doi.org/10.5194/hess-22-5387-2018.

Defra (Department for Environment Food and Rural Affairs) (2010). Natural Environment Adapting To Climate Change. The report can be viewed at: https://assets.publishing.service.gov.uk/media/5a79011 4e5274a2acd18b748/pb13323-natural-environment-adaptation-100326.pdf.

El-Shinawi, A., and Naymushina, O. (2015). Geotechnical aspects of flood plain deposits in south east Aswan City, Egypt. *ARPN: Journal of Engineering and Applied Sciences*, 10(8), 3490–3497.

El Shinawi, A., Ibrahim, R. A., Abualigah, L., Zelenakova, M., and Elaziz, M. A. (2021). Enhanced adaptive neuro-fuzzy inference system using reptile search algorithm for relating swelling potentiality using index geotechnical properties: A case study at el sherouk city, Egypt. *Mathematics*, 9(24). https://doi.org/10.3390/math9243295.

El Shinawi, A., Ramadan, F., and Zelenakova, M. (2021). Appraisal for the environment, weathering and provenance of Upper Cretaceous-Lower Tertiary shales, Western Desert, Egypt. *Acta Montanistica Slovaca*, 26(1), 171–184. https://doi.org/10.46544/AMS.v26i1.1

El Shinawi, A., Kuriqi, A., Zelenakova, M., Vranayova, Z., and Abd-Elaty, I. (2022). Land subsidence and environmental threats in coastal aquifers under sea level rise and over-pumping stress. *Journal of Hydrology*, 608, 127607. https://doi.org/10.1016/j.jhydrol.2022.127607.

Erkens, G., Bucx, T., Dam, R., De Lange, G., and Lambert, J. (2015). Sinking coastal cities. *Proc. IAHS*, 372, 189–198. https://doi.org/10.5194/piahs-372-189-2015.

Gambolati, G., and Teatini, P. (2015). Geomechanics of subsurface water withdrawal and injection. *Water Resources Research*, 51, 3922–3955.

Gleick, P. H. (1998). Climate change, hydrology, and water resources. *Reviews of Geophysics*, 27, 329–344.

Gogu, R., Carabin, G., Hallet, V., Peters, V., and Dassargues, A. (2001). GIS-based hydrogeological databases and groundwater modelling. *Hydrogeology Journal*, 9(6), 555–569.

Guzy, A., and Malinowska, A. A. (2020). Assessment of the impact of the spatial extent of land subsidence and aquifer system drainage induced by underground mining. *Sustainability*, 12(19), 7871. https://doi.org/10.3390/su12197871.

IPCC. (2001). Climate change 2001: Impacts, adaptations, and vulnerability: Contribution of working group II to the third assessment report of the intergovernmental panel on climate change. In Mc Carthy, J. J., Canziani, O. F., Leary, N. A., Dokken, D. J., and White, K. S. Cambridge University Press, Cambridge and New York, NY.

IPCC. (2007). An assessment of the intergovernmental panel on climate change. Adopted section by section at IPCC Plenary XXVII (Valencia, Spain, 12–17 November 2007), represents the formally agreed statement of the IPCC concerning key findings and uncertainties contained in the Working Group contributions to the Fourth Assessment Report.

IPCC. (2013). Climate change 2013: The physical science basis. Contribution of Working Group I to the Fifth Assessment Report of the Intergovernmental Panel on Climate Change [Stocker, T. F., Qin, D., Plattner, G.-K., Tignor, M., Allen, S. K., Boschung, J., Nauels, A., Xia, Y., Bex, V., and Midgley, P. M. (Eds.).]. Cambridge University Press, Cambridge, and New York, NY, 1535pp.

Khedr, M. (2017). Challenges and issues in water, climate change, and food security in Egypt. In Negm, A. M. (Eds.), *Conventional Water Resources and Agriculture in Egypt*. The Handbook of Environmental Chemistry, vol 74. Springer, Cham. https://doi.org/10.1007/698_2017_67.

Kramarenko, V. V., El-Shinawi, A., Matveenko, I. A., and Shramok, A. A. (2016). Clay swelling of quaternary and paleogene deposits in the south-eastern flanks of West Siberian iron ore basin. *IOP Conference Series: Earth and Environmental Science*, 33, 012041. https://doi.org/10.1088/1755-1315/33/1/012041.

Li, Z., Sun, Z., Tian, Y., Zhong, J., and Yang, W. (2019). Impact of land use/cover change on Yangtze River delta urban agglomeration ecosystem services value: Temporal-spatial patterns and cold/hot spots ecosystem services value change brought by urbanization. *International Journal of Environmental Research and Public Health*, 16(1), 123. https://doi.org/10.3390/ijerph16010123.

Lu, Y., Wu, P., Ma, X., et al. (2019). Detection and prediction of land use/land cover change using spatiotemporal data fusion and the Cellular Automata–Markov model. *Environmental Monitoring and Assessment*, 191, 68. https://doi.org/10.1007/s10661-019-7200-2.

Maps—UNESCO Land Subsidence International Initiative. https://www.landsubsidenceunesco.org/mapsOrg/maps/. Accessed 2 June 2020.

Mascaro, G. (2018). Hydrologic Cycle. In: White, W.M. (eds) Encyclopedia of Geochemistry. *Encyclopedia of Earth Sciences Series*. Springer, Cham. https://doi.org/10.1007/978-3-319-39312-4_158.

MATLAB and Mapping Toolbox Release. (2019a). Mapping toolbox user's guide—map_ug.pdf. https://www.mathworks.com/products/mapping.html. Accessed 27 March 2020.

Nabih, M., and El Shinawi, A. (2020). Qualitative analysis of clay minerals and swelling potential using gamma-ray spectrometry logs: A case study of the Bahariya Formation in the Western Desert, Egypt. *Applied Radiation and Isotopes*, 166, 166. https://doi.org/10.1016/j.apradiso.2020.109384.

Naymushina, O., Shvartsev, S., Zdvizhkov, M., and El-Shinawi, A. (2010). Chemical characteristics of swamp waters: A case study in the Tom River basin, Russia. Water-Rock Interaction. *Proceedings of the 13th International Conference on Water-Rock Interaction*, WRI-13, 955–958.

Prasad, A. S., and Francescutti, L. H. (2017). Natural disasters. In Quah, S. R. (Ed.), *International Encyclopedia of Public Health* (Second Edition). Academic Press, 215–222. ISBN 9780128037089. https://doi.org/10.1016/B978-0-12-03678-5.00519-1.

Rujoiu-Mare, M.-R., Olariu, B., Mihai, B.-A., Nistor, C., and Săvulescu, I. (2017). Land cover classification in Romanian Carpathians and Subcarpathians using multi-date Sentinel-2 remote sensing imagery. *European Journal of Remote Sensing*, 50(1), 496–508. https://doi.org/10.1080/22797254.2017.1365570.

Runoff and Peak Flow- Institute for Transportation. (2020). https://intrans.iastate.edu/app/uploads/sites/15/2020/03/2B-4.

Sarkar, S., and Maity, R. (2021). Global climate shift in 1970s causes a significant worldwide increase in precipitation extremes. *Science Reports*, 11, 11574. https://doi.org/10.1038/s41598-021-90854-8.

Shinawi, A. E., Mésáros, P., and Zeleňáková, M. (2020). The implication of petrographic characteristics on the mechanical behavior of middle Eocene limestone, 15th May City, Egypt. *Sustainability (Switzerland)*, 2(22), 1–15. https://doi.org/10.3390/su12229710.

Shinawi, A. E., Zeleňáková, M., Nosair, A. M., and Abd-Elaty, I. (2022). Geo-spatial mapping and simulation of the sea level rise influence on groundwater head and upward land subsidence at the Rosetta coastal zone, Nile Delta, Egypt. *Journal of King Saud University – Science*, 34(6). https://doi.org/10.1016/j.jksus.2022.102145.

Sposób, J. (2011). Water balance in terrestrial ecosystems. In Gliński, J., Horabik, J., and Lipiec, J. (Eds.), *Encyclopedia of Agrophysics*. Encyclopedia of Earth Sciences Series. Springer, Dordrecht. https://doi.org/10.1007/978-90-481-3585-1_267.

Sprenger, M., Stumpp, C., Weiler, M., Aeschbach, W., Allen, S., Benettin, P., Dubbert, M., Hartmann, A., Hrachowitz, M., Kirchner, J. W., et al. (2019). The demographics of water: A review of water ages in the critical zone. *Reviews of Geophysics*, 57, 800–834.

Sweet, W., Hamlington, B., Kopp, R., Weaver, C., Barnard, P., Bekaert, D., and Pendleton, M. (2022, February). Global and regional sea level rise scenarios for the United States: Updated mean projections and extreme water level probabilities along U.S. coastlines. National Oceanic and Atmospheric Administration, National Ocean Service, Silver Springs, Maryland. Retrieved May 3, 2022, from https://oceanservice.noaa.gov/hazards/sealevelrise/noaa-nos-techrpt01-global-regional-SLR-scenarios-US.pdf.

Tabari, H. (2020). Climate change impact on flood and extreme precipitation increases with water availability. *Science Reports*, 10, 13768. https://doi.org/10.1038/s41598-020-70816-2.

Taiz, L., and Zeiger, E. (2009). *Plant Physiology*, 4th Edition. Sinauer, Sunderland, 719.

Taylor, K. E., Stouffer, R. J., and Meehl, G. A. (2012). An overview of CMIP5 and the experiment design. *Bulletin of the American Meteorological Society*, 93, 485–498.

Trajkovic, S., Kisi, O., Markus, M., Tabari, H., Gocic, M., and Shamshirband, S. (2016). Hydrological hazards in a changing environment: Early warning, forecasting, and impact assessment. *Advances in Meteorology*. https://doi.org/10.1155/2016/2752091.

UNDRR (United Nations Office for Disaster Risk Reduction). (2017). Terminology of disaster risk reduction. https://www.undrr.org/drr-glossary/terminology.

U.N. ISDR. (2004). Living with risk: A global review of disaster reduction initiatives United Nations, Geneva.

U.N. ISDR. (2011). Killer year caps deadly decade – Reducing disaster impact is "critical" says top UN Disaster Official United Nations, Geneva, Switzerland.

Visser-Quinn, A., Beevers, L., Lau, T., and Gosling, R. (2021). Mapping future water scarcity in a water abundant nation: Near-term projections for Scotland. *Climate Risk Management*, 32, 100302. https://doi.org/10.1016/j.crm.2021.100302.

Whittow, J. B. (1984). *Dictionary of Physical Geography*. Penguin, London.

WMO (World Meteorological Organization). (2012). Management of flash floods – A tool for integrated flood management, Version 1.0, 2012. World Meteorological Organization, Geneva. https://library.wmo.int/viewer/50629/download?file=ifmts_16_en.pdf&type=pdf&navigator=1.

WMO (World Meteorological Organization). (2023). Characteristics of tropical cyclones. World Meteorological Organization, Geneva. https://wmo.int/content/characteristics-of-tropical-cyclones.

Zamani Nuri, A., Farzaneh, M. R., Fakhri, M., Dokoohaki, H., Eslamian, S., and Khordadi, M. J. (2013). Assessment of future climate classification on Urmia Lake basin under effect of climate change. *International Journal of Hydrology Science and Technology*, 3(2), 128–140.

Part III

South Asia

12 Climate Change Impacts on Water Resources and Sediment Yield in South Asia

*Shahriar Rahman, Sanzida Rahman,
Aziza Ahsan, and Saeid Eslamian*

12.1 INTRODUCTION

Climate change will impact South Asian economies, societies, and environments, threatening development and poverty. Bangladesh, Bhutan, northern India, and Nepal experience more natural disasters than globally. India and Pakistan, the two largest economies, have water shortages in many locations, including major cities (Hirji et al., 2017).

Water resource management is crucial for population growth, livelihoods, and resilience, as 23.7% of the global population has just 4.6% of sustainable water sources (Ahmed and Suphachalasai, 2014). Water shortages will affect one-third of South Asia, North and Sub-Saharan Africa, South America, and the Middle East. (Basu and Shaw, 2013). Climate change will worsen the shortage of water supplies for food, drinking water, everyday necessities, hydropower, industry, and severe hydrological events (Fathian et al. 2016). Climate change will unevenly influence surface and groundwater availability, demand, and quality. Indo–Gangetic Basin aquifers are estimated to store 30,000 km^3, more than 100 times the storage capacity of all South Asian dams, tanks, and reservoirs. Its massive natural reserve is 20 times the combined annual average flows of the Indus, Ganges, and Brahmaputra (Hirji et al., 2017).

Groundwater, which powered the green revolution and brought millions out of poverty, is crucial to the region's economy and will become much more important as economies grow, populations increase, surface water supplies become less dependable, and the climate changes. Groundwater accounts for 40% of water usage but only 13% of renewable water, and irrigation uses 60–80% of water. South Asia, the world's biggest groundwater abstractor, pumps 62% of its 555 billion cubic meters (bcm) of yearly renewable groundwater. Policymakers still overlook groundwater reserves. Surface water infrastructure receives the bulk of water policy financing, whereas groundwater management receives significantly less. This chapter discusses South Asian water resource management, sediment output, and transportation under climate change (Hirji et al., 2017).

12.2 CLIMATE CHANGE IMPACTS ON WATER RESOURCES AND SEDIMENT YIELD IN SOUTH ASIA

By 2050, South Asia's population is expected to expand by about 2.2 billion (Vinke et al., 2017). By the end of the twenty-first century, the global mean atmospheric temperature is forecasted to rise by 1.8–4.0°C, depending on the greenhouse gas (GHG) scenario (IPCC, 2007; Shrestha et al., 2013). The most current measures of total monsoon rainfall in South Asia suggest a decline over the last several decades.

DOI: 10.1201/9781003473398-15

Heavy floods, droughts, declining snowpack, and glacier losses in the Himalayas that alter river flow are just a few of the climate concerns and consequences that pose a danger to people in South Asia (Vinke et al., 2017). In coastal parts of South Asian nations, the consequences of sea-level rise, tropical storm strength, and droughts are intensified (Vinke et al., 2017). For instance, a cyclone such as SIDR in Bangladesh would result in an average storm with a return period of 10 years and a sea level increase of around 27 cm by 2050 (NASA, 2007; Wassmann et al., 2009).

Moreover, droughts impacted more than 50% of India's cropland. Crop productivity is decreased by the weight of salinity from brackish groundwaters or salt-damaged soils (Wassmann et al., 2009). Climatic change substantially impacts the movement of sediments across watersheds to 13 harbors and canals excavated for water or flood control (Dahl et al., 2018). Watersheds produce millions of tonnes of silt each year, reducing water storage (Azim et al., 2016). Research has shown how climate change affects sediment production, streamflow, and soil erosion. The sediment output ranged from −0.7 to 13.7% due to variations in precipitation from −0.7 to 17.8% and watershed temperatures from 0.03 to 2.4°C (Azim et al., 2016; Michael et al., 2005; Zhu et al., 2008). The yield of suspended sediment and streamflow do not always vary due to climate change in the same way. According to studies, the temperature change will disrupt precipitation patterns, cause snow to melt earlier than usual, and ultimately change the water cycle in the bay (Azim et al., 2016).

12.3 WATER RESOURCES OF SOUTH ASIA

Afghanistan, Pakistan, the Maldives, India, Bangladesh, Nepal, Bhutan, Sri Lanka, and Burma comprise South Asia. An overview of the water resources of the nine South Asian countries is provided in this section.

12.3.1 Overview of Water Resources: Afghanistan

Afghanistan is situated in central Asia and has a semi-arid climate. Twenty-five percent of its territory is 2500 m above sea level with snow and glaciers, 18% is deserts–plainlands–forests, and 17% is rivers (Ahmad and Wasiq, 2004; Habib, 2014). The rivers originated from snow melting in the Hindu Kush and Pamir mountains (Uhl, 2006). Afghanistan receives an average of 300 mm per year of rainwater; however, the eastern part has a different rainfall pattern of partial monsoon effect (Frenken, 2013). Its rivers have a storage capacity of around 55 bcm. The water flow splits into five main river basins: Helmand, Kabul (Indus), Amu Darya, Hari Rud–Murghab, and Northern flowing rivers (Table 12.1). Kabul and the Amu Darya contribute nearly two-thirds of the surface water, covering 25% of river basins.

TABLE 12.1

Summary of the Five Major Rivers Hydrological Data in Afghanistan

River basin	Water capacity (billion m³/year)	Water (%)	Catchments area (km²)	Catchments area (%)
Kabul (Indus)	21.6	26	54,000	11
Helmand	9.3	11	190,000	41
Hari Rud – *Murghab*	30.6	4	40,000	12
Amu Darya	48.1	57	309,000	14
Northern flowing rivers	18.8	2	1,100	11
No Drainage Area				10

Source: Aawar et al., 2019; Alim, 2006; Aquastat, 2010.

The accessible surface water in Afghanistan is 2480 m³/capita/year, higher than in Iran (1430 m³/capita/year). The Ministry of Irrigation, Water Resources, and Environment (MIWRE) reported that approximately 47% of the water from total precipitation is released out of the country due to transboundary issues (Alim, 2006). The three major groundwater aquifers are the Great Southern Basin, the Central Highland Region, and the Northern Plain. These aquifers commonly comprise gravel, sand, clay, siltstone, marl, and sandstone containing gypsum and salt (Alim, 2006). The estimation of groundwater recharge is near 15 billion m³. Both fresh and saline water are found throughout the country; saline water is available in the northern part. The total water storage capacity is 75 bcm annually, where 57% is surface water and 18% is groundwater. In all, 32 dams and reservoirs are being built around the nation. Thirty million people (70% urban and 29% rural) have water resources. However, just 45.5% have clean drinkable water. The majority of water is needed for irrigation, followed by drinking, sanitation, industrial, and power generation. Water has become a large and critical problem in urban and rural locations.

12.3.2 CHALLENGES OF WATER RESOURCES

The absence of water resource management rules and regulations, the presence of conventional irrigation systems, the absence of international climate change networks, poor water resource management, a lack of public awareness, a substandard sanitation system, and industrial development are Afghanistan's greatest challenges. The US Geological Survey (USGS) reported that Afghanistan would face water scarcity and could lose half its figure in the next 50 years (USGS, 2010). Excessive water extraction and damage to many water infrastructures during the civil war are other reasons that hamper the water management system. Moreover, the growing population needs more water in the future. Around 84.6% of irrigation water comes from rivers and streams, 7.9% from springs, 7% from Karezes (kanats), and 0.5% from shallow and deep wells. Already people suffer from long-term drought, lowering groundwater, rough land, and loss of forests and wild animals. All the conventional groundwater irrigation schemes have been diminished or entirely dried up. For instance, around 60–70% of karezes are not usable and 85% of shallow wells are dead from over-pumping aquifers (Qureshi, 2002). Inappropriate forms of irrigation, such as watering by pond and furrow, exceed water requirements. Besides this, civil war and a lack of adequate water control have directly damaged approximately 27–36% of the overall irrigation infrastructure in Afghanistan. Recent climate change creates pressure on the refilling of groundwater in semi-arid regions. People are not aware of the water's value due to their limited knowledge. That is the major drawback of properly utilizing domestic and irrigation water resources. Moreover, less precipitation and water consumption practices result in low groundwater recharge.

12.3.3 CLIMATE CHANGE IMPACTS ON WATER RESOURCES

In 2018–2019, less precipitation induced extreme drought (Jones, 2020). Lack of rain and snow threatens the agriculture sector, and thousands of wells have already dried up (Akbari, 2018). White, Tanton, and Rycroft estimated that if river flows decrease by 10 or 20% as expected, the fertile land will decrease by 40–49% (White et al., 2014). This scenario predicts that most people will live in poverty within the next 50 years. The Kabul region is home to approximately 7 million people, of whom 80% lack access to potable water and 95% lack sanitation facilities. Increasing temperature is melting glaciers faster and causing floods by bursting a natural dam adjacent to a glacial lake. In July 2018, a landslide destroyed many houses, schools, and farmland generated by downstream water forces. Snow melting in spring and summer was a blessing for the entire country to keep the weather cool, but extreme melting is dangerously causing floods. In mountain areas, high erosion rates are apparent during floods. The large volume of sediment creates a variety of significant water resource management issues eroded and transported downstream, such as reservoir siltation, turbine destruction, the decline in water quality, and the transport of chemical contaminants (Ali and De Boer, 2010).

12.3.4 Sediment Transport and Yield

The spatially dispersed relationship between sediment supply and annual erosion rates led to the development of annual sediment yield forecasts. The basin's mean annual sedimentation rates were determined by aggregating these predicted sediment outcomes.

The long-term sediment discharge of 195.1 metric tonnes per year (Mt year^{-1}) river in Besham Qila, determined by Ali and De Boer (2007) near a basin outlet, and the sediment storage of 192.2 Mt year^{-1}, based on the Tarbela Dam Project hydrographic research (Ali and De Boer, 2007; Tarar, 2006). The upper Indus River basin's sediment delivery rate of 0.28 is similar to the upper Yangtze River's (Ali and De Boer, 2007). Dividing the expected yearly average sediment output by the number of basin cells yielded 1110 tonnes per square kilometer per year (t km^2 year^{-1}). Unlike other mountain drainage basins, the Upper Yangtze generates 524 t km^2 year^{-1} of silt annually (Ali and De Boer, 2007). The Ganges–Brahmaputra sediment outputs are 604 t km^2 year^{-1} (Wasson, 2003), the Bolivian Andes 1400 (Saavedra, 2005), and the Yellow River basin 2232 (Xu and Cheng, 2002). The projected monthly mean sediment output distribution pattern of estimated monthly sediment outputs at Besham Qila matches observed Indus River sediment data.

12.3.5 Remarkable Treaties

Afghanistan shares water with Iran, Pakistan, Kyrgyzstan, Tajikistan, Turkmenistan, and Uzbekistan via its main rivers. Afghanistan and the former Soviet Union reached international agreements regarding the Amu Darya transboundary water utilization and quality. Afghanistan and Pakistan have not reached a water agreement concerning the Kabul River (King and Sturtewagen, 2010). Afghanistan and Iran share the Helmand River, on which only Afghanistan has recently signed an official agreement. Protocol 1 of the Helmand River Treaty allocated mutual Helmand River administrators to these two nations in 2003.

12.3.6 Good Practices and Lessons Learned

Major water resource management projects are concerned with irrigation systems and their improvement. The major farmland irrigation schemes are described below:

> *Lower Kokcha Irrigation and Hydropower Project:* This project started before the 1980s, but the physical work was abandoned because of war. By 2020, the plan will be implemented through 96,000 ha of restoration cycles and the irrigation of 37,000 ha of additional land (Klemm and Shobair, 2010).
>
> *Kelagay Irrigation and Hydropower Project:* The project is situated in the province of Baghlan on the Kunduz River, one of Amu Darya's tributaries. The project's main goals are to increase local employment opportunities, secure a source of irrigation water for 43,250 ha of existing farmland, produce 60 MW of hydroelectricity, provide effective water irrigation for 25,365 ha, encourage fishing, and establish processing industries (Klemm and Shobair, 2010).
>
> *Upper Amu or Lower Panj Irrigation and Hydropower Project:* The precise area that will be irrigated for this project has not yet been determined. However, it is anticipated that the region that might be irrigated would include more than 500,000 ha of existing and new land and up to 1,000 MW of installed hydropower (Klemm and Shobair, 2010).
>
> *NATO's ENVSEC (Environment and Security) Initiative:* This initiative emerged out of a project proposal titled "Strengthening Transboundary Cooperation on the Management of Water Resources between Afghanistan and Tajikistan in the Upper Amu Darya River Basin" that was submitted in 2007 by UNEP/GRID Arendal/Norway on behalf of the Tajik Agency for Hydro-Meteorological and Environmental Monitoring in Dushanbe (Klemm and Shobair, 2010).

Agromet Project (2004): The Agromet Project built a network of agrometeorological data by setting up 102 stations and collecting precipitation amounts, snowfall, and other parameters. Over 150 monthly reports and seasonal reports were produced by the Agromet project (Mack et al., 2014).

Dam Construction: Over 60 dams with a combined capacity of more than 2 km³ and 109 diversion schemes are intended to meet the water resource requirements of the Kabul and Helmand Basins. The main purposes of these infrastructures are to ensure water supply, flood control, hydroelectric power generation, reserve water, and recreation (Stewart, 2016).

Afghanistan needs to reinvent proper hydro-meteorological stations for a sustainable cross-border water solution that disappeared due to the civil war. Afghanistan has limited technical capacities, and creating a well-structured organization for transboundary water issues is necessary. For state confidence and mutual faith, the predictability and transparency of policy planning are crucial. The 30-year knowledge deficit in Afghanistan can be met by sharing experts, training, and mutual information such as geospatial mapping and remote sensing. Afghanistan and its neighbors may also want to start a multi-stakeholder dialogue to build trust and make plans for cross-border water and water security guards between governments. In conclusion, a synopsis is that a joint scientific and technical assessment could be conducted to enhance the compilation, assessment, and evaluation of hydrological data, an informal meeting of scientists from Afghanistan and its neighbors on the importance of developing river basin–based hydrological mechanisms.

12.4 OVERVIEW OF WATER RESOURCES: PAKISTAN

Pakistan's main water sources include rainfall, glaciers, and groundwater. With an average annual rainfall of 125 to 750 mm, the monsoon season (June to September) receives around 70% of the yearly precipitation. The Sindh plains receive a lot of rain in July and August, albeit the rain falls less heavily inland than it does at the coast. Moreover, the annual precipitation in South Punjab and North Sindh is modest, coming in at less than 152 mm (Kahlown and Majeed, 2003). On the other hand, the mean annual precipitation in several places, such as Jhelum, Rawalpindi, Attock, and Mianwali, has surpassed that range. Around 3% of the mountains in the Upper Indus Basin — or 35431.037 square Kms— are in Pakistan Ahmed et al. (2007). The Indus, Beas, Chenab, Jhelum, Sutlej, and Ravi are six major rivers that flow into Pakistan's mountainous areas, along with three smaller ones (Haro, Soan, and Siran). The Indus has several western tributaries, including Kunar, Punj, and Kora. Various minor rivers, including the Gomal Kurram, Tai, Kohat Tank, and others, join from the right bank of the Indus River. The Indus River serves as Pakistan's main source of surface water. The Indus River network annually holds an average of 154 million acre feet (MAF) of water. The most significant freshwater resource in Pakistan is the Indus Plain, which spans from the foothills of the Himalayas to the Arab Sea.

Pakistan is blessed with wide-ranging unconfined storage of water. Its availability has increased by 3% in the past years. Over 500,000 tube wells are installed to pump groundwater and figure over 41.6 MAF (Ali et al., 2009). Though Pakistan has more glaciers than any other country except polar regions, this country faces water security issues (Sleet, 2019).

12.4.1 CHALLENGES OF WATER RESOURCES IN PAKISTAN

Pakistan's main water resource management problems are due to climate change, population growth, and improper water allocation. Global warming is causing glacier retreat events that cause floods. Furthermore, less precipitation affects stressing groundwater levels and is unequally distributed in different regions within the country. Besides, it is responsible for fluctuating surface water quality, drought, ecosystem degradation due to saltwater intrusion in the local area, waterlogging, and reduction of reservoir capacity for sedimentation. Many cities in Pakistan already struggle with a lack of water, and the situation worsens due to insufficient sewage and commercial wastewater systems. The outdated irrigation system is inefficient in Pakistan, where agriculture accounts for 93% of the nation's water requirement. It necessitates more groundwater abstraction, resulting in a major decline in groundwater levels. The Indus River Basin has consistently lost up to 1000 m^3 of water over the last 50 years, making it the second-most stressed aquifer in the world. Now this country takes out around 61 km^3 of water every year from aquifers that are beyond acceptable limits. Another challenge for water resources in Pakistan is industrialization.

Water is contaminated by untreated industrial wastes, basically from textiles and tanneries. The two biggest industries located in Karachi release heavy metals, oil, and other toxic materials directly into the rivers (Sleet, 2019). Pakistan promotes industrial growth to secure foreign trade. On the other hand, the irrigation system is also blamed for water contamination by washing and mixing pesticides and fertilizer into the river. Global climate change is associated with the hydrological cycle, which impacts on directly ground and surface water. Snow and ice melting is a critical event in Pakistan for water availability and hydropower generation (Ali et al., 2009).

12.4.2 Climate Change Impacts on Water Resources in Pakistan

According to the International Commission for Snow and Ice (ICSI), by 2035, there is a high probability of flash floods and Indus River overflow due to the Himalayan glaciers' fast pace of retreat (Ali et al., 2009). In the last two decades, the number of flood incidents per decade has risen significantly in Pakistan. Pakistan faced a severe freshwater shortage during the drought period between 1998–2002. The surrounding Indus Basin is most affected, notably the Punjab, Baluchistan provinces, and four Sindh Province districts. Punjab and Sindh of Pakistan are facing severe waterlogging and salinity problems. Since 1998, waterlogging has affected 9.1 Mha, while severe waterlogging affected 4.9 Mha, i.e., 33.7% and 18.2% of farmland, respectively. Arsenic is naturally high in the Indus Basin. A study showed that 50–60 million Pakistani inhabitants (especially Lahore and Hyderabad residents) are in danger of arsenic contamination (Sleet, 2019). Excessive groundwater extraction has also been accompanied by increased salinity. Glaciers and snow melting at high velocity may carry sediments downstream. It has mainly resulted from soil erosion, deforestation, and inappropriate land-use problems. The dams and reservoir capacity are going down for sedimentation.

12.4.3 Sediment Transport and Yield

In Pakistan, the reasons behind soil erosion are the high-intensity rainfall, falling water force for steep slopes, and fragile soil, and around 16 Mha areas are exposed to soil erosion. In 2010, sediment yields were between 8.15 and 12.31 tonnes per hectare per year (t/ha/yr) in the Dhrabi watershed (Iqbal et al., 2012). The nearby gully catchment produces 4.1 t/ha/yr of sediment, less than 12.31 t/ha/yr, which shows that terraces have a lot of potential to stop erosion. The runoff coefficients for these catchments are thought to be between 0.09 and 0.75. The annual amount of sediment in the Indus River is thought to be 435 million tonnes, and every day, 500,000 tonnes of sediment are added to Tarbela Reservoir (Iqbal et al., 2012).

July and August provide 80–85% of the sediment load to the Shyok, Gilgit, Hunza, and Indus Rivers. April and July sediment output peaks in Karora's Gorband River are comparable. Snow and ice melting limit sediment runoff in the primary Indus tributary. A study of sediment transport's size-frequency characteristics found that the Basin's effective release range is between 1.5 and 2.0 and 5.5 and 6.0 for the average release and decreasing downstream (Ali and De Boer, 2007). For Pakistan's Naran Watershed, the effect of climate change on sediment yield at each site was looked at. The results show seasonal snow and monsoon discharges are expected to rise (Azim et al., 2016). This would help bring in more sediment each year. Since 1974, the reservoir of the Tarbela Dam has had a trapping efficiency of 89%, and sediment deposition has decreased by 27.6% to the reservoir's storage capacity of 14.3 $\times$ 10^9 m^3 (Ali and De Boer, 2007). Lack of flushing arrangements, the sediment still stuck there, covering a large portion of the dam.

The Mangla watershed is in the northeast of Pakistan. It is 33,421 km long, about 53% of the Jhelum River basin (Aslam and Yoshimura, 2017). In April–June, the north and northeast basins were higher for sediment yields because rainfall patterns vary from medium to low altitudes in these areas. Rawal is an artificial reservoir near Islamabad, and its catchment area is 275.2 km^2. Using the Revised Universal Soil Loss Equation and GIS, we can predict that soil erosion will range from 0.1 to 28 t/ha/yr in this area (Nasir et al., 2006). Though Fateh Jang watershed has 1–10% steeper slopes, its soil erosion rate is low, 9–26 t/ha, an advantage of trees where the barren land erosion rate is very high, 17–41 tonnes per hectare (t/ha) (Iqbal et al., 2012). Sarah did another study in 2010 and found that the Rawal Lake watershed has a high risk of erosion (26%). This is because the steep slope rains a lot (>1000 mm), and there aren't many plants there (Bashir et al., 2013). Pothohar Plateau is the largest adjoining drylands where gully erosion affects 1.21 farmland out of 2.2 Mha and only 0.61 Mha of land is suitable for cultivation (Iqbal et al., 2012). Vulnerable to soil erosion having a 19 t/ha/yr average annual soil loss, with the maximum erosion occurring in river channels and hilly areas (70–208 t/ha/yr). The average sediment yield is 4.3 t/ha/yr, with a maximum of 148 t/ha/yr (Ullah et al., 2018). The actual gross reservoir storage volume was estimated to be roughly 18.37 MAF, however, sedimentation reduced its efficiency by around 26% (Ahmed et al., 2007).

12.4.4 REMARKABLE TREATIES

India and Pakistan signed the Inter-Dominion Agreement on May 4, 1948, to ensure a constant water supply for agriculture. The World Bank mediated the 1960 signature of the Indus Basin Water Treaty (IBWT) to resolve the water conflict between India and Pakistan. India is granted access to the waters of the Ravi, Beas, and Sutlej Rivers in the east, while Pakistan is granted access to the waters of the Indus, Khyber, and Indus Rivers in the west (Chenab, Jhelum, and Indus). The proper execution of the Indus Water Treaty with India and an agreement with Afghanistan might boost Pakistan's water allotment. The four major hydroelectric projects producing tension between the two nations are the Ratle Hydroelectric Plant, Baglihar Dam, Kishanganga Hydroelectric Plant, and Salal Hydroelectric Power Station. At least nine rivers run across shared territory between Afghanistan and Pakistan, but no agreement exists for their administration. The Kabul River originates in Pakistan, which may generate energy for Pakistan and Afghanistan (Tariq et al., 2020).

12.4.5 GOOD PRACTICES AND LESSONS LEARNED

Pakistan's National Environment Policy provides a comprehensive mechanism for resolving environmental issues related to freshwater and coastal water pollution. "Clean Drinking Water for All by 2008" was the primary objective of the Clean Drinking Water Programme, which was approved in July 2004. Pakistan Wetlands Programme (PWP) is a Federal Environment Ministry initiative that the Worldwide Fund for Nature in Pakistan has implemented since July 2005. This initiative seeks to reduce poverty while protecting the biodiversity of Pakistan's internationally significant wetlands (GoP, 2007).

In 1955, when Pakistan was experiencing an acute power shortage, construction of the Warsak Dam on the River Kabul near Peshawar began. Pakistan constructed two great dams: Mangla, which has a 5.88 MAF gross storage capacity; Terbala, with an 1162 MAF storage capacity; and replacement works for the Indus Basin. There are numerous dams, and smaller irrigation schemes have also been implemented.

12.5 OVERVIEW OF WATER RESOURCES: THE MALDIVES

The Maldives is geographically located in the Indian Ocean at the confluence with the Arabian Sea. The primary source of freshwater is rainwater; other sources are groundwater, surface water, and human-made lakes. The average precipitation rate in the south of the Maldives is higher than that of the north — some freshwater lakes, known as *Kulhis*, meet water demand for the community. Insufficient rainfall causes a shortage of groundwater (Chachibaia, 2016). The usable water resource is limited as each island's freshwater lens is small. Precipitation, soil permeability, island size, and variations in sea level all affect how big the freshwater lens is. People must use groundwater for their everyday requirements.

Access to clean water is a crucial issue for Maldivians as the population is increasing — the insufficiency of the freshwater lens to sustain a large population with increased water intake per capita. Towards the conclusion of the dry season, the groundwater lens on around half of the populated islands is nearly empty, forcing the people to seek government water to satisfy their daily requirements. The declining volume of groundwater lenses is due to a smaller amount of recharges caused by increased impervious surfaces and rainwater capture by harvesting systems and increased pumping. Groundwater lens contamination primarily occurs from septic tank sewerage.

12.5.1 CHALLENGES OF WATER RESOURCES

Groundwater resources in the Maldives are depleting fast because of the shallow depth of the water table. There is a high chance of groundwater contamination from human activities on the surface because of improper household sanitation. On the outer side of the capital city Male, water and sewerage conditions are inferior. Almost 90% of people are out of the water supply network, and only 10% get this facility. The same scenario in the sewerage system; only 25% is covered by the sewerage network. Besides, water resources are vulnerable to increased salinity due to over-extraction. The soil properties of the Maldives are mostly very permeable. If water extraction remains the same in the dry season, rainfall cannot recharge the ground as well (Bailey et al., 2015).

Furthermore, the rising sea levels in the last decade, as a result of global warming, increased the possibilities of saline intrusion. Consequently, freshwater lenses are increasingly exhausted, and seawater may sometimes penetrate the periphery of the islands. Thus, it poses a growing threat to the groundwater supplies of the Maldives.

12.5.2 CLIMATE CHANGE IMPACTS ON WATER RESOURCES

The Maldives is a small nation, consisting of 26 atolls southwest of Sri Lanka, and its 1192 islands consist mostly of coral reefs and sand bars (Derler, 2019). This formation makes shallow aquifers and contains freshwater scarcity (Derler, 2019). This nation is anticipated to be the most susceptible to climate change since its greenhouse gas emissions rate is less than 0.01% of the global average. Several major challenges are merging with climate change to cause water shortages. The Maldives contains small isolated islands which are affected by tidal inundation. The shoreline is too close to human communities, public services, and vital infrastructure. As a result, the effects of sea-level rise, floods, beach erosion, storm surges, and large waves are already felt in infrastructure. Human activities are seriously raising the intensity of the nation's vulnerability to overcrowding of several islands, improper infrastructure, and deforestation on the beach. It is quite challenging to fulfill the water demand in these areas. Recent studies published by the International Climate Change Council indicate that the global sea level will rise between 300 and 1000 mm by 2100, depending on various emission scenarios. Less rainfall affects to fill of large catchments in which water is used in the dry season. Natural disasters have recently affected water resources (Stojanov et al., 2017). In 2004, the Maldives suffered an enormous tsunami that devastated many islands.

12.5.3 Sediment Transport and Yield in the Maldives

Sand export rates are lower during the northeast monsoon, although fore-reef places are more similar. At Vabbinfaru, wave-driven currents and wind waves carry sediments from the leading edge of the windy bank, which flows from Vabbinfaru Island to the north and south of the reef, and mix them with sediments from the protected edges. Gravitational settling and animal migration may become more significant under these circumstances. Vabbinfaru's sand exports were high, around 113,438 kg.m^{-1}.y^{-1} (where kg.m^{-1}.y^{-1} is the total platform export considering the width of the trap in meter (m), ranging from 6 to 123 kg.m^{-1}.y^{-1}, with an average annual reef front export rate of 52.2 kg.m^{-1}. Hubbard et al., discovered a range of 14 to 151 kg.m^{-1}.y^{-1} sediment was being carried over an open-shelf reef in Cane Bay, St. Croix (Hubbard et al., 1981). Results from the past demonstrated that exporting sand from outside rivers was a keyway for massive sand reservoirs to expand. Sand exports vary from season to season over the platform perimeter, with lower export rates in places more exposed to waves (Morgan and Kench, 2014).

12.5.4 Treaties, Exclusive Projects, and Lessons Learned

The Maldives has no transboundary treaty with the countries of its maritime boundaries. Using the Reverse Osmosis (RO) process, the United Nations Office for Project Services (UNOPS) and the United Nations Development Programme (UNDP) are using money from the Adaptation Fund to pay for a project. Three islands, Mahibadhoo, Gadhoo, and Ihavandhoo have already built water plants that make water and get it to people. After the 2004 Indian Ocean Tsunami, these plants were brought to the country. The main difference is that the current system uses solar energy instead of diesel. Solar power minimizes costs and makes rainwater cleaner and affordable to all. Seawater for desalinization is extracted from underground by well into the RO plants. At the same time, rainwater harvested from public roofs is piped into the filtration and disinfection system. Then desalinated seawater and filtered rainwater are collected into the enormous main tank and combined for delivery to local users. A backup generator power source is set up for an emergency case to prevent disturbances.

An initiative for improving water quality by the Government of the Maldives includes a pledge to establish policies and introduce programs to tackle the islands' water shortages during dry periods. Innovative and viable water supply techniques may be found to assist populations affected by climate change. Such integrated water production and delivery give the Maldives more confidence in managing the risks and uncertainties associated with climate change (Millar, 2002).

12.6 OVERVIEW OF WATER RESOURCES: SRI LANKA

Rainfall is the major water source in Sri Lanka, and the annual average rainfall ranges from 900–5000 mm. Based on rainfall patterns, two distinct climate zones are assigned: wet (20%) and dry zone (80%). Average rainfall in the wet zone is over 1900 mm per year and less than 1500 mm per year in the dry zone. Cumulative rainfall in Sri Lanka is, on average, 121 km^3 per year, and runoff is 50 km^3 per year. This country is blessed with over 103 rivers and more than 12,000 reservoirs, each consisting of more than 1000 km^2 basin area. The largest catchment is *Mahaweli* River, with 10,600 km^2 draining 4009 million m^3 per year, and the largest discharge point is *Kaluganga* River draining 4032 million m^3 per year, with a catchment area of 2688 km^2 (Withanachchi et al., 2014). Groundwater is another primary water resource in Sri Lanka. In this country, six types of aquifers are distributed: coastal sand, deep confined, lateritic (cabook), alluvial shallow karstic, and shallow regolith aquifer of the Hard Rock Region. Around 8000 km^3 per year is projected to be the total renewable groundwater resource (Kijne et al., 2003).

12.6.1 Challenges of Water Resources

Considering climate change, economic development pressures, a growing population, and increasing water demand, Sri Lanka's sustainable water resource management is a challenge (Senevirathne, 2015). Water-related issues in Sri Lanka are primarily a result of many human activities. Feces, industrial waste, and pesticides highly contaminate the water in urban areas. There is a high fluoride concentration in groundwater in certain arid zones and a high iron concentration in rugged, stony, alluvial regions. Another significant issue affecting groundwater supplies in Sri Lanka is excessive use, particularly through tube wells. In addition, the leading causes of marine pollution are oil accidents, ship refuse disposal, and coral and sand extraction. Except for irrigation and hydroelectric systems, the general condition of Sri Lanka's water supply is inadequate (Bandara, 2003).

12.6.2 Climate Change Impacts on Water Resources

Sri Lanka (65,610 km^2) is located in the southern Indian Ocean, bordered by coastal lowlands and a central hill range. Climate change has an influence on water security by raising air temperature, modifying evaporation and precipitation rates, and affecting stream flows and sea-level rise. Sea breezes and heavy humidity moderate the warm climate of Sri Lanka. The increasing frequency of floods and droughts affects agronomy, fisheries, and forest composition; land use patterns change rapidly and on hydropower generation.

Climate change impacts are forcing food price instability, the outburst of diseases, high prices of electricity, trade imbalance, and housing problems. Around 45% of rainfall water is lost through evaporation and transpiration, 35% flows in rivers and streams, and the remaining 20% penetrates through the earth's surface. Rainwater flows into the river along 4500 km, and the water volume is 127–130 billion m^3. The central highlands contribute to hydropower and irrigation development. In the northern and eastern portions of the arid region's central highlands, water capacity is limited, resulting in water scarcity. The arid zone of Sri Lanka will encounter severe economic consequences due to climate change (Eriyagama et al., 2010).

12.6.3 Sediment Transport and Yield

In Sri Lankan River basins, measurement or assessment of suspended sediment loads is crucial. Continuous monitoring of SSCs for at least one hydrological year is preferred for the calculation of a representative yearly total suspended load. Small rivers in low-lying areas are more susceptible to temporal variation, such as shifts in land use caused by seasonal crops, and the delivery of sediment is comparatively greater than that of large rivers (Illangasinghe and Hewawasam, 2015). Sedimentation of inland reservoirs was regarded as a critical issue (Hewawasam et al., 2003; Illangasinghe and Hewawasam, 2015). Current river discharge testing by the Irrigation Department, Sri Lanka, is based on river stage and discharge rating curves. The sub-water shoreline of the Upper Mahaweli has the largest sediment yield (Dias et al., 2019; Diyabalanage et al., 2017). The Mahaweli watershed is Sri Lanka's most significant due to its size and economic significance (Dias et al., 2019).

12.6.4 Exclusive Projects and Achievements

The primary objective of this ambitious plan was the generation of hydropower, the provision of water to arid lands for the development of agricultural production, the control of flooding, and the construction of infrastructures for landless and destitute families. Significant programs are enumerated in Table 12.2.

12.6.5 Treaties and Lessons Learned

Sri Lanka does not have any water treaty with its neighboring countries. Established planning methods, such as water protection planning, can be expanded to include monitoring for climate change

TABLE 12.2
List of Exclusive Projects in Sri Lanka

Program name	Implementing agency	Program outputs	Reference
Mahaweli Development Project (MDP) (1960-1985)	MDP	Construction of Victoria (the largest dam), Randenigala and Kotmale Dams	(Fernando and Nizar, 2019)
Minipe Left Bank Canal Rehabilitation Project (MLBCRP) Upper Elahera Canal Project (UECP) North Western Province Canal Project (NWPCP)	MWSIP	Providing uninterrupted water infrastructure for irrigable land in the North-Western Province, permitting *Yala* and *Mahaa* to farm during both growing seasons.	
Moragahakanda-Kalu Ganga development project (2007)		Moragahakanda dam (official name Kulasinghe Reservoir) is constructed across the Amban River at Elahera which is the 1st initiative for Mahaweli development	(Padma, 2019)

threats and impacts. Scenario-based methods that consider several different climate futures are suggested because of the "data gap" in most developing countries and difficulties in downscaling climate forecasts at the basin scale and below. Because of uncertainties about climate change impacts in water resource planning, "robust uncertainty" should be the approach to technical choices, for example, suitable for various conditions like rainfall and runoff. This means that the reliability of various sources should be given more consideration, for instance, placing boreholes and deeper wells in more efficient aquifers, promoting large springs, and enhancing health security. However, shallow wells can be utilized as a water supply option for local families and communities.

12.7   OVERVIEW OF WATER RESOURCES: INDIA

All river basins in India flow 1869 km^3 annually (Cassen, 2016). Notwithstanding their high precipitation rates, the Ganga, Brahmaputra, and Barak catchments account for only around a third of the country's surface water reserves (Table 12.3) and 60% of the region's (Mohile, 2006). 431 km^3 of rechargeable groundwater is estimated (Cassen, 2016; Kumar et al., 2005).

12.7.1   Challenges and Climate Change Impacts on Water Resources

Due to irrigation, hydroelectricity, and urban and industrial water needs, India has water resource management issues (Jain et al., 2017). Water, food, and healthcare are threatened by changing temperatures. India has seen several disastrous extremes in recent decades due to rising hot and cold days. The 2016 drought impacted 330 million people in 10 states and cost US$100 billion (Jain, 2019). Changes in precipitation pattern, surface temperature, groundwater recharge, river surface timing, and evaporation may impact freshwater availability (Goyal and Surampalli, 2018; Madhusoodhanan et al., 2016). India's 5-year precipitation pattern shifts from 6–15% of the typical (Singh and Kumar, 2018). Northern India lost 2 cm of groundwater every year. In South India, pumping and precipitation fluctuations enhanced soil water storage by 1–2 cm per year from 2002 to 2013. Southwest monsoons cause 80% of India's yearly precipitation between June and September (Singh and Kumar, 2018). Several key river basins have had fewer and more frequent rainy days. Precipitation will rise by 30% in the short future (2040–2069) and by 50% during the long term

TABLE 12.3

Yearly Average Flow and Utilizable Flow of Rivers

Name of river basin	Flow per year (Avg.) in km³	Utilizable flow in km³
Subarnarekha	12.37	6.81
Mahanadi	66.88	49.99
Cauvery	21.36	19
Indus	73.31	46
Brahmni–Baitarani	28.48	18.3
Godavari	110.54	76.3
Ganga	525.02	250
Ganga–Brahmaputra sub-basin	629.05	24
Ganga–Meghna (Barak) sub-basin	48.36	
Kaveri	55.33	12.3
Kachchh and Saurashtra including Luni (west flowing)	15.1	14.98
South of Tapi (west flowing)	200.94	36.21
Godavari and Krishna (East flowing)	1.81	13.11
Mahanadi and Godavari (East flowing)	17.08	
south of Cauvery (East flowing)	6.48	
Krishna and Pennar (East flowing)	3.63	
Pennar and Cauvery (East flowing)	9.98	16.73
Tapi	14.88	14.5
Krishna	69.81	58
Narmada	45.64	34.5
Pennar	6.32	6.86
Sabarmati	3.81	1.93
Mahi	11.02	3.1
Rivers draining into Bangladesh	8.57	
Rivers draining into Myanmar	22.43	

Source: Jain, 2019; Jain et al., 2017.

(2070-2099) (Goyal and Surampalli, 2018; Lacombe and McCartney, 2014). In the previous 15 years (2000–2014), India's average annual temperature has climbed 0.25°C from 25.06°C. Average maximum and lowest temperatures rose by 0.28°C and 0.22°C, respectively, from 2000 to 2014 (Jain et al., 2017). Throughout 41 years, daily Sutlej River records in Northern India at three measured stations (Kasol, Sunni, and Rampur) dropped (1970–2010). The basin has great hydropower and agricultural potential. The average yearly river streamflow declined significantly, and the Ravi River declined somewhat. Due to snow and glacier melting fluctuations, the Chenab increased statistically between 1961 and 1995. The historical annual streamflow (1982–2012) of the Gomti River (30,437 km²), a Ganges River tributary in Northern India, decreased at four-gauge rainfall sites (Neemsar, Sultanpur, Jaunpur, and Maighat). On the peninsula of India, stream flows in the Mahanandi River Basin (catchment area of 141,589 km²) dropped by 3388 million m3 between 1972 and 2007. Due to a rise in floods between 2075 and 2100 and a 32.5% fall in the average runoff in April between 2050 and 2075, the Mahanadi's peak runoff increased by 38% in September (Goyal and Surampalli, 2018).

From northwest to south India, drylands encompass 106 km² (arid, semi-arid, sub-moist) (Goyal and Surampalli, 2018). Drought impacts many Indians. India saw severe droughts in 1965, 1972, 1979, late 1987, and late 2010. West Rajasthan, Saurashtra, and Kutch experienced 31 droughts from 1875 to 2004 (130 years) (Kane-Potaka et al., 2018). Jammu and Kashmir experienced 28 droughts, and Gujarat, 27 (Goyal and Surampalli, 2018). Indian agricultural and Himalayan rivers such as the Indus, Ganges, and Brahmaputra affect glacier changes. Between 1960–2004, 11 basins in the Indian Himalayas lost 0.2–0.7% of their glaciers, averaging 0.2–1.4 km^2 (Bolch et al., 2012).

12.7.2 SEDIMENT TRANSPORT AND YIELD

In recent years, reservoir sediment deposition has been identified as a significant issue. Reservoir deposition diminishes storage capacity and creates significant issues in the event of stabilization and operation of the dam. Due to deposition and entrapment at various locations in the upstream basin (Table 12.4), it was observed that not all eroded and conveyed sediments reached the outflow of the basin (Goyal and Surampalli, 2018).

The production of sediments in any region results from the interaction between meteorological factors and the land surface. The study of soil erosion and sediment measurements is cumbersome because several factors affect erosion rates. Accurate field measurements for sediments, soil resources, period and intensity of rainfall, the density of vegetation, and farming practices, are needed. Govind Sagar reservoir's sediment yield rate was found to be 700 m^3/year, which is in a very low range (Subramanian, 1993). The Indian River's annual sediment load is just over 1.2 trillion tonnes and around 10% of the world's sedimentary flows to the oceans. In its sediment release, Indian rivers exhibit marked seasonal and spatial variability (Javed et al., 2016).

12.7.3 GOOD PRACTICES

India has made many good examples of water resources management. Some major programs are listed in Table 12.5.

12.7.4 REMARKABLE TREATIES

Water in much of India remains politically disputed. The region faces water scarcity and agricultural problems and the situation will exaggerate owing to expansion in industrialization and the demands for energy and water. India has many transboundary rivers, with Pakistan (the Indus), Bangladesh (Teesta, Brahmaputra), Nepal, Bhutan, and China (UNDP, 2010).

TABLE 12.4
Sediment Yield of Rivers in India

River name	Area of basin (km^2)	Sediment (10^6 ton/yr)	River run-off (10^6 m^3/yr)
Ganga	750,000	329	493,000
Brahmaputra	580,000	597	510,000
Narmada	87,900	70	46,700
Mahi	25,500	9.7	11,000
Kaveri	66,300	1.5	21,500
Indus	970,000	100	238,000
Mahanadi	41,000	15.7	54,500
Godavari	313,000	170	92,200
Brahmani	28,200	20.4	16,300
Irrawady	430,000	265	422,000
Tapi	49,000	25	18,000
Krishna	251,400	4	32,400

Source: Dutta et al., 2017.

TABLE 12.5

Good Practices on Water Resource Management in Different States of India

Project name	Implementing agency	Main objectives	Mission outcomes
Mission Kakatiya, Telangana	Telangana State Government, Telangana	To restore small water supplies and improve the construction of minor irrigation infrastructure. To implement this project, the feed channels have been redefined; and irrigation channels have been repaired.	Increased soil groundwater level and capacity for water retention. In agricultural soil preparation, techniques such as perforation blending reduced the use of chemical fertilizers and increased water retention.
Narmada (Sanchore)	Govt. of Rajasthan, Rajasthan	The goal was to efficiently manage water use, drainage, and planting trees carefully. Especially a broad range of plants which are resistant to salinity to soil fertility.	Output and salinity of the soil have both dropped Losses in agriculture and property losses have been greatly reduced. Increased food production.
Har Khet ko Pani	Ganga Rejuvenation, Government of India, Chittoor, Andhra Pradesh & Ministry of Water Resources, River Development	To include rehabilitation and promotion of crop diversification in conventional water structures. To work under the National Rural Job Guarantee Scheme (MGNREGA).	The water table and irrigation capacity have been strengthened. Recovery and refurbishment of water bodies can lead to efficient water use.
Mulching– Harvesting many benefits in cardamom	The Indian Cardamom Research Institute (ICRI), Western Ghats	To analyse the soil fertility on the farm.	It successfully helped maintain the plantation with compact clumps for 17 years.
Pani panchayat	Department of Water Resource Orissa, Orissa	The primary aim was to strengthen the state water resource planning and growth process. The main goals of the activity were to help and ensure that everyone who used water got an equal amount.	The institutional structure helps to control the resource sustainably and efficiently.
Bhungroo-Groundwater injection well	Government of Gujarat, Gujarat	This system injects and stores excess rainfall water underground.	It has contributed to the protection of food. This structure has minimized the pressure on women to make them the primary owners and experts.
Community-managed drinking water supply program	Gujarat, Water and Sanitation Management Organization	This cooperative approach to the rural water supply program is controlled by a demand to supply the village population with sufficient and safe water.	This action covered approximately 76.84% of rural households in Gujarat since 2014. In the communities, particularly for girls, substantial improvements have been observed in collecting long-distance water, reducing waterborne diseases, and overall improvements in health.

(Continued)

TABLE 12.5 CONTINUED

Good Practices on Water Resource Management in Different States of India

Project name	Implementing agency	Main objectives	Mission outcomes
Mazhapolima initiative	Kerala, Thrissur District Administration	During the rainy season, roof rainwater is brought through pipes with a sand filter to fill the water tank at the end of an open well.	The money expended previously on drinking water is now used for constructing self-supported buildings on the rainwater roof.
Meeting water requirements through innovation	Gujarat, Swajaldhara	The tanks were installed around a perennial spring at an altitude of 120 meters from the main village under this program.	Those villagers could get water from their village and no longer had to rely on a single well that was open to meet their regular water requirements.
Adaptive water management	Rajasthan, Jal Bhagirathi Foundation, Village Communities	The money which was raised through the support of every family in the village is used for the renovation of water channels and tree plantation to increase the inflow of water to sustain the catchment.	The Jal Sabha has a sustainable financial resource to periodically hold the talaab through the coupon system. The town was able to adjust to evolving weather patterns and repeated droughts.
Birkha Bawar	Jodhpur, Rajasthan, Umaid Heritage Real Estate	The project emphasizes sustainable stormwater management in the housing complex, apart from stormwater collection and conservation. The rainwater from open areas is collected through natural paths and from the housing roof connected by drainage pipes.	The arrangement has enhanced the value of the complex by perfectly mixing good architectural design and well-preserved green areas in a low-precipitation environment.
Johads, Haryana	Uttar Pradesh, Rajasthan	Johads are simple barriers to mud and rubble installed over a slope to avoid rainwater. Johads obtain water from the Mozzone that eventually reaches the groundwater.	Results in improved access to water and was an agricultural, animal husbandry resource, etc.
Artificial glaciers	Ladakh, Himachal Pradesh, Chewang Norphel	This plan uses artificial glaciers and canals to move water from the main glacial stream to a small area of shade outside the village, where it will stay frozen.	A significant increase in agricultural production and thus the local population's income.
Dhara–Vikas enhancing rural water security in drought-prone areas	Sikkim, Department of Rural Development	The key objective was to ensure water protection in areas where water is minimal. To enhance mountainous ecosystem hydrology to the population as a water tower and to manage catastrophe risk by reducing landslides and floods.	The recharging of the lakes and the revival of as many as 50 Springs in Sikkim have had a big effect.
Project Bhujal watershed rejuvenation	Bundelkhand, Uttar Pradesh, Anandana	The Bhujal project primarily focused on rejuvenating the tanks, generating 100 million liters of water storage capacity.	After this procedure, the strength of the cultivation has increased to 30%, and the groundwater has increased significantly to 2–5 m.

Source: Aayog, 2017; Morris and Fan, 1998.

Indus Water Treaty: The World Bank mediated the Indus Water Treaty, which Pakistan and India signed on September 19, 1960. The treaty specifies both countries' obligations and rights over the Indus River. From April 1, 1948, India began taking water out of canals that ran into Pakistan. David Lilienthal, a former executive of the Tennessee Valley Authority United States, was in charge in 1951. To develop and administer the Indus River plan jointly with the World Bank's advice and assistance, he suggested that India and Pakistan negotiate a deal. India and Pakistan built dams and link canals (e.g., the Jhelum River Mangla Dam and the Indus River Tarbela Dam). As the Treaty only permits India to utilize 20% of the total Indus water, it clearly outlines what the two nations are doing and are not doing. India constructed the Bhakra Dam on the Satluj, the Pong and Pandoh Dam on the Beas, and the Thein (Ranjitsagar) Dam at Ravi to harness the eastern rivers' waters that were exclusively given to it (Zeitoun and Warner, 2006).

India–China Water Treaty: China and India are competing for resources along the Brahmaputra River, which flows through regions of Asia prone to territorial conflict. China's Yarlung Zangbo river flows 2900 km from Tibet to Bangladesh and India to the Ganges and Gulf of Bengal (Akhter, 2015). India primarily depends on Tibetan water since it has access to a third of its water sources. There are several reasons why India and China are facing a water crisis, including continuous border disputes over Arunachal Pradesh and China's upstream dam construction (Lovelle, 2016). Flash floods were caused by a dam failure in Tibet in June 2000 in Arunachal Pradesh. Due to their rapid population increase, China and India are experiencing severe water shortages. India lacks a large enough reservoir to store water since China has an abundance of ice, groundwater, and surface water, and water is distributed unevenly in certain areas (Amarasinghe et al., 2005; Mohile, 2006).

12.7.5 Lesson Learned

The problem before the whole of humanity and not just for India is to be seen as climate change. If this serious problem is ignored locally, it will later be the most harmful and deadly problem globally. For a developing country such as India, climate change poses incredibly difficult challenges. However, India wants no restrictions on its prospects for development. It also wants to be seen as a global force emerging. While its new status can best serve the former, the latter will require it to be a leader in addressing the most significant global climate change challenges. Climate change's consequences on India's diversified economy and people are little studied. India should address its adaptation and mitigation methods, which include increasing protective capacity and reducing the frequency and severity of catastrophic environmental disasters.

12.8 OVERVIEW OF WATER RESOURCES: BANGLADESH

Bangladesh's low riparian zone has three major rivers: the Ganges, Brahmaputra, and Meghna. Transboundary water and precipitation distribution determine a country's water management and productivity. Bangladesh's dry and wet seasons affect its agroecosystems and social operations. Bangladesh's principal water sources include the linked rivers, estuaries, canals, the floodplain, wetlands, intertidal lands, and water, *haor* (vast depression areas with seasonal water), *baor* (oxbow lake), *beels* (depressions with stagnant water), ponds, lakes, and groundwater aquifers. (Lovelle, 2016).

Surface Water: Surface flow is the most significant part of Bangladesh's water supply (Ahmed and Roy, 2007). Bangladesh's water supplies are available in about 89% (1210 km^3/year), while only 9% and 2% are sufficient for precipitation and groundwater, respectively.

Rainfall: Bangladesh, except for Eastern Hills, has an annual precipitation level of 284 km^3. The precipitation is spread across the country unevenly. They range from 1500–5500 mm in the northwest, with a mean rainfall of 2200 mm (Rahman et al., 2019). The average rainfall was 2360 mm between 1960 and 1997 (Rahman, 2015).

Transboundary Flow: The nation's 808 rivers span 24,000 km. Fifty-four Indian and 3 Myanmar rivers cross into Bangladesh. The Ganges, Brahmaputra, and Meghna rivers drain 108, 0.58, and 0.09 million km^2, respectively. Peak flows from the Brahmaputra, Ganges, upper Meghna, and lower Meghna are 100,000, 75,000, 20,000, and 160,000 m^3/s, respectively (Ahmed and Roy, 2007).

Groundwater: Groundwater is one of Bangladesh's leading irrigation and household water sources (Rahman and Rahman, 2010). The yearly recharge volumes and the characteristics of the soil storage influence Bangladesh's soil water availability. Bangladesh has three primary aquifers at three depths (from a few meters to 2500 m). Most of Bangladesh's regions are comprised of alluvial and deltaic sediments deposited by three major rivers. These alluvial deposits have primarily created an unregulated aquifer in most of the land. Bangladesh's quaternary alluvium is a large aquifer with relatively strong transmission and storage properties. In Bangladesh, the total water supply is approximately 25.750 million m^3, of which about 14.150 million m^3 in 1996–1997 were used (Ahmed and Roy, 2007).

12.8.1 Challenges of Water Resources

Unregulated extractions, expanding urban demand, climate change, land-use changes, and environmental requirements are the nation's biggest issues. Soil water is heavily used for irrigation, agriculture, and industry. Groundwater irrigates northwest (97%), northcentral (84%), southwest (76%), and northeast (45%). Bangladeshi farmers need 33 km^3 of water annually. Northwest (11.4 km^3) has the largest irrigation water requirement. The northeast and southeast have the lowest irrigation water requirement (4.3 km^3). Domestic and industrial water consumption is 2.7 km^3 and 0.08 km^3, respectively (Hossain et al., 2013; MoWR, 2004).

12.8.2 Impacts of Climate Change on Water Resources

Bangladesh is significant in terms of the prospective effects of climate change resulting from global warming. Precipitation and evaporation will increase, groundwater resources will be reduced, and projected transboundary flows during the dry season can be reduced by 2050. The dry season is considerably longer than usual, and the demand for crop water will rise as high as 25% above current needs by 2050. Surge depths would rise, and by 2050 an increase of 0.44 m will make drainage even more congested (Rahman, 2015). This will significantly increase drainage requirements and flood time over peak flood levels. Nevertheless, these results are difficult to predict. The increase in sea levels caused by global warming, the rivers and the floodplains continue to gather sediment, and the Ganges Basin subsides influence sea levels to the land level (Rahman et al., 2019).

In and around the metropolis of Dhaka and Chittagong, the water quality is poor. Coastal groundwater is susceptible to seawater intrusion (MoWR, 2004). Due to upstream withdrawals, the decrease in dry season runoff exacerbates the salt intrusion. In addition to affecting water quality, water contamination also affects bioaccumulation. Approximately 95% of domestic and industrial and 70% of irrigation water sources are derived from groundwater. Arsenic contamination is the primary issue with groundwater. The population, particularly in the Lower Ganges plain, has been subjected to arsenic contamination because of an increasing reliance on groundwater for drinking and irrigation. The contamination of Bangladesh's groundwater with arsenic affects approximately 35 million people and 37 districts. An estimated 35 million individuals are at risk from arsenic in their potable water (Ahmad et al., 1994).

Flooding (flash flooding, river flooding, rainwater flooding, and storm-induced coastal flooding) in Bangladesh is common, and two-thirds of the country is inundated yearly (Gupta et al., 2005). The four most destructive in recent history were the 1987, 1988, 1998, and 2004 floods, with the worst record in 1998. Fluvial erosion is a normal cause of forced resettlement and landlessness. River erosion involves altering the path, and forming new channels in flooding (Mirza and Dixit,

1997). According to the Bangladesh Water Development Board (BWDB), riverbank erosion affects 1200 km of land. In the last 125 years, 42 intense cyclones have struck coastal regions, 14 of which have occurred in the previous 25 years (Ahmed, 2006; MoWR, 1999).

12.8.3 Sediment Transport and Yield in Bangladesh

Three-quarters of the Ganges and Brahmaputra rivers drain into the Bengal Basin, depositing 1 million tonnes of silt (Rahman, 2015). From India, China, Nepal, and Bhutan, Bangladesh's Ganges, Jamuna, Meghna, and Meghna rivers carry around 1 trillion tonnes of silt to the Delta annually. The Delta's long-term sediment allocations account for one-third of the sediment carried by those rivers on the flood plain, tidal plain, and sub-aqueous delta, creating vertical accretion and lateral advancement. A 16-km-thick fluvial-deltaic layer was formed in the Bengal Basin during the Paleogene due to the high sediment burden of the rivers, annual overbank flooding, and subsidence of the basin. Approximately one-third of the annual burden of sediments is deposited in the river floodplain. However, the dispersed sediment burden has not been investigated for a brief duration (moist and dry seasons within one year) (MoWR, 2004). Using such point positions could result in unrepresentative sediment assessments, as sediment could be transported upstream and deposited downstream of the data collection site. Periodically, the sediment loads for both main streams are measured at a single measuring station along each channel (Rice, 2007; Rice, 2010).

12.8.4 Exclusive Project

Bangladesh Delta Plan 2100 (BDP 2100): A water-central, techno-economic long-term plan for the Bangladesh Delta. BDP 2100 is Bangladesh Delta's long-term, stable, holistic sustainable development plan. The BDP 2100 aims at effective and sustainable water and flood protection in Bangladesh while addressing natural disasters and climate change through adaptive and systematic strategies (Sarker et al., 2011). The BDP 2100 has more specific goals, such as protecting against floods and disasters caused by climate change, making water safer improving the quality of water use, and protecting and preserving management (BPC, 2017). BDP 2100 included both the national level (for example, risk of flooding, freshwater supplies) and the unique hotspot strategies (river/estuary, droughts, flash flood, hilly, urban and coastal zones) (BPC, 2017, 2019).

12.8.5 Remarkable Treaties between Bangladesh and India

Most of the rivers in Bangladesh, including the largest rivers, are transboundary with India. Ganges Water Sharing and the Teesta Treaty are two important deals between Bangladesh and India.

> *Ganges Treaty:* India–Bangladesh signed the Ganges Water Sharing Treaty in 1996. Ganges Treaty allocates surface waters at the Farakka Dam near the India–Bangladesh boundaries. The treaty is concerned with the real flows of different levels and not with the accurate flow of 75% as provided for in previous agreements. The fundamental formula is that the two countries share 50:50 in the lean season (BPC, 2019).
>
> *Teesta Treaty:* Bangladesh and India signed the Teesta Treaty in April 2017 to establish equitable water-sharing policies between the two nations. This treaty aims to nourish these two countries and maximize the advantages of the common river system (Tabassum, 2003).

12.8.6 Lesson Learned

Since the early 1990s, several studies have raised awareness of climate change's potentially detrimental effects on numerous sectors of the economy (Rahaman, 2020). Policymakers need scientific climate change understanding in their language and timeframes. To better understand and inform policymakers, climate change research should be supported. Climate change adaptation is more likely to be included in the sector's present and anticipated operations by policymakers, planners, and administrators. National and international professionals and analysts must communicate their knowledge with those who make choices and plans on the ground more effectively.

12.9 OVERVIEW OF WATER RESOURCES: NEPAL

Nepal has 6000 rivers, 224 km² of annual gross average water flow per capita, and 9000 m³ of water per capita (Table 12.6). The four principal water systems are the Mahakali, Gandaki, Karnali, and Koshi Rivers (Huq, 2002). The nation's available surface water is estimated at 225 billion m³ annually (Aryal and Rajkarnikar, 2011).

Glaciers occupy around 3.6% of Nepal's total land area (WECS, 2005). It disclosed 3252 glaciers covering 5323 km² and the ice reserve valued at 481 km³. There are 779 glaciers in the Koshi Basin, an area of 1,409,84 m², and an overall 152.06 km³ ice reserve. In the Gandaki River Basin, 1025 glaciers occupy 203,015 m² and an estimated 191.39 km³ ice reserve (Mool et al., 2001).

Nepal has many closed waterways, including reservoirs, swamps, dams, and other small wetlands (Table 12.7). The National Lake Protection Committee has recently published a study documenting 5358 lakes (including 2323 glacial lakes) in Nepal. Lakes are also scattered over the east–west longitudinal range at various heights. Nine Ramsar sites are found in Nepal. Ghodaghodi Lake

TABLE 12.6
Surface Water Resources in Nepal

River basin origin	Area of the catchment in km²	Discharge per year (Annual)	Average discharge
Himalayas	105,573	157 km³/year	4979 m³/s
Siwalik Zone	23,150	53 km³/year	461 m³/s
Mountains and hills	17,000	14.5 km³/year	1682 m³/s
Total	145,723	224.5 km³/year	7122 m³/s

Source: Shrestha et al., 2018; Suhardiman et al., 2015.

TABLE 12.7
Total Water Resources in Nepal

Name of the river	Total length in km	Area of drainage Nepal	Area of drainage Total
Karnali	507	41,890	44,000
West Rapti	257	6,500	6,500
Bagmati	163	3,700	3,700
Kankai	108	1,330	1,330
Mahakali	223	5,410	15,260
Babai	190	3,400	3,400
Narayani	332	28,090	34,960
Sapta Koshi	513	31,940	60,400
Others		24,921	24,921
Total		**147,181**	**194,471**

Source: Aryal and Rajkarnikar, 2011.

District, Beeshazari, and Associated Lakes, Jagadishpur, Koshi Tappu, Rara Lake, Mai Pokhari, Phoksundo Lake, and Gosaikunda Lakes are some of Nepal's significant lake locations. Other large lakes in Pokhara (Kasky district) are Phewa, Begnas, and Rupa. Nepal has plenty of groundwater. It is estimated that the country's renewable water capacity is 12 km^3. Besides Terai, towns such as Kathmandu depend heavily on groundwater resources for their day-to-day water requirements (Aryal and Rajkarnikar, 2011).

12.9.1 Challenges of Water Resources in Nepal

Despite abundant water, Nepal confronts numerous challenges in managing its water resources (WEPA, 2010). Water scarcity is a significant issue in rural watersheds of central Nepal and the Hindu Kush – Himalayas. Water scarcity during the year's arid months poses a particular threat to agriculture and domestic use. Untreated waste has contaminated the surface water and refuses from densely populated residential areas (Shrestha et al., 2015).

12.9.2 Impacts of Climate Change on Water Resources in Nepal

Climatic change disturbs the natural cycle, affecting the water supply. Changes in climate, particularly in mountainous regions, would significantly impact water supplies (Merz et al., 2003). Nepal in the south and arctic conditions in areas above 5000 m in the north have a tropical to subtropical climate in the Ganga plains. Climate change in Nepal has not only affected agriculture and drinking water. It involves the retreat of glaciers, floods, slopes and sedimentation from heavy plummeting events, and increased dry-season flows with possible threats to water and energy supply in lean-season conditions. Additional impacts include increased warming days, predictable precipitation patterns, seasonal changes, drought and storms, and declining natural water supplies. The survey showed that climate change had a significant effect, such as water drying up, decreased water volume and number of springs and decreased crop productivity (Merz et al., 2003).

12.9.3 Sediment Transport and Yield

Sediment yield is a dynamic phenomenon of glacial, fluvial erosion, weathering and land sliding (Basyal, 2017). The catchment area, topography, slope, rainfall, and terrestrial features are highly influential factors (Gajmer, 2014). The average sediment was transported 40.904 $\pm$ 12.453 megatons (Mt) per year from the highest of the Himalayas in the hydropower reservoir (Giordano and Wolf, 2003). The major Narayani River tributaries transport sediments across the High Himalayas and Mahabharata highlands upstream from the major Boundary Thrust. Recent research calculated the annual suspended load flow at the Narayani watershed scale to be 130 Mt/year and the late Holocene erosive flow to be 110–184 Mt/year based on 10 Be cosmogenic nuclides in river sediments. Suspended sediment concentration (SSC) ranged from 0.012–0.588 kg/m^3 at Manahara River, 0.021–0.163 kg/m^3 at Bagmati, 0.082–0.600 kg/m^3 at Dhobi Khola and 0.042–0.916 kg/m^3 at Bishnumati River (Aquastat, 2010).

12.9.4 Good Practices and Projects

Since 2008, the Irrigation and Water Resources Management Project (IWRMP) has focused on enhancing agricultural productivity, sustaining chosen irrigation systems in Nepal, and building institutional capacity for improved water resource management. 415,200 water consumers of selected Farmer-Managed Irrigation Systems (FMIS) spanning 268.59 km^2, mostly in hilly regions, benefit. The project also aims to implement four Agency Managed Irrigation Systems (AMIS) and considerable institutional reforms throughout 230 km^2 (Islam et al., 1999). The project has four key components: (i) irrigation, (ii) infrastructure, (iii) development and (iv) enhancement: increase irrigation water service in selected mountain, hill, and Terai systems of three western regions (40 districts), and improve and extend groundwater irrigation in Terai. At the field level, the initiative has achieved its goals of boosting agricultural output, water efficiency, and water user management of irrigation systems. A Water Resources Information Center (WRIC) and Water Resource Information System (WRIS) are active and publicly accessible. The research project has improved the physical efficiency of selected surface water systems and strengthened Water User Associations (WUAs) institutionally and financially to improve irrigation services.

In western Nepal, the Digo Jal Bikas project focuses on the urgent need to resolve the management and production of water supplies in west Nepal. This research project seeks to enhance sustainable water resources in the Western Region of Nepal by including stakeholders in policymaking that balance economic development, social justice, and healthy, resilient ecosystems and inform decision-makers. The initiative targets basins and sub-basins in the growing Mid-Western and Far-Western Regions of Nepal (Giordano and Wolf, 2003).

12.9.5 Remarkable Treaties

Mahakali Treaty: India calls the Ganges' greatest tributary, *"Sarada"*, the Mahakali. On February 12, 1996, Nepal and India signed the Mahakali Treaty, which comprised the Sarada barrage, Tanakpur barrage, and Pancheshwar Multipurpose Project, addressing integrated construction, which took effect on June 5, 1997. From a procedural standpoint, the Mahakali Treaty comprises three different agreements. It binds all parties to comply with the Mahakali Treaty principles. The convention supports the concepts of fair usage and equitable benefit sharing. Therefore, before any project begins on the Mahakali River, either party must reach an agreement. The project is intended to achieve the total net profit as agreed between the parties. The advantages for all parties in developing the project shall be evaluated in terms of power, irrigation, flood control, etc. A good example of conflict resolution on foreign rivers is the Mahakali Treaty. It offers reasonably advanced methods to solve conflicts (Gajmer, 2014).

12.9.6 Lesson Learned

Glacier-dominated mountains provide water to many populations. Global climate change presents distinct difficulties to downstream ecosystems. Glacier-dominated regions across the globe continue to bring unique challenges to civilization as they adapt to global climate change and lead populations to consider and prepare for it. From the tropical south to the alpine north, Nepal has several climates. Food and forest policy have major climatic impacts. In many locations, residents have responded to stress independently of government response. Floods, aridity, drought, and forest fires caused by climate change might greatly impact local food and livelihood systems.

12.10 OVERVIEW OF WATER RESOURCES: BHUTAN

Bhutan is a small land of 47,000 km^2 in the Eastern Himalayas (Rahaman, 2009). Bhutan also has an abundance of water supplies and its main Himalayan rivers in the north. Bhutan's greatest water sources are glaciers, rivers and basins, glaciates, and high-altitude wetlands. Bhutan has four major rivers: Drangme Chhu, Punatsang, Sankosh, and Wangchhu. Each one is swiftly moving down the Himalayas to join the Brahmaputra River in India, south through the Duars, and finally, Bangladesh, where it joins the enormous Ganges. Water in rivers, reservoirs, springs, rivers, and ponds is accessible locally (NEC, 2016). Approximately 58,930 m^3 of annual freshwater per cup indicates the country is not suffering from water stress (Singh and Karki, 2004).

India's biggest river system, Drangme Chhu, runs southwest of Arunachal Pradesh.

The Great Himalayan Range feeds Puna Tsang Chhu (a 320-km Puna) in northwest Bhutan. They join the Puna Tsang Chhu near Punakha, flowing south to West Bengal, India. Wang Chhu's

370-kilometer affluence is in Tibet (NEC, 1992). The Wang Chhu drains the Ha, Paro, and Thimphu valleys west and southeast of central Bhutan. In the Grand Himalayan Area, the Mo Chu or Sankosh River is also known as the upper reaches of Punakha. A 9900 km² basin drains the main river and has at least 53.80 m³/second annual water flow. The Pho Chu and Wangdi Phodrang Tang Chu are 20 km downstream from Punakha. Bhutan's greatest river system, the Manas, with its primary branches and affluents, is 3200 km long. In eastern Bhutan, the big Manas or Gong River drains approximately 18,300 km². Amochhu is the country's smallest river system. The 358 km long river Amochhus comes from Tibet's Chumbi Valley. Amochhu Basin includes the Haa, Samze, and Chhukha districts Districts. It encompasses 2298 km², almost 6% of the nation's land area. The Wangchhu Basin covers 4596 km² (11% of Bhutan) and comprises the Haa, Paro, Thimphu, and Chukha.

One of the main rivers in Bhutan is Punatsangchhu. Bhutan's biggest basin is 9645 km², 25% of its overall area. The Punatsangchhu Basin is located in the districts of Gasa, Punakha, Wangduephodrang, Dagana, and Sarpang. The Mangdechhu basin is made up of Bumthang, Trongsa, Zhemgang, and Sarpang (Singh and Karki, 2004). Its 7380 km² land area is nearly 19% of the nation. Following Punatsangchhu, the Drangmechhu basin is the second-largest body of water. It takes up 8457 km² or 22% of all of Bhutan. Bhutan has more than 59 lakes that cover an area of about 42.5 km² (Table 12.8). The majority of the country is more than 3500 m above sea level.

Bhutan's main natural resources are freshwater and glaciers (FAO, 1978). Approximately 1.6% of Bhutan's total land area is covered by glaciers, snow, and ice between 4050 and 7230 m above sea level. In Bhutan and the downstream riparian states, the glaciers comprise around 10% of the total land area and constitute major sustainable freshwater sources (FAO, 1978). In Bhutan, there are 2674 ice lakes. Due to the vulnerability of their Himalayan climate, glacial lake outburst floods (GLOFs) pose a significant risk to downstream populations, as they contribute considerably to the country's water resources (Savada, 1991).

12.10.1 CHALLENGES OF WATER RESOURCES IN BHUTAN

Climate change intensifies glacial lake outburst flooding, flash floods and landslides, wind storms, forest fires, and seasonal watersheds (NEC, 2016).

12.10.2 IMPACTS OF CLIMATE CHANGE ON WATER RESOURCES

Bhutan has a fragile mountain environment, and in this scenario, climate change poses grave threats (GEF, 2019). Increased landslides, glacial lake eruptions, and detrimental effects on agriculture, biodiversity, and water supplies are at risk. Climate change can make the country extremely sensitive to water scarcity. Increasing rainfall in existing reservoirs can increase runoff, increase soil erosion, and accelerate sedimentation. Hydroelectric power and urban water supply would suffer from lower average snow-fed streams, higher peak flows, and sediment discharge (NEC, 1992).

TABLE 12.8

Lakes of Bhutan with Their Surface Area

Name of lakes	Location	Surface area in km²
Buli	Shamgong	0.02
Ho Ko Tso	Punakha	0.6
Gulandi	Diapham	0.01
Luchika	Wangdi Phodrang	0.025

Source: NEC, 2016.

Increases in global warming will cause glaciers to withdraw, glacial lakes to become increasingly voluminous, and eventually cause glacial lake outburst flooding (GLOFs) with possible disasters (Singh and Karki, 2004). Some glaciers in Bhutan have been observed to retreat by about 20–30 cm annually. The latest GLOF in Bhutan was held in the Punakha–Wangdue valley on October 7, 1994. Part of the lagoon of Lugge Tsho in eastern Luana devastated the downstream valley (NEC, 2016). This event affected 91 households (Singh and Karki, 2004). Watermills have been washed out, dry ground is destroyed, and pasture land has caused significant harm to the local people (Choden, 2009).

12.10.3 SEDIMENT TRANSPORT AND YIELD

The Himalayan River is complex in sediment transport, and sampling sediments by those rivers is always difficult. River sediment loads vary widely annually. Most sediment loads in the monsoon months are transported. The Himalayan Rivers cannot achieve accurate and consistent sediment rating equations (Singh and Karki, 2004). The flux rate and sediment concentration changes are fast and unexpected (Choden, 2009). No published data on sediment transport studies are available in Bhutan. Sediment yield in the Himalayan rivers is discussed, and some contrast with the Punatsangchhu River is made regarding the factors that influence sediment transport (Tiwari and Sharma, 2012).

The Punatsangchhu River Basin is one of Bhutan's three major river basins. Sediment movement is influenced by hydrology and climate, geology, land use, and topography. Transport by sediment is dependent on flow, so topography influences sediment transport. Due to its steep topography, soil erosion is an obvious process in Bhutan. The influence of soil erosion on river sediment load is substantial. The correlation between the river and sediment concentration for 2007 was very good, with an R^2 value of 0.8 for river Punatsangchhu. Average release and concentration for the 1993 and 1996–2008 record periods were correlated with 0.53. The correlation with an R^2 value of 0.13 was very low for some years. This shows that the fluvial release does not predict the concentration of sediments in the Punatsangchhu River. Most of the sediments are moved even during monsoon months, accounting for about 90% of the total load shuttled in a year. The sediment is supplied by melting water originating from glaciers, processes such as mass migration and mud flux, debris torrents, and erosive canal flush from the landslide, moraine, and glacier-dammed lakes during disastrous outbreaks (Choden, 2009).

12.10.4 REMARKABLE TREATIES

The successful collaboration between Bhutan and India over the last 3 years has benefited both nations significantly in transboundary water management. Bhutan's size and small population have brought great advantages to the nation. In comparison, while India's benefits were also substantial, they were not a "player" because of their size and population, such as Bhutan. The Indian–Bhutan cooperation is an excellent example of how transboundary rivers can be managed within a global framework for collaborative development, which uses water as a driver for economic growth and poverty reduction in an impoverished region (Choden, 2009).

12.10.5 LESSON LEARNED

The Kingdom of Bhutan's mountains may not seem a clear place for learning about climate change. South Asia and the wider world have to learn a lot from Bhutan. The negotiations aimed to agree on maintaining global warming at the pre-industrial level by a maximum of 2°C.. According to the British Energy and Environment Intelligence Unit, Bhutan declared in 2009 that it would stay carbon neutral and set the most aggressive COP21 emission reduction pledges. Since its huge forests capture carbon dioxide, it's carbon neutral. However, it will not be easy to remain neutral as industrial and transport emissions grow quickly. It needs to retain its tree cover vigorously and find ways to expand economically in a carbon-neutral or reduced fashion.

12.11 OVERVIEW OF WATER RESOURCES: MYANMAR

12.11.1 WATER RESOURCES OVERVIEW

Myanmar has plentiful water resources. The catchment area of the river basins in Myanmar is approximately 737,800 km² (Biswas, 2011). Myanmar, Bangladesh, India, China, Thailand, and Laos share the Andaman Sea and the Bay of Bengal. Bangladesh, India, China, Thailand, and Laos share the Andaman Sea and the Bay of Bengal with Myanmar. Myanmar has a total water supply of 1577 km³. Several kinds of water resources are available here, including groundwater, surface water, and glaciers; ice caps are also excellent sources of water (Thein, 2019; WEPA, 2020).

Myanmar has four major rivers named the Ayeyarwady, Chindwin, Salween River (Thanlwin), and Sittaung; these are the major surface water resources and are crisscrossed by many other rivers. The Ayeyarwady, Sittaung, Chindwin, Mekong, and Thanlwin Rivers (Salween) normally stream from north to south. However, the Myittar River is the only one in Myanmar that flows from south to north. Lakes, tanks, ponds, springs, falls, and dams are important water sources. The four main rivers are 4654 km long. The annual potential of eight rivers in the surface water is 1,081,885 km³ and a total area of 737,800 km² (WEPA, 2020). The four Main Delta Rivers with their 59 tributaries (rivers), 581 barrages and 208 water pumps, 36 large hydro-power (dumps), and 69 springs and 38 lakes are all available water sources (Table 12.9).

Over 60 reservoirs (> 1 Mm³) have been installed in irrigation plans by the Government of Myanmar. These wide reservoirs are rated at 7760 Mm³ total storage capacity (2018). A valuable source of water provides villagers with runoff collected in small storage sites (< 1 Mm³) (*Myanmar Integrated Water Resources Management-Strategic Study*, 2014). The overall capacity for water storage is today estimated to be about 1.020 Mm³ in about 2000 small reservoirs (Boori et al., 2017). Myanmar has 11 distinct aquifer types based on its stratigraphic unit (Meel et al., 2014). The consistency and quantity of the aquifers differ depending on their lithology and deposition (Table 12.10).

Groundwater is mainly found in carboniferous-permanent calestone. It is in the eastern part of the area on the Mesozoic and Precambrian strata. Eocene rocks in the east of Rakhine and Chin states contain groundwater. Basin water is primarily brackish, and freshwater is rarely found in this region. Groundwater is present in quaternary-age alluvial beds and is fresh and brackish in some sections (Zaw et al., 2018).

TABLE 12.9

Potential Sources of Surface Water

River basin	Area (km²)	Surface water annual (km³)
Ayeyarwady (Upper)	193,300	227.92
Ayeyarwady(lower)	95,600	85.80
Thanlwin (in Myanmar)	158,000	257.92
Bago River	5,300	8.02
Bilin River and other rivulets	8,400	31.17
River in Thanintharyi Division	40,600	130.93
Rivers in Rakhaing State	58,300	139.25
Chiundwin	115,300	141.29
Mekong (in Myanmar)	28,600	17.63
Sittoung	34,400	41.95
Total	**737,800**	**1,081.88**

Source: Thein, 2019; WEPA, 2020.

TABLE 12.10
Potential Ground Water Resources in Myanmar

River basin	Area (km²)	Ground water annual (km³)
Ayeyarwady (Upper)	193,300	92.599
Ayeyarwady(Lower)	95,600	153.249
Thanlwin (in Myanmar)	158,000	74.779
River in Thanintharyi Division	40,600	39.278
Rivers in Rakhaing State	58,300	41.774
Chiundwin	115,300	57.578
Mekong (in Myanmar)	28,600	7.054
Sittoung	34,400	28.402
Total	**737,800**	**494.71**

Source: WEPA, 2020; Zaw et al., 2018.

12.11.2 Challenges of Water Resources in Myanmar

Though Myanmar has more abundant water resources than many other countries in South Asia from an environmental point of view, water supplies face many significant threats. The challenges are caused by sedimentation, waste, contamination, climate change, deforestation, landscape changes, and urban growth. In the cities of Homalin and Monywa, near mining activities, changes and declining water quality heavy metals were detected, including plumage and mercury. The Ayeyarwady River's water quality has deteriorated during the previous decade. The river's season may be linked to high arsenic levels of 30 ppb and cyanide values of 0.14 mg/L. Siltation, at 360 million t/y, is the world's third most significant, owing mostly to mining activities, deforestation, soil conservation, and over-exploitation of land. Besides, the Ayeyarwady Delta already shows initial indications of major changes such as mangrove exploitation, overfishing, riverbank erosion, and water quality deterioration due to seawater intrusion.

Groundwater is today heavily contaminated and polluted by mining and industry, waste disposal, and agricultural drainage water fertilized. Rivers and their affluents face declining water quality due to mining, industrial disposal, and flood erosion. Sedimentation and sand bars are an obstacle to navigation, and the composition of the rivers varies in the long term. In some locations, groundwater quality is declining, and the coastal areas have been shut down; mostly, water-bearing strata have been unconfined (Zaw et al., 2018). Many groundwater supplies in the coastal region are either saline – or primarily polluted by arsenic – natural (Thein, 2019).

12.11.3 Impacts of Climate Change on Water Resources

Climate change in Myanmar will impact the economic, productive, social, and environmental sectors (Meel et al., 2014). From 2001 to 2100, the Myanmar climate change predictions are as follows (Shrestha et al., 2020):

- From December to May, the nation's average temperature rises, particularly in the north and central regions, with the greatest increases occurring in the north and central regions.
- Increase in rainfall during the monsoon season, including an increase across the country from March to November (especially in northern Myanmar) and a decrease from December to February.
- More severe weather conditions, including cyclones/strong winds, high floods/storms, heavy rainfall, and drought.

Global climate change seriously affected Myanmar and, along with an increased risk of flooding and droughts, the daily temperature has risen significantly in the entire region (Than and Maung, 2017). In the dry areas, delta areas, hilly zones, and some coastal areas, Myanmar has been experiencing water scarcity (Myint, 2018). Owing to the drier and hotter weather conditions during the dry season, the shallow groundwater table (seasonal shallow aquifer) falls, and the surface water supplies, such as reservoirs, wetlands, and rainwater storage tanks, are almost dry. Besides, the present water infrastructure, such as dams and reservoirs, is limited due to reduced rainfall inputs and high evaporation processes. Overexploitation of groundwater by increased human activities threatens the sustainability of groundwater supplies (MCCA, 2019). The shorter monsoon cycle would cause significant amounts of rain to decline for short periods, resulting in flooding, water pollution, and riverbank erosion. The progressive receding of the glaciers will impact the Himalayan region of Myanmar, which provides significant quantities of water to many regions of Myanmar (Than and Maung, 2017).

Seawater intrusion is common in coastal areas, especially in the river delta. The amount of salt infiltration upstream varies depending on the water outlet discharge rate. The Ayeyarwady Delta coastal waterways are one of the most vulnerable areas due to marine water intrusion. The aquaculture sector faces several challenges, including unforeseen and uneven precipitation patterns, floods and drought in the main agricultural areas, intensive growing activity, and illegal hydrology in watershed regions. The lowlands, such as the Deltaic Ayeyarwady/Yangon Region, are becoming more vulnerable to sea-level rise (Myint, 2018). Increased anomalous weather events such as extreme rainfall or drought would significantly affect the hydropower capacity of many of Myanmar's rivers (Myint, 2018).

A rise in severe weather will disrupt existing transport infrastructure dramatically and raise maintenance costs. Cyclone Nargis, which struck the South Coast of Myanmar in 2008, cost US$36 million due to infrastructure damage as well as the loss of lives. Cyclone Nargis estimated damages/losses to the industrial sector in Myanmar at US$ 1.814 billion (MCCA, 2019).

12.11.4 SEDIMENT TRANSPORT AND YIELD IN MYANMAR

The Myanmar rivers are known to send sediment, carbon, and nutrients to the sea, but their exact contribution is unknown (NAPA, 2012). Myanmar's major river is the Ayeyarwady (IFC, 2017). Often soil erosions have impacted it, resulting in significant sediment production (Aung and Aye, 2017a). It has the world's fifth-highest sediment discharge river (Anthony et al., 2019), with an annual average watershed sediment output of 272.8 Mt (Table 12.11).

TABLE 12.11

Estimated Sediment Yields from the Ayeyarwady Sub-catchments Based on the Sources of Minerals and Geochemical and Geochronological Signatures

Catchment	Sediment yield (tones km^{-2})
Mytinge	200
Mali	1300
Chindwin	1700
Nmai	1300
Sweli	1,000–1,500
Taping	1,000–1,500
Total	**1000 (approximate)**

Source: Aung and Aye, 2017b.

From 2003 to 2010, the average amount of sediment each year was 43,759 t, 67,569 t, 48,220 t, 54,589 t, 65,987 t, 57,934 t, 46,368 t, and 66,159 t, respectively (IFC, 2017).

12.11.5 EXCLUSIVE PROJECTS AND PRACTICES

Myanmar's National Water Policy for watersheds, water bodies, fisheries and sanctuaries, groundwater aquifers, and inshore and marine water is the first comprehensive water policy in the country. The national policy for integrated water resource management aims to create, distribute, and manage Myanmar's water resources in an integrated and inclusive manner. It also seeks to contribute significantly to the struggle against poverty, green growth, and the nation's sustainable development by ensuring that all socially responsible individuals have equal access to clean water. The Myanmar National Water Framework Directive and the accompanying Water legislation employ a "top-down" methodology. The National Water Policy acknowledges the current situation and recommends a structure for developing legislation and systems, as well as an action plan with a unified national perspective, which includes the Myanmar National Water Framework Directive (MNWFD) (Aung and Aye, 2017b). Myanmar has made significant efforts to extend the use of gravity-fed canals and reservoirs, river pumping systems, and groundwater systems for irrigation (Meel et al., 2014).

12.11.6 REMARKABLE WATER TREATIES AND DAMS

Myanmar has a more or less favorable water supply situation (IWMI, 2015). There are ten transboundary rivers in Myanmar such as the Mekong, Pakchan, Salween, Ganges–Brahmaputra–Meghna, Ayeyarwady, Kaladan, Karnaphuli, Sangu, Matamuhuri, and Naf which create a relationship between different countries like China, Thailand, Bangladesh, India, Nepal, Vietnam, and Laos (Meel et al., 2014). The Salween River flows through three countries on the Thai–Myanmar border in the Himalayan plateau (FAO, 2011; WEPA, 2020). In China, it is known as the Nu River, in Myanmar as the Thanlwin River, and in Thailand as the Salween River (Bailey and Bryant, 2005; Magee and Kelley, 2009). China, Myanmar, and Thailand have unilateral plans to build dams and project construction along the Salween River. A joint technical committee, consisting principally of power-firm representations, was set up between Thailand and Myanmar in 1989.

Eight major hydropower dam projects were selected in 1992 by the Japanese Electronic Power Development Corporation (EPDC), some entirely in Myanmar and others in parts of the Salween River Basin. Thai companies have been encouraged to invest in Myanmar, and Thailand has agreed to create a cross-frontier bridge to facilitate commerce and tourism. It has begun the Salween River hydroelectric dam project. Thailand agreed to import water from Myanmar to satisfy its irrigation and drinkable water needs and as a power source. Since December 2002, Myanmar's military has discussed the possibility of constructing large structures on the Salween River with the Electricity Production Authority of Thailand. Thailand and Myanmar resolved in August 2004 to construct five dams in the Salween River Basin as a cooperative enterprise, beginning with the Tasang Dam (FAO, 2011). The heads of the Mekong and Salween Rivers are several transboundary rivers. Several dams have been built or are planned for the river (FAO, 2011).

There are 13 dams on the Chinese side of the River Nu and 7 on the Myanmar side of the Salween River. In the Myanmar–Thai borderland, where the Salween River forms a natural border, it is anticipated that the Weigyi Dam and Dagwin Dam will have seven dams on the Myanmar and Thai sides (Magee and Kelley, 2009). Hatgyi Dam is in the Karen State in Myanmar, but it is adjacent to the Thai province of Mae Hong Son (Krongkaew, 2004). The Hatgyi Dam will be constructed on the Salween River. The site is in the Kayin State of Myanmar, near the Thai border (Magee and Kelley, 2009). The lagoon created by the dam reaches the Thai boundary (Krongkaew, 2004). The 100-m-high dam should have a capacity of 1200 MW (Magee and Kelley, 2009). The dam is a multinational partnership effort comprising Thai, Chinese, and Burmese firms (McCully, 1996).

12.11.7 Lesson Learned

Myanmar is a water-rich nation with several rivers, reservoirs, and dams. In Myanmar's dry zone (Sagaing, Mandalay, and Magway areas), populations struggle to get safe drinking water year-round despite abundant precipitation. The rivers are sources of food, water, and economic opportunities for the native people of Myanmar. The production and the effects of hydropower would lead to a dispute between riparian countries, riparian civil unrest, and local violence. Unlike neighboring countries, most rivers are still untouched and safe. Myanmar has the potential to avoid the errors made by others and to utilize its water supplies for sustainable growth completely.

12.11.8 Transboundary Water Resources in South Asia

South Asia's rivers and aquifers make its populations and its states interdependent. Around 1.8 billion people live in seven countries (Bangladesh, Bhutan, India, Afghanistan, Pakistan, Nepal, the Maldives, and Sri Lanka). India, Bangladesh, Pakistan, Bhutan, Nepal, and Afghanistan have 21 major rivers. The Indus, Ravi, Beas, Sutlej, Jhelum, and Chenab Rivers form the Indus Basin, which connects China to Bhutan, Bangladesh, India, and Nepal (Biswas, 2011; Magee and Kelley, 2009).

The Koshi, Gandaki, and Mahakali rivers join India and Nepal. The Brahmaputra, Ganges, and Teesta are the main rivers in India, Bangladesh, and Bhutan. The Kabul Basin is shared between Pakistan and Afghanistan. The Naf River links Bangladesh and Myanmar with one another. Transboundary Himalayan rivers flow through Bhutan, Nepal, India, and Bangladesh. They offer a golden chance to improve living standards in the countries with the most people in the world (Williams, 2018). The effects of climate change on the Himalayan habitats and freshwater glacial lakes in the area need everyone to work together. Sri Lanka and the Maldives must work together on this. South Asia's transboundary rivers, like the Indus and the Ganges– Brahmaputra–Meghna (GBM), and the communities that depend on them are going through changes that have never happened before; this is because of the rapid growth of water, power, food, and trade infrastructure in the area and around them (Hill, 2013).

Transboundary river systems provide many ecosystem resources throughout political frontier countries and societies. Governments and companies invest in numerous regional trade and connectivity projects, many directly and indirectly connected to common rivers and riparian resources. For instance, Nepal, Bangladesh, and India work together to improve inland water transport (IWT) on the Ganges and Brahmaputra rivers as part of the multi-modal regional connectivity trade. South Asian water scarcity may seem ironic and paradoxical because the Himalayas and the Hindu Kush mountains, which separate this area from the rest of Asia, have huge freshwater stocks (Verghese, 1997). This water runs along 20 key streams that have sustained South Asian cultures and economies over the past decades (Williams, 2018).

12.12 DISCUSSION

Over 20% of climate-vulnerable countries of the world are in South Asia. Natural catastrophes have harmed 700 million people in the last decade. The countries of South Asia have already been affected by climate change, and it is necessary to take measures to mitigate its negative effects. South Asian nations must, for instance, prepare for the consequences of global warming in the Himalayas, which provide water to Bangladesh's coastal areas for 1 billion people, even as they combat the human causes of climate change. Carbon dioxide (CO_2) concentrations, one of the principal GHGs, have steadily increased in South Asia because of accelerated industrialization and other human activities. India and Pakistan emit the most CO_2, whereas the Maldives emit the most per capita. Because most South Asians continue to rely on agriculture and rural economic activity for any changes to their land, productivity, and insecurity would cause millions more to fall into extreme poverty (Joy and Paranjape, 2007). Due to their geo-climate, socioeconomic-demographic histories, and reliance on agriculture and pastoral livelihoods, South Asian nations are extremely susceptible to climate change. Some academics say climate change renders South Asia very sensitive to food and fiber, biodiversity, water resources, coastal ecosystems, human health, and land degradation. Over the past 200 years, the global average temperature has been measured at 1.5°C. The normal greenhouse gas requirements in the atmosphere should be between 450 and 550 ppm (parts-per-million) CO_2 as the key explanation for the increase in temperature.

Furthermore, researchers believe that the earth's temperature will continue to rise due to the impacts of greenhouse gases and the release of large quantities of frozen gases from the Arctic and Antarctic (Fallesen et al., 2019). South Asia will warm by 2–6°C by 2100. The future of several South Asian regions, especially the Maldives, Bangladesh's south coast districts, and Indian and Sri Lankan coasts, is uncertain (Sivakumar and Stefanski, 2010). Besides destroying lives and habitats, many economic interests, livelihoods, and human interests would be affected. Evidence suggests that a single meter increase in sea level in India alone will result in a US$1.259 million welfare loss (Fallesen et al., 2019).

12.13 CONCLUSIONS

With the recent disastrous floods, cyclones, storms, droughts, variations in monsoon timing, and anomalies in precipitation patterns, South Asian nations are now facing a wide range of climatic changes. In recent years, a lot has been stated to explain why climate change is a significant issue for the nations of South Asia:

Floods and other extreme events: Seven extreme floods afflicted Bangladesh between 1984 and 2017, and 10,655,145 people endured more devastation in 2007. South Asian nations are the most susceptible to flooding worldwide. More dangerous floods in 2005 resulted in the death of more than 1000 people in Mumbai, India. Almost 2000 deaths in Pakistan were caused by floods in 2010, and 20 million homeless people lost US$10 billion in Pakistan. The total number of people who suffered from the floods was 4,626,078 in Sri Lanka between 2005 and 2014. Bhutan suffered from three big floods in 2009, 2011, and 2013 (Islam and Sultan, 2009). Another alarming predictor of climate change can be steady and rising events like hurricanes, cyclones, earthquakes, droughts, and more. Cyclones are most damaging in the lowly coastal regions, particularly Bangladesh and India. Strong waves are causing significant damage and killing thousands of people. In 1970, over 300,000 lives were killed in Bangladesh by a cyclone. For landing disturbances, Bhutan has been affected by landslides every year. The Orissan population will never forget the super cyclone of 1999, with an enormous toll of approximately 10,000 lives lost, land, and crops worth billions, and the ecological balance of coastal areas being devastated by the destruction of mangroves (Hasnat et al., 2018). Pakistan and India's 1999 and 2000 droughts caused water table collapse and crop failures. South Asia, particularly India, experienced an extended period of heatwaves.

Things are shifting due to global warming, either in the northeastern or in other areas historically regarded as areas with the heaviest rainfall in the world. In many South Asia areas, the precipitation rate has changed slowly but distinctly, and many will increase manyfold soon. In several parts of South Asia, there is a shift in the rainy season.

Rising sea level for glacier melting and lack of freshwater: The Himalayan ice mass melts quickly over the global average and threatens this area.

In 1947, sea levels climbed 1.4 mm and 3.9 mm per year in the southeast and southeast central zones closest to Bangladesh. By 2050, the sea level will increase by 45 cm, impacting 10–15% of the land and 35 million people. The Indian Ocean started rising in the 1960s and will rise 15–38 cm by 2050. Sri Lanka's coastline is just 1 m above sea level. The Maldives average 1.5 m above sea level and reach a high of 2 m. Sea-level rise reduces the drinking water supply and is responsible for land degradation, impacting aquaculture (Fallesen et al., 2019). The Maldives is threatened by freshwater depletion due to salinity. Desertification has caused water shortages in Afghanistan, India, and Pakistan.

Pollutions and Diseases in South Asia: Since 1990, South Asia's GHG emissions have increased by 3.3% yearly due to increased industrial activity, cars, and brick kilns (Islam and Sultan, 2009). Pakistan pollutes the air the most, followed by India. Cars, factories, and brick kilns pollute Bangladesh's air. Bhutan's main air pollutant is forest fires. Pollution is affecting Nepal's air quality. Sri Lanka's main air pollutants are cars, trash, and biomass.

Climate change may cause thousands of South Asians to die from diseases. According to the World Bank, malaria, dengue, cholera, and hepatitis are the most common diseases in urban and rural South Asia. Their main causes include flooding, water toxicity, water pollution, and logging. Due to climate change, Assam, West Bengal, Bihar, Orissa, Andhra Pradesh, Madhya Pradesh, Tamil Nadu, and practically every area of Bangladesh and Pakistan will face comparable occurrences of a severe toll from these illnesses in the future years.

REFERENCES

Aawar, T., Khare, D., and Singh, L. (2019). Identification of the trend in precipitation and temperature over the Kabul River sub-basin: A case study of Afghanistan. *Modeling Earth Systems and Environment*, 5(4), 1377–1394. https://doi.org/10.1007/s40808-019-00597-9.

Aayog, N. (2017). Selected best practices in water management. SMARTNET, Ministry of Housing and Urban Affairs, Government of India. https://smartnet.niua.org/sites/default/files/resources/bestpractices-in-water-management.pdf. Accessed 14 April 2023.

Ahmad, M., and Wasiq, M. (2004). *Water Resource Development in Northern Afghanistan and its Implications for Amu Darya Basin*. The World Bank, Washington, DC.

Ahmad, Q. K., Ahmad, N., and Rasheed, K. S. (1994). *Resources, Environment, and Development in Bangladesh: With Particular Reference to the Ganges, Brahmaputra, and Meghna Basins*. Academic Publishers, Kolkata.

Ahmed, A., Iftikhar, H., and Chaudhry, G. (2007). Water resources and conservation strategy of Pakistan. The Pakistan development review, Pakistan Institute of Development Economics, No., 997–1009.

Ahmed, A. M. M. M., and Roy, K. (2007). Utilization and conservation of water resources in Bangladesh. *Journal of Developments in Sustainable Agriculture*, 2(1), 35–44.

Ahmed, A. U. (2006). Bangladesh climate change impacts and vulnerability: A synthesis: Climate change cell. Department of Environment, Dhaka.

Ahmed, M., and Suphachalasai, S. (2014). *Assessing the Costs of Climate Change and Adaptation in South Asia*. Asian Development Bank, Metro Manila.

Akbari, M. Z. (2018). Afghanistan: Drought and climate change effects. *The Daily Outlook Afghanistan*. Kabul, Afghanistan. http://outlookafghanistan.net/topics.php?post_id=20167. Accessed 14 April 2023.

Akhter, M. (2015). The hydropolitical cold war: The Indus waters treaty and state formation in Pakistan. *Political Geography*, 46, 65–75.

Ali, G., Hasson, S., and Khan, A. M. (2009). Climate change: Implications and adaptation of water resources in Pakistan. Global Change Impact Studies Centre, Islamabad.

Ali, K. F., and De Boer, D. H. (2007). Spatial patterns and variation of suspended sediment yield in the upper Indus River basin, northern Pakistan. *Journal of Hydrology*, 334(3–4), 368–387.

Ali, K. F., and De Boer, D. H. (2010). Spatially distributed erosion and sediment yield modeling in the upper Indus River basin. *Water Resources Research*, 46(8), 1–16.

Alim, A. K. (2006). Sustainability of water resources in Afghanistan. *Journal of Developments in Sustainable Agriculture*, 1(1), 53–66.

Amarasinghe, U., Sharma, B. R., Aloysius, N., Scott, C., Smakhtin, V., and De Fraiture, C. (2005). *Spatial Variation in Water Supply and Demand Across River Basins of India* (Vol. 83). IWMI, Colombo.

Anthony, E. J., Besset, M., Dussouillez, P., Goichot, M., and Loisel, H. (2019). Overview of the monsoon-influenced Ayeyarwady River delta, and delta shoreline mobility in response to changing fluvial sediment supply. *Marine Geology*, 417, 106038.

Aquastat, F. (2010). A global information system on water and agriculture. Food and Agriculture Organization of the United Nations (FAO), Rome.

Aryal, R. S., and Rajkarnikar, G. (2011). *Water Resources of Nepal in the Context of Climate Change*. Government of Nepal, Singha Durbar, Kathmandu.

Aslam, M. H., and Yoshimura, K. (2017). Sediment yield in Jhelum River Basin with and without climate change impact in Pakistan. Paper presented at the Proceedings of the Society of Civil Engineering B1.

Aung, S. W., and Aye, N. (2017a). Sediment yield simulation in upper Ayeyarwady Basin. *American Scientific Research Journal for Engineering, Technology, and Sciences*, 27(1), 405–418.

Aung, S. W., and Aye, N. (2017b). Sediment yield simulation in upper Ayeyarwady Basin. *American Scientific Research Journal for Engineering, Technology, Sciences*, 27(1), 405–418.

Azim, F., Shakir, A. S., and Kanwal, A. (2016). Impact of climate change on sediment yield for Naran watershed, Pakistan. *International Journal of Sediment Research*, 31(3), 212–219.

Bailey, S., and Bryant, R. (2005). *Third World Political Ecology: An Introduction*. Routledge, London.

Bailey, R. T., Khalil, A., and Chatikavanij, V. (2015). Estimating current and future groundwater resources of the Maldives. *Journal of the American water resources Association*, 51(1), 112–122.

Bandara, N. (2003). Water and wastewater related issues in Sri Lanka. *Water Science Technology*, 47(12), 305–312.

Bashir, S., Baig, M., Ashraf, M., Anwar, M., Bhalli, M., and Munawar, S. (2013). Risk assessment of soil erosion in Rawal watershed using geoinformatics techniques. *Science International*, 25(3), 583–588.

Basu, M., and Shaw, R. (2013). Water policy, climate change and adaptation in South Asia. *International Journal of Environmental Studies*, 70(2), 175–191.

Basyal, R. (2017). Climate change impact in water resources: A case of Nepal. Climate Action Tracker, Berlin. Retrieved from Climatetraker. org: http://climatetracker. Accessed 25 May 2019.

Biswas, A. K. (2011). Cooperation or conflict in transboundary water management: Case study of South Asia. *Hydrological Sciences Journal*, 56(4), 662–670.

Bolch, T., Kulkarni, A., Kääb, A., Huggel, C., Paul, F., Cogley, J. G., Frey, H., Kargel, J. S., Fujita, K., and Scheel, M. (2012). The state and fate of Himalayan glaciers. *Science*, 336(6079), 310–314.

Boori, M. S., Choudhary, K., Evers, M., and Paringer, R. (2017). A review of food security and flood risk dynamics in Central Dry Zone area of Myanmar. *Procedia Engineering*, 201, 231–238.

BPC. (2017). Bangladesh Delta Plan 2100. Bangladesh Planning Commission Government of the People's Republic of Bangladesh Sher-e-Bangla Nagar, Dhaka 1207, Bangladesh www.plancomm.gov.bd.

BPC. (2019). Integrated assessment for the Bangladesh Delta Plan 2100 – Analysis of selected interventions, Dhaka.

Cassen, R. (2016). *India: Population, Economy, Society*. Palgrave Macmillan, London.

Chachibaia, K. (2016). Protecting (scarce) fresh water in the Maldives. www.climatechangecentre. net_pdf_waterresources.

Choden, S. (2009). Sediment transport studies in Punatsangchu River, Bhutan. MSc Thesis, Lund University, Sweden.

Dahl, T. A., Kendall, A. D., and Hyndman, D. W. (2018). Impacts of projected climate change on sediment yield and dredging costs. *Hydrological Processes*, 32(9), 1223–1234.

Derler, Z. (2019). How three islands have inspired the Maldives to fight water shortages. *Climate Home News*.

Dias, B., Udayakumara, E., Jayawardana, J., Malavipathirana, S., and Dissanayake, D. (2019). Assessment of soil erosion in Uma Oya catchment, Sri Lanka. *Journal of Environmental Professionals Sri Lanka*, 8(1), 39–51.

Diyabalanage, S., Samarakoon, K., Adikari, S., and Hewawasam, T. (2017). Impact of soil and water conservation measures on soil erosion rate and sediment yields in a tropical watershed in the Central Highlands of Sri Lanka. *Applied Geography*, 79, 103–114.

Dutta, D., Das, R., and Mazumdar, A. (2017). Assessment of runoff and sediment yield in the Tilaya Reservoir, India using SWAT Model. *Asian Journal of Water, Environment Pollution*, 14(3), 9–18.

Eriyagama, N., Smakhtin, V., Chandrapala, L., and Fernando, K. (2010). Impacts of climate change on water resources and agriculture in Sri Lanka: A review and preliminary vulnerability mapping (Vol. 135). IWMI, Colombo.

Fallesen, D., Khan, H., Tehsin, A., and Abbhi, A. (2019). South Asia needs to act as one to fight climate change. *World Bank Blogs*, Washington DC.

FAO. (1978). A report prepared for the Feasibility Study for Development of Inland Fisheries Project. http://www.fao.org/3/L8853E/L8853E02.htm. Accessed 20 May 2019.

FAO. (2011). *AQUASTAT Transboundary River Basins – Salween River Basin*. Food and Agriculture Organization of the United Nations (FAO), Rome.

Fathian, F., Dehghan, Z., and Eslamian, S. (2016). Evaluating the impact of changes in land cover and climate variability on streamflow trends (Case study: Eastern subbasins of Lake Urmia, Iran). *Journal of Hydrology Science and Technology*, 6(1), 1–26.

Fernando, S., and Nizar, M. (2019). Sri Lanka's water resource development: Diverting water to the dry zone. http://www.ft.lk/opinion/Sri-Lanka-s-water-resource-development-Diverting-water-to-the-dry-zone/14-690486.

Frenken, K. (2013). Irrigation in Central Asia in figures: AQUASTAT survey-2012 (1020–1203). Italy.

Gajmer, B. (2014). Nepal: Irrigation and water resource management. The World Bank, Vol. 11, Washington, DC.

GEF. (2019). Coming to grips with water: How Bhutan is overcoming water challenges magnified by the onset of climate change, Global Environment Facility(GEF), https://www.thegef.org/news/coming-grips-water-how-bhutan-overcoming-water-challenges-magnified-onset-climate-change (accessed on 29 February 2024).

Giordano, M. A., and Wolf, A. T. (2003). Sharing waters: Post-Rio international water management. Paper presented at the Natural resources forum.

GoP. (2007). The international water resources management policies and actions and the latest practice in their environmental evaluation and strategic environmental assessment. Ministry of Environment, Government of Pakistan (GoP), Pakistan.

Goyal, M. K., and Surampalli, R. Y. (2018). Impact of climate change on water resources in India. *Journal of Environmental Engineering*, 144(7), 04018054.

Gupta, A. D., Babel, M. S., Albert, X., and Mark, O. (2005). Water sector of Bangladesh in the context of integrated water resources management: A review. *International Journal of Water Resources Development*, 21(2), 385–398.

Habib, H. (2014). Water related problems in Afghanistan. *International Journal of Educational Studies*, 1(3), 137–144.

Hasnat, G. T., Kabir, M. A., and Hossain, M. A. (2018). Major environmental issues and problems of South Asia, Particularly Bangladesh. In Hussain, Chaudhery Mustansar (Ed.), *Handbook of Environmental Materials Management*. Springer, Cham, 1–40.

Hewawasam, T., von Blanckenburg, F., Schaller, M., and Kubik, P. (2003). Increase of human over natural erosion rates in tropical highlands constrained by cosmogenic nuclides. *Geology*, 31(7), 597–600.

Hill, D. P. (2013). Trans-boundary water resources and uneven development: Crisis within and beyond contemporary India. *South Asia: Journal of South Asian Studies*, 36(2), 243–257.

Hirji, R., Nicol, A., and Davis, R. (2017). *South Asia Climate Change Risks in Water Management: Climate Risks and Solutions-Adaptation Frameworks for Water Resources Planning, Development, and Management in South Asia*. World Bank, Washington, DC.

Hossain, M., Rahman, M., and Tamim, M. (2013). Coping with new challenges in water resources management in Bangladesh. International Symposium for Next Generation Infrastructure October 1-4, 2013, Wollongong, Australia.

Hubbard, D. K., Sadd, J. L., and Roberts, H. H. (1981). The role of physical processes in controlling sediment transport patterns on the insular shelf of St. Croix, US Virgin Islands. Paper presented at the Proceedings of the Fourth International Coral Reef Symposium, Manila.

Huq, S. (2002). Lessons learned from adaptation to climate change in Bangladesh. Climate Change Team The World Bank Washington DC USA October 2002.

IFC. (2017). Baseline assessment report geomorphic and sediment transport-strategic environmental assessment of the hydropower sector in Myanmar, International Finance Corporation (IFC), retrieved from https://www.ifc.org/content/dam/ifc/doc/mgrt/chapter-3-sea-baseline-assessment-geomorphology-and-sediment-transport.pdf (accessed on 29 February 2024).

Illangasinghe, S., and Hewawasam, T. (2015). Introducing surface sampling threshold factor for suspended sediment transport: Model development using Sri Lankan tropical highland river basins. *Hydrology Research*, 46(1), 136–155.

IPCC. (2007). The AR4 synthesis report, section 2 (4) attribution of climate change. Intergovernmental Panel on Climate Change (IPCC).

Iqbal, M. N., Oweis, T. Y., Ashraf, M., Hussain, B., and Majid, A. (2012). Impact of land-use practices on sediment yield in the Dhrabi Watershed of Pakistan. *Journal of Environmental Science Engineering*, 1(3A), 406–420.

Islam, A., and Sultan, S. (2009). Climate change and South Asia: What makes the region most vulnerable? MPRA Paper 21875, University Library of Munich, Germany.

Islam, M. R., Begum, S. F., Yamaguchi, Y., and Ogawa, K. (1999). The Ganges and Brahmaputra rivers in Bangladesh: Basin denudation and sedimentation. *Hydrological Processes*, 13(17), 2907–2923.

IWMI. (2015). Improving water management in Myanmar's dry zone for food security, livelihoods and health. International Water Management Institute (IWMI), Vol. IWMI Reports 229586, Colombo.

Jain, S. K. (2019). Water resources management in India–challenges and the way forward. *Current Science*, 117(4), 569–576.

Jain, S. K., Nayak, P., Singh, Y., and Chandniha, S. K. (2017). Trends in rainfall and peak flows for some river basins in India. *Current Science*, 112(8), 1712–1726.

Javed, A., Tanzeel, K., and Aleem, M. (2016). Estimation of sediment yield of Govindsagar Catchment, Lalitpur District, (U.P.), India, using remote sensing and GIS techniques. *Journal of Geographic Information System*, 8(5), 595–607.

Jones, S. (2020). In Afghanistan, climate change complicates future prospects for peace. https://www.national-geographic.com/science/2020/02/afghan-struggles-to-rebuild-climate-change-complicates/. Accessed 29 May 2019.

Joy, K., and Paranjape, S. (2007). Understanding water conflicts in South Asia. *Contemporary Perspectives*, 1(2), 29–57.

Kahlown, M. A., and Majeed, A. (2003). Water-resources situation in Pakistan: Challenges and future strategies. In: *Water Resources in the South: Present Scenario and Future Prospects, Commission on Science and Technology for Sustainable Development in the South*, Islamabad, 21–39.

Kane-Potaka, J., Wani, S., and Pillai, L. (2018). The social and environmental value of CSR investments in agriculture including the approach and value of science backed solutions. International Crops Research Institute for the Semi-arid Tropics (ICRISAT), Hyderabad.

Kijne, J. W., Barker, R., and Molden, D. J. (2003). *Water Productivity in Agriculture: Limits and Opportunities for Improvement* (Vol. 1). IWMI & CABI, Colombo.

King, M., and Sturtewagen, B. (2010). Making the most of Afghanistan's river basins: Opportunities for regional cooperation. EastWest Institute, New York.

Klemm, W., and Shobair, S. (2010). The Afghan part of Amu Darya Basin. Impact of Irrigation in Northern Afghanistan on Water Use in the Amu Darya Basin. Food Agriculture Organization, Italy.

Krongkaew, M. (2004). The development of the Greater Mekong Subregion (GMS): Real promise or false hope? *Journal of Asian Economics*, 15(5), 977–998.

Kumar, R., Singh, R., and Sharma, K. (2005). Water resources of India. *Current Science*, 89(5), 794–811.

Lacombe, G., and McCartney, M. (2014). Uncovering consistencies in Indian rainfall trends observed over the last half century. *Climatic Change*, 123(2), 287–299.

Lovelle, M. (2016). Co-operation and the Brahmaputra: China and India water sharing. Future Directions International. http://www.futuredirections.org.au/publication/co-operation-and-the-brahmaputra-china-and-india-water-sharing/. Accessed 26 March 2019.

Mack, T. J., Chornack, M. P., Vining, K. C., Amer, S. A., Zaheer, M. F., and Medlin, J. H. (2014). Water resources activities of the US Geological Survey in Afghanistan from 2004 through 2014. US Department of the Interior, US Geological Survey.

Madhusoodhanan, C., Sreeja, K., and Eldho, T. (2016). Climate change impact assessments on the water resources of India under extensive human interventions. *Ambio*, 45(6), 725–741.

Magee, D., and Kelley, S. (2009). Damming the Salween river. In Molle, Francois, Sokhem, Pech, Foran, Tira, and Kakonen, Mira (Eds.), *Contested Waterscapes in the Mekong Region: Hydropower, Livelihoods and Governance*. Routledge Taylor and Francis, London, 115–140.

MCCA. (2019). Impact of climate change and the case of Myanmar. https://myanmarccalliance.org/en/climate-change-basics/impact-of-climate-change-and-the-case-of-myanmar/.

McCully, P. (1996). *Silenced Rivers: The Ecology and Politics of Large Dams*. Bloomsbury Publishing, New york.

Meel, P., Leewis, M., Tonneijck, M., Leushuis, M., Groot, K., Jongh, I., and Nauta, T. (2014). Myanmar integrated water resources management strategic study. Consortium Royal Haskoning DHV.

Merz, J., Nakarmi, G., and Weingartner, R. (2003). Potential solutions to water scarcity in the rural watersheds of Nepal's Middle Mountains. *Mountain Research Development*, 23(1), 14–18.

Michael, A., Schmidt, J., Enke, W., Deutschländer, T., and Malitz, G. (2005). Impact of expected increase in precipitation intensities on soil loss—Results of comparative model simulations. *Catena*, 61(2–3), 155–164.

Millar, J. (2002). The ability of the Maldives to cope with freshwater scarcity via the adaptive capacity of its political economy. Occasional Paper, No. 44, 53.

Mirza, M., and Dixit, A. (1997). Climate change and water resources in the GBM basins. *Water Nepal*, 5(1), 71–100.

Mohile, A. (2006). Inter-basin transfer of water in India: Prospects and problems. *Water Energy International*, 63(4), 34–53.

Mool, P. K., Wangda, D., Bajracharya, S. R., Kunzang, K., Gurung, D. R., and Joshi, S. P. (2001). Inventory of glaciers, glacial lakes and glacial lake outburst floods. Monitoring and early warning systems in the Hindu Kush-Himalayan Region: Bhutan. International Centre for Integrated Mountain Development (ICIMOD);United Nations Environment Programme (UNEP), Kathmandu.

Morgan, K., and Kench, P. (2014). A detrital sediment budget of a Maldivian reef platform. *Geomorphology*, 222, 122–131.

Morris, G. L., and Fan, J. (1998). *Reservoir Sedimentation Handbook: Design and Management of Dams, Reservoirs, and Watersheds for Sustainable Use.* McGraw Hill Professional.

MoWR. (1999). National water policy. Ministry of Water Resources Government of the People's Republic of Bangladesh, Dhaka.

MoWR. (2004). National Water Management Plan (NWMP): Main report. Dhaka.

Myint, K. M. (2018). How will climate change impact Myanmar? https://www.myanmarwaterportal.com/news/667-how-will-climate-change-impact-myanmar.html.

NAPA. (2012). Myanmar's National Adaptation Programme of Action (NAPA) to climate change. In No. E. C. Committee (Ed.), National Coordinating Body (Vol. 128). Ministry of Environmental Conservation and Forestry, Myanmar.

NASA. (2007). https://www.nasa.gov/mission_pages/hurricanes/archives/2007/h2007_sidr.html.

Nasir, A., Uchida, K., and Ashraf, M. (2006). Estimation of soil erosion by using RUSLE and GIS for small mountainous watersheds in Pakistan. *Pakistan Journal of Water Resources*, 10(1), 11–21.

NEC. (1992). Bhutan towards sustainable development in a unique environment. National Environmental Secretariat, Planning Commission.

NEC. (2016). WATER- securing Bhutan's future. Royal Government of Bhutan, Bhutan.

Padma, S. (2019). Sri Lanka in 2018. *Asian Survey*, 59(1), 108–113.

Qureshi, A. S. (2002). *Water Resources Management in Afghanistan: The Issues and Options* (Vol. 49). IWMI, Colombo.

Rahaman, M. M. (2009). Principles of transboundary water resources management and Ganges treaties: An analysis. *International Journal of Water Resources Development*, 25(1), 159–173.

Rahaman, M. M. (2020). Hydropower development along Teesta river basin: Opportunities for cooperation. *Water Policy*, 22(4), 641–657.

Rahman, M. R. (2015). Water resources. In Bangladesh National Conservation Strategy (2016-2031). Dhaka, Bangladesh: Ministry of Environment and Forests (MoEF).

Rahman, S., Islam, M. S., Khan, M. N. H., and Touhiduzzaman, M. (2019). Climate change adaptation and disaster risk reduction (DRR) through coastal afforestation in South-Central Coast of Bangladesh. *Management of Environmental Quality: An International Journal*, 30(3), 498–517. https://doi.org/10.1108/MEQ-01-2018-0021.

Rahman, S. H., and Rahman, S. (2010). Climate change and water resources. https://www.researchgate.net/publication/235707918_Climate_Change_and_Water_Resources. Accessed 24 June 2020.

Rice, S. K. (2007). Suspended sediment transport in the Ganges-Brahmaputra, river system, Bangladesh. Master's thesis, Texas A & M University, Texas.

Rice, S. K. (2010). Suspended sediment transport in the Ganges-Brahmaputra, river system, Bangladesh. Texas A & M University, USA.

Saavedra, C. (2005). Estimating spatial patterns of soil erosion and deposition of the Andean region using geo-information techniques: A case study in Cochabamba, Bolivia. Wageningen University and Research.

Sarker, M. H., Akter, J., Ferdous, M., and Noor, F. (2011). Sediment dispersal processes and management in coping with climate change in the Meghna Estuary, Bangladesh. Paper presented at the Proceedings of the Workshop on Sediment Problems and Sediment Management in Asian River Basins.

Savada, A. M. (1991). *Bhutan: A Country Study River Systems.* GPO for the Library of Congress, Washington.

Senevirathne. (2015). Sustainable water resources management in Sri Lanka: Present situation and way forward. https://rwsn.blog/2015/02/20/sustainable-water-resources-management-in-sri-lanka-present-situation-and-way-forward/#:~:text=There%20are%20103%20river%20basins,the%20Asian%20and%20Pacific%20countries. Accessed 5 May 2020.

Shrestha, B., Babel, M., Maskey, S., Van Griensven, A., Uhlenbrook, S., Green, A., Akkharath, I. J. H., and Sciences, E. S. (2013). Impact of climate change on sediment yield in the Mekong River basin: a case study of the Nam Ou basin, Lao PDR. *Hydrology and Earth System Sciences*, 17(1), 1.

Shrestha, N., Lamsal, A., Regmi, R. K., and Mishra, B. K. (2015). Current status of water environment in Kathmandu Valley, Nepal. Water and Urban Initiative Working Paper Series Number 03 — April 2015, United Nations University.

Shrestha, S., Imbulana, N., Piman, T., Chonwattana, S., Ninsawat, S., and Babur, M. (2020). Multimodelling approach to the assessment of climate change impacts on hydrology and river morphology in the Chindwin River Basin, Myanmar. *Catena*, 188, 104464.

Shrestha, S. R., Tripathi, G. N., and Laudari, D. (2018). Groundwater resources of Nepal: An overview. In Mukherjee, A. (Ed.), *Groundwater of South Asia*. Springer. Springer, Singapore, 169–193.

Singh, N., and Karki, S. (2004). Disclosure, legislative framework and practice on integrated water resources management in Bhutan. Microsoft Word - Bhutan Second Draft no photos.doc (iucn.org)

Singh, U. K., and Kumar, B. (2018). Climate change impacts on hydrology and water resources of Indian River basins. *Current World Environment*, 13(1), 32.

Sivakumar, M. V., and Stefanski, R. (2010). Climate change in South Asia. In Lal, R., Sivakumar, M., Faiz, S., Mustafizur Rahman, A., Islam, K. (Eds.), *Climate Change and Food Security in South Asia*. Springer, Dordrecht, 13–30.

Sleet, P. (2019). Water resources in Pakistan: Scarce, polluted and poorly governed. Future Directions International Pty Ltd., Nedlands, WA.

Stewart, A. K. (2016). Chapter 9 – Dams in Afghanistan. In Shroder, J., and Ahmadzai, S. J. (Eds.), *Transboundary Water Resources in Afghanistan*. Elsevier, Boston, 213–268.

Stojanov, R., Duží, B., Němec, D., and Procházka, D. (2017). Slow onset climate change impacts in Maldives and population movement from islanders' perspective. USA.

Subramanian, V. (1993). Sediment load of Indian rivers. *Current Science*, 64, 928–930.

Suhardiman, D., Clement, F., and Bharati, L. (2015). Integrated water resources management in Nepal: Key stakeholders' perceptions and lessons learned. *International Journal of Water Resources Development*, 31(2), 284–300.

Tabassum, S. (2003). Indo-Bangladesh treaty on Farakka Barrage and international law application. *Pakistan Horizon*, 56(3), 47–62.

Tarar, R. N. (2006). Performance of tarbela dam project. Paper presented at the Pakistan Engineering Congress, 69th Annual Session Proceedings Report.

Tariq, M. A. U. R., van de Giesen, N., Janjua, S., Shahid, M. L. U. R., and Farooq, R. (2020). An engineering perspective of water sharing issues in Pakistan. *Water*, 12(2), 477.

Than, Z., and Maung, M. T. (2017). Climate change and groundwater resources in Myanmar. *Journal of Groundwater Science Engineering*, 5(1), 59–66.

Thein, M. (2019). Water resources: Myanmar. Yangon, Myanmar.

Tiwari, H., and Sharma, N. (2012). A case study of Wangchu (Bhutan) reservoir sedimentation using HEC-RAS. Sedimentation in Reservoirs, 122–126.

Uhl, V. W. (2006). Afghanistan: An overview of ground water resources and challenges. *Ground Water*, 44(5), 626.

Ullah, S., Ali, A., Iqbal, M., Javid, M., and Imran, M. (2018). Geospatial assessment of soil erosion intensity and sediment yield: A case study of Potohar Region, Pakistan. *Environmental Earth Sciences*, 77(19), 705.

UNDP. (2010). Good practices in water security. UNDP, United Nations Development Programme (UNDP).

USGS. (2010). Afghanistan's Kabul Basin faces major water challenges. *ScienceDaily*, 17 June. www.science-daily.com/releases/2010/06/100616142351.htm.

Verghese, B. G. (1997). Water conflicts in South Asia. *Studies in Conflict Terrorism*, 20(2), 185–194.

Vinke, K., Martin, M. A., Adams, S., Baarsch, F., Bondeau, A., Coumou, D., Donner, R. V., Menon, A., Perrette, M., and Rehfeld, K. J. R. E. C. (2017). Climatic risks and impacts in South Asia: Extremes of water scarcity and excess. *Regional Environmental Change*, 17(6), 1569–1583.

Wassmann, R., Jagadish, S., Heuer, S., Ismail, A., Redona, E., Serraj, R., Singh, R., Howell, G., Pathak, H., and Sumfleth, K. (2009). Climate change affecting rice production: The physiological and agronomic basis for possible adaptation strategies. *Advances in Agronomy*, 101, 59–122.

Wasson, R. (2003). A sediment budget for the Ganga–Brahmaputra catchment. *Current Science*, 1041–1047.

WECS. (2005). National water plan for Nepal. Government of Nepal, Kathmandu.

WEPA. (2010). State of water resources: Nepal. Awaji.

WEPA. (2020). State of water environmental issues: Myanmar. http://www.wepa-db.net/policies/state/myanmar/myanmar.htm#:~:text=Myanmar%20is%20a%20country%20endowed,constitute%20national%20water%20resources%20annually.

White, C. J., Tanton, T. W., and Rycroft, D. W. (2014). The impact of climate change on the water resources of the Amu Darya Basin in Central Asia. *Water Resources Management*, 28(15), 5267–5281.

Williams, J. M. (2018). Stagnant rivers: Transboundary water security in South and Southeast Asia. *Water*, 10(12), 1819.

Withanachchi, S. S., Köpke, S., Withanachchi, C. R., Pathiranage, R., and Ploeger, A. (2014). Water resource management in dry zonal paddy cultivation in Mahaweli River Basin, Sri Lanka: An analysis of spatial and temporal climate change impacts and traditional knowledge. *Climate*, 2(4), 329–354.

Xu, J., and Cheng, D. (2002). Relation between the erosion and sedimentation zones in the Yellow River, China. *Geomorphology*, 48(4), 365–382.

Zaw, T., Moe, A. K., Ko, M., and Swe, Y. M. (2018). Groundwater resources of Myanmar. In Abhijit Mukherjee (Ed.), *Groundwater of South Asia*. Springer, Singapore, 153–168.

Zeitoun, M., and Warner, J. (2006). Hydro-hegemony–a framework for analysis of trans-boundary water conflicts. *Water Policy*, 8(5), 435–460.

Zhu, Y.-M., Lu, X., and Zhou, Y. (2008). Sediment flux sensitivity to climate change: A case study in the Longchuanjiang catchment of the upper Yangtze River, China. *Global Planetary Change*, 60(3–4), 429–442.

13 Detection of Land Surface Temperature in the Northwest Part of Bangladesh

A Remote Sensing Based Approach

Arshib Imtiaj, Md Hasibul Hasan, Md. Mostafizur Rahman, Md. Salah Uddin, and Saeid Eslamian

13.1 INTRODUCTION

Urban areas are now the preferable living place for about half of the world's population, this is expected to reach two-thirds by 2030 (Zhang, 2016). The increasing statistics of global urbanization have had an adverse effect on the micro climate due to the rise in air and surface temperatures (Makvandi et al., 2019; Souza et al., 2016). All over the world, this land use land cover (LULC) change, especially vegetation, is improvised by impervious cover, for instance, concrete buildings, pavements, or asphalt (Makvandi et al., 2019; Fathian et al., 2015). This results in energy absorption and radiation that may accelerate local and global problems such as biological hazards (air pollution and urban heat islands), concentration of greenhouse gases, and deterioration of public health and wellbeing (Lai and Cheng, 2010). However, the increasing temperature of urban areas in the modern era is a burning issue in many developed and underdeveloped countries. Cities such as London (UK), Bolzano (Italy), Madrid (Spain), and Nice (France) in Europe (dos Santos, 2020; Hellings and Rienow, 2021); Kumasi (Ghana) and Freetown (Sierra Leone) in Africa (Buo et al., 2021; Mustafa et al., 2021); and Tokyo (Japan), Beijing (China), Delhi (India), Mumbai (India), and Kuala Lumpur (Malaysia) in Asia (Dutta et al., 2019; Hua and Ping, 2018; Jain et al., 2021; Liu et al., 2021; Siddique et al., 2020) are being affected rigorously by land surface temperatures. Also, in less developed nations such as Bangladesh, the effect is not negligible, particularly in the mega-city of Dhaka.

According to WB (2013), "Bangladesh will be the most affected country in South Asia by an expected 2°C rise in the world's average temperatures in the next decades, with rising sea levels and more extreme heat" (Mirza, 2011). On the other hand, northern parts of Bangladesh have been facing drought problems and a reduction in water bodies (Hoque et al., 2020). Past experience has revealed that the northwestern part of Bangladesh has gone through critical changes such as landscape degradation, river erosion, land use changes, and extensive unplanned human activities (Kafy et al., 2021b; Mim and Zamil, 2020). Besides, quick urbanization in recent decades has attracted people to migrate to this area, with the city Rajshahi no exception (Kafy et al., 2020; Rahman, 2010) as it is one of the warmest regions in northwestern Bangladesh because of its geographical location and socioeconomic conditions. This massive rate of haphazard infrastructural development with its construction materials is affecting the urban climate, which is influencing district temperature rise (Dewan et al., 2021). This is due to the rural–urban migration from the nearby disaster-prone areas (river erosion, flood, and drought). Nowadays migrants prefer Rajshahi rather than Dhaka because of its low cost of living, employment opportunities, and available municipal services (Hasan et al.,

2017; Hasan et al., 2022; Rahaman et al., 2021). As a result, vegetation and water reserves are gradually reduced with the increase of settlements (Kafy et al., 2021a, 2021b).

During the last decade, several research studies regarding LST detention have been conducted, however, most of the studies were Dhaka-based (Ahmed et al., 2013; Imran et al., 2021). Several early studies have addressed land surface temperature and urban heat islands (Roy et al., 2021). However, little importance has been given to the capture sensitivity analysis and the most responsible factors behind the change in land surface temperature (Ferdous and Rahman, 2018) in the northwestern part of Bangladesh. Besides, temperature mapping with a physical survey is relatively expensive, whereas the use of remote sensing (RS) and geographic information systems (GIS) methods in land temperature detection are being used a lot by researchers for their simplicity and cost-effectiveness (Dey et al., 2021; Habibie et al., 2021).

This chapter will deal with investigating the relationship between land surface temperature variations over land cover by using the normalized difference vegetation index (NDVI), normalized difference water index (NDWI), and normalized difference built-up index (NDBI) techniques to identify the comfort zones within the study area. In addition, a comprehensive assessment of sensitivity has been carried out to explore which component is more sensitive to heat. The approach of remote sensing along with GIS has been followed in bringing up the scenario where satellite images (Landsat-5 TM and Landsat 8 OLI) from different years (1995, 2005, and 2019) were used. The findings thus provide a guide to the concerned authorities in the development policies for better land zoning ways of alleviating warming effects.

13.2 STUDY AREA

Rajshahi district is predominantly an agricultural area (BBS, 2011) and is well-known throughout Bangladesh as an educational city. It is located between the latitudes of 24°07′ and 24°43′ north and the longitudes of 88°17′ and 88°58′ east. The district's total area is 2,425.37 km², with a total of 633,758 houses and 2,595,197 people (BBS, 2011) This district is bordered on the north by Naogaon, on the south by the Padma River and Kushtia district, on the east by Natore, and on the west by Chapainawabganj, while Rajshahi City Corporation (RCC) is located on the banks of the Padma River (Figure 13.1).

The main sources of income are agriculture (59.35%), non-agricultural laborers (3.36%), industry (0.99%), commerce (14.25%), transportation and communication (4.36%), service (8.97%), construction (1.45%), and others (6.74%) (BBS, 2011). Although the district is not industrially developed, it has experienced a significant increase in unplanned housing activity in recent years. RCC is presently provoked by several commercial activities such as expanding market areas, rapidly developing restaurant businesses, and so on, which are impacting the urban land use activity. As a result of the increased construction rate to accommodate this high demand, the city is losing its green spaces. The climate in Rajshahi is classed as tropical, where the summer is much rainier than the winter (BBS, 2011). During the winter season, the minimum and maximum mean temperatures range from 9°C to 14°C. The minimum and highest mean temperatures in the summer range from 25.5°C to 38.7°C. In April, the humidity level was around 77%, and in July, it was over 88% (BBS, 2011).

13.3 METHODOLOGY

A systematic approach was adopted to prepare the LULC map from satellite images and analyze the sensitivity with different indices. First, this study processed satellite images through an image processing system where the radiometric and geometric corrections were used. The supervised classification technique was adopted for the LULC map which needs prior to accuracy assessment, to check whether the result is acceptable or not. This study also considered three different normalized indices to understand the land use and land cover scenario. To calculate the land surface temperature, three different steps namely, conversion of the digital number (DN) values to spectral radiance,

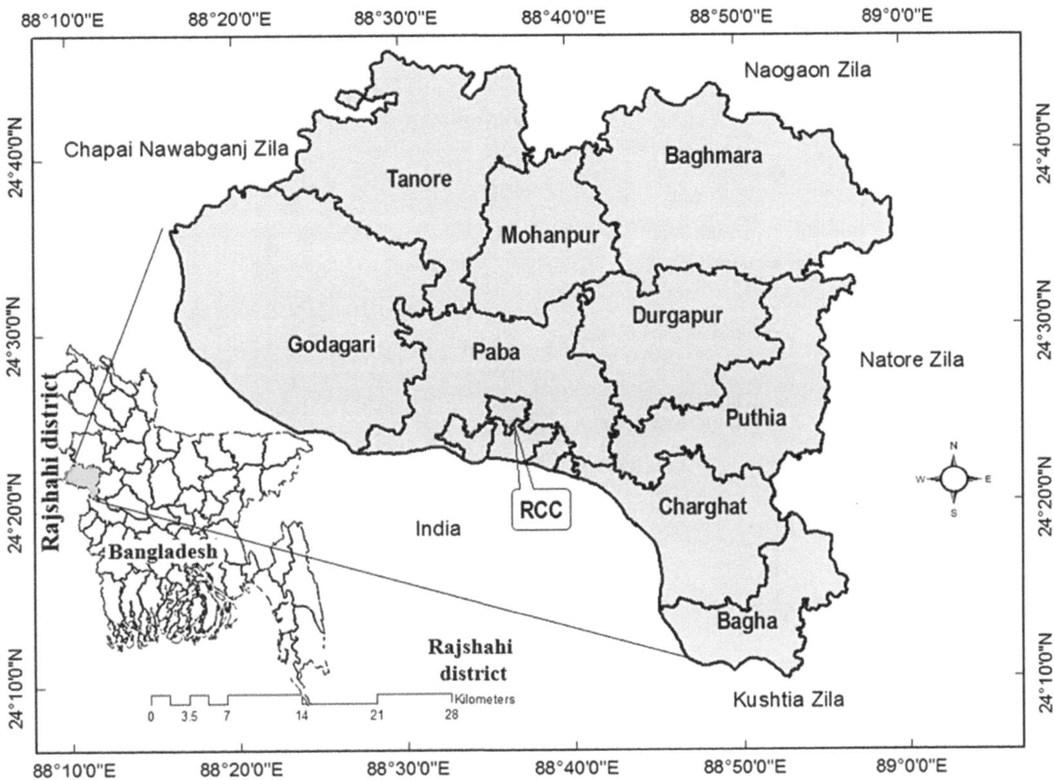

FIGURE 13.1   Study area map.

then spectral radiance to top of atmosphere (TOA) brightness temperature, and land surface emissivity (LSE) were used (Figure 13.2).

13.3.1   Remote Sensing Data

The satellite images were acquired from the United States Geological Survey (USGS) with a 30 m resolution and had cloud coverage of less than 10%. Landsat 5 TM and Landsat-8 OLI-TIRS images were considered for the years of 1990, 2006, and 2018 respectively that were taken by using 138/43 path and row (Table 13.1).

13.3.2   Image Pre-preprocessing

Both geometric and radiometric corrections have been conducted to process the satellite images for further analysis. The geometric correction is confined to the process of geo-referencing of Landsat images with the Universal Transverse Mercator (UTM) projection system that was ratified by availing some known ground points. In this study, 25 ground points were identified from Google Earth to transfer images into the Bangladesh Transverse Mercator (BTM) map projection system, having the datum of D_Everest_1830 from the World Geodetic System (WGS84).

On the other hand, the reflectance sensor data is influenced by many factors (e.g., atmospheric absorption and scattering, in-sensor factors, data processing procedures, etc.) (Teillet, 1986). Moreover, these effects need to be addressed to avoid the task becoming complicated (Mas, 1999; Turker and Asik, 2002). A radiometric correction needs to capture the absolute surface reflectance digital number (DNs) generated from the satellite (Chavez, 1996). The study followed (López et al., 2016) for the radiometric correction of the satellite images.

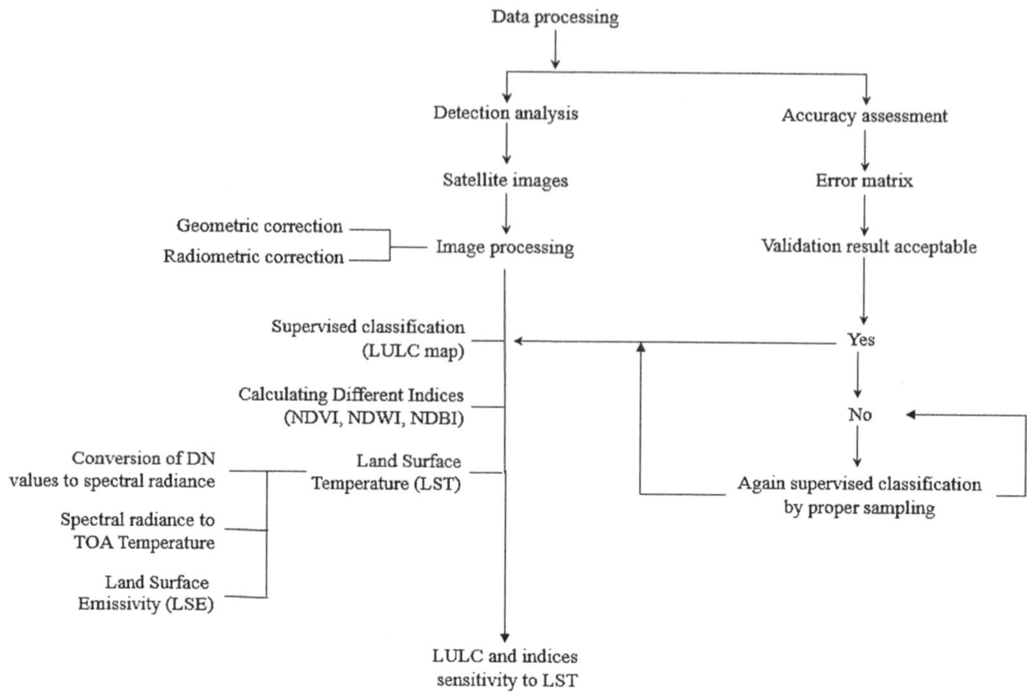

FIGURE 13.2   Flow chart of the data analysis.

TABLE 13.1

Details on Collected Satellite Image

Sensors name	Date of data acquired (yy-mm-dd)	Time	Path	Row
Landsat 4–5 Thematic Mapper (TM)	1995-04-18	03:39:25 PM	138	43
Landsat 4–5 Thematic Mapper (TM)	2005-04-13	04:17:43 PM	138	43
Landsat 8 Operational Land Imager (OLI)	2016-04-11	04:30:12	138	43

13.3.3 Calculating Different Indices

The study considered three different normalized indices to understand the land use and land cover scenario of the study area. These are the normalized difference vegetation index (NDVI), normalized difference water index (NDWI), and normalized difference built-up index (NDBI) respectively. NDVI is the most commonly used vegetation index to indicate a plant's photosynthetic activity. On the other hand, NDWI and NDBI show water bodies and built-up areas accordingly. These three indices were calculated by using Equations (13.1), (13.2), and (13.3) given below (NASA, 2015):

$$NDVI = \left(NIR - Red \right) / \left(NIR + Red \right) \tag{13.1}$$

$$NDWI = \left(Green - NIR \right) / \left(Green + NIR \right) \tag{13.2}$$

$$NDBI = \left(SWIR1 - NIR \right) / \left(SWIR + NIR \right) \tag{13.3}$$

Where, bands 3, 4, 5, and 6 are green, red, NIR, and SWIR1 for Landsat 8 OLI, similarly bands 4 and 3 are NIR and red for Landsat TM 4-5.

13.3.4 Deriving Land Surface Temperature (LST)

To detect land surface temperature, calculating the digital number (DN) into the top of atmosphere (TOA) brightness temperature is essential, followed by the conversion of TOA brightness temperature into land surface temperature.

13.3.4.1 Conversion of DN Values to Spectral Radiance

LST is calculated from the thermal band where bands 10 and 11 are the thermal bands for Landsat 8 and band 6 is the thermal band for Landsat TM 4-5. The digital numbers (DNs) of bands 6, 10, and 11 were converted to spectral radiance using the Equations (13.4) and (13.5) given below: (NASA, 2015)

$$L_{\lambda} = M_L.Q_{cal} + A_L \quad \text{(For Landsat 8 OLI)} \tag{13.4}$$

$$L_{\lambda} = \frac{V}{225\left(R_{max} - R_{min}\right) + R_{min}} \quad \text{(For Landsat TM 4-5)} \tag{13.5}$$

Where L_{λ} = Spectral radiance; M_L = Radiance multiplicative scaling factor for the band; A_L = Radiance additive scaling factor for the band; and Q_{cal} = pixel value of band 10 and 11 in DN. Similarly, V represents the DN of band 6; R_{max}, 1.896(mW×cm−2×sr−1) and R_{min}, 0.1534(mW×cm−2×sr−1).

13.3.4.2 Spectral Radiance to TOA Temperature

Then spectral radiance of bands 10 and 11 of Landsat 8 and band 6 of Landsat TM 4-5 were converted to top of atmosphere (TOA) brightness temperature where TOA was converted to degrees Celsius as it was derived in kelvin. The formula is shown below: (NASA, 2015)

$$BT = \frac{K_2}{L_{\lambda}\left\{\left(\frac{K_1}{L_{\lambda}}\right) + 1\right\}} \tag{13.6}$$

Where, BT refers to top of atmosphere brightness temperature, in Kelvin; L_{λ} = Spectral radiance; $K1$ = Thermal conversion constant for the band; and $K2$ = Thermal conversion constant for the band.

13.3.4.3 Calculate Land Surface Temperature

To investigate LST, the value of land surface emissivity (LSE) is needed along with TOA temperature. The LST and LSE of Landsat 8 OLI and Landsat TM 4-5 can be calculated by the following formula (Weng et al., 2004):

$$P_v = [(NDVI - NDVI_{min}) / (NDVI_{max} - NDVI_{min})]^2 \tag{13.7}$$

$$e = 0.004\, P_v + 0.986 \tag{13.8}$$

$$LST = BT / 1 + \lambda * (BT / p) * ln(e) \tag{13.9}$$

Where, e = Land surface emissivity; P_v = Proportion of vegetation; λ = wavelength of emitted radiance; P = h*c/s (1.438*10⁻² m K); h = Planck's constant (6.626×10⁻³⁴ J s); s = Boltzmann Constant (1.38 ×10⁻²³ J/K) and c = Velocity of Light (3 ×10⁸ m/s).

TABLE 13.2
Land Cover Types

Land cover type	Description
Built-up Area	Mainly all infrastructure. Such as residential, commercial, mixed-use, and industrial areas, settlements, road networks, pavements, etc.
Water Body	Ponds, wetlands, marshy land, low land, and swamps.
Vegetation	Trees, natural vegetation, mixed forest, gardens, parks, and playgrounds.
Agricultural	Agricultural lands, vacant land, open space, and crop fields,

13.3.5 LAND COVER CLASSIFICATION

There are various types of classification methods available to identify the LULC status from RS data. Broadly, the methods are supervised and unsupervised classification approaches. The unsupervised classification method is based on the iterative self-organizing data analysis technique algorithm (Lu et al., 2004). On the other hand, supervised classification is based on training sample analysis, which is decided by researchers/experts, and this type of classification method is more precise compared to others to see the accuracy obtained by the MLC method. Considering this benefit, the supervised classification method is the most popular and most widely used by many researchers (Fonji and Taff, 2014; Islam et al., 2015, 2018; Islam and Sarker, 2016). Having this consideration, the study used supervised classification with the maximum likelihood classification (MLC) algorithm for LULC determination by using ArcGIS 10.3 software. The types of LULC found from the analysis are listed in Table 13.2.

Initially, the corrected images (geometric and radiometric corrected) of each date are compiled into a single layer using three bands to create a false color composite (FCC) image to distinguish different LULC types like agricultural land, vegetation, and settlement.

13.3.6 ERROR MATRIX

The error matrix is one of the best ways to assess the accuracy of satellite images for land use land cover analysis as it aims to assess the effectiveness of sampling for land cover classification (Hasan et al., 2020; Hasan et al., 2023; Rwanga and Ndambuki, 2017). For the error matrix, a satellite image of April 2019 has been selected where 160 ground points were chosen. Google Earth was used as a reference source for ground truthing. Table 13.3 presents the error matrix of the land cover classified image of 2019 that shows the relationship between ground truth data and the corresponding classified data. The columns of the table represent produced value (map makers' point-of-view, selecting a sample for the land cover) and the rows express observed value (user's point-of-view, how accurately land class is presented on the ground). For accuracy assessment, the study considered the producer's and user's accuracy, commission, and omission errors, overall accuracy, and Kappa coefficient followed by (Rwanga and Ndambuki, 2017) that have been derived from the error matrix.

13.4 RESULTS

13.4.1 ACCURACY OF THE STUDY

The land covers derived from the analysis and the land covers from Google Earth are well-matched with better accuracy. The result shows that the producer Accuracy (PA) ranged from 100% to 0.63% whereas the user accuracy (UA) was found at 98% to 80% (Table 13.4). Water bodies and vegetation

TABLE 13.3
Error Matrix

Land use class	Water body	Built up area	Agricultural land	Vegetation	Row total
Water body	40	0	1	0	41
Build up area	0	40	10	0	50
Agricultural land	0	0	25	3	28
Vegetation	0	0	4	37	41
Column Total	40	40	40	40	160

TABLE 13.4
Accuracy Assessment

Land class	UA	PA	CE	OE	OA	Kappa
Water body	0.98	1	0.02	0	0.89	0.85
Built up area	0.80	1	0.20	0		
Agricultural land	0.89	0.63	0.11	0.37		
Vegetation	0.90	0.93	0.10	0.08		

are found to be more accurate both for PA and UA. High commission error (CE) in this study is found for built-up areas (20%) and high omission error (OE) for agriculture (37%). That means that overestimation was incorrectly included for the built-up area while sampling, on the other hand, underestimation was evaluated for agricultural land cover. Besides, overall accuracy is found at 89% and the Kappa coefficient is measured at 0.85 that illustrates preferable accuracy for each land cover classification in its field of interest according to Fleiss et al. (2004) (Fleiss et al., 2004).

13.4.2 Temperature Difference between Urban and Rural Areas

RCC is considered an urban area, whereas the other 9 Upazilas are supposed to be rural areas. These classifications have been done on the basis of some criteria (e.g., land cover type, population size, and municipality services). The temperature difference between urban and rural areas is identified in the study area due to their land cover differences. For instance, the amount of concrete and bitumen and the existence of greenery and water bodies. The maximum land cover of the urban area is the built-up area where a plethora of vegetation, agricultural land, or water bodies are found in the rural part of Rajshahi or vice versa.

Figure 13.3 illustrates that the temperature difference between urban and rural areas varies by 1° to 2° for each type of land cover, where for every land class, the rural area temperature is lower than the urban due to the absorption of heat by built-up areas. More specifically, the mean temperature in built-up areas is 31.87°C and in rural areas it is 32.82°C. Similarly, the mean temperatures are 30.39°C and 31.35°C for vegetation, 29.90°C and 31.61°C for agricultural land, as well as 29.77°C and 30.89°C for aquatic land.

From Figure 13.4, it has been found that the southern and southwestern parts of the urban area are concentrated on built-up areas, and the northern and northeastern parts are covered by other land use types (Figure 13.4, A2). On the other hand, high temperatures have been observed in the southern and southwestern areas and lower temperatures in other parts of the city (Figure 13.4, A1). Consequently, a proportionate relationship exists between surface temperature and LULC. Likewise, Godagari, Paba, Durgapur, Puthia, and Charghat Upazilar rural areas seem to be hotter

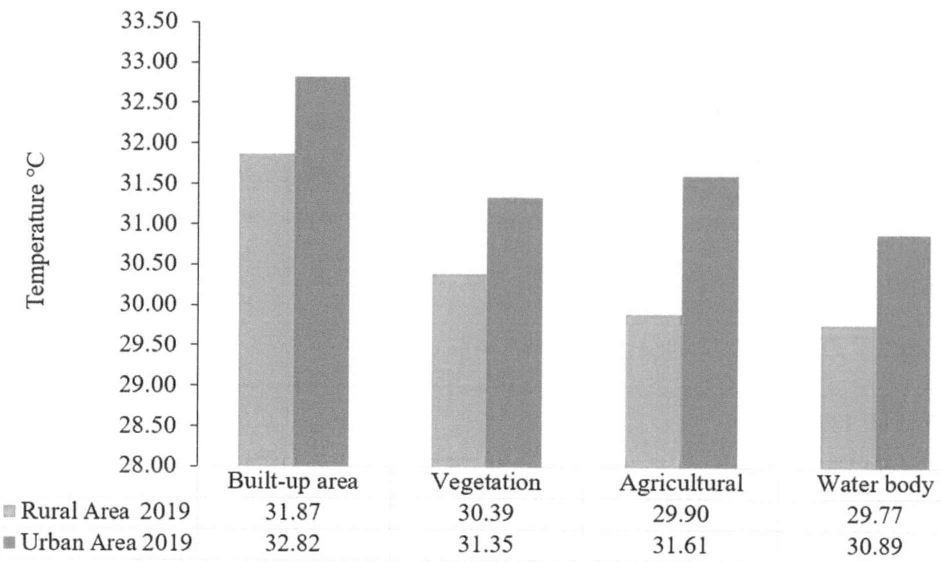

	Built-up area	Vegetation	Agricultural	Water body
Rural Area 2019	31.87	30.39	29.90	29.77
Urban Area 2019	32.82	31.35	31.61	30.89

FIGURE 13.3 LULC-wise temperature difference between urban and rural areas in 2019.

than the other Upazilas (Figure 13.4, B2) where there is a corresponding relationship between surface temperature and LULC as in urban areas (Figure 13.4, B1).

13.4.3 Analysis of Comfort Zone

The temperature variation from 1995 to 2019 has been identified both for urban (RCC) and rural (Other Upazilas except RCC) areas to understand the comfort zone (Brändle et al., 2015). To do that, temperatures are divided into four classes, where temperatures up to 15°C are considered a cool area, 16–25°C as a comfort zone, 26–32°C are treated as warm, and more than 32°C as an uncomfortable area. This study found that the temperature values ranged from 23°C to 40°C. Figure 13.5 illustrates that maximum areas of Rajshahi district are in the warm zone all over the study period, having no cool zone. A small portion of a comfortable zone was found in Baghmara Upazila in 1995, whereas most of the areas of Tanore, Godagari, Paba, and Charghat were found to be uncomfortable areas until 2005. But in the year 2019, the spatial distribution of comfort zones showed a reduction in uncomfortable hot zones whereas Rajshahi City Corporation gave the inverse result.

Table 13.5 illustrates that the warm zone decreased by 88.57 km² in 2005 and again increased to 200.15 km² in 2019. However, the area of comfort zone was in a decreasing trend all over the study period, which decreased to 0.84 km² in 2005 and no comfort zone was available in the year of 2019. Though the very hot zones decreased in 2019, RCC followed a reverse trend in this case.

13.4.4 Relation between LST and Different Indices

Figure 13.6 shows the temperature difference between various Upazilas of Rajshahi in the year 2019, where the highest minimum temperature was found in Rajshahi City Corporation and the lowest in Puthia. Figure 13.7 shows that Rajshahi City Corporation, treated as an urban area, is found built-up concentrated (395.001 km²) where a small amount of agricultural and vegetated areas, as well as water bodies, are available. Besides, a large amount of agricultural land is found in Godagari, Tanore, Mohonpur, and Baghmara Upazila, where the maximum vegetation is in Bagha Upazila.

NDVI's highest value varies from 0.74 to −0.19, whereas NDWI and NDBI vary from 0.65 to −0.33 and 0.33 to −0.74. The NDVI value is found high in several areas (e.g., Mohanpur, Baghmara,

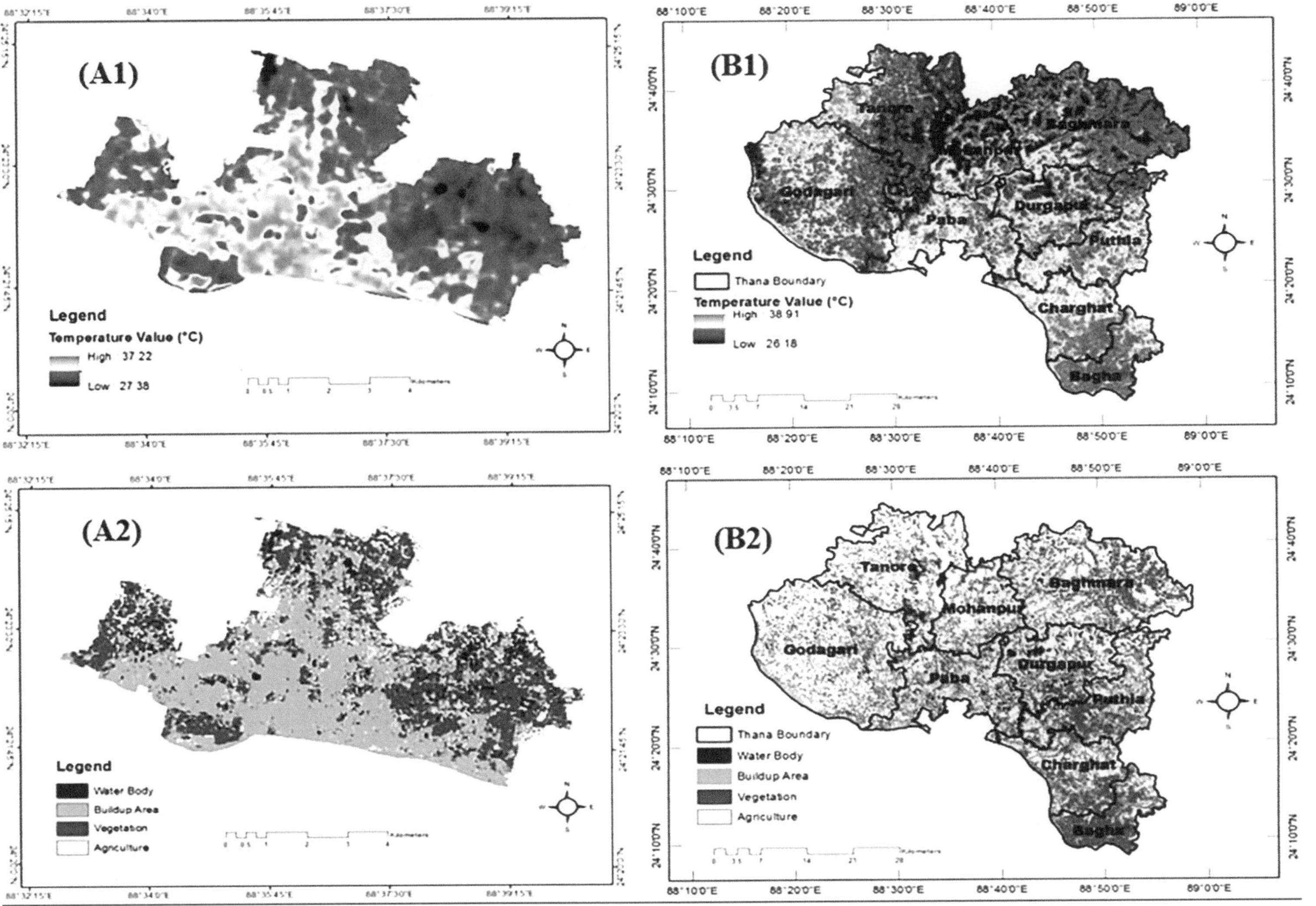

FIGURE 13.4 LST and land cover of urban and rural area, (A1) LST of urban area, (A2) Land cover of urban area, (B1) LST of rural area, (B2) Land cover of rural area.

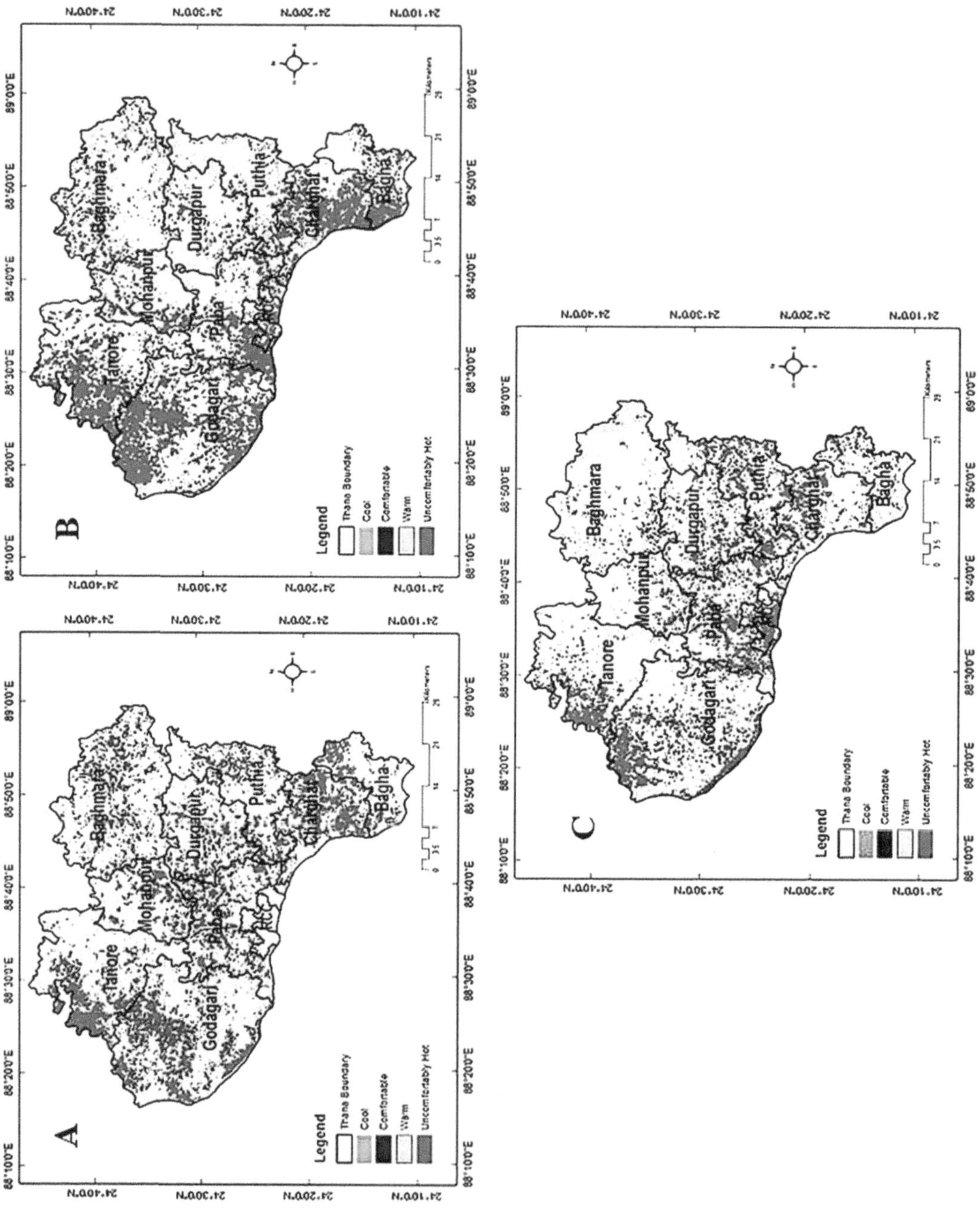

FIGURE 13.5 Year-wise comfort zone scenario of Rajshahi district, (A) 1995, (B) 2005, (C) 2019.

TABLE 13.5

Area Distribution in km² of Comfort Zone of Rajshahi District in 1995, 2005, and 2019

Temperature category	Area (km²), 1995	Area (km²), 2005	Area (km²), 2019
Cool	–	–	–
Comfortable	6.15	0.84	–
Warm	1684.16	1595.59	1793.74
Uncomfortably hot	509.39	603.28	405.97

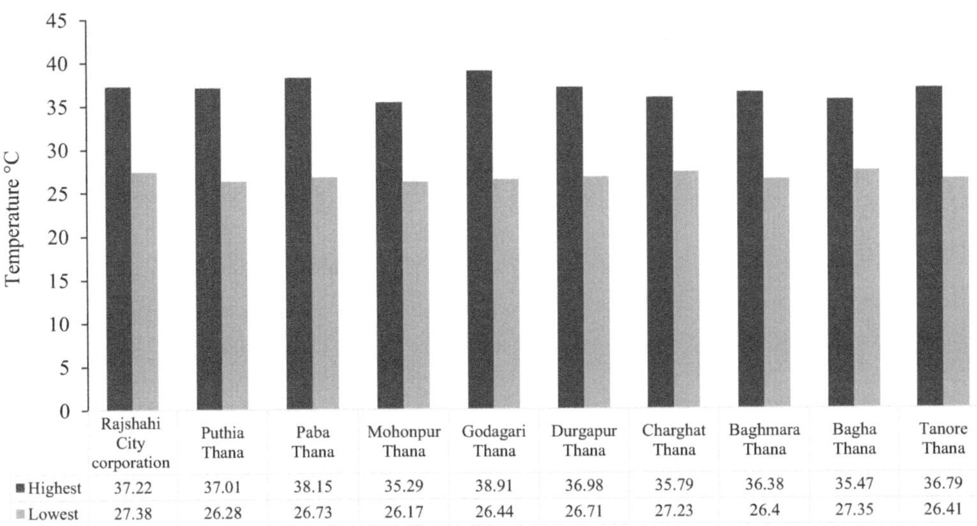

	Rajshahi City corporation	Puthia Thana	Paba Thana	Mohonpur Thana	Godagari Thana	Durgapur Thana	Charghat Thana	Baghmara Thana	Bagha Thana	Tanore Thana
Highest	37.22	37.01	38.15	35.29	38.91	36.98	35.79	36.38	35.47	36.79
Lowest	27.38	26.28	26.73	26.17	26.44	26.71	27.23	26.4	27.35	26.41

FIGURE 13.6 Highest and lowest temperature difference of different upazila of Rajshahi district in 2019.

Tanore, Godagari Upazila) having low temperatures. Similarly, RCC and Godagari Upazila fall under lower NDWI values with higher temperatures. With the increase in NDWI value, the presence of water bodies increases and temperature decreases, as well as NDVI (with respect to vegetation). On the other hand, with respect to NDBI, RCC, and the adjacent areas of Paba, Puthia, and Charghat that are close to the city corporation area, the northwest portion of Godagari and Tanor are found to have high values of NDBI and temperature.

13.4.5 Indices Sensitivity to Temperature

A simple application of the correlation coefficient can be exemplified by using data from a sample of 348 points that were selected from different places in the RCC shown in Figure 13.8. All of the indices (e.g. NDVI, NDWI, and NDBI) are continuous and skewed with the response to temperature. From Pearson correlation analysis between indices and LST, the resulting correlation coefficients are −0.80 (NDVI), −0.88 (NDWI), and 0.88 (NDBI) respectively. In this case, all of the indices are strongly correlated with temperature, i.e., high positive and negative correlation. Where two indices (NDVI and NDWI) correlation coefficients are similar and lead to the same conclusion, the NDBI reveals different statistical conclusions. According to Hinkle et al. (2003), the correlation coefficient between 0.70 to 0.90 or −0.70 to −0.90 has a high positive or negative correlation where the above

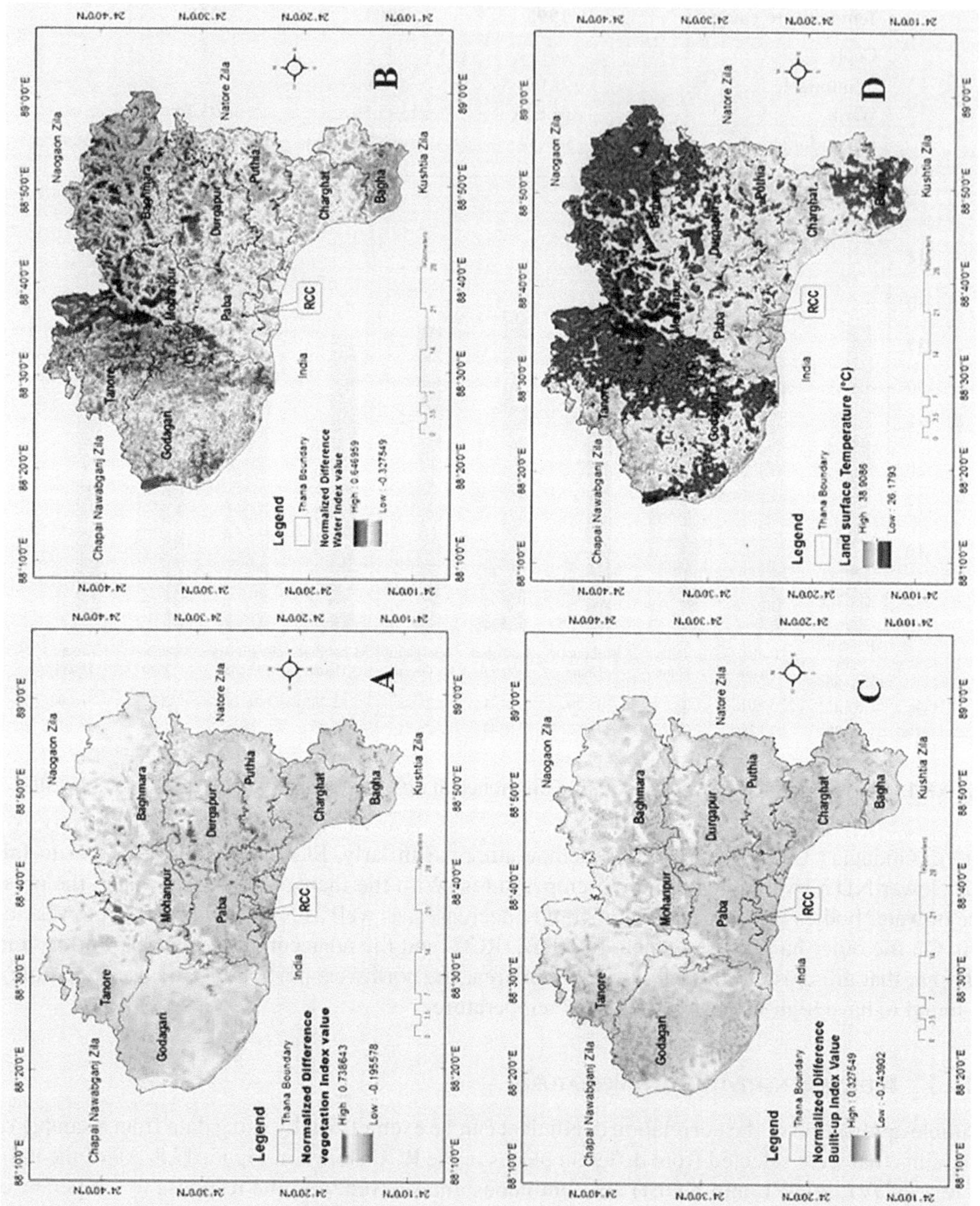

FIGURE 13.7　Indices comparison with LST, (A) NDVI, (B) NDWI, (C) NDBI, and (D) LST.

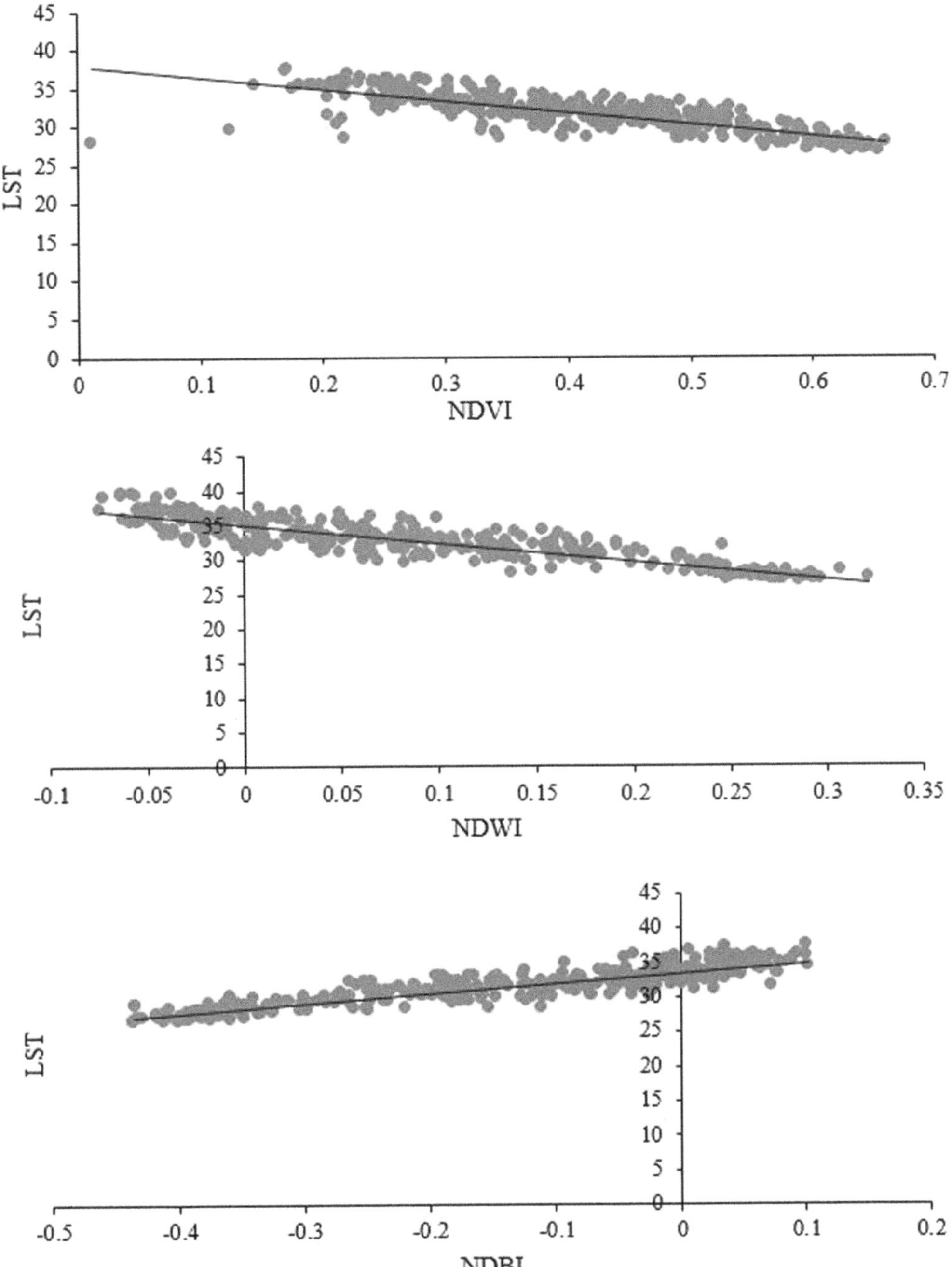

FIGURE 13.8 Correlation coefficient analysis between LST and three different indices (NDVI, NDWI, and NDBI).

or below value range demonstrates a very high or moderate correlation. In this study, the correlation coefficient value between NDWI, NDVI, and temperature was found to be −0.81 and −0.88 respectively, which depict a highly negative correlation. On the contrary, the NDBI value is found at 0.88, which means it is highly positively correlated with temperature. Therefore, with an increase in the NDBI value, temperature increases, and with a decrease in the NDBI value, temperature decreases. Reverse situations have been found in the case of NDWI and NDVI.

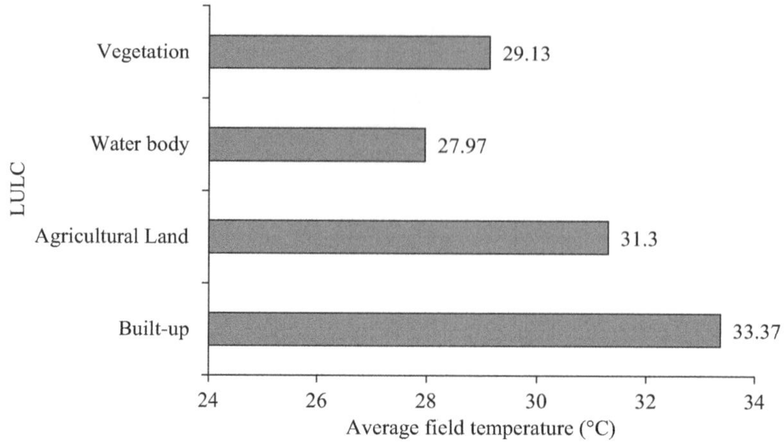

FIGURE 13.9   Average field temperature of the study area.

To understand the above relationship, a field survey was conducted in RCC on the different land covers to cross-check the sensitivity to temperature. Figure 13.9 shows the temperature value for four different land covers where built-up areas have the highest value (33.37°C). On the other hand, the value is 29.13°C for vegetation, 27.97°C for water bodies, and 31.30°C for agricultural land.

13.5   DISCUSSIONS

The study was carried out to determine the consequences of land cover changes on land surface temperature in Rajshahi district for the year 2019. Furthermore, chronological changes in temperature have been found in the years 1995, 2005, and 2019. Previous studies regarding LULC and LST (Ahmed et al., 2013; Imran et al., 2021; John et al., 2020; Kafy et al., 2020; Raja, 2012) were helpful in leading as well as comparing the results of the study. High temperatures in built-up areas, and contrariwise sloping temperatures in agriculture, vegetation, and water bodies have been fabricated in this study, which coincides with previous research (Ahmed et al., 2013; Imran et al., 2021; John et al., 2020). Dey et al. (2021) reported that in Rajshahi district, the buildup area has increased by 14.78% while greenery space has decreased by about 6.27% since 1990. Kafy et al. (2021b) added that about 3.35 % of the area in Rajshahi will be more than 35°C due to the conversion of urban areas by 2039. Nevertheless, the study results show that 18% (396 km^2) of the study area was built-up areas in 2019, among which almost 50% of the built-up area was in RCC. These concrete-concentrated urban areas led to the increase in LST, while Imran et al. (2021) revealed similar results in Dhaka, Bangladesh.

The southern part of Rajshahi (especially the RCC area) experienced substantial growth in settlement areas, leading to the LST. Urban expansion occurred due to population growth in Rajshahi city due to a higher birth rate and migrated people from rural areas to urban areas for the sake of better livelihood (Kafy et al., 2021c, 2021d). Sometimes this is due to natural or anthropogenic disasters (river erosion, flood, drought, scarcity of employment, etc.). As Kafy et al. (2021a) argued, most of the water bodies and greenery turned into settlements from 1990 to 2019.

On the other hand, with respect to the comfort zone of Rajshahi, the uncomfortable zone was decreased by about 4.70%, though the warm area was increasing at an alarming rate (9.06%) from 1995 to 2019 (Figure 13.10). The study also found that the number of comfortable zones (between 15–25°C) decreased by 0.3% where RCC (urban areas) gradually turned into warm to uncomfortably hot areas. A corresponding trend of temperature shifting was identified by Imran et al. (2021) in Dhaka city of Bangladesh where the temperature increased by 0.24°C per year, while Pal and Ziaul (2017) found an increase of 0.11°C per year in West Bengal, India.

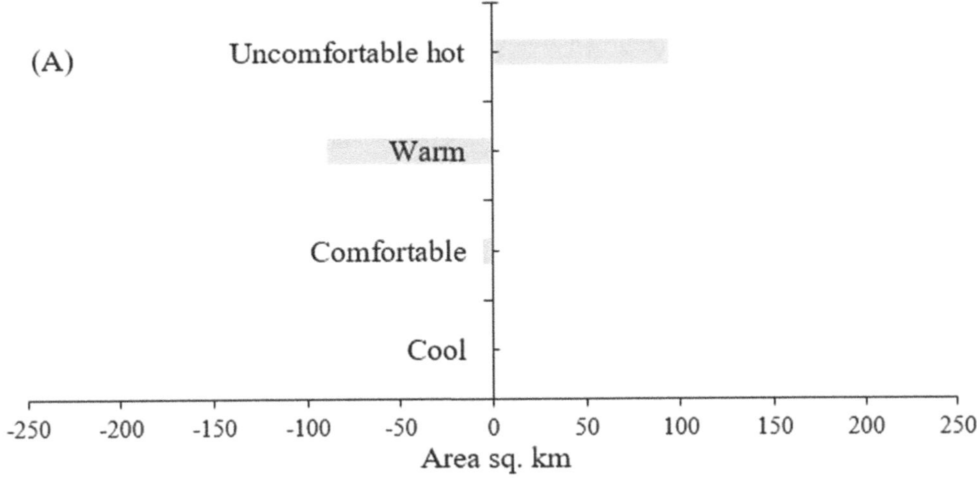

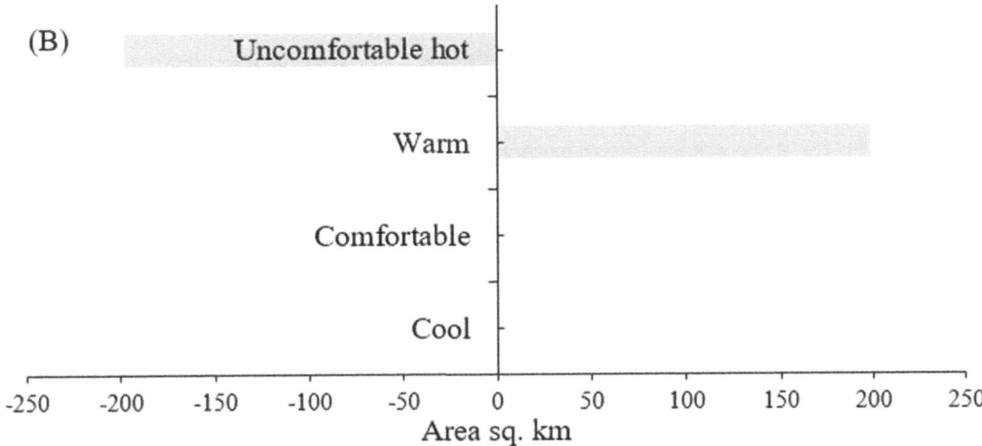

FIGURE 13.10 Gain and loss of comfort zone (A) 1995 to 2005 and (B) 2005 to 2019.

Further, regression analysis of the study resulted in a positive accordance between LST and NDBI but a negative kinship between LST and the other two indices (NDVI and NDWI). The analysis shows that LST increased with the increase in built-up areas (NDBI) or vice-versa. On the other hand, LST decreased when vegetation (NDVI) and water bodies (NDWI) increased in trend or vice-versa. Likewise, Imran et al. (2021) identified similar results from the perspective of LST and different land cover indices that testify to the LST sensitivity to indices.

The findings of this chapter thus suggest that urbanization is the forward driver for LULC and LST respectively. However, without pledging a better decentralization policy, it is hard to control the unplanned built-up areas in Rajshahi because they gradually increase their major opportunities and services (e.g., education, medical facilities, jobs) (Farhana and Mannan, 2018).

13.6 CONCLUSIONS

The study reveals the relationship between the land cover and surface temperature in the Rajshahi district of Bangladesh, where the temperature difference between the concrete concentrated (urban) and agriculture, vegetation, and water concentrated (rural) has been clearly identified. This relationship is further clearly identified by linear regression analysis between different indices of NDVI,

NDWI, NDBI, and LST. The temperature difference between the urban and rural areas is almost 1°C to 2°C. Most of the land in the urban area is wrapped in rock or asphalt, and these construction materials are the main reasons for increasing urban temperature, especially in RCC. However, the highest temperature difference between the urban and rural areas was found in the agricultural land (1.71°C), where 1.01°C in the built-up area, 0.98°C in vegetation, and 1.12°C in the water body. Besides, to find a comfortable zone, it is notable that the hot zone (15°C to 25°C) has been on an increasing trend from 1995 to 2019, irrespective of LULC. It is partially supposed to be the reason for LST due to climate change, though further research ought to be needed to confirm it. From the findings of the study, policymakers in Rajshahi ought to be concerned about urbanization and future urban growth. An environmentally friendly proper guideline is thus needed both for horizontal and vertical expansion where green roofs, climate-friendly construction materials, and extensive tree plantation may reduce the urban microclimate warming effect.

ACKNOWLEDGMENTS

The author expresses gratitude to the Department of Urban and Regional Planning at Rajshahi University of Engineering and Technology (RUET) for their assistance during the data collection phase and their ongoing support throughout the research process. Their contributions were instrumental in ensuring the successful completion of this chapter. The authors also extend their appreciation to the Bangladesh Meteorological Department (BMD), the United States Geological Survey (USGS), and the Survey of Bangladesh for their assistance in supplying the required data for this study at no cost. The data provided by these organizations were crucial in conducting the research and verifying the results. The authors are grateful for their cooperation and support throughout the study.

DATA AVAILABILITY

The datasets that were created and analyzed during this study are available upon request to the corresponding author. To ensure transparency and reproducibility, the corresponding author will respond to requests for the data in a prompt manner. Making the data available to interested parties can facilitate scientific progress and encourage other researchers to expand on the conclusions of this research.

REFERENCES

Ahmed, B., Kamruzzaman, M. D., Zhu, X., Rahman, M., and Choi, K. (2013). Simulating land cover changes and their impacts on land surface temperature in Dhaka, Bangladesh. *Remote Sensing*, 5(11), 5969–5998.

BBS. (2011). *District Statistic*. Bangladesh Bureau of Statistics. The Government of People's Republic of Bangladesh.

Brändle, J. M., Langendijk, G., Peter, S., Brunner, S. H., and Huber, R. (2015). Sensitivity analysis of a land-use change model with and without agents to assess land abandonment and long-term re-forestation in a Swiss mountain region. *Land*, 4(2), 475–512.

Buo, I., Sagris, V., Burdun, I., and Uuemaa, E. (2021). Estimating the expansion of urban areas and urban heat islands (UHI) in Ghana: A case study. *Natural Hazards*, 105(2), 1299–1321. https://doi.org/10.1007/s11069-020-04355-4.

Chavez, P. S. (1996). Image-based atmospheric corrections-revisited and improved. *Photogrammetric Engineering and Remote Sensing*, 62(9), 1025–1035.

Dewan, A., Kiselev, G., and Botje, D. (2021). Diurnal and seasonal trends and associated determinants of surface urban heat islands in large Bangladesh cities. *Applied Geography*, 135, 102533.

Dey, N. N., Al Rakib, A., Kafy, A. A., and Raikwar, V. (2021). Geospatial modelling of changes in land use/land cover dynamics using Multi-layer perception Markov chain model in Rajshahi City, Bangladesh. *Environmental Challenges*, 4, 100148.

dos Santos, R. S. (2020). Estimating spatio-temporal air temperature in London (UK) using machine learning and earth observation satellite data. *International Journal of Applied Earth Observation and Geoinformation*, 88(December 2019), 102066. https://doi.org/10.1016/j.jag.2020.102066.

Dutta, D., Rahman, A., Paul, S. K., and Kundu, A. (2019). Changing pattern of urban landscape and its effect on land surface temperature in and around Delhi. *Environmental Monitoring and Assessment*, 191(9). https://doi.org/10.1007/s10661-019-7645-3.

Farhana, D. K. M., and Mannan, D. K. A. (2018). Socioeconomic impact of regional rural urban migration: A revisit of slum dwellers of Rajshahi City Corporation. Available at SSRN 3098838.

Fathian, F., Prasad, A. D., Dehghan, Z., and Eslamian, S. (2015). Influence of land use/land cover change on land surface temperature using RS and GIS techniques. *International Journal of Hydrology Science and Technology*, 5(3), 195–207.

Ferdous, J., and Rahman, M. T. U. (2018). Temporal dynamics and relationship of land use land cover and land surface temperature in Dhaka, Bangladesh. *Proceedings of the 4th International Conference on Civil Engineering for Sustainable Development (ICCESD 2018)*, 9–11.

Fleiss, J. L., Levin, B., and Paik, M. C. (2004). The measurement of Interrater agreement. In *Statistical Methods for Rates and Proportions*, 598–626. https://doi.org/10.1002/0471445428.ch18.

Fonji, S. F., and Taff, G. N. (2014). Using satellite data to monitor land-use land-cover change in North-eastern Latvia. *SpringerPlus*, 3(61), 1–15. https://doi.org/10.1186/2193-1801-3-61.

Habibie, M. I., Noguchi, R., Shusuke, M., and Ahamed, T. (2021). Land suitability analysis for maize production in Indonesia using satellite remote sensing and GIS-based multicriteria decision support system. *GeoJournal*, 86(2), 777–807.

Hasan, M. H., Sayeeda, N., and Hossain, A. K. M. S. (2017). Developing fire adaptation strategies at community level: A case study of Shreerampur Slum in Rajshahi City Corporation. *International Conference on Planning, Architecture and Civil Engineering. Rajshahi University of Engineering and Technology*, Rajshahi.

Hasan, M. H., Hossain, M. J., Chowdhury, M. A., and Billah, M. (2020). Salinity intrusion in southwest coastal Bangladesh: An insight from land use change. In Haque, A., Chowdhury, A. (Eds.), *Water, Flood Management and Water Security Under a Changing Climate*. Springer, Cham, 125–140. https://doi.org/10.1007/978-3-030-47786-8_8

Hasan, M. H., Chowdhury, M. A., and Wakil, M. A. (2022). Community engagement and public education in northwestern part of Bangladesh: A study regarding heritage conservation. *Heliyon*, 8(3), e09005.

Hasan, M., Newton, I. H., Chowdhury, M. A., Esha, A. A., Razzaque, S., and Hossain, M. J. (2023). Land use land cover change and related drivers have livelihood consequences in coastal Bangladesh. *Earth Systems and Environment*, 7, 1–19.

Hellings, A., and Rienow, A. (2021). Mapping land surface temperature developments in functional urban areas across europe. *Remote Sensing*, 13(11). https://doi.org/10.3390/rs13112111.

Hinkle, D. E., Wiersma, W., and Jurs, S. G. (2003). *Applied Statistics for the Behavioral Sciences*. Houghton Mifflin, Boston.

Hoque, M. A. A., Pradhan, B., and Ahmed, N. (2020). Assessing drought vulnerability using geospatial techniques in northwestern part of Bangladesh. *Science of the Total Environment*, 705, 135957. https://doi.org/10.1016/j.scitotenv.2019.135957.

Hua, A. K., and Ping, O. W. (2018). The influence of land-use/land-cover changes on land surface temperature: A case study of Kuala Lumpur metropolitan city. *European Journal of Remote Sensing*, 51(1), 1049–1069. https://doi.org/10.1080/22797254.2018.1542976.

Imran, H. M., Hossain, A., Islam, A. K. M. S., Rahman, A., Bhuiyan, M. A. E., Paul, S., and Alam, A. (2021). Impact of land cover changes on land surface temperature and human thermal comfort in Dhaka city of Bangladesh. *Earth Systems and Environment*, 5, 1–27. https://doi.org/10.1007/s41748-021-00243-4

Islam, W., and Sarker, S. C. (2016). Monitoring the changing pattern of land use in the Rangpur City corporation using remote sensing and GIS. *Journal of Geographic Information System*, 8(4), 537–545. https://doi.org/10.4236/jgis.2016.84045.

Islam, G. M. T., Islam, A. S., Shopan, A. A., Rahman, M. M., Lázár, A. N., and Mukhopadhyay, A. (2015). Implications of agricultural land use change to ecosystem services in the Ganges delta. *Journal of Environmental Management*, 161, 443–452. https://doi.org/10.1016/j.jenvman.2014.11.018.

Islam, K., Jashimuddin, M., Nath, B., and Nath, T. K. (2018). Land use classification and change detection by using multi-temporal remotely sensed imagery: The case of Chunati wildlife sanctuary, Bangladesh. *The Egyptian Journal of Remote Sensing and Space Science*, 21(1), 37–47. https://doi.org/10.1016/j.ejrs.2016.12.005.

Jain, S., Roy, S. B., Panda, J., and Rath, S. S. (2021). Modeling of land-use and land-cover change impact on summertime near-surface temperature variability over the Delhi–Mumbai Industrial Corridor. *Modeling Earth Systems and Environment*, 7(2), 1309–1319. https://doi.org/10.1007/s40808-020-00959-8.

John, J., Bindu, G., Srimuruganandam, B., Wadhwa, A., and Rajan, P. (2020). Land use/land cover and land surface temperature analysis in Wayanad district, India, using satellite imagery. *Annals of GIS*, 26(4), 343–360.

Kafy, A. A., Al Faisal, A., Sikdar, S., Hasan, M., Rahman, M., Khan, M. H., and Islam, R. (2020). Impact of LULC changes on LST in Rajshahi district of Bangladesh: A remote sensing approach. *Journal of Geographical Studies*, 3, 11–23.

Kafy, A. A., Naim, N. H., Khan, M. H. H., Islam, M. A., Al Rakib, A., Al-Faisal, A., and Sarker, M. H. S. (2021a). Prediction of urban expansion and identifying its impacts on the degradation of agricultural land: A machine learning-based remote-sensing approach in Rajshahi, Bangladesh. In Singh R. (Ed), *Re-envisioning Remote Sensing Applications*. CRC Press, Boca Raton, 85–106.

Kafy, A. A., Al Faisal, A., Al Rakib, A., Roy, S., Ferdousi, J., Raikwar, V., Kona, M. A., and Al Fatin, S. M. A. (2021b). Predicting changes in land use/land cover and seasonal land surface temperature using multi-temporal landsat images in the northwest region of Bangladesh. *Heliyon*, 7(7), e07623. https://doi.org/10.1016/j.heliyon.2021.e07623.

Kafy, A. A., Rahman, A. N. M. F., Al Rakib, A., Akter, K. S., Raikwar, V., Jahir, D. M. A., Ferdousi, J., and Kona, M. A. (2021c). Assessment and prediction of seasonal land surface temperature change using multi-temporal Landsat images and their impacts on agricultural yields in Rajshahi, Bangladesh. *Environmental Challenges*, 4, 100147.

Kafy, A. A., Al Rakib, A., Akter, K. S., Rahaman, Z. A., Faisal, A.-A., Mallik, S., Nasher, N. M. R., Hossain, M. I., and Ali, M. Y. (2021d). Monitoring the effects of vegetation cover losses on land surface temperature dynamics using geospatial approach in Rajshahi city, Bangladesh. *Environmental Challenges*, 4, 100187.

Lai, L. W., and Cheng, W. L. (2010). Urban heat island and air pollution-an emerging role for hospital respiratory admissions in an urban area. *Journal of Environmental Health*, 72(6), 32–35.

Liu, F., Hou, H., and Murayama, Y. (2021). Spatial interconnections of land surface temperatures with land cover/use: A case study of Tokyo. *Remote Sensing*, 13(4), 1–26. https://doi.org/10.3390/rs13040610.

López-Serrano, P. M., Corral-Rivas, J. J., Díaz-Varela, R. A., Álvarez-González, J. G., and López-Sánchez, C. A. (2016). Evaluation of radiometric and atmospheric correction algorithms for aboveground forest biomass estimation using Landsat 5 TM data. *Remote Sensing*, 8(5), 369.

Lu, D., Mausel, P., Brondízio, E., and Moran, E. (2004). Change detection techniques. *International Journal of Remote Sensing*, 25(12), 2365–2401. https://doi.org/10.1080/0143116031000139863.

Makvandi, M., Li, B., Elsadek, M., Khodabakhshi, Z., and Ahmadi, M. (2019). The interactive impact of building diversity on the thermal balance and micro-climate change under the influence of rapid urbanization. *Sustainability*, 11(6), 1–20. https://doi.org/10.3390/su11061662.

Mas, J. F. (1999). Monitoring land-cover changes: A comparison of change detection techniques. *International Journal of Remote Sensing*, 20, 139–152. https://doi.org/10.1080/014311699213659.

Mim, M. A., and Zamil, K. M. S. (2020). GIS-based surface water changing analysis in rajshahi city corporation area using ensemble classifier. *Proceedings of International Joint Conference on Computational Intelligence*, 39–47.

Mirza, M. M. Q. (2011). Climate change, flooding in South Asia and implications. *Regional Environmental Change*, 11(1), 95–107.

Mustafa, E. K., Abd El-Hamid, H. T., and Tarawally, M. (2021). Spatial and temporal monitoring of drought based on land surface temperature, Freetown City, Sierra Leone, West Africa. *Arabian Journal of Geosciences*, 14(11). https://doi.org/10.1007/s12517-021-07187-z.

NASA. (2015). *Landsat 8 Data User Handbook*. National Aeronautics and Space Administration.

Pal, S., and Ziaul, S. K. (2017). Detection of land use and land cover change and land surface temperature in English Bazar urban centre. *The Egyptian Journal of Remote Sensing and Space Science*, 20(1), 125–145.

Rahaman, M. A., Hossain, M. I., Kamal, A., and Chowdhury, A. M. (2021). Climate-induced displacement and human migration landscape in Bangladesh. In *Handbook of Climate Change Management*. https://doi.org/10.1007/978-3-030-22759-3_255-1.

Rahman, M. M. (2010). Factors of economic transformation in sub-urban areas of Rajshahi City, Bangladesh. *Journal of Life and Earth Science*, 5, 47–55.

Raja, D. R. (2012). GIS-based spatial simulation of impacts of urban development on changing both land cover area and land surface temperature in Dhaka city. Dissertation, Bangladesh University of Engineering and Technology, Bangladesh.

Roy, B., Bari, E., Nipa, N. J., and Ani, S. A. (2021). Comparison of temporal changes in urban settlements and land surface temperature in Rangpur and Gazipur Sadar, Bangladesh after the establishment of city corporation. *Remote Sensing Applications: Society and Environment*, 23, 100587.

Rwanga, S. S., and Ndambuki, J. M. (2017). Accuracy assessment of land use/land cover classification using remote sensing and GIS. *International Journal of Geosciences*, 8(4), 611.

Siddique, M. A., Dongyun, L., Li, P., Rasool, U., Khan, T. U., Farooqi, T. J. A., Wang, L., Fan, B., and Rasool, M. A. (2020). Assessment and simulation of land use and land cover change impacts on the land surface temperature of Chaoyang District in Beijing, China. *PeerJ*, 2020(3), 1–26. https://doi.org/10.7717/peerj.9115.

Souza, D. O. de, Alvalá, R. C. dos S., and Nascimento, M. G. do. (2016). Urbanization effects on the microclimate of Manaus: A modeling study. *Atmospheric Research*, 167(January), 237–248. https://doi.org/10.1016/j.atmosres.2015.08.016.

Teillet, P. M. (1986). Image correction for radiometric effects in remote sensing. *International Journal of Remote Sensing*, 7, 1637–1651. https://doi.org/10.1080/01431168608948958.

Turker, M., and Asik, O. (2002). Detecting land use changes at the urban fringe from remotely sensed images in Ankara, Turkey. *Geocarto International*, 17, 47–52. https://doi.org/10.1080/10106040208542243.

WB. (2013). *Warming Climate to Hit Bangladesh Hard with Sea Level Rise, More Floods and Cyclones.* World Bank, Washington, DC.

Weng, Q., Lu, D., and Schubring, J. (2004). Estimation of land surface temperature–vegetation abundance relationship for urban heat island studies. *Remote Sensing of Environment*, 89(4), 467–483.

Zhang, X. Q. (2016). The trends, promises and challenges of urbanisation in the world. *Habitat International*, 54(November 2015), 241–252. https://doi.org/10.1016/j.habitatint.2015.11.018.

14 Heavy Metals on the Sediments of the Ship-Breaking Zone and Impact on Fish Diversity in Coastal Bangladesh

Prabal Barua and Saeid Eslamian

14.1 INTRODUCTION

Ship-breaking (SB) is usually defined as the entire or partial dismantling of vessels in order to recover materials such as steel and rubbish for recycling and reuse. It includes an intensive extent of challenging activities, starting with the removal of all equipment and gear to cutting down and recycling the vessel structure. Ship-scrapping activities are held responsible for polluting the adjacent environment A scrapping vessel generates an oversized amount of waste and unsafe substances, alongside many assets [1–5], and valuable materials such as steel and wood, which are recovered for reuse or recycling purposes [2, 6–9]. Ship-breaking may be a challenging process, thanks to the structural complexity of ships and therefore there are many environmental, safety, and health hazards involved [10]. The history of ship-breaking is as nearly old as shipbuilding. There are approximately 50,000 ships within the world's waters [11]. After 25–30 years, the value of re-investment to accumulate this certificate (Certification of Fitness (CoF) for Ship) is no longer profitable. As a result, about 700 ships (15–25 million deadweight tons) are sold per annum to at least one Asian scrapyard. Most ship-breaking yards are now operating within South Asian countries thanks to lower labor costs and fewer stringent environmental regulations in regards to handling the disposal of lead paint and other toxic substances [2].

There are two sayings that are particularly suitable to the SB industry, which are "nothing lasts forever", and "waste not, want not". Even the world's best-designed and most loved vessels reach the end of their useful trading life. Because when a ship reaches the end of its service, there remains only one commercial option – to be sold for scrap. And there are plenty of people interested in turning obsolete ships into profit – a nice symbolic relationship. There used to be another saying relevant to ship-breaking, "out of sight, out of mind" this, however, is no longer the case. Thus, the location of ship-scrapping activities was transient and moved from one region to another due to the high demand for scrap material and labor costs from the beginning of this practice. This activity was initiated in the United States and Europe, especially Spain, Portugal, and Italy; and continued from 1945 to 1970 (Figure 14.1) [12].

But the activity radically shifted from the United States and Europe to Far-East Asia, especially South Korea, China, and Taiwan during the early part of the 1970s (Figure 14.2) [12].

Taiwan was leading in the practice during the 1980s and presently they have almost stopped the scrapping practice. Since then Bangladesh, China, India, Pakistan, and some other developing countries have engaged in the ship-scrapping business (Figure 14.3).

DOI: 10.1201/9781003473398-17

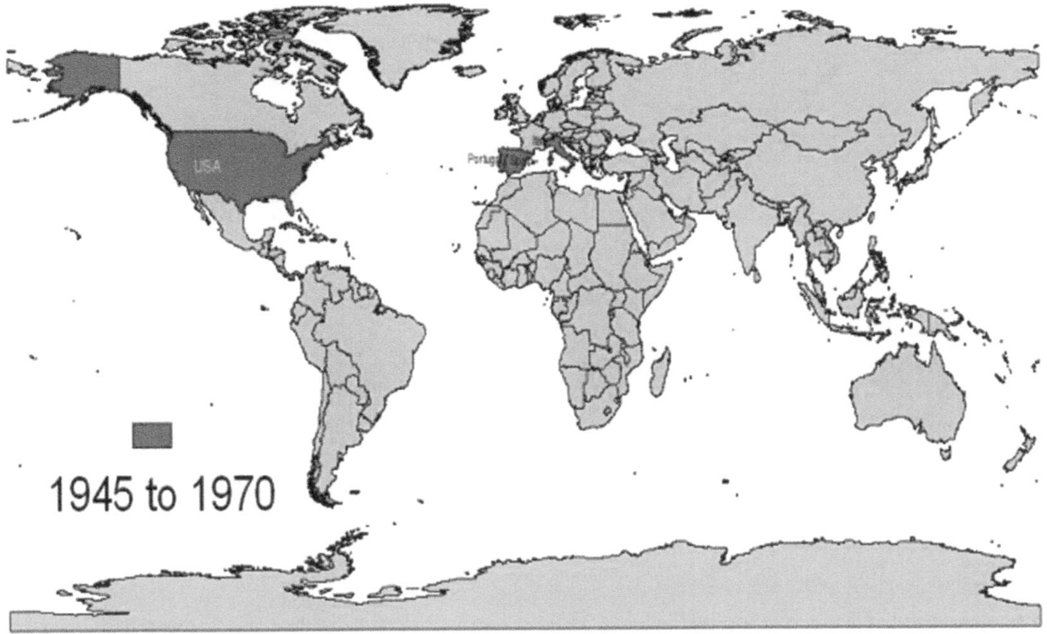

FIGURE 14.1 Ship-breaking countries of the First World (1945–1970).

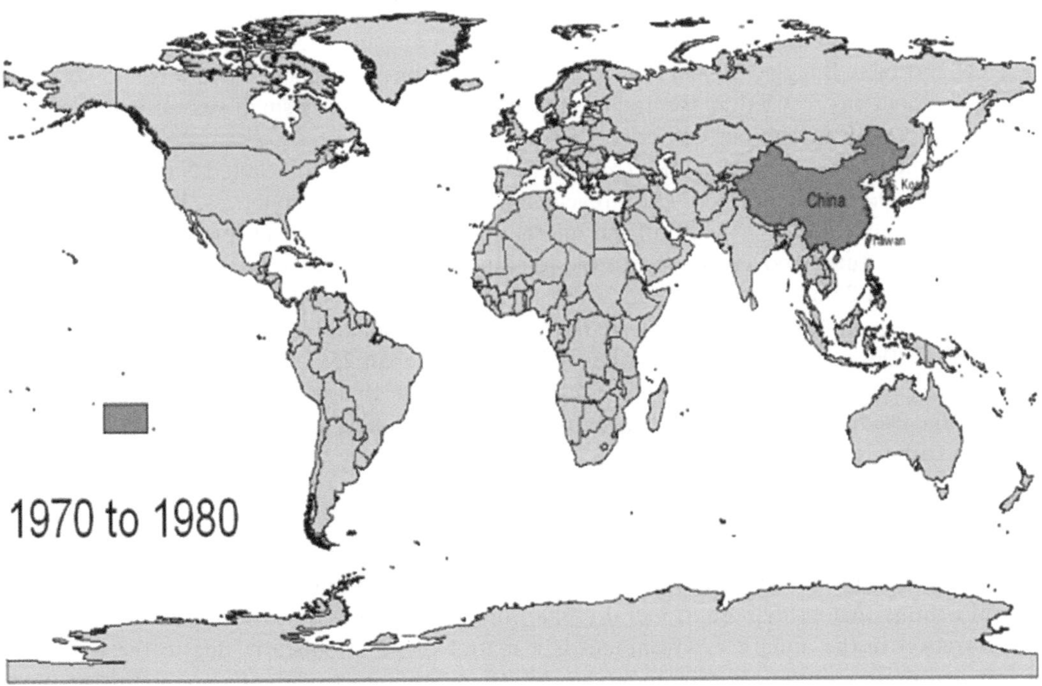

FIGURE 14.2 Migration of ship-breaking to Far-East Asia (1970–1980).

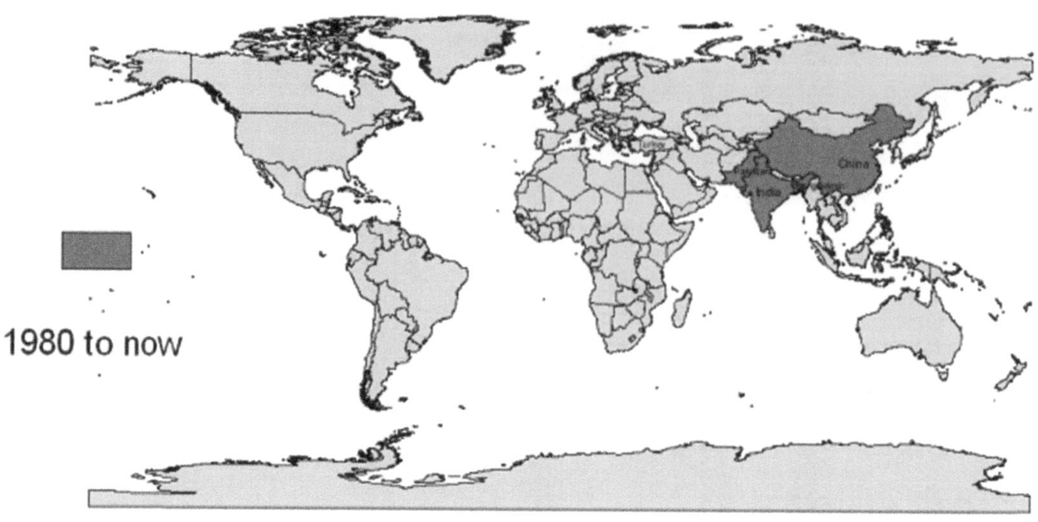

FIGURE 14.3 Ship-breaking countries from 1980 to now.

Before 1994, China was the pioneer in ship-scrapping activities, but they stopped the activity when the government, people, and breakers became aware of the adverse impact on the environment, particularly the coastal environment. India started scrapping activity vigorously since the reduction of activity by China. But in 1997, when a major explosion occurred due to the scrapping of a large oil tanker and caused the deaths of an unknown number of employees, the government did not allow the ship scrapper to scrap the ship without having a gas-free certificate. This regulation led to a rapid decline in tanker demolishing in India. As a result, the tanker scrapping practice in India has shifted to Bangladesh and since then Bangladesh has been dismantling large tankers along its coast without any restriction. But India is presently scrapping only small vessels with a gas-free certificate from the appropriate authority.

A report titled *Review of Maritime Transport 2019* published by the United Nations Conference on Trade and Development (UNCTAD) revealed the data that three countries, Bangladesh, India, and Pakistan, account for 70–80% of the international recycling market for ocean-going vessels with China and Turkey covering most of the remaining market. According to the report, in 2018, India dismantled 25.6% of ocean-going vessels, Pakistan 21.5%, Turkey 2.3%, and China 2%. In 2018, China, Japan, and South Korea were the top countries globally for ship production, representing together 90% of shipbuilding activities (China 40%, Japan 25%, and South Korea 25%). The five recycling countries have a large appetite for scrap metal. Bangladesh, Pakistan, and to a large extent India, use the steel from recycled ships in mills where steel is rerolled so that it can be used directly in urban construction [13–14].

14.2 POLLUTANTS FROM SHIP-BREAKING

The large number of toxic heavy metals is discharged by anthropogenic activities as well as by natural actions that contribute to metal contamination in aquatic environments. Heavy metal concentration in the aquatic environment is a matter of great concern, due to the toxicity of metals and their accumulation in aquatic habitats. A large part of the heavy metal input ultimately accumulates in the estuarine zone and consequently in the continental shelves, since these areas are important sinks for suspended marine and associated land-derived contaminants [4, 7, 15–23].

The metal contaminants in aquatic systems usually remain either in solution or suspension form and finally tend to settle down to the bottom or are taken up by the organisms. The coastal zone receives an outsized amount of metal pollution from agricultural and industrial activity. Heavy metals have contaminated the aquatic environment in the present century due to intense industrialization and urbanization. Elevated chemical concentrations were due to ship-scrapping activities, antifouling sources, and sewage disposal. Several other organic and inorganic contaminants have been linked to ship-breaking activities (Table 14.1) [4, 7, 15–23].

Heavy metal contamination of the environment has been occurring for centuries, but its extent has increased markedly in the last 50 years due to technological developments and increased consumer use of materials containing these metals (Table 14.2).

Soil contamination becomes a priority when the concentration of chemicals within the soil is enough to induce health risks, either by direct exposure or from secondary pollution of life-sustaining resources such as water [7, 14, 24–26]. Contamination of heavy metals within the sediments provides a temporal indication of the aquatic environment condition and acts as a serious reservoir for metals, though some sediment also can act as a source of contaminants.

14.3 SHIP-BREAKING IN BANGLADESH

Bangladesh has emerged as a big player in the global ship-breaking recycling industry thanks to cheap labor, government support, high demand for recovering materials from ships, and the absence of strict environmental regulation [7, 23]. In the period from 2016 to 2018, Bangladesh dismantled the most ships in terms of gross tonnage within the world [27] because the country remains the foremost productive dumping field for end-of-life ships laden with toxins. There is large-scale knowledge of the irreparable damage, due to dirty and dangerous practices around the coastal tidal mudflats, yet the benefit is the only necessity for many shipyard businessmen when selling their ships for breaking. The ship-breaking industry of Bangladesh has captured the global market by dismantling around 47.2% of world vessels.

In 1980, Turkey held the top position in ship breaking while in 1990 it shared the position with China and South Korea. Though India and China held the position for two decades, Bangladesh has grabbed it for the first time. As per the report, scrapping is a segment of the maritime supply chain dominated by developing countries due to several factors, including lower labor costs, a high proportion of utilization of steel from recycled ships for domestic manufacturing, and, at times, weak enforcement of environmental regulations. Moreover, several countries have tightened regulations pertaining to ship demolition. This move is linked to the anticipation of the entry into force of the IMO Hong Kong International Convention for the safe and environmentally sound recycling

---

TABLE 14.1

Pollutant Outbreak from Ship-Breaking Activities

Pollutants	Materials containing pollutants
Asbestos	Hanger liners, mastic under insulation, cloth over insulation, cable, lagging and insulation on pipes and hull, adhesive
Polychlorinated biphenyls	Rubber products such as hoses, plastic foam insulation, cables, silver paint and habitability paint
Lead	Lead and chromate paint, lead ballast, batteries, generators and motor components
Excess noise	With grinding, hammering, metal cutting and other activities
Fire	Ignited insulation, matting, lagging and residual fuel, and from lubricants and other flammable liquids
Others	Heavy metals in ship transducers, ballast and paint coatings; mercury in fluorescent light tubes, thermometers, electrical switches, light fittings and fire detectors

TABLE 14.2

Showing the Effects of Toxic Metals in Marine Biota

Pollutants	Organisms	Effects
Heavy metals	Fish	• At 1 µg-cd/1 earlier hatching occurs • Increase mortality • Reduction of the body's defense system
	Coelenterates	• At 1 µg-cd/1 ctenophores lose growth and survivability • Irregular cell division
	Mollusk	• At 5 µg-cd/1 Crassorstrea virginia gets slightly delayed development • Delayed the maturation system
	Crustaceans	• Increased mortality and delayed development • Effects occur on the shell development • Irregular cell division
	Sea birds	• Mortality increase • Reduction of the body's defense • Retardation of growth • Loss of breeding capacity • Reduction of shell thickness of eggs
	Benthos	• Irregular structure • Acute toxic condition at the bottom • Retardation of growth

TABLE 14.3

Total Ship Scrapping in Bangladesh and Global Contribution for Ship-Breaking

Year	Scrapping Ship	Amount of LDT (MT) handled	Global Contribution (%)
2000	73	892,756	10
2001	152	1,909,055	28
2002	84	1,519,735	22
2003	88	1,088,338	14
2004	145	1,333,667	40
2005	94	840,927	45
2006	187	132,0170	57
2007	103	774,065	42
2008	172	1,660,212	49
2009	175	2,192,751	38
2010	107	1,296,831	35
2011	150	1,896,102	40
2012	230	3,456,380	33
2013	170	2,250,456	30
2014	172	2,110,120	22
2015	195	2,396,280	25
2016	222	6,240,100	33
2017	197	6,456,150	36
2018	185	2,396,077	36
2019	236	7,849,569	45

Source: Present Study.

of ships in 2019, as well as a European Union regulation, which has been in force since December 21, 2018. According to the Bangladesh Ship Breakers and Recyclers Association (BSBRA), 221 ships were brought for demolition in 2015, 250 ships in 2016, 214 ships in 2017, and 221 ships in 2018. From the different literature and documents, it is found that scrapping ships in Bangladesh is increasing sharply compared to other countries globally. In 2000, Bangladesh scrapped a total of 73 ships, weighing 892,756 MT, a global share of 10%, while in 2006 the number of ships scrapped increased to 87, a global share of 57%. In 2019, Bangladesh scrapped 236 ships, weighing 7.8 million MT, a global contribution of 45% (Table 14.3) [5, 7, 10, 14, 28–30].

Ship-breaking activities have generated thousands of direct and indirect employment opportunities for the foremost marginalized people, who are in desperate need of a livelihood in Bangladesh. Moreover, ship-breaking and recycling industries (SBRIs) are a part of the nucleus of an industrial symbiosis involving light engineering and developing industrial sectors, in countries such as Bangladesh [14].

Ship-breaking activities offer direct employment opportunities for about 40,000 people and nearly 200,000 are also engaged in different businesses related to ship-breaking activities in Bangladesh [31–32]. The ship-breaking industry is a major cause of marine pollution in Bangladesh through the contamination of the coastal soil along with seawater which subsequently impairs the ecological setup. Though the ship-breaking industry (SBI) is vital for the national economy of Bangladesh, it has become perilous due to its effect on the environment, adverse effects on biodiversity, and human health impacts. Very few measures have been taken so far in the sustainable development of the adjacent fishing sector, which is the worst victim of the ship-breaking industry. Thus, while the ship-breaking industry has a positive aspect because of its employability and promising economic side, it also poses a serious threat to the fishing sector due to its health hazards and serious environmental pollution risk [7, 28, 33]. The contamination of soil in developing countries (e.g., Bangladesh) poses higher risks thanks to rapid industrialization and therefore the lack of stringent regulations. Countries that partake in the ship-breaking and recycling industry (SBRI) on open beaches are particularly liable for the contamination of various spheres of the environment [2, 7, 15, 24, 34].

Contamination of aquatic systems from heavy metals has been a grave problem worldwide. In regards to the economic importance of the coastal regions and the adverse effects of metal pollution on living resources, this chapter focuses on assessing the changing patterns of the buildup of the nine heavy metals (Fe, Mn, Cr, Ni, Zn, Pb, Cu, Cd, Mg and Hg) within the core sediments over the span of 40 years from 1980 to 2019 and its impact on fish diversity round the ship-breaking zone of Bangladesh.

14.4 PROFILE OF THE SHIP-BREAKING AREA

The area of ship breaking zone of Bangladesh is located in Sitakund Upazila (sub-districts) of the Chittagong district. The eastern side of this sub-district is hilly, and the Bay of Bengal lies on the western side. The zone is situated approximately 20 km southwest of Chittagong city on the coast of the Bay of Bengal in Bangladesh. For the collection of sediments and fish and fishing communities, the whole ship-breaking area was divided into six sampling stations, according to some characteristics such as on the basis of the importance of fishing, proximity to a village, canals or ditches, and density toward the ship breaking zone. Thus total investigated area was separated into (a) the ship-breaking zone (reference sites 1–6) sites in and around ship-breaking activities and (b) reference site 7 which is far away from the ship-breaking area for comparative analysis (Table 14.4). Sandwip, in the east, has been considered as the control site (site 7) because it is diagonally opposite and away from the ship-breaking yards and the water and soil qualities are apparently free from pollutants as revealed from earlier studies.

The geographical location of the ship-scrapping zone is between latitude 22°25′ and 22°28′N, and longitude 91°42′ and 91°45′E (Figure 14.4).

TABLE 14.4
Geographical Location of the Study Areas

Station No.	Station Name	Location
1	Sonaichari	lat. 910 65" E and long. 220 56" N
2	Sitalpur	lat. 910 68" E and long. 220 52" N
3	Madambibirhat	lat. 910 70" E and long. 220 49" N
4	Bhatiari	lat. 910 73" E and long. 220 44" N
5	Salimpur	lat. 910 73" E and long. 220 44" N
6	Kumira	Lat 910 41"E and 220 84" N
7	Sandwip	Lat 91042//E and 22029//N

14.5 METHODOLOGY OF THE STUDY

Soil samples from the inter-tidal zone were collected throughout the year 2019 during the high tide from the seven sampling stations namely Salimpur, Fauzdarhat, Sonaichari, Kumira, Madambibirhat, Sitalpur, and Sandwip Island. All samples were collected during the pre-monsoon season, monsoon season, and post-monsoon season. Collected samples were then air dried, sieved to remove stones and organic parts, and made into a composite for each sampling location by taking an equal amount from harmonized samples for three depths. The standard solution was prepared before every determination of the analysis of the present work. After that 100 g from each soil sample was packed and preserved at room temperature before being transported to BCSIR (Bangladesh Council of Scientific and Industrial Research) laboratories, Chattogram, Bangladesh. Collected sediment samples were analyzed by the atomic absorption spectrophotometric measurement (AAS) (the model is 3300, Thermo Scientific, Designed in the UK, Made in China) using standard analytical procedure (Table 14.5).

The collection method, preservation process, and heavy metals analysis of sediment samples have not changed since 1980. The authors review the findings of different studies related to heavy metals concentration in sediments around the ship-breaking area of the Chittagong coastal area from 1980 to 2015 in the archives of the different departments of the University of Chittagong, Bangladesh. The authors then assess the changing pattern of heavy metals in sediments of ship-breaking and non-ship-breaking zones between the present study finding and values of earlier 40 years (1980–2019) record on selected parameters.

The long-term trends are interpreted as signals of climate change and their possible implications to the fisheries biodiversity are briefly outlined in this first-order analysis. The status of fish diversity for the 25 km adjacent area to the ship-breaking area known as the Chittagong coastal area was collected by the application of the Rapid Participatory Rural Appraisal (RRA/PRA) technique. Participatory appraisal involved a series of qualitative multidisciplinary approaches to learn about local-level conditions and local fishermen's perspectives. A total of 200 questionnaire surveys, personal interviews of 250 fishermen, and 20 semi-structured interviews were conducted in 10 fishing villages, with fish landing being the center of the study area for the purpose of primary data collection.

14.6 RESULTS

14.6.1 Changing Pattern of Heavy Metals in Sediments

Coastal zones are considered as a geographic space of interaction between terrestrial and marine ecosystems, which is of great importance for the survival of a broader variety of animals, plants, and marine diversity species according to the environmental point-of-view [3, 14, 35]. So, coastal

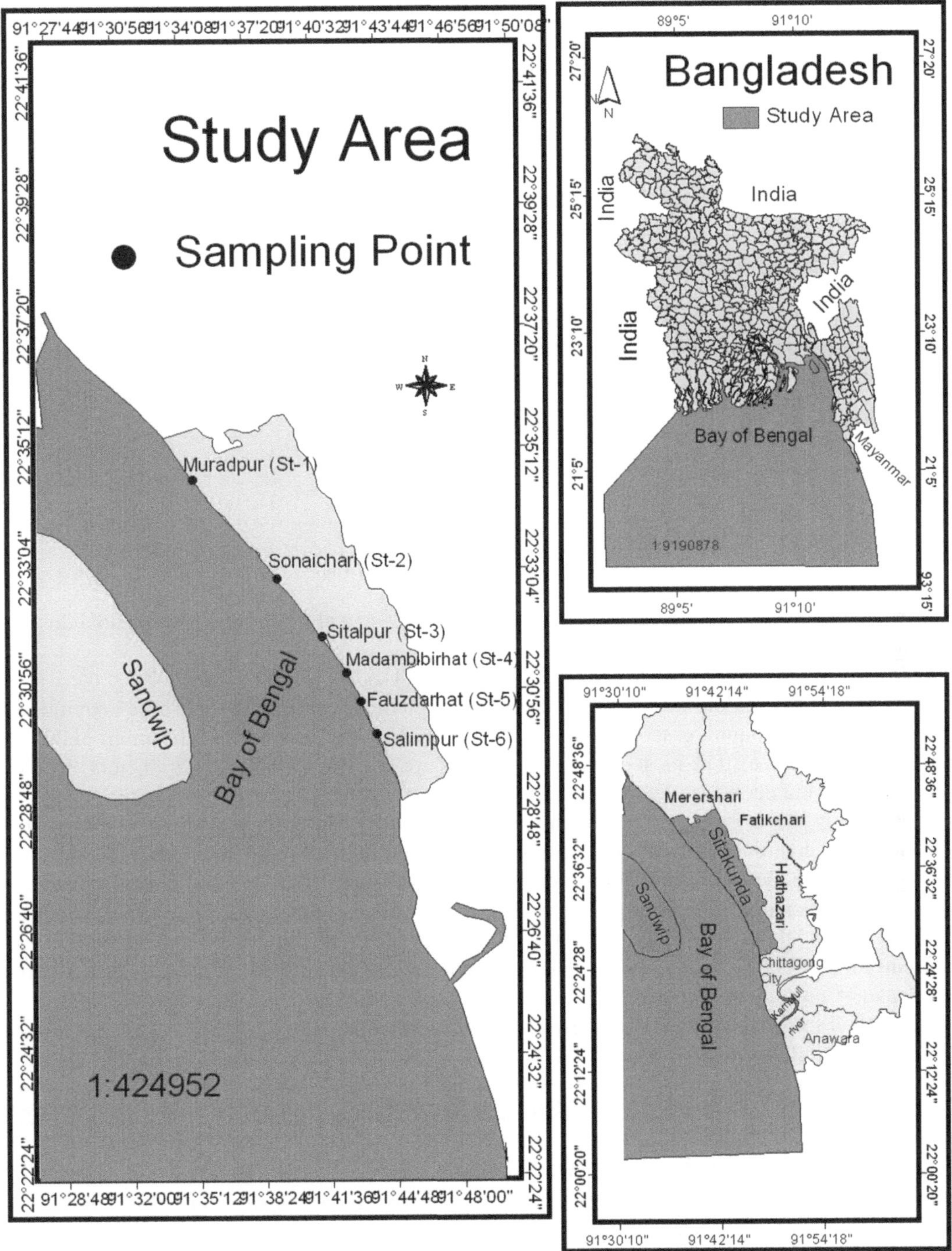

FIGURE 14.4 Map of ship-breaking areas indicated by black spots with Sandwip as the control area.

pollution has been significantly increasing over recent years, and expanding environmental problems have been recorded in many developing countries such as Bangladesh. Different industrial and urban activities in the coastal areas introduce large amounts of trace metals into the coastal environment responsible for permanent disturbances in coastal ecosystems, leading to ecological

TABLE 14.5

Spectral Lines Used in Emission Measurements and the Instrumental Detection Limits for the Elements Measured Using AAS

Elements	Wavelength (nm)	Instrumental detection limit (mg/l)
Fe	248.3	0.0043
Pb	217.0	0.013
Cr	357.9	0.0054
Co	240.7	0.01
Cd	228.8	0.0028
Mn	279.5	0.0016
Ni	232.0	0.008
Zn	213.9	0.0033
Cu	324.8	0.0045
Al	309.3	0.028

and environmental damage and constituting a potential risk to a number of flora and faunal diversity such as humans, through food chains [35–36].

Operational management of the ship-breaking industry in Bangladesh is so poor that these activities produce huge environmental pollution in the coastal areas of Bangladesh. These activities discharge a number of liquid, gaseous, and solid pollutants that are hazardous to the environment and human beings [23, 26]. The most common pollutants are oil, physiochemical parameters, asbestos, heavy metals, and persistent organic pollutants.

Iron (Fe) is the most important transition metal in all living organisms and the most widely used metal, accounting for over 90% of worldwide metal production. It is common that primary sources of Fe are the land, runoff, waste disposal, and industrial releases into the environment [7]. Fe that comes from ship breaking settles down the mud of shallow waters. There it consumes dissolved water and thus depletes the oxygen of the water, which is so important to all aquatic life. The concentration of Fe was found in the highest levels in the Madambibirhat and Sitalpur areas which are the most prominent ship-breaking areas while concentrations were lowest in the control site Sandwip over the 40 years. Quantitatively, the Fe concentrations rose by 3.82 to 6.55 times over the 40 years in the ship-breaking zone and 1.08 to 3.17 times higher levels than the standard range of Fe concentration 27000 µg.g^{-1} [37]. On the other hand, the level of Fe recorded in the control area of Sandwip Island rose by 3.20 times from 1980 to 2019 but the concentration of Fe was found as 4,200 µg.g^{-1} which was 0.142 times lower than the standard concentration (Table 14.6).

The authors found that Manganese (Mn) concentration rose 2.10 to 4.56 times in the ship-breaking area; in the control area, this rate of increase was found to be 2.87 between 1980–2019. On the other hand, during the present study, the Mn level was found to be 2.77 times to 8.63 times higher than the standard range of Mn concentration 1.17 µg.g^{-1} in the ship breaking area which was proposed by IAEA [37]. The authors assess the increasing trend of Mn level in the control area to 1.96 times higher than the standard range of Mn levels in sediment (Table 14.6). During the result of the present study, the value of nitrogen (Ni) increased by 4.54 to 9.17 times in the affected sites, compared to an 8.8 times increase at Sandwip over the 40 years (Table 14.6). Comparatively Ni level in the study area of the ship-breaking zone was found at a higher range of 1.05 to 1.37 times than the standard level [37]. On the other hand, the Ni level was found to be 0.19 times lower in the control area than the IAEA proposed standard level of Ni in sediment samples.

TABLE 14.6

Concentration of Heavy Metals in the Ship-Breaking Zone and Control Area over the 40 years

Station	1980	1985	1990	1995	2000	2004	2007	2010	2015	2019	Standard concentration of selected metal in Sediments (IAEA, 1990)	Times of Increases over the 40 years	Comparative result than standard level
Concentration of Fe(µg/g)													
Station-1	12,456	15,854	18,850	21,500	22,552	30,450	41361	45565.8	52,456.9	57,560		4.21	+1.94
Station-2	20,520	28,450	35,450	48,040	50,657	55,870	64,895	70,900	85,556.6	94,590		4.17	+3.17
Station-3	18,980	26,408	31,870	45,350	49,450	54,875	60,580	64,560	72,654	80,760	27,000	3.82	+2.69
Station-4	10,390	12765	15800	18,210	20,445	25,654	35,216	37,654	44,345	50,879		4.27	+1.64
Station-5	4,456	7,980	10,650	12,120	15,450	18,650	22,932	25,564	29,213	35,890		6.55	+1.08
Station-6	6,450	8,390	10,654	13,500	15,800	19,800	20,971	22,450	35,653	40,240		5.52	+1.32
Station-7	1,200	1,487	2,010	2587	2,765	3,265	3,393	3,545	3,845	4,200		3.20	-0.142
Concentration of Mn(µg/g)													
Station-1	2.40	2.80	3.40	3.90	4.65	6.05	6.89	7.90	8.25	8.90		3.43	7. 05
Station-2	3.80	4.20	4.90	5.90	6.50	7.10	8.25	9.50	9.89	10.25		2.6	8.45
Station-3	4.80	5.90	6.00	7.09	8.16	9.00	9.80	10.02	10.10	10.80		2.108	8.63
Station-4	2.00	2.90	3.80	4.90	5.40	7.00	8.25	8.50	9.12	11.23	1.17	4.56	7.79
Station-5	1.00	1.20	1.75	1.90	2.10	2.50	2.64	3.50	3.95	4.30		3.95	3.37
Station-6	0.90	1.10	1.01	1.50	1.90	2.00	2.32	2.90	3.25	4.26		3.61	2.77
Station-7	0.8	0.9	1.0	1.20	1.30	1.50	1.8	2.0	2.3	2.8		2.87	1.96
Concentration of Ni (µg/g)													
Station-1	12.9	15.9	20.7	35.8	40.9	45.8	48.96	55.90	60.65	68.50		5.31	1.22
Station-2	14.8	20.8	28.9	35.8	48.8	52.8	55.8	60.8	65.60	70.45		4.76	1.26
Station-3	16.9	22.9	34.7	50.8	55.5	58.5	60.9	64.5	68.32	76.80		4.54	1.37
Station-4	8.5	14.9	21.8	24.8	30.6	32.8	35.12	58.5	62.43	67.45	56.1	7.93	1.20
Station-5	6.8	10.9	15.9	20.8	28.1	35.7	45.12	50.7	55.98	60.55		8.90	1.08
Station-6	6.4	11.6	14.8	16.8	22.7	28.6	35.36	40.65	48.45	58.69		9.17	1.05
Station-7	1.2	1.7	1.9	2.1	2.5	3.2	3.98	5.8	8.5	10.56		8.8	-0.19
Concentration of Cu (µg/g)													
Station-1	8.66	10.9	15.9	20.71	25.33	28	30.67	36.77	40.90	50.90		5.88	1.54
Station-2	12.55	15	20	31.9	38.9	42.7	46.9	51.55	59.56	66.70		5.31	2.02
Station-3	16.8	20.77	28.9	38.07	45.75	50	52.8	73.50	80.65	90.45		5.38	2.74
Station-4	12.66	15.19	22	28.9	30.55	35.1	39.85	45.22	50.55	58.90	33.0	4.65	1.78
Station-5	7.8	10	13.76	15.7	17.8	20.9	21.05	35.66	42.45	50.45		6.47	1.53
Station-6	9.15	13	16.87	19.08	22.76	27.65	28.01	34.88	40.34	49.45		5.40	1.50
Station-7	0.8	0.9	1.2	1.5	1.8	1.9	2.05	2.5	3.0	4.0		5	-0.12

(Continued)

TABLE 14.6 CONTINUED

Concentration of Heavy Metals in the Ship-Breaking Zone and Control Area over the 40 years

Station	1980	1985	1990	1995	2000	2004	2007	2010	2015	2019	Standard concentration of selected metal in Sediments (IAEA, 1990)	Times of Increases over the 40 years	Comparative result than standard level
Concentration of Cr (μg/g)													
Station-1	19.90	33.80	39.80	54.80	70.70	75.90	78.36	85.70	92.33	96.70		4.86	1.25
Station-2	22.00	34.90	40.60	56.50	72.90	92.80	90.50	96.60	99.30	100.15		4.55	1.29
Station-3	22.80	35.70	40.80	55.00	70.80	88.80	95.60	98.70	102.35	109.40		4.80	1.42
Station-4	20.90	30.80	35.70	50.80	65.90	79.00	86.72	90.50	98.43	105.40	77.2	5.04	1.36
Station-5	18.90	25.90	30.60	45.80	50.40	60.60	68.35	70.90	79.00	85.40		4.52	1.11
Station-6	15.34	20.80	28.00	35.90	50.80	60.80	69.89	75.50	85.34	90.40		5.90	1.17
Station-7	4.75	5.00	5.90	7.80	10.70	15.80	19.00	22.40	23.45	28.45		5.99	-0.37
Concentration of Pb (μg/g)													
Station-1	45.7	66	89.8	100.8	130.9	140.5	147.83	152.9	160.44	165.34		3.62	7.25
Station-2	58.8	75.9	98.79	120.8	142.9	150.8	160.4	168.60	172.33	175.89		2.99	7.71
Station-3	65.5	84.9	110.85	135.9	150.8	160.8	175.65	185	190.44	196.90		3.00	8.63
Station-4	40.66	48.9	58.5	78.76	92.89	100.9	122.03	135.9	145.50	150.45	22.8	3.70	6.60
Station-5	20.2	22.9	25	28.8	30.5	35.8	36.78	50.50	60.66	65.45		3.24	2.87
Station-6	22.5	25.6	29.8	32.7	35.9	40.5	41.57	59.45	65.90	70.23		3.12	3.08
Station-7	5.8	6	6.2	7	7.3	7.5	8.82	12.54	15.44	16.45		2.84	-0.72
Concentration of Cd (μg/g)													
Station-1	0.19	0.25	0.39	0.59	0.6	0.8	0.94	1	1.45	1.70		8.95	8.94
Station-2	0.4	0.5	0.61	0.79	0.8	1	1.2	1.5	1.9	2.2		5.5	11.58
Station-3	0.4	0.6	0.7	0.85	0.9	1.1	1.5	1.6	2.0	2.8		7.0	14.74
Station-4	0.15	0.25	0.39	0.50	0.75	0.79	0.83	0.9	1.2	1.7	0.19	11.33	8.94
Station-5	0.20	0.25	0.30	0.38	0.40	0.48	0.57	0.60	1.20	1.50		7.5	7.90
Station-6	0.10	0.25	0.30	0.38	0.45	0.50	0.59	0.65	0.7	1.2		12.0	6.31
Station-7	0.02	0.03	0.05	0.07	0.09	0.1	0.13	0.16	0.17	0.18		9.0	-0.94
Concentration of Zn (μg/g)													
Station-1	38.75	56.9	80	95.65	110.6	128.8	142.85	150.2	158.90	165.20		4.26	1.74
Station-2	42.89	65.5	84.9	100.3	120.65	130.3	148.9	162	170.56	175.67		4.09	1.85
Station-3	50.7	75.4	90.8	110.2	128.7	148.8	165.5	185.4	195.90	199.34		3.93	2.10
Station-4	35.8	45.7	58.8	75.8	85.8	100.8	102.05	115.5	125.90	135.86	95	3.80	1.43
Station-5	25.79	30.2	50.7	69.8	75.8	80.65	83.78	99.8	110.50	117.54		4.56	1.24
Station-6	32.8	39.9	59.7	78.9	85.77	115.8	119.86	125.8	130.89	138.20		4.21	1.45
Station-7	7.1	10.2	15.3	18.1	19.9	21.8	22.22	24.8	26.90	29.90		4.21	-0.31

Copper (Cu) is intimately related to the aerobic degradation of organic matter in sediments. The main sources of Cu in the coastal waters are antifouling paints and this metal entered into the water body through industrial effluents containing $CuSO_4$ used in metal plating and fishing operations [7, 37]. IAEA sets up the standard range of Cu level in sediment samples as 33 $\mu g.g^{-1}$. The authors found that Cu level in the study areas during the present analysis time were 49.45 to 90 $\mu g.g^{-1}$. 45 $\mu g.g^{-1}$ which provides that the Cu level increased 4.65 to 6.7 times higher compared to the result of sediment analysis in the ship-breaking area during 1980 to 2019. All the ranges of Cu level increased 1.50 to 2.74 times higher compare to the standard level of 33.00 $\mu g.g^{-1}$ (Table 14.6). Although Cu level in sediments of control area increased five times over the years, the concentration of Cu is still below the standard range of the IAEA [37]. Despite the existence of a number of detoxifying and storage systems for Cu, it is the most toxic metal after mercury and silver, to a wide spectrum of marine life, hence its value in antifouling preparations [8, 12, 23, 38].

Chromium (Cr) distribution in the ship-breaking area was very complex in nature. The concentration of Cr did not follow any regular pattern of distribution. The concentration of Cr varied from 85.40 to 109.40 $\mu g.g^{-1}$ in the ship-breaking zones, while it was 28.45 $\mu g.g^{-1}$ in the control site. But the recommended value of Cr is 77 $\mu g.g^{-1}$ [37]. It was also found that Cr concentration rose by 4.52 to 5.90 times over the last 40 years and the rate of increase compared to the standard level was 1.11–1.42 times but less range in the control area (Table 14.6).

The authors found the lead (Pb) concentration in the sediment sample of the ship-breaking zone rose from 65.45 $\mu g.g^{-1}$ to 196.90 $\mu g.g^{-1}$ and for Sandwip Island this range was 16.45 $\mu g.g^{-1}$ proved that the Pb concentration of affected areas was found to be 2.99 to 3.70 times higher range than recommendation level 22.8 $\mu g.g^{-1}$. Although the range of this trace metal rose 2.84 times over the 40 years where the level of this metal was found lower than the standard range in the control area (Table 14.6).

Cadmium (Cd) is one of the most toxic metals measured in this study [39]. The present level of Cd concentration in the study area becomes much higher than the recommended level of trace metals in sediment and increased 7.5 to 12 times within 40 years in the ship-breaking zone (Table 14.6). On the other hand, the Cd level increased by 9 times in the control zone but was still found in a lower range than the standard level of Cd concentration. The higher concentration could be attributed to the combined effects of oil and oil spillage, petroleum hydrocarbons from ships, tankers, mechanized boats, etc.

The concentration of Zinc (Zn) was found higher range in all the study areas adjacent to the ship-breaking area compared to the IAEA levels [37]. Zn concentration in the sediment samples increased from 3.80 to 4.56 times higher over the 40 years in ship-breaking area and this result indicated a higher range compared to the standard level (Table 14.6). it was also found that Zn concentration in the control area increased 4.21 times over the study period but the range of this metal was found at a lower level than the standard value of Zn in sediment.

14.6.2 Impact of Heavy Metals Toxicity on Fish Diversity

The coastal environment of Bangladesh is highly dynamic and ecologically assorted with critical terrestrial and aquatic surroundings such as mangrove forests, tidal flats, and wetlands. These enriched ecosystems presently exist at risk because of environmental impacts and human interruption. The wastes from ship-breaking, including oils and persistent organic pollutants (POPs), come into the Bay of Bengal. This, the world's longest bay, is home to a range of marine biodiversity, including lots of vulnerable and endangered species, and pollution has been recognized as one of three major transboundary problems influencing its marine ecosystem.

It is found that fish species from the inland and marine water bodies of Bangladesh declined gradually over the last two decades and the catch of fish declined about 40% compared to the past 20 years. Research indicated that the major responsibilities for fish decline from the river are the increased fishing pressure and habitat destruction due to coastal pollution [7, 22, 25, 32, 40–41].

TABLE 14.7

ESBN Catch Composition of Commercial Fish Species (%) for the last 15 years. (2005–2010) and (2011–2019) at the Coastal Area of Chittagong

Sl. No.	Scientific Name of the species	English name	Local name	Year of Harvesting (Average)		
				2005–2010	201–2015	2016-2019
1	Harpodon nehereus	Bombay duck	Loitya mach	40.56	35.40	32.30
2	Acetes sp.	Sergastid shrimp	Gura chingri	15.560	10.20	10.09
3	Arius sp.	Carfish	Kata mach	2.04	1.50	1.40
4	Coilia sp.	Anchoby	Alua mach	1.832	1.34	1.50
5	Metapenaeus Monoceros	Speckled/ Brawn shrimp	Harina chingri	3.282	2.90	2.40
6	Trypauchen vagina		Lal chewa	0.85	0.50	0.20
7	Penaeus sp.	Shrimp	Chingri (Bagda, Dorakata, White)	3.354	1.859	1.390
8	Lepturacanthus savala	Ribbon fish	Churi mach	3.50	2.80	2.00
9	Johnius sp.	Black jew fish	Kala poa	2.392	2.110	1.990
10	Thryssa sp.	Hairpin anchoby Hamilton's	Phasa/Phaisya	1.234	1.150	1.09
11	Polynemus paradiseus	Long thread tassel fish	Risshsha Topshe	0.245	0.14	0.12
12	Sillago domina	Lady fish	Hundra mach	0.324	0.250	0.190
13	Tenualsola ilisha	Hilsha shad	Ilish/ Jati ilish	0.232	0.10	0.20
14	Mystus gulio	Catfish	Guilda	0.124	0.12	0.09
15	Pampus sp.	Pomfret	Rupchanda	0.016	0.010	0.009
16	Lates calcarifer	Giant seaperch/ Seabass	Koral/Bhetki	1.450	0.96	0.25
17	Mugil sp.	Mullet	Bata	2.709	2.00	1.80
18	Scomberomorus guttatus	Spanish mackerel	Maitya	1.790	1.2	1.10
819	Alepes djeddaba	Djeddaba crevalle	Moori	0.450	0.230	0.160
20	Argyrops spinier	Longspine Sea Bream	Lal Datina	1.890	1.50	1.10

Source: Marine fisheries dept. Chittagong, December, 2019.

Different studies show that metal concentrations in estuarine water are relatively higher due to the rapid acceleration of the industrial sector. Metal concentrations are higher in fish than in water and sediment. Elevated levels of trace metals are highly detrimental to fish and human mechanisms, as shown by different studies. Oil pollution is responsible for environmental deterioration due to its adverse effects on estuarine biota, fish and shellfish, phytoplankton, and zooplankton. Industrialization is needed for the development of the country. However, it should be eco-friendly for effective and sustainable development and for the protection of the environment (aquatic). The toxic chemical pollutants such as mercury (Hg), Pb, Cd, cyclic olefin copolymer (COC), and dissolved oxygen (DO) were found higher than the environmental quality standard (EQS) value which is dangerous for the entire aquatic ecosystem and public health [7, 42–43]. Excessive discharge of nitrogen, sulfur, and phosphorus compounds in the water system can cause eutrophication [7, 17, 20, 44].

Aquatic pollution, soil erosion, siltation, decline of wetlands, and biodiversity, as well as correct management issues, are feasible causes for the reduction of fishery resources [20, 45]. The fishery resources in the coastal area of Chattogram seem to be affected by the ship-breaking activities as revealed by increased fishing efforts, reduced catch per unit effort (CPUE) per boat/day, reduced species diversity, and increased amount of trash fish destruction and collection of fish fry. There are 10 fishery landing centers in the Chattogram coastal area and the Department of Marine Fisheries Survey of Chattogram records data every month. From the study of 15 years of data (2005–2019) it is found that only a few of the 20 commercial species are high in catch numbers (*Harpadon nehereus, Acetes* sp. *Arius* sp., and *Coilia sp.*) dominated by *Harpodon nehereus* (35.40%), *Acetes* sp. (10.56%), and the rest are negligible (Table 14.7).

Notably, in comparison to the past 10 years of data [40], species diversity in Set Bag Net (SBN) catch is reduced significantly, and some species (30 species) which are not found in recent years in the coastal area of Chittagong according to the statement of fishermen and market survey. The

TABLE 14.8

Fish Species, Which Are Not Found in SBN Catches at Sitakund Coastal Area, Chittagong (Adjacent to Ship-Breaking Area) Compared to the Availability 30 years Ago

Sl. No.	Scientific name of fish species	Local name
1	Osteogenous staenocephalus	Aspisoa katamach
2	Scolopsis vosmeri	Nemipscol mach
3	Eleotris fusca	Dora bailla
4	Uranoscopus guttatus	Faton mach
5	Dendrophysa russelli	Kala poa
6	Bahaba chaptis	Chapti mach
7	Pomadasys opercularis	Grunti mach
8	Polynemus sextarius	Kala tailla
9	Gobuis sadanandio	Nandi bailla
10	Gobuis melanosome	Kalthu bailla
11	Sphyraena fosteri	Khika mach
12	Sphyraena obtusata	Khika mach
13	Carangoides melampygus	Bungda muri
14	Saurida elongate	Tiktiki mach
15	Pricanthus macracavthus	Prica mach
16	Pricanthus tayenus	Prica mach
17	Cynoglossus macrolepidotus	Lamba kukurjib
18	Epinephelus lanceolatus	Bole coral
19	Otolithoides brunneus	Lombu fish
20	Cybium guttatuam	Maitta
21	Coilia ramkorati	Olua
22	Sphyraena forsteri	Dharkuta
23	Escualosa thoracata	Hichiri Machh
24	Scomberooides commersonianus	Chapa Kori
25	Priacanthus tayenus	Pari Machh
26	Polynemus paradiseus	Hriska mach
27	Carangoides malabaricus	Lahmuri mach
28	Anodontostoma chacunda	Koiputi mach
29	Arius thalassinus	Kata mach
30	Apocryptes serperaster	Dora chau mach

present study also analyzed the catch per unit effort (CPUE) of fish per boat/day from 15 years of data (2005–2019) for both SBN (Set Bag Net) and ESBN (Estuarine Set Bag Net) catch. The results show a significant decrease in CPUE per boat/day with many fold increase of boats from 2005 to 2019 (Table 14.9).

So, the decrease in catch and availability of fish is not only due to pollution, or destruction of post larvae (PL) fish but also other factors are responsible such as the increase of the number of boats and catching, demands for fish, etc. During the present survey pre-set questionnaire survey of 250 fishermen from different categories was done by visiting 10 fishing villages several times. The horizontal expansion of the ship-breaking yards has posed a threat not only to the diversified coastal resources but also to the adjacent coastal inhabitants, especially the fishing communities. In the course of this research work, some important information has been obtained. The studied fishing communities were found to have a lack of access to primary education and health facilities, which are basic needs. Communication and drinking water supply were observed to be satisfactory, however, the sanitation status was found to be very poor. A participatory rural appraisal (PRA) study showed that about 90% of them were local and full-time fishermen and 10% were migratory of different districts including Bhola, Barisal, Mymensingh, and others. The fisher folk are dissatisfied with different NGO activities working in this area. No government aid was found to be available for the welfare of fishing communities.

It was found that about 80% of the fishermen had either nets or boats or both of them. They used both mechanized and non-mechanized boats and some traditional fishing crafts (called Dinghi) for fishing. Among the fishing nets Sets Beg Net (Behundi Jal) and Gill Net (Ilisha Jal) were found to be widely used. Though the gear is available for fishing, they catch a very small amount with every effort. They equally reported that the fish catch had reduced by more than half from the previous time. The fishermen also reported that while they catch fish at sea they face piracy. They also face the musclemen, middlemen, and swindlers when they return with fewer amounts of fish. These criminals snatch away the fish forcefully. The fishermen are exploited by the dealers through a practice called "Dadon" (earnest money which the fisherman needs to pay back at an exorbitant price with interest). It is made obligatory that the price of fish is determined by the middleman. They are

TABLE 14.9

Last 15 CPUE Rates of Mechanized Boats in Chattogram Coastal Area

Sl. No.	Year	Landing center	Observed boats	Amount of fish (mt)	Yield day/boat (kg)
1	2005	14	1030	6142.56	139.23
2	2006	12	1151	5010.42	88.08
3	2007	12	1230	4980.80	82.26
4	2008	10	1350	4100.90	65.31
5	2009	10	1450	3950.90	59.76
6	2010	10	1600	3700.30	55.68
7	2011	10	1500	3500.00	53.56
8	2012	10	1500	3200.89	52.50
9	2013	10	1420	3000.45	50.45
10	2014	10	1400	2800.45	48.59
11	2015	10	1380	2600.45	45.00
12	2016	10	1350	2500.56	43.45
13	2017	10	1320	2460.45	41.90
14	2018	10	1200	2270.89	40.90
15	2019	10	1050	2005.20	35.90

forced to sell fish to the lenders at a very nominal or very cheap price. The middlemen indulge in maintaining miscreants to exercise their authority over them [7, 10, 46].

From this study, it was observed that about 75% of fishermen believed ship-breaking activity was creating problems for them, the rest did not respond. Also, 90% of the villagers of the vicinity were anxious about their existence in the future. They complained about suffering from respiratory difficulties, high-decibel sounds, fumes, toxic chemicals, and skin diseases. It is clear from the sociological survey that coastal fish species diversity has reduced due to ship-breaking activities. The fishermen's hereditary profession is now in a vulnerable position due to the ship-breaking activity meaning they have less access to credit, etc. But the most interesting thing is that they are not conscious of their rights and deprivation. The increasing ship-breaking activity is depleting the fishery resources which simultaneously decreases catch per unit effort (CPUE). So this trend is provoking the fisher folk to change their livelihoods as they never feel comfortable and safe (Table 14.9).

Adverse effects on fish, both in natural and under laboratory conditions, were reported due to heavy metal toxicity that may include reduced fertility, problematic reproduction, hatching delay, damaged kidneys, slower growth and development, organ deformities, abnormal behavior, and even death. The most common deformities can be seen in the vertebral column, swim bladder, cephalic region, fins, and lateral line. Moreover, fish are expected to have damaged gills, gut, and sensory systems under elevated levels of toxic metals [7, 10, 29]. The impact of toxicity of heavy metals can influence the individual growth rates, physiological functions, reproduction, and mortality of fish diversity. The metals reach fish bodies through three possible ways: gills, digestive tract, and body surface [7, 10].

The influence of the release of waste materials from ship-breaking activities on aquatic plants, mangroves, salt marshes, fishes, and benthic invertebrates has been documented [7, 10, 37, 46–48]. Deteriorations of stocking density by 50–100% have been recorded for Bombay ducks (*Harpadon nehereus*), Hilsa (*Tenualosa ilisha*), prawns, and other fish species in water bodies near ship-breaking zones [13, 47]. Heavy metal pollutants were found at significantly higher than the maximum permissible level in suspended particulate matter, marine sediments, fishes, and other seafood, from locations near the ship-breaking zone of Bangladesh [3, 13, 23, 29, 32, 46, 49].

14.7 CONCLUSIONS

The ship-breaking activities contaminate the coastal soil and the seawater environment and thus change their ecological settings. The problem is mainly associated with the discharge of ammonia, burned oil spillage, floatable grease balls metal rust (iron), and various other disposable refuse materials together with high turbidity of seawater. Ship-breaking activity has become of paramount importance within the economic context of poverty-stricken Bangladesh. This book chapter demonstrated that there has been an increment of two to eight times of selected heavy metals from the findings of 1980 to 2019. The pattern of heavy metals concentration in the ship-breaking area followed a pattern of Fe>Pb>Cr>Mn>Zn>Ni>Cu>Cd>Hg. The authors also found that 30 species of marine fishes became irregular and threatened with extinction than they were highly available 40 years ago. The authors recommended that if this ship-breaking sector takes some eco-friendly approaches in compliance with the principles of the blue economy and overcomes challenges, this may be an enormous and sustainable industry in the future. The pollution of aquatic ecosystems due to heavy metals is rising at an alarming rate in Bangladesh. Ship-breaking activities are considered a profitable industry with significant contributions to respective national economies but with a hefty environmental cost due to the contamination of soil and water. The ship-breaking activities contaminate the coastal soil and seawater environment and thus change their ecological settings. As a result, the growth and abundance of fish may seriously be affected. Ship-breaking has become a major contributor to the development of Bangladesh's national economy. Considering the positive role of ship-breaking in the national economy, ship-breaking practices can not be stopped. Rather a sustainable approach should be taken to minimize the negative consequences of ship-breaking activities in our coastal zone. However, the following steps may be taken for the sustainable practice of ship-breaking activities in the Chittagong coastal area:

- The government should formulate and implement a national policy and principles for safe and sustainable ship-breaking after consultation with relevant organizations, employers, and workers.
- A gas-free certificate (in the true sense) must be obtained before any ship is broken. Oil must be removed and the oil tanks must be thoroughly cleaned either chemically or manually and the ship breakers must obtain a tank clearance certificate from the Mercantile Marine Department before beaching.
- Vessels must pump out the maximum possible quantity of oil at the anchorage before beaching. All the oily sludge, rags, rust, sawdust, etc. must be removed and disposed of at a safer place.
- Vessels causing marine pollution by spill over-flow or dumping of oil or oily sludge, etc. will be liable to be prosecuted under Bangladesh Marine Pollution Ordinance.
- A systematic and periodic inspection of the whole yard should be done before a certificate of compliance is issued by the Department of Environment (DoE) and Department of Shipping to control pollution during ship-breaking.
- Waste reception facilities with safe management for hazardous materials are to be established under the control of a competent authority as an independent body (experts from all the relevant organizations).
- The sea shall be kept undisturbed as far as practicable for the healthy growth of marine biodiversity and human health. Because, many of the ship-breaking components are highly toxic, persistent, and carcinogenic in nature and they prove fatal for the aquatic food chain and human health.

REFERENCES

1. Ashraful, M. A. K. (2003). Trace metals in littoral sediments from the North east coast of the Bay of Bengal along the ship breaking area, Chittagong, Bangladesh. *Journal of Biological Science*, 10(3), 1050–1057.
2. Abdullah, H. M., Mahboob, M. G., and Biruni, A. A. (2010). Drastic expansion of ship breaking yard in Bangladesh: A cancerous tumor to the coastal environment. *Regional Environmental Change*, 80(5), 150–175.
3. Hasan, A. H., Selim, H. H., Reza, S. H., Kabir, M. H., and Siddique, A. M. (2020). Accumulation and distribution of heavy metals in soil and food crops around the ship breaking area in southern Bangladesh and associated health risk assessment. *Applied Sciences*, 20(3), 45–65.
4. Neşer, G. H., Kontas, A. O., Ünsalan, D. H., Uluturhan, E. L., and Altay, O. L. (2015). Heavy metals contamination levels at the Coast of Aliaga (Turkey) ship recycling zone. *Marine Pollution Bulletin*, 64(5), 882–887.
5. Rahman, M. S., Saha, N. Y., and Molla, A. H. (2014). Potential ecological risk assessment of heavy metal contamination in sediment and water body around Dhaka export processing zone, Bangladesh. *Environmental Earth Science*, 71(5), 2293–2308.
6. Bhattacharjee, S. M. (2009). From Hong Kong to Basel: International environmental regulation of ship-recycling takes one step forward and two steps back. *Trading and Law Development*, 1(1), 193–230.
7. Barua, P., Rahman, S. H., and Molla, M. (2017). Heavy metals influence in sediments and its impact on Macrobenthos at ship breaking area of Bangladesh. *Asian Profile*, 45(2), 50–65.
8. Rahman, M. M., Asaduzzaman, M. L., and Naidu, R. O. (2013). Consumption of arsenic and other elements from vegetables and drinking water from an arsenic-contaminated area of Bangladesh. *Journal of Hazard Materials*, 26(3), 1056–1063.
9. Sujauddin, M. H., Koide, R. H., Komatsu, T. O., Hossain, M. M., Tokoro, C. T., and Murakami, S. H. (2015). Characterization of ship breaking industry in Bangladesh. *Journal of Material Cycle and Waste*, 17(3), 72–83.

10. Hossain, M. M., and Islam, M. M. (2007). Ship breaking activities and its impact on the coastal zone of Chittagong, Bangladesh: Towards sustainable management. Chittagong, Bangladesh. *Coastal Management*, 34(3), 45–65.
11. US EPA. (2007). *Framework for Metals Risk Assessment.* U.S. Environmental Protection Agency, Washington.
12. Luoma, S. N. (1990). Processes affecting metal concentrations in estuarine and coastal marine sediments. *Coastal Pollution*, 30(3), 45–65.
13. Bhuyan, M. S., Bakar, M. A., Akhtar, S. H., Hossain, M. B., and Ali, M. M. (2018). Heavy metal contamination in surface water and sediment of the Meghna River, Bangladesh. *Environmental Monitoring and Management*, 8(2), 273–279.
14. Das, J., and Shahin, M. A. (2020). Ship breaking and its future in Bangladesh. *Journal of Ocean and Coastal Economics*, 6(2), 60–80.
15. Abdullah, H. M., Mahboob, M. G., Banu, M. R., and Seker, D. G. (2012). Monitoring the drastic growth of ship breaking yards in Sitakunda: A threat to the coastal environment of Bangladesh. *Environmental Monitoring and Assessment*, 1855, 3839–3851.
16. Alam, S. M., and Faruque, A. B. (2014). Legal regulation of the ship breaking industry in Bangladesh: The international regulatory framework and domestic implementation challenges. *Marine Pollution*, 47(3), 46–56.
17. Barua, P., Mitra, A., Banerjee, K., and Chowdhury, M. S. N. (2011). Seasonal variation of heavy metal concentration in water and oyster (*Saccostrea cucullata*) inhabiting central and western sector of Indian Sundarbans. *Environmental Research Journal*, 5(3), 121–130.
18. Hossain, M. S., Khan, Y. S. A., Chowdhury, S. R., Saifullah, S. M., Kashem, M. B., and Jabbar, S. M. A. (2004). Environment and socio-economic aspects: A community based approach from Chittagong coast, Bangladesh. *Jahangirnagar University Journal of Science*, 27(3), 155–176.
19. Jabber, S. M. A. (1994). Impact of ship scrapping on the coastal environment of Bangladesh. *Journal of NOAMI*, 6(3), 55–75.
20. Khan, Y. S. A., and Talukder, A. B. M. (1998). Pollution in the coastal water of Bangladesh. *Journal of NOAMI*, 10(1), 11–25.
21. Muhibbullah, M. (2003). Health hazards and risks vulnerability of ship breaking workers: A case study on Sitakunda ship breaking industrial area of Bangladesh. *Journal of Geographical and Regional Planning*, 2(8), 172–184.
22. Mostofa, K. M. G., Liu, C. Q., Vione, D. H., Gao, K. H., and Ogawa, H. H. (2013). Sources, factors, mechanisms and possible solutions to pollutants in marine ecosystems. *Environmental Pollution*, 182(3), 461–478.
23. Sujauddin, M. J., Koide, R. T., Komatsu, T. L., Hossain, M. M., and Tokoro, C. T. (2017). Ship breaking and the steel industry in Bangladesh: A material flow perspective. *Journal of Indian Ecology*, 21(3), 191–203.
24. Ali, M. M., Ali, M. L., Islam, M. S., and Rahman, M. Z. (2016). Preliminary assessment of heavy metals in water and sediment of Karnaphuli River, Bangladesh. *Environmental Nanotechnology, Monitoring Management*, 5(3), 27–35.
25. Reddy, M. S., Mehta, B. H., Dave, S. L., Joshi, M. L., and Karthikeyan, L. L. (2007). Bioaccumulation of heavy metals in some commercial fishes and crabs of the Gulf of Cambay, India. *Current Science*, 92(5), 1489–1491.
26. Umair, S. M., Bjorklund, A. B., and Petersen, E. E. (2015). Social impact assessment of informal recycling of electronic ICT waste in Pakistan using UNEP SETAC guidelines. *Research Constant*, 95(3), 46–57.
27. Nøst, T. H., Halse, A. K., Randall, S., Borgen, A. R., and Schlabach, M. Y. (2015). High concentrations of organic contaminants in air from ship breaking activities in Chittagong, Bangladesh. *Environmental Science Technology*, 49(19), 11372–11380.
28. Sinha, S. (1998). Ship scrapping and the environment-the buck should stop!. *Maritime Pollution Management*, 25(4), 397–403.
29. Mitra, A., Banerjee, K., Ghosh, R., and Ray, S. K. (2011). Bioaccumulation pattern of heavy metals in the shrimps of the lower stretch of the River Ganga. *Mesopatamian Journal of Marine Science*, 25(2), 1–14.
30. Hossain, K. A. (2015). Overview of ship recycling industries in Bangladesh. *Journal of Environmental Analytical Toxicology*, 5(5), 297–302.
31. UNEP. (2005). *Technical Guidelines for the Environmentally Sound Management for Full and Partial Dismantling of Ships.* UNEP.

32. Zakaria, N. M. G., Ali, M. T., and Hossain, K. A. (2012). Underlying problems of ship recycling industries in Bangladesh and way forward. *Journal of Naval Architecture and Marine Engineering*, 9(2), 91–102.

33. Siddique, N. A., Parween, S. T., Quddus, M. M. A., and Barua, P. (2009). Heavy metal pollution in sediments at ship breaking area of Bangladesh. *Asian Journal of Water, Environment and Pollution*, 6(3), 7–12.

34. Ahmed, J. U., and Goni, M. A. (2010). Heavy metal contamination in water, soil, and vegetables of the industrial areas in Dhaka, Bangladesh. *Environmental Monitoring and Assessment*, 166(1–4), 347–357.

35. Mehedi, Y. A. (1994). Comparative pen picture of pollution status of the coastal belt of Bangladesh with special reference to Halishahar and Ship Breaking area, Chittagong, Bangladesh. *Aquatic Pollution*, 45(4), 45–65.

36. Habib, A. (2001). *Delipara: An Obscure Fishing Village of Bangladesh*. Community Development Center (CODEC), Chittagong.

37. IAEA. (1990). *Guidebook on Applications of Radiotracers in Industry*. Technical Report Series No. 316. International Atomic Energy Agency, Viena.

38. Das, B., Ahmed, Y. S., and Sarkar, M. A. K. (2003). Trace metal concentration in water of the Karnafully River Estuary of the Bay of Bengal. *Pakistan Journal of Biological Sciences*, 5(3), 607–608.

39. Neff, J. M. (2008). Estimation of bioavailability of metals from drilling mud barite. *Integrating Environmental Assessment Management*, 4(2), 184–193.

40. Aijun, A. J., Kawser, A. B., Yong, H. X., Xiang, Y. H., and Rani, S. B. (2016). Heavy metal accumulation during the last 30 years in the Karnaphuli River estuary, Chittagong, Bangladesh. *Water Management*, 5(3), 20–34.

41. Deb, A. K. (1995). Ship breaking in the south-east coast of Bangladesh: Pros and Cons. *Aquatic Management*, 35(4), 45–66.

42. Aktaruzzaman, M. M., Chowdhury, M. A., Fardous, Z. M., Alam, K. M., and Hossain, M. S. (2014). Ecological risk posed by heavy metals contamination of ship breaking yards in Bangladesh. *International Journal of Environmental Research*, 8(3), 469–478.

43. Babul, A. R. (2000). Study on ship breaking industry: Bangladesh perspective. *Coastal Management*, 20(3), 50–75.

44. Bhuiyan, M. A. H., Dampare, S. B., Islam, M. A., and Suzuki, S. H. (2015). Source apportionment and pollution evaluation of heavy metals in water and sediments of Buriganga River, Bangladesh, using multivariate analysis and pollution evaluation indices. *Environmental Monitoring Assessment*, 187(3), 4075–4075.

46. Jordao, C. P., Pereira, M. G., Bellato, C. R., Pereira, J. L., and Matos, A. T. (2003). Assessment of water systems for contaminants from domestic and industrial sewages. *Environmental Monitoring and Assessment*, 79(3), 75–100.

45. Khan, M. A. A., and Khan, Y. S. A. (2008). Trace metals in littoral sediments from the north east coast of the Bay of Bengal along the ship breaking area, Chittagong, Bangladesh. *Journal of Biological Sciences*, 3(2), 1050–1057.

47. Martin, J. A. R., Arana, C. D., Ramos, M. J. J., Gil, C. L., and Boluda, R. H. (2015). Impact of 70 years urban growth associated with heavy metal pollution. *Environmental Pollution*, 196(3), 156–163.

48. Naser, H. A. (2013). Assessment and management of heavy metal pollution in the marine environment of the Arabian Gulf: A review. *Marine Pollution Bulletin*, 72(5), 6–13.

49. Salarijazi, M., Abdolhosseini, M., Ghorbani, K., and Eslamian, S. (2016). Evaluation of quasi-maximum likelihood and smearing estimator to improve sediment rating curve estimation. *International Journal of Hydrology Science and Technology*, 6(4), 359–370.

15 Environmental Impacts on Sediment Load and Bank Erosion in River Channels in Bangladesh

Shafi Noor Islam, Sandra Reinstädtler, Albrecht Gnauck, and Saeid Eslamian

15.1 INTRODUCTION

Generally, river systems play an important role in the development of economic growth, cultural interactions, and human civilization. Historically, riverbanks have always been settlement areas for human civilizations and their developing culture (BWDB, 1971; Milliman et al., 1989). Water is required for domestic composition, sanitary use, agricultural production, and the protection of biodiversity ecology and landscape ecosystems (Islam, 2007, 2007a). From the perspective of developing countries, river systems and hydrologic situations are very different (Adnan, 1991; Biswas, 2000).

In Bangladesh, freshwater demand on rivers has increased highly during the last century due to extensive population growth (Islam and Gnauck, 2007a; Gleick, 1993). The location of Bangladesh is in South Asia between 20°34′ to 26°38′ north latitude and 88°01′ to 92°42′ east longitude, with an area of 147.570 km² (Sarkar, 2008; Islam and Gnauck, 2011, 2012). It has a high population of about more than 140 million people (BBS, 1993) with a very low per capita Gross National Product (GNP) of US\$ 670. Three major types of landscapes form Bangladesh: floodplains (80%), terraces (8%), and hills (12%). Excepting the eastern hilly region, almost all of the country lies in the active delta of three major rivers, the Ganges, the Brahmaputra, and the Meghna.

The Ganges–Brahmaputra–Meghna (GBM) river system is one of the major river systems in Bangladesh. Together with the Karnafuli River system, the Surma–Kushiyara River system, and the Tista River system they form a network (BWDB, 1972; Milliman et. al., 1989). The size of the GBM river system drainage basin is about 1.76 million km² of which 62% in India, 18% in China, 7.5% in Nepal, 4% in Bhutan, and 7.5% in Bangladesh (Elahi et al., 1998; Islam and Gnauck, 2007a, 2008). The GBM rivers contribute strongly to the formation of a deltaic landscape and valuable ecosystems with high biodiversity. The Ganges–Brahmaputra–Meghna (GBM) delta covers a major portion of Bangladesh just above the Bay of Bengal in the southwest region with a coastline of about 710 km (Ahmed and Falk, 2008). This delta is a tract of vast alluvial flat roughly resembling the Greek letter "Δ" (Delta) and commences at the offtake of the Bhagirathi near Gaur of Malda district in West Bengal, India (Rob, 1998).

The GBM delta is one of the largest and most fertile deltas in the world and occupies the lower part of the Bengal Basin in the South Asian region (Goodbred, 2004; Islam and Gnauck, 2008). The delta is about 360 km wide along the Bay of Bengal. Strong influences of the alluvial land exist by depositional activities of the rivers Ganges, Brahmaputra, and Meghna and their various distributaries and other large and small streams of the region. Sediment transport is an essential dynamic river phenomenon. It influences directly water quality by erosion and deposition of

DOI: 10.1201/9781003473398-18

"

biomass, chemicals, and suspended stony material. Benthic habitats and still-water areas are necessary for biotic life and are endangered areas. Active currents transport mineral materials washed out of soil from urbanized and agricultural catchments as well as nutrients discharged by domestic, agricultural, and industrial wastewater (Lijklema et al., 1986). Silt deposits of the delta cover an area of about 105,000 km² (Goodbred, 2004).

The deltaic coastal landscapes of the Ganges mouth area are undergoing rapid hydro-morphological and ecological changes due to natural geomorphic processes and human influences on the landscape. The river courses in the delta are broad, active, and dynamic. They carry a vast amount of 6 million m³/s water and sediments 2.4 billion tons annually in the Bay of Bengal (Anwar, 1988). The landscapes of the deltaic region encompass strong aesthetic, cultural, biological, and geological values (Zube, 1986). The coastal landscape has continuous erosion and depositional action. River deltas with low-flow conditions act always as accumulator of their watershed analogous to lakes and reservoirs (Lijklema et al., 1986). The Ganges River (Padma) and Brahmaputra River (Jamuna) carry fresh water and sediments to the catchment area downstream.

At present, the development of the GBM delta depends strongly on the hydrological and transport regimes of these three river systems and influences the developing national economy and changing environment quality (Elahi et al., 1998; Islam, 2001). Transport of clay materials, sandy and stony sediments of different sizes, and bank erosion in the GBM river basin strongly influence the population development of people living in the delta area. Silt and clay material are brought in by normal flow and flooding of the rivers annually. The hydrodynamic development of the delta region is rapid but the economic growth trends are slow. It is essential to rapidly increase this economic growth and the total economic development of the deltaic region.

The objectives of this chapter are to present a short overview of selected topics of hydro-morphological changes in the delta and environmental processes taking place within the delta. Another one is to get an understanding of future changes in the GBM delta caused by environmental, hydrological, and anthropogenic impacts on the river channel system of the GBM rivers system, which contribute to the changes in the deltaic floodplain landscape in Bangladesh. Therefore, the protection of such highly valuated deltaic floodplain ecosystems should be the focus of governmental and management decisions (Islam, 2001; Islam and Gnauck, 2008; Khan and Islam, 2008).

15.2 THE GANGES–BRAHMAPUTRA–MEGHNA RIVER DELTA

The location of the Ganges–Brahmaputra Meghna river delta is at the latitude of 24°40′N and 88°02′E longitude. The longitudinal extension of the delta is from 88°E to 91°50 E longitude (Rob, 1998) (Figure 15.1). The major portion of the Bengal Basin and the GBM delta is flat with quaternary sediments eroded from higher lands on three sides and deposited by the Ganges, Brahmaputra, and Meghna (GBM) rivers and their tributaries and distributaries. The total population living in the delta area which is covering 150 million people and includes five major cities: Calcutta, Khulna, Jessore, Faridpur, and Barisal (Islam, 2001; Goodbred, 2004). A popular calling of Bangladesh is "the land of rivers". One of the neighboring countries is Myanmar, sharing a mountainous borderline of 288 km in length in the southeast. The coastal zone of Bangladesh comprises the delta and is under the process of active delta development and morphological changes by the Ganges–Brahmaputra–Meghna river systems (BWDB, 1972). The major ecosystems by distributaries are levee, delta front (channel, distributary mouth bar, distal bar, etc.) that are in turn modified by the effects of wave and wave influence processes, tidal processes, and to an extent semi-permanent currents, oceanic currents, and wind effects.

The Bengal Delta is extremely complex and dynamic, it is likely that three different delta types exist here, namely, tide-dominated, tide-river-dominated, and tide-wave-dominated (Anwar, 1988). A broad deltaic plain subject to frequent flooding, and a small hilly and terrace region crossed by swiftly flowing rivers. The temperature is 23°C to 40°C; the annual precipitation range is 1474 mm to 2265 mm. The dry months have an average rainfall between 20 mm to 65 mm and the average

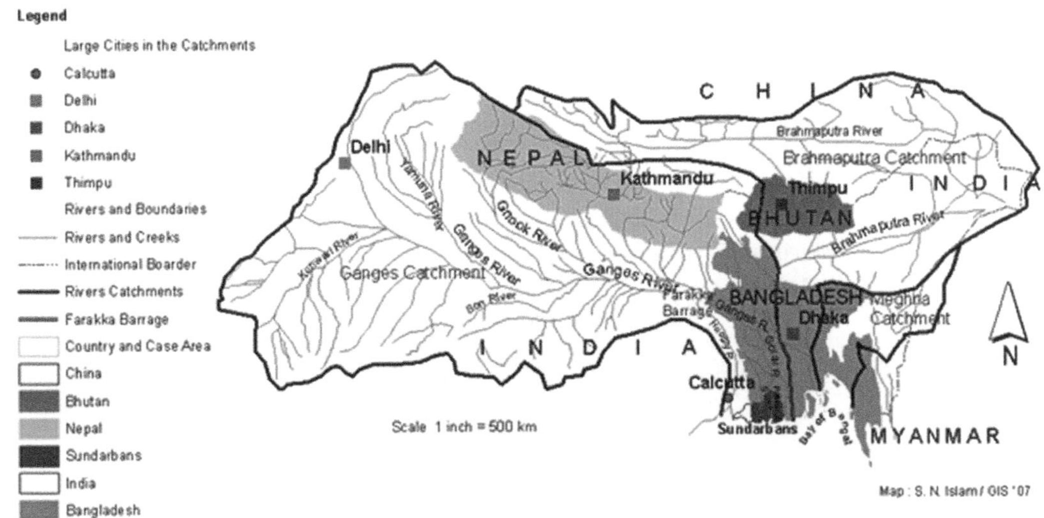

FIGURE 15.1 Geographical location of Bangladesh and Ganges-Brahmaputra-Meghna River catchments areas.

Source: Islam, and Gnauck, 2008.

evapotranspiration rate is 15 mm to 90mm/month (Islam and Gnauck, 2008). The country extends 900 km from north to south and 600 km from east to west (Islam and Gnauck, 2008). On the south, it is a highly irregular deltaic coastline of about 710 km, fissured by many rivers and streams flowing into the Bay of Bengal. The territorial waters of Bangladesh extend 12 nautical miles, and the exclusive economic zone of the country is 200 nautical miles. Altitudes up to 105 m above sea level occur in the northern part of the plain, most elevations are less than 10 m above sea level. Elevations decrease in the coastal south, where the terrain is generally at sea level. With such low elevations and numerous river waters, concomitant flooding is a predominant physical feature. The only exceptions to Bangladesh's low elevations are the Chittagong hills in the southeast, the low hills of Sylhet in the northeast, and the highlands in the north and northwest. The Chittagong hills rise steeply to narrow ridgelines, generally no wider than 36 m, 600 to 900 m above sea level. The highest elevation in Bangladesh exists at Keokradong in the southeastern part of the hills at 1046 m. West of the Chittagong hills is a broad plain, cut by rivers draining into the Bay of Bengal, which rises to a final chain of low coastal hills, mostly below 200 m, that attain a maximum elevation of 350 m. West of these hills is a narrow, wet coastal plain located between the cities of Chittagong (north) and Cox's Bazar (south).

15.3 THE RECENT GANGES–BRAHMAPUTRA– MEGHNA DELTA IN BANGLADESH

The development of the Ganges–Brahmaputra–Meghna delta has been continuing since it began some 125 million years ago (Ma), after the fragmentation of the Gondwanaland since the early cretaceous (Islam and Gnauck, 2008). The area of the GBM rivers delta is 105,000 km² (Hanna, 2001; Goodbred, 2004). The average elevation of the delta in Khulna Barisal, the southern part of Faridpur, and the eastern part of Noakhali district is less than 2 m (Rob, 1998; Islam, 2006, 2008, 2016).

The river courses in the delta are carrying water discharge, 29,692 m³/s on average into the Bay of Bengal. The discharge is 100,000 m³/s during the flood time and 6041 m³/s during the low water season (Islam and Gnauck, 2008). Figure 15.2 shows the present day Ganges-Brahmaputra-Meghna delta where four major categories are distinguished.

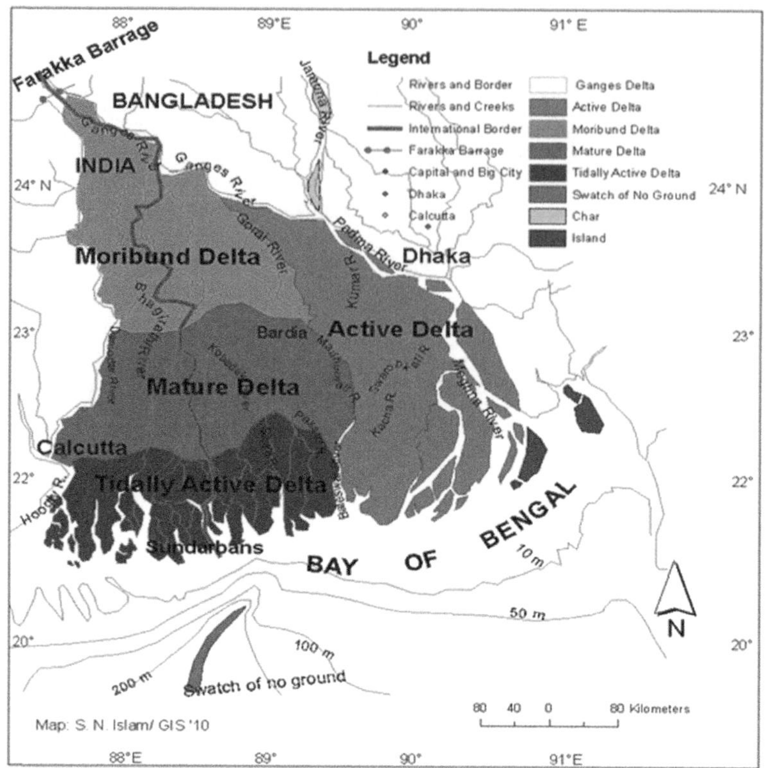

FIGURE 15.2   Present day Ganges-Brahmaputra-Meghna delta.

Source: Map modified from Islam and Gnauck, 2008.

15.3.1   THE MORIBUND DELTA

In the western part of the deltaic plain, its borders are laying on the east by the Gorai–Madhumati River comprising the areas from northern Jessore and the greater Kushtia district. The moribund delta lies between latitudes of 21°15′N and 24°40′N and longitudes of 88°0′E and 90°0′E. It covers an area of 29,500 km² in Bangladesh and includes Mushidabad (the east portion of the Bhagirathi district), a part of Nadia of India, and Meherpur, Chuadanga, Kushtia as well as the northern part of greater Jessore district of Bangladesh (Islam, 2006, 2016; Islam and Gnauck, 2008; Goodbred, 1999). As a result, land building has caused and the delta has grown moribund up to this limit (Islam, 1995). The deltaic area lacks the important effect of the flooding from the Ganges and is characterized by rivers that are now abandoned or drying up. Due to the drastic fall in the flow of the Gorai River, at least 7 of the 15 rivers dependent on it are now nearly dead while 8 others are following. The nearly dead rivers are Hisna, Kaliganga, Kumar, Kamkumra, Harihar, and Chitra. These have lost connections with their parent streams. The rivers having now no offtake from the main stream cannot bring down enough water and silt, even in flood, and being confirmed within high levees, are not in a position to inundate the entire area (Brammer, 1969).

15.3.2   THE MATURE DELTA

The mature delta comprises the saline non-forested areas in the northern part of the Khulna (Islam, 1995). The mature delta lies between the latitudes of 23°16′N to 20°10′N and longitudes of 87°50′E to 89°70′E. The mature delta covers an area of approximately 28,500 km² in Bangladesh and in West Bengal of India. On average, it includes Murshidabad, Nadia, and 24 Pargana of India and Meherpur, Kushtia, Chuadanga, Jhenaidah, Magura, Jessore, Norail, and the northern part of Khulna, Sathkhira and Bagerhat districts of Bangladesh (Islam, 2006; Islam and Gnauck, 2008). In the mature delta, although the land is suitable for cultivation recent salinity intrusion is a major handicap, especially in the early part of the summer (March–May) in the greater Khulna and Jessore districts. Due to high salinity in the mature delta, the agricultural cropping intensity is very low in this area (Islam, 1995). The urbanization growth develops in the Ganges mature delta. The major cities such as Calcutta, Khulna, Bagerhat, and Satkhira are located in the mature delta (Figure 15.2).

15.3.3 THE ACTIVE DELTA

The Ganges–Brahmaputra active delta lies between the latitudes of 23°70′N and 21°55′N and the longitudes of 89°10′E and 90°50′E and covers an area of about 26,500 km². The area is about 300 km long in the north–south direction and almost 100 km and 130 km wide in the upper and middle reaches of the active delta respectively. The active Ganges–Brahmaputra delta includes the Rajbari, Faridpur, Madaripur, Shariatpur, Gopalgonj, Pirojpur, Jhalokhati, Barisal, Borguna, Patuakhali, Bhola, and Lakshmipur districts of Bangladesh (Islam and Gnauck, 2008). The area of the Ganges–Brahmaputra active delta is about 22,500 km², which forms the deltaic lands per excellence (Islam, 1995, 2006; Islam and Gnauck, 2008; Goodbred, 1999). The non-saline tidal and estuarine char-lands and islands constitute a sort of transitional region to the saline delta. However, in the southern part of the Barisal and Patuakhali districts, salinity is also a handicap and precludes cultivation of crops other than paddy during the rainy season (Islam, 1995).

15.3.4 THE TIDALLY ACTIVE DELTA

The southwestern part of the Ganges delta is a coastal mangrove area and tidally active delta (Islam and Gnauck, 2008). The present tidally active delta lies between the latitude of 2210′N to 21°10′N and the longitude of 87°50′E to 89°60′E. Coastal tidal waves and currents dominate hydrodynamic activities. The tidally active delta actually is the southern part of the Sundarbans region. The area is approximately 20,000 km². It includes the southwestern part of the coastal region especially the major portion of the Bagerhat, Khulna, and Sathkhira districts of Bangladesh and the South 24 Pargana district of West Bengal, India (Islam, 2006; Islam and Gnauck, 2008). The establishment of the present-day northern limit of the Sundarbans was at ca. 3000 Years BP. The coastline retreated from the Khulna region at about 1800 Years BP. The northern limit of this last transgression is unknown (Gupta, 1981; Islam, 2001). According to Khan and Islam (2008) the lower deltaic coastal floodplain has five assemblages recognized in the lower coastal deltaic plain in five different depositional environments such as levee or levee complex, bill or depression, abandoned meander belt, inter-distributary, and estuarine plain (Khan and Islam, 2008). The thickness of the Holocene sediments ranges from 30 m to 70 m in the deltaic plain, usually floored by the Pleistocene period stiff clays with the exception of the abandoned meander belt deposit where Holocene channel sand deposited directly on the Pleistocene sand (Khan and Islam, 2008). Besides this, coastal subsidence of 3 mm/a on average, occurs in the coastal region in Bangladesh which is in general the tidally active and the southern part of the lower active delta (Goodbred and Kuehl, 2000; Khan and Islam, 2008). A rise of relative sea level movements and the sustainability of the present mangrove vegetation of Bangladesh could be predicted (Islam, 2001). The study would therefore suggest that most of the mangroves along the Bangladesh coast and the tidally active delta would collapse under sea level scenarios where rates of rise up to 1.5 cm per year are predicted by IPCC (Milliman et al., 1989).

15.4   DATA AND INFORMATION BASE

Primary and secondary data sources serve as databases. Different governmental and non-governmental organizations and agencies added a broad information base on river systems in Bangladesh, sedimentation processes, and environmental impacts on the GBM river delta. Additional information stems from *char-land* studies on the Ganges–Padma River basin and close observations of Ganges River and Jamuna River characteristics (Islam, 2006). Besides this, an in-depth investigation of water salinity modeling on 16 rivers in the Sundarbans region gives fundamental information on chemical transport mechanisms and settling procedures. Physical observations of rivers and river-related conversations with local people especially in the southwestern region and active deltaic region in Bangladesh contribute to the information base. Data was also collected from the local people in the GBM rivers catchment areas during field data collection in 2003 and a short visit to Bangladesh in 2008 and 2021. Collected water quality reports from IWM, Dhaka, BCAS, BIWDB, Department of Environment and Forestry, and The Center for Environmental and Geographic Information Services (CEGIS) Dhaka, as well as relevant published research works in Bangladesh and outside of the country together with papers in journals and books in Bangladesh and outside of the country, form the secondary database. Data processing, analysis, and visualization take place by software packages MS Excel, and VISIO 32. For mapping of river sedimentation, water, and environmental pollution investigations, ArcGIS 10.3 software is a helpful tool.

15.5   FLOODS, RIVERBANK EROSION, AND SEDIMENTATION IN THE GBM DELTA RIVER CHANNELS

Watershed erosion and bank erosion contribute mainly to sediment loading of the rivers in the GBM delta. For some long-term average sediment loads, the Universal Soil Loss Equation (USLE) (Wischmeier and Smith, 1978) is a helpful tool for the calculation and estimation of the amount of settled material (Mills et al., 1985).

The GBM river system is a unique and complicated river system in Bangladesh. The channel system and the sedimentation processes are also unique in the Bangladesh river system process. The average rainfall in Bangladesh generates annually only 100 million acre-feet of water whereas 1100 million acre-feet of water comes from outside of Bangladesh. Thus, about 90% of the water carried by the river systems, namely, the Brahmaputra, the Ganges, the Meghna, and other smallest rivers, are the inflow from outside of the country, these rivers carry water from an area of about 960,000 km^2 of which only 7.6% lies in Bangladesh, Water enters Bangladesh through three major channels while the main and major discharge also takes place through these three major channels. Finally through Meghna River channel the main discharge is drained into the Bay of Bengal. The river system has evolved to carry the normal flow of water generated in the catchment area. Bangladesh falls in the region of very heavy rainfall. About 80% of the rainfall occurs during the five months from May–September. The Ganges–Brahmaputra–Meghna river catchment areas and the percentage covered by the countries are located in the downstream and within the catchment areas. This is the widest river system in the country flowing from North to south. The discharge during the rainy season is enormous, the highest recorded flood was in August 1988, and the river flow discharge was 98,600 m^3/s (Islam, 1995; Islam et al., 2014).

15.5.1   FLOODS IN THE GBM DELTA

The reasons for massive floods in Bangladesh are the following:

- Snow melting in the Himalayan region
- Hydrographic and geomorphic changes in the Brahmaputra River basin.

Annually, an estimated 2.4 billion tons of sediments are carried by the river systems of Bangladesh by floods that reduce the water-carrying capacity of the rivers. It is also creating massive floods (Coleman, 1969; Anwar, 1988).

The primary cause of floods in Bangladesh is rainfall in the catchment areas of the rivers. Situated in the monsoon belt with the Himalayas in the north, Bangladesh falls in a region of very heavy rainfall. About 80% of the rainfall occurs during the five-month period from May to September. Especially the newly formed lands (so-called *char-lands*) are repeatedly affected by massive floods, with concomitant riverbank erosion because of the shift of river channels (Ahmed, 1956; Hassan et al., 2000). The newly formed char-lands and eroded riverbanks are inhabited by some of the most desperate and vulnerable people in the country. These people will never have abandoned fragile riverbanks and char-lands because of flooding (Haque, 1985). However, the average rainfall in Bangladesh generates annually only 100 million acre-feet of water whereas 1100 million acre-feet of water comes from outside of Bangladesh. Thus, about 90% of the water carried by the GBM river systems and other smaller rivers is the inflow from outside of the country.

The rivers carry water from an area of about 960,000 km² of which only 7.5% lies in Bangladesh. Water enters Bangladesh through three major channels but the discharge takes place through one major channel. The river system has evolved to carry the normal flow of water generated in the catchment area. The catchment of the mighty Brahmaputra–Jamuna River is about 583,000 km² of which 293,000 km² is in Tibet, 241,000 km² in India, and only 47,000 km² in Bangladesh. This is the widest river system in the country flowing from north to south. The discharge during the rainy season is enormous, the highest recorded flood was in August 1988 and the river flow discharge was 98,600 m³/s (Islam, 1995). High rates of forest depletion in the GBM catchment areas tend to aggravate the floods in downstream areas. Unplanned development of constructions such as roads, railways, embankments, polders, etc. create barriers to the flow of water and aggravate the floods. In the 20th century, 18 major floods were recorded in 1900, 1902, 1907, 1918, 1922, 1954, 1955, 1956, 1962, 1963, 1968, 1970, 1971, 1974, 1984, 1987, 1988, 1998, and 2008, of which 3 (1987, 1988, and 1998) were catastrophic floods. According to the statistical evidence of Bangladesh in both 1987 and 1988, the country experienced disastrous floods. The 1987 flood is estimated a 30–70 year event, which affected 75,300 km², almost 40% of the total area of the country. The 1988 flood inundated about 82,000 km² and about 60% of the area with an estimated recurrent period of about 50–100 years (Hussain, 1993; Elahi, 1998; Hussain, 1987). August has always been the month when widespread flooding occurred. Floods from May to July are usually high time in the catchments of Brahmaputra–Jamuna and the Meghna, from August to October due to the combined flows of that river and the Ganges. As a result, the flow of the Brahmaputra–Jamuna is more erratic than that of the Ganges. The gradient of the Jamuna River averages 1:11,850 that is slightly more than that of the Ganges. During the monsoon, the Jamuna River widens in some parts up to 20 km and transports between 60,000 to 100,000 m³/s water (FAP 24, 1996a).

No sophisticated system is so far available in Bangladesh to assess the damage due to floods. Considering historical data and information on floods, the most devastating floods occurred in 1988, 1998, 2005, 2008, 2016, 2018, 2021, and 2022 (Figure 15.3a and b).

The official sources indicate the damage of crop acreage at 4.93 million Bangladeshi Taka (Bangladesh Taka (TaKa.); US$1 = 95 Taka). Consequently, rice production lowered by 2.4 million tons per year (Miah, 1988; Islam, 1993, 1995). Estimates of the average annual flow of the Jamuna River at Bahadurabad (the point where the Brahmaputra enters into Bangladesh) are to be 501 million acre-feet. The damage caused by the 1987 flood was staggering, according to one estimate, the total loss was about TaKa. 3000 million (over US$ 95 million).

15.5.2 Riverbank Erosion in the GBM Delta

The two Himalayan Rivers, the Ganges and the Brahmaputra, are among the most sediment-loaded rivers in the world (Islam et al., 1999; Milliman and Meade, 1983). Almost every year,

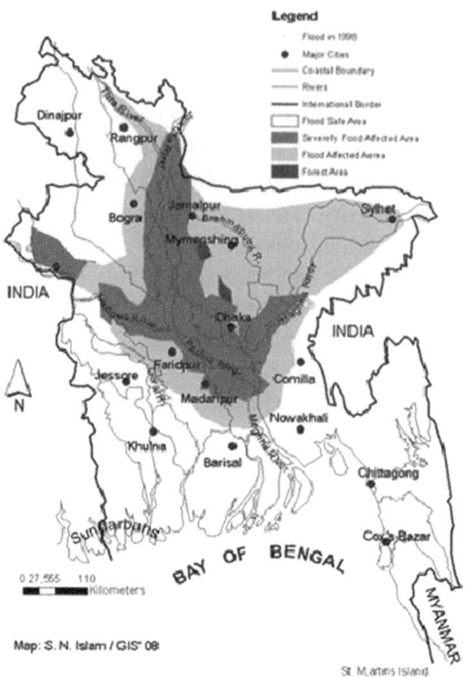

FIGURE 15.3A Historical flood inundation and bank erosion patterns in 1998 in the Ganges-Brahmaputra-Meghna River basins in the Bengal Delta in Bangladesh.

Source: Islam and Gnauck, 2008.

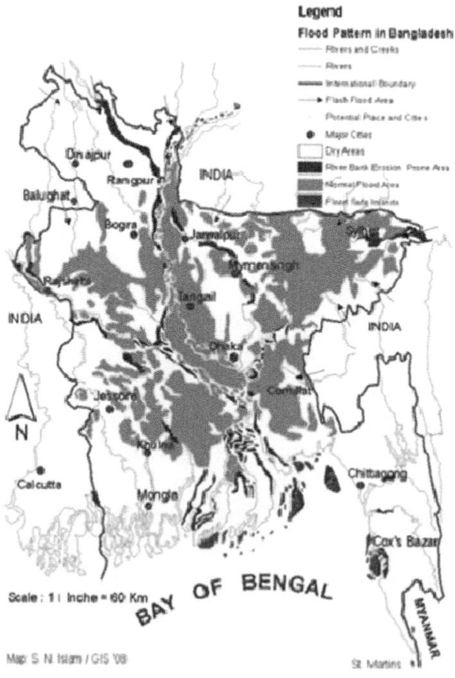

FIGURE 15.3B Historical flood inundation and bank erosion patterns in 2008 in the Ganges-Brahmaputra-Meghna River basins in the Bengal Delta in Bangladesh.

Source: Islam and Gnauck, 2008.

the Ganges and Brahmaputra Rivers in Bangladesh transport 316 and 721 million tons of sediment respectively. These high loads of suspended sediment reflect the very high rate of denudation in their drainage basins. The role of the Ganges and Brahmaputra rivers in mechanical denudation of the basin areas are together estimated 365 mm 10^3 yr^{-1} (Islam et al., 1999). The Brahmaputra River is naturally a braided river, within the braided belt of the Jamuna River. There are many *chars* of different sizes. An assessment of the 1992 dry season Landsat images shows that the Jamuna contained 56 large island *chars*, each longer than 3.5 km. There are an additional number of 226 small island *chars*, varying in length between 0.35 and 3.5 km. This includes sandy areas as well as vegetated *chars*. In the Jamuna River the period between 1973 and 2000, *chars* have consistently appeared in the reaches opposite to the "Old Brahmaputra" offtakes, north and east of Sirajganj and in the southernmost reach above the confluence with the Ganges (EGIS, 2000).

In the rainy season, it brings down around 13 million tons of sediment per day (Coleman, 1969; Hanna, 2001) which helps to format the *riverine islands* in the Brahmaputra (Jamuna) – Padma River channels. In Bangladesh from 1981 to 1993, about 729,000 people were displaced by riverbanks and *riverine island* erosion in the river basin and the catchment areas. Floods are more or less a recurring phenomenon in Bangladesh and often have been within tolerable limits. However, occasionally they become devastating. Each year in Bangladesh about 26,000 km², 18% of the country is flooded. During severe floods, the affected area may exceed 55% of the total area of the country. In an average year, 844,000 million m³ of water flows into the country during the humid period (May to October) through the three main rivers namely the Ganges, the Brahmaputra–Jamuna, and the Meghna. This volume is 95% of the total annual inflow.

The river Brahmaputra (Jamuna) is one of the three main rivers of Bangladesh. The river is a major contributor to the building up of the delta through siltation and rapid sedimentation process in the river basin and the river catchment areas. There is evidence of severe bank erosion and rapid rates of bank line retreat along the Brahmaputra–Jamuna River. River bank erosion and sedimentation process is a natural phenomenon of river characters but very often anthropogenic activities and actions can have significant impacts on the rate of morphological changes as well as downcutting accretion and environmental pollution are natural processes for an alluvial river. The development of sand mining, infrastructure building on the river bank, bank revetment, construction of reservoirs, and land use alterations have changed the natural geomorphological dynamics of rivers (Lane and Richards, 1997; Fuller et al., 2011; Li et al., 2007), these type of geomorphological features are occurring due to the rapid suspended loads and particles sedimentation.

15.5.3 Sedimentation in the GBM Delta

Sedimentation processes depend on the water flow velocity, slope of the riverbed, and particle size (Graf, 1985; Vanoni, 1977). Internal and imported particulate material settles to lower river segments. Current activities form the riverbeds in downstream areas. They are getting ridged and higher the width of channels. This is the main reason for reducing the width of rivers. The maximum amount of sediment deposits on the riverbed. While dissolved organic and inorganic water ingredients are distributed over the completely cross-sectional area of the river, the transport of particulate material of different sizes takes place at the river bottom by rolling, sliding, or suspending movement. Sedimentation and land-forming patterns depend on the offtake of the river. For instance, the mean annual peak discharge of the Gorai River in the late 1990s is about 4500 m³/s. Its bed material varies from silts to fine sands. The annual average sediment transport is about 50 million tons (for the last three decades). The mean size of bed materials at the mouth of the Gorai River is 0.17 mm (Delft Hydraulics and DHI, 1996). Table 15.1 gives some information on flow velocity and particle size dependencies.

Sediments and particulate materials within the water column settle in downstream river segments and deposit on the surface of the riverbed (Ambrose et al., 1993). For settling velocities, the

TABLE 15.1

River Flow Velocity and Particle Size

Flow velocity (cm/s)	Diameter of particle size (mm)	Denotation
<10	< = 0.002	clay
<10	0.002–0.05	silt
10	0.05–0.3	Fine-grained sand
17	0.3–0.6	middle-grained sand
25	0.6–2	coarse-grained sand
50	2–6	fine-grained gravel
75	6–20	middle-grained gravel
150	20–60	coarse-grained gravel
>150	60–80	stones
200	>80	big stones
300	>180	log of stones

Source: After Eisma, 1993.

application of Stoke's relationship is a helpful tool. In dependence of particle size, the following equation is useful (Ambrose et al., 1993; Batchelor, 2000)

$$v_s = 8.64 \ g/18\mu \ (r_p - r_w) \ d^2_p$$

where: v_s = Stokes velocity for particle size with diameter d_p and particle density r_p (m/day), g = gravity acceleration (equal 981 cm/sec^2), μ = absolute viscosity of water (equal 0.01 poise or g/cm^3 sec at 20 °C), r_p = density of particles (g/cm^3), r_w = density of water (equal 1.0 g/cm^3), d_p = particle diameter (mm). Table 15.2 gives an impression of Stokes settling velocities at 20 °C (Ambrose et al., 1993; Bachelor, 2000).

Table 15.3 displays estimations of sediment loads of the Ganges–Padma river system in Bangladesh obtained from different information bases. Major portions of the sediments flow through the river channels of the Ganges–Brahmaputra–Meghna rivers and the river estuaries in southern Bangladesh (Tables 15.4 and 15.5).

TABLE 15.2

Dependence of Settling Velocities at 20°C Particle Diameter and Density

	Particle Density (g/cm³)	Particle Density (g/cm³)	Part. Density (g/cm³)	Part. Density (g/cm³)
Diameter of particles (mm)	1.80	2.00	2.50	2.70
0.001	0.04	0.05	0.07	0.08
0.002	0.15	0.19	0.28	0.32
0.005	0.94	1.20	1.80	2.00
0.01	3.80	4.70	7.10	8.00
0.02	15.00	19.00	28.00	32.00
0.05	94.00	120.00	180.00	200.00
0.3	300.00	400.00	710.00	800.00

Source: Modified from Ambrose et al., 1993.

TABLE 15.3

Estimates of the Sediment Load of the Ganges–Padma River in Different Sources

Susp. Sediment (Million Tons/yr)	Sampling station	Period of Investigation	Reference
328	Calcutta	1981	Abbas and Subramanian, 1984
729	Farakka	1981	Abbas and Subramanian, 1984
403	-	1981	
485	Harding Bridge	1958–1962	Coleman, 1969
520	Harding Bridge	1966–1969	BWDB, 1972
375	-	1960–1967	NEDECO, 1967
680	Harding Bridge	1966–1967	Milliman and Meade, 1983
1600	Bengal Delta	1968	Coleman, 1969
350-600	Harding Bridge	1980–1986	
316	Bahadurabad	1980, 1983–1986, 1988–1995	Islam et al., 1999

Source: After Islam et al., 1999.

TABLE 15.4

Incoming Sediment Load in River Basins in Bangladesh

Rivers	Sediment load from secondary information base (Million Tons/year)	Sediment load calculated by Rahman et al., (2018) (Million Tons /year)
Ganges River	260–680	150–590
Brahmaputra River	390–1160	135–615
Upper Meghna River	6–12	No information
Total	1000–2400	Average 500

TABLE 15.5

Sediment Load of the Brahmaputra River as Estimated in Different Studies

Sampling Stations	Observation period	Suspended Sediment (Million Tons yr⁻¹)	References
Pandu, Assam	1955–1979	402	Goswami, 1985
not known	1987	710	Subramanin, 1987
Bahadurabad	1982–1988	650	Hossain, 1987
Bangladesh	1960	750	NEDECO, 1967
Bahadurabad	1958–1969	617	Coleman, 1969
Bahadurabad	1967–1969	541	BWDB, 1972
Bahadurabad	1966–1967	1157	Milliman and Meade, 1983
Bengal Delta	1968	800	Holeman, 1968
Bahadurabad	1989–1994	721	Islam et al., 1999

Source: After Islam et al., 1999.

The channel stability is often threatened and a threat to the environment and local ecosystem services as well as food insecurity. However, the current capacity of the GBM delta by means of transport, erosion, sedimentation, and accumulation of dissolved and particulate material is fundamental for riverine ecology. Nevertheless, the strong pollution of riverine ecosystems by clay and silt is an effective cause of agricultural crops and intensive fisheries production in the river basins of the major rivers in Bangladesh.

15.6 ENVIRONMENTAL IMPACTS TO RIVER CHANNELS IN THE GBM DELTA

River, estuary, and floodplain sedimentation in Bangladesh. The sediments from the Ganges–Brahmaputra–Meghna systems discharge into the Bay of Bengal through the Lower Meghna mouth turn clockwise, part of the sediments deposit in the ocean, and the rest of the sediments re-enter into the western estuarine systems. During cyclonic events, this equilibrium breaks down and a bulk amount of sediments re-enters into the system through the estuary mouths. Sedimentation, as expected, is the maximum in the unprotected regions of the coast, but not inside polders. Among the rivers and estuaries, maximum sedimentation takes place in the Sundarbans system. In the Meghna River estuary region, particles settle by distinct sedimentation processes in the region between Sandwip and Urir char that will eventually join these two landforms. Strong sedimentation comes off in the Pyra port region in the mouth of the Tetulia River, in the Kutubdia channel, and in the Moheshkhali channel.

15.6.1 IMPACTS DUE TO SEDIMENTATION IN RIVER CHANNELS

The Gorai River flows through the part of the delta which was formed about 500 years ago, at present geomorphological landscape pattern is changeable and unstable (EGIS, 2000). The Gorai River is the distributary of the Ganges River, which is the main river, and upstream water source in the Sundarbans mangrove wetland region. Annual floods inundate the entire *char-land*. The upstream and downstream reaches of the Gorai River basin are continuously regimenting due to particle flows during the high flooding season. The rate of sedimentation is very high. Figure 15.1 presents a conceptual model of environmental impacts in the Padma–Jamuna river system. The inhabitants of *char-land* have to live with flood or to migrate to a safe place. Floods and riverbank erosions are the most threatened natural phenomena for the Char–Janajat people, and almost 6,000 people are facing flood problems in the *char-land*. In the Purba Khas Bandarkhola mouza there are 3,000 people living in the past. Some negative impacts derive from the conceptual model.

Figures 15.4 and 15.5 shows impacts of massive floods (in 1998, 2008, 2020 and 2022) on the *char-land* in the Padma-Jamuna river channel in the Ganges deltaic region. The particles settling down of the particles on the riverbed create multiple types of pollution and negative impacts on river water. Because of organic and inorganic particles and other settling material, water quality, and soil quality are degraded resources that are the major threats to the local terrestrial ecosystem and food security as well as health insecurity. Especially, chemical and pharmaceutical pollution of the riverbed is a threat to fishing and water use by people. By comparison, only about 187,000 million m³ of stream flow is generated by rainfall inside the Char–Janajat location which is one the largest chars (riverine islands) in the Padma River basin (Jansen et al., 1994). The deforestation intensity in the upstream Himalayas increases the deposition of sediments in the downstream areas and consequently changes the river morphology frequently (Jansen et al., 1994).

The actual allocation of Ganges water at the Farakka Barrage point in India has been causing serious concern to Bangladesh due to the reduced availability of flows during the past 47 years (Figure 15.6) (Islam, 2007a). The abnormal reduction of Ganges flows caused excessive siltation, the elevation of bed levels, and consequent reduction in the flood discharging capacity of the channels, silting of the Ganges thereby blockage of the Gorai river offtake (Anwar, 1988; Islam and Gnauck, 2009).

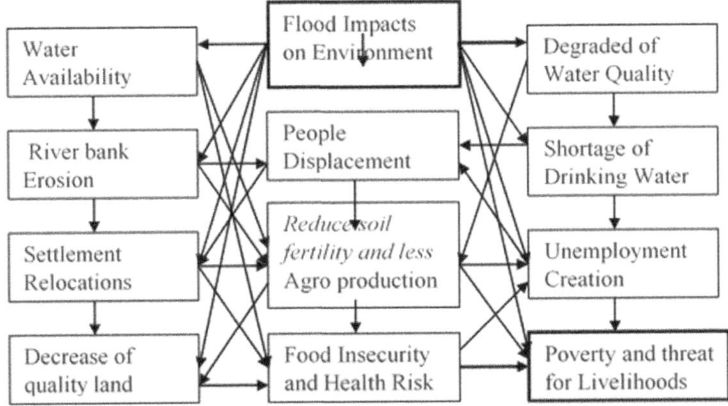

FIGURE 15.4 A conceptual model on environmental impacts due to floods and erosion in the Padma–Jamuna River channels.

FIGURE 15.5 Sedimentation process on the Gorai River Basin (Left) and Saline water flowing at the coastal region in Shatkhira Region near to the Sundarbans mangrove forest area in Bangladesh 2003 and 2021.

Source: Islam, 2003.

FIGURE 15.6 Scenarios of River Basin area (Gorai River) in 2003.

Source: Islam, 2003.

15.6.2 RIVER CHANNEL MIGRATION

Dynamic physical processes of rivers can cause channels in areas to move laterally, or "migrate" over time. The area within which a river channel is likely to move over a period is known as the channel migration zone. Channel migration can occur gradually as a river erodes one bank and deposits sediment along the other. The natural meander patterns of Padma river channels are the result of the dissipation of energy from flowing water and the transportation of sediment. Channel migration also can occur abruptly as the river channel shifts (or avulses) to a new location (Islam, et al. 1999; Rahman et al., 2018). Avulsions are usually unpredictable events that occur during high flood flows when the existing channel cannot transport all of the water and sediment supplied to it (Salarijazi et al., 2016). The highest rate of channel migration generally occurs when steep rivers flow out of foothills into flatter floodplains (Figure 15.7). Deltaic sediments of quaternary formation characterize the riverine floodplain land (Islam et al., 1999; Khan and Islam, 2008). The island of Padma River channel is located in the Ganges active delta. This delta has come into existence by the devotional activities of the three mighty rivers in the region the Ganges, the Brahmaputra, and the Meghna (Islam and Gnauck, 2008). Figures 15.7, 15.8, and 15.9 show the large char geomorphic condition and landslide pattern in 1995 and in 2008.

The Ganges–Padma River channel may shift laterally by more than 300 m annually (compare figures 15.7, 15.8 and 15.9). This process makes char-land (riverine island) erosion a devastating natural hazard in Bangladesh that causes people to be displaced for a short time or more cases permanently (Haque, 1985). The whole char is called Char–Janajat (Figures 15.7, 15.8 and 15.9) and it is unstable and prone to annual flooding and uncertain of rural livelihoods. The whole char (Char–Janajat) area was 84.09 km^2 and the area of Purba Khas Bandarkhola was measured 6.7 km^2. However, satellite images and GIS mapping show the gradual reduction of island areas in the river basin of the Ganges–Padma River (Islam and Gnauck, 2008; Khan and Islam, 2008; Islam et al. 1999).

The Char–Janajat (island) is extremely vulnerable to erosion, accretion, and flood hazards due to frequent annual floods and sedimentation. The recent analysis of the satellite images (from 1995 to 2008 and 2020) indicates that over 80% of the area within the river channel had been char-lands of char–Janajat in the Ganges–Padma River channel. The same analysis shows that about 80% of the char-lands persisted between 1 and 10 years while only 20% lasted for 20 years or more in the Ganges–Padma river channel (Rahman et al., 2018).

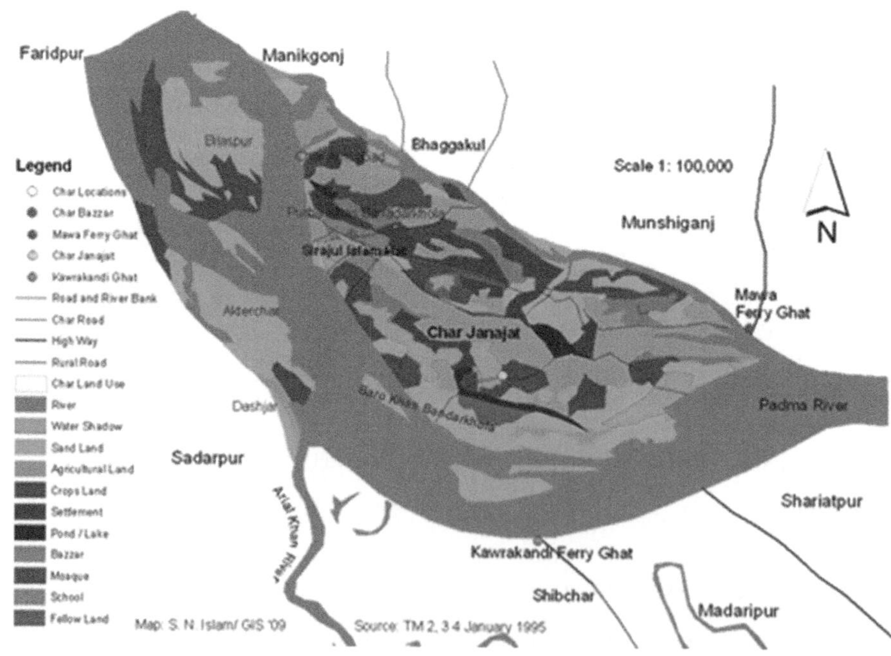

FIGURE 15.7   River channel and sedimentation and channel migration at the Ganges-Padma River basin.

Source: Islam, S. N., 2005.

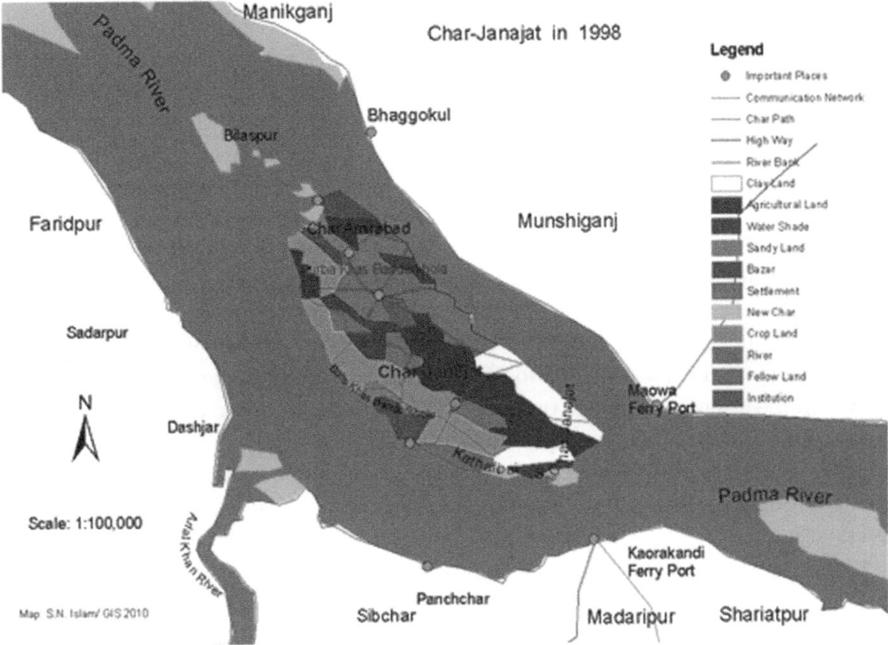

FIGURE 15.8 River channel and Sedimentation and channel migration at the Ganges-Padma River basin.
Source: Islam, 2005.

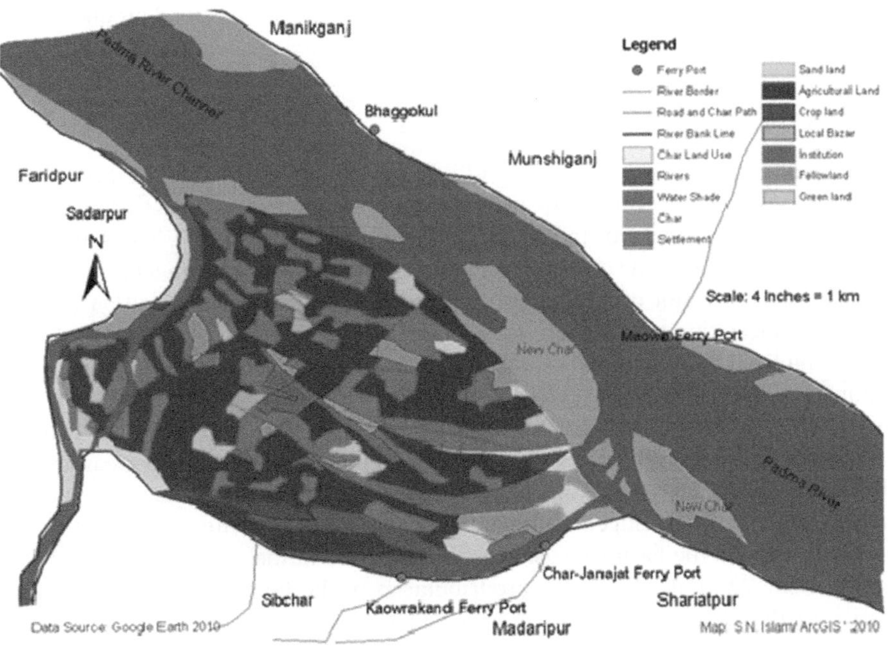

FIGURE 15.9 River channel and Sedimentation and channel migration at the Ganges-Padma river basin.
Source: Islam, 2012.

15.6.3 Agricultural Productivity and Cropping Pattern Changes

The Ganges river floodplain, the catchment area, and the peat basins in the study area are non-saline zones. Soil salinity range was between 8644 dS/m during the dry season (November–May) (EGIS, 2000; Islam and Gnauck, 2008). Since the Ganges River water withdrawal of groundwater due to drought, massive floods, and agricultural irrigation, saline groundwater from the south intrudes into the freshwater aquifer in the north (Islam, 2021). This salinization process and polluted water from the Ganges basin as well as the huge sedimentation in the basin and in the floodplain areas are overwhelming the local agro-ecological systems (Islam and Gnauck, 2009; 2009a). The actions of humans in this aquatic environment contribute mainly to the reduction of ecosystem services. The UN report stated that almost 50% of the land had been highly degraded by anthropogenic and natural hazards influences (Vitousek, 1997). By 2050, the number of countries will face water stress and hazardous water that will rise to 20, after 600 million people. The future of water stress and scarcity will have serious impacts on the socio-economic improvement of the countries. It has probably adversely affected their food production levels. It is likely the precipitation alone, which would guide up from only 23% of affected land masses worldwide (Islam and Gnauck 2009; 2009a).

Floods and drought by themself do not trigger an emergency, but long-lasting floods and drought become an emergency due to their severe impact on agriculture and the lives of the drought victims directly and indirectly related to life, water, air, forest, soil, food, and agricultural crop production. In 1974, Bangladesh experienced severe floods and a severe drought in 1975. This flood affects about 47% of the area and 53% of the population through floodwater, polluted water, and siltation of agricultural lands. River basins are silted by sedimentation, which has reduced the agricultural production in river catchment areas in Bangladesh. Monsoon rainfall, although very heavy compared to surrounding areas, suffers from extreme variability, often exceeding 60% (Rahman et al., 2018). The fact that the lesser the amount of rainfall, the higher the variability holds good in the north-western and southwestern parts of Bangladesh (Islam, 1995). Massive floods and polluted sedimentation processes in Bangladesh rivers are affecting agricultural food production and human life in general in the four major river basins and catchment areas in Bangladesh.

15.6.4 Degraded River Floodplain Ecosystems

From a gross estimate, Goodbred and Kuehl (2000) assumed that one-third of incoming sediments serve as depositions in the floodplain. In a later study, researchers found out that about 23% to 47% of incoming sediments were deposited on the floodplain (WARPO, 1998). The same study reveals the fact that sedimentation in the coastal floodplain is spatially variable. In addition, sedimentation in this region is also a result of largely anthropogenic and climatic interventions. Due to spatial variability of sedimentation and impacts due to intervention, sediment management in the region is particularly complicated. Most of the estuaries and rivers in the western part lost their conveyance due to sedimentation causing large-scale waterlogging (WARPO, 1998).

The most of the River basis in Bangladesh are already started to lose their natural flows and this is the reason over sedimentation process are running in the large rivers basin areas and the most of the small channels has been blocked due to over sedimentation. Through the sedimentation process the river beds are becoming ridged and natural channel also becoming narrow and some cased silted the water flows in the basins areas. Most of the rivers in the northern region of Bangladesh, including the Tista, Atrai and Karotoya, are tributaries of the Jamuna. The most important of the distributaries are the Dhaleswari, the Buriganga, and the Lakhya, the rivers that serve the important inland ports of Narayanganj and Dhaka, but recent scenario these rivers capacity of water flowing is reducing and the water quality degrading very rapidly. The Brahmaputra-Jamuna drains an area of 360,00 sq. miles (930,000 Sq. km) Peak flows of up to 2,500.000 cusecs are reached between mid-June and early September when this river carrying 2.4 billion tons of Sediments and 60% of sediments are settling down on the river beds and restructuring the river's bed morphology due to huge amount of upstream water and water loads or particles which are settling down on the River beds (Schulze and Islam, 2012; Jansen et al., 1994).

The combined drainage area of the Ganges-Brahmaputra-Meghna is over 800,000 Sq. miles (2.1 million sq.km). It covers parts of India, Bhutan, Nepal, and China. Only 7.5 % of the drainage area is within Bangladesh. Prior to 1975 and the opening of the Farakka Barrange, some 1240 million cusecs of water flowed through the country's three great rivers, equivalent to about six times all the rainfall received within the country (Jansen et al., 1994). The shortage of water in the basin areas and flooding in the catchment areas with huge particles in water which is continuously degrading the water quality and soil quality and fertility in floodplain areas. The sea level rise and saline intrusion in the southern coastal floodplain of Bangladesh is facing a serious ecosystem problems, as a result the ecosystem services are under threat in the rivers basins and floodplain areas in Bangladesh. Based on the study findings there are some strategic policies towards control and manage sedimentation and erosion control that could make balance ecosystem services in the riverine floodplain area.

15.7   CONCLUSIONS

The four types of ecosystem services are functioning in the riverine catchment areas in Bangladesh as well as any riverine catchment and basin areas of the world. In this chapter, the Ganges–Brahmaputra–Meghna rivers deltaic region has identified and investigated the following ecosystem services such as supporting services, provisioning of services, regulating services, and cultural services. All these services are now under threat in the Gorai River catchment area and the coastal region in Bangladesh. The settlement relocation in the riverine islands and basin areas, people displacement, rural livelihood, and annual food are interlinked in riverine Bangladesh. Almost every year more than 30,000 *choura people* of Char–Janajat are facing and struggling against floods and *char-land* erosion. The study finding shows that the flood and *char-land* erosion are the main reasons for the cyclic dislodgment of people and settlement relocation in the same *char* and outside of the *char* within 90 km² range in the Padma River channel in Sibchar Upazila of Madaripur District. The Purba Khas Bandarkhola Mouza is one of the vulnerable unstable *char-land* in Bangladesh where the *choura people* cyclically move and migrate because of unstable *char-land* and uncertain livelihood. The finding also lays out that the mouza changes its shape and size by erosional and accretion processes due to devastating floods. Therefore, the *char* inhabitants have been displaced a maximum of 17 times and a minimum of 6 times within 50 years of *char* life that is a threat to *char* livelihood and sustainable *char-lands* development.

The floods in the *char-lands* erosion is a challenge for managing *char* settlements, cultural landscape protection, agricultural cropping systems, crop biodiversity, and riverine ecology of Char–Janajat Island. It is necessary to find an alternative approach or adaptation strategies for sustainable livelihood in the *char-lands* of Padma River channels in the Ganges–Brahmaputra delta in Bangladesh.

The findings of this chapter lead to the following remarks that may serve as basics for implementation in Char–Janajat of Sibchar upazila in Madaripur district. It can be an alternative approach to solve the *char-land* erosion, deposition, people displacement, and settlement relocation problems in the river basins in different regions in Bangladesh.

The government, NGOs, and social organizations need to make an effort to develop methodologies mechanisms, policy-making, validate tools, impact assessment indicators, and instruments for flood and *char-land* erosion protection and management at the micro level in the *char-lands* in the river channels.

Long-term *char-land* data analysis, visualization, mapping, planning, and modeling for future flood forecasting, char-land erosion control, and effective management measures need the application of Geographical Information Systems (GIS), Remote Sensing techniques as well as the application of statistical and hydro-ecological procedures.

For the protection of the *char-land* people's displacement, settlements, and their properties, the construction of hydro-engineering dams or embankments is necessary.

Besides this, the implementation of nonstructural means such as flood monitoring services, forecasting, and warning systems is an alternative solution to *char-lands* problems with high transparency for the local people.

Community participation, local adaptation strategy, awareness education, and applied research on floods, *char-land* erosion, and agricultural cropping systems should be incorporated into the national development agenda. Local settlers would be equal partners in the *char-land* development for safer settlement and livelihood by incorporating indigenous knowledge, skills, and capacities for ecological management and a sustainable livelihood in *char-land* that could be the reality in the riverine environmental context in Bangladesh. The practical alternative should be to promote and strongly support local, regional, and national level programs and initiatives to enable societies to become resilient to the negative impact of *char-land* erosion in the river channel. An interdisciplinary integrated *char-lands* settlement sustainability and migration policy and strategies development plan should be developed based on the result of flood impacts, *char-land* erosion, traditional cropping system, *char* people indigenous adoption skills should include and consider the problem based and project-based management plan of river's bar or *char-lands* in the river channels in Bangladesh.

ACKNOWLEDGMENTS

The chapter incorporates ideas and findings of research I have jointly carried out with my colleagues and Professor Gnauck, and the work carried out by myself on flood mapping, erosion, and deposition through sediment load in the river basins in Bangladesh. I acknowledge their contributions and remain thankful for their support as well as thanks to the dwellers who gave me a lot of information concerning sediment load and environmental issues.

REFERENCES

Abbas, N., and Subramanian, V. (1984). Erosion and sediment transport in the Ganges river basin (India). *Journal of Hydrology*, 69(1–4), 173–182.

Adnan, S. (1991). *Flood, People and Environment*. Research and Advisory Services, Dhaka.

Ahmed, N. (1956). The pattern of rural settlement in east Pakistan. *Geographical Review*, 46(3), 388–398.

Ahmed, R., and Falk, G. C. (2008). Bangladesh: Environment under pressure. *Geographische Rundschau International Edition*, 4(1), 13–19.

Ambrose, R. B. Jr., Wool, T. A., and Martin, J. L. (1993). *The Water Quality Analysis Simulation Program, WASP 5. Part A: Model Documentation*. Environmental Research Laboratory, Athens, GA.

Anwar, J. (1988). Geology of coastal area of Bangladesh and recommendation for resource development and management In *National Workshop on Coastal Area Resource Development and Management, Part II. CARDMA, Dhaka, Bangladesh* (pp. 36–56).

Batchelor, G. K. (2000). *Introduction to Fluid Mechanics*. Cambridge University Press, Cambridge, MA.

BBS. (1993). *Bangladesh Population Census, 1993*. Bangladesh Bureau of Statistics, Dhaka.

Biswas, A. K. (2000). Water for urban areas of the developing world in the twenty first century. In Uttio, J. I., and Biswas, A. K. (Eds.), *Water Resource Management and Policy Water for Urban Areas-Challenges and Perspective*. UNU, Tokyo, 1–23.

Brammer, H. (1996). Geographical complexities of detailed impact assessment for the Ganges-Brahmaputra-Meghna Delta of Bangladesh. In Warrik, R., Barrow, E. M., and Wigley, T. M. (Eds.), *Climate and Sea Level Change*. Cambridge University Press, Cambridge, 263–275.

BWDB (Bangladesh Water Development Board) (1972). Sediment investigations in Main Rivers of Bangladesh, 1968 & 1969. *BWDB Water Supply Paper No. 359*. BWDB, Bangladesh.

Coleman, J. M. (1969). Brahmaputra river: Channel process and sedimentation. *Sedimentary Geology*, 3, 129–239.

Delft Hydraulics and DHI. (1996). *Bed Form and Bar Dynamics in the Main Rivers of Bangladesh, FAP 24 River Survey Project, Special Study Report 9*. Flood Plan Coordination Organization, Dhaka.

Eisma, D. (1993). *Suspended Matter in the Aquatic Environment*. Springer, Berlin.

Elahi, K. M., Das, S. C., and Sultana, S. (1998). Geography of coastal environment: A study of selected issues. In Bayes, A., and Mohammad, A. (Eds.), *Bangladesh at 25, an Analytical Discourse on Development*. University Press Limited, Dhaka, 336–368.

Environmental and Geographical Information Services (EGIS). (2000). *Bangladesh Water Development Board-Environmental Baseline of Gorai River Restoration Project*. (EGIS I, EGIS II and EGIS III) Environment and GIS support project for water sector planning, Ministry of Water Resources, Government of the Peoples Republic of Bangladesh. EGIS, Dhaka, 190.

FAP-24. (1996). *River Survey Project Morphology of Gorai Off-Take: Special Report No. 10*. WARPO, Flood Action Plan, Dhaka.

Fuller, I. C., Basher, L., Mardern, M., and Massey, C. (2011). Using morphological adjustments to appraise sediment flux. *Journal of Hydrology (New Zealand)*, 50(1), 59–79.

Gleick, P. (1993). Water and conflict. *International Security*, 18, 79–112.

Goodbred, S. L., and Kuehi, S. A. (1999). Holocene and modern sediment budgets for the Ganges–Brahmaputra river system: Evidence for highstand dispersal to flood-plain, shelf, deep-sea depocenters. *Sedimentary Geology*, 27(6), 559–562.

Goodbred, S. L., and Kuehl, S. A. (2000). The significance of large sediment supply, active tectonics, and estuary on margin sequence development: Late quaternary stratigraphy and evolution of the Ganges–Brahmaputra delta. *Sedimentary Geology*, 133(3–4), 227–228.

Goodbred, S. L., and Nicholls, R. (2004). Towards integrated assessment of the Ganges–Brahmaputra delta. In *Proceedings of the 5th International Conference on Asian Marine Geology, and 1st Annual Meeting on Asian Marine Geology, and 1st Annual Meeting of IGCP475 Delta and APG Mega Delta*. 13 February, 1–15.

Goswami, U., Sarima, J. N., and Patgiri, A. D. (1999). River channel changes of the Subansiri in Assam, India. *Geomorphology*, 30, 227–244.

Graf, W. H. (1971). *Hydraulics of Sediment Transport*. McGraw Hill, New York.

Gupta, H. P. (1981). Palaeoenvironments during holocene time in Bengal Basin, India as reflected by palynostratigraphy. *Journal of Palaeosciences*, 27(2), 138–159.

Hanna, S. W. (2001). *Facing the Jamuna River-Indigenous and Engineering Knowledge in Bangladesh*. Bangladesh Resource Centre for Indigenous Knowledge (BARCIK), Bersha (Pvt) Ltd, Nilkhet, Dhaka.

Haque, M. (1985). *Impact of Riverbank Erosion in Kazipur, an Application of Landsat Data*. REIS workshop, Jahangirnagar University, Dhaka.

Hassan, M., Haque, M. S., and Saroar, M. (2000). Indigenous knowledge and perception of the *Charland* people in cropping with natural disasters in Bangladesh. *Grassroots Voice – A Journal of Resources and Development*, III, 34–44.

Holeman, J. N. (1968). The sediment yield of major rivers of the world. *Water Resource Research*, 4, 737–747. https://doi.org/10.1029/WR004i004p00737.

Hossain, M. M. (1987). A simplified sediment equation from dimensional consideration. *Journal of the Indian Water Resources Society*, 7(2), 35–44.

Hussain, M. M. (1993). Economic effects of riverbank erosion: Some evidence from Bangladesh. *Disasters*, 17, 25–32.

Islam, A. (1995). *Environment, Land Use, and Natural Hazards in Bangladesh*. University of Dhaka, Dhaka.

Islam, M. R., Begum, S. F., Yamaguchi, Y., and Ogawa, K. (1999). The Ganges and Brahmaputra rivers in Bangladesh: Basin denudation and sedimentation. *Hydrological Processes*, 13(17), 2907–2923. https://doi.org/10.1002/(sici)1099-1085(19991215)13:17<2907::aid-hyp906>3.0.co;2-e.

Islam, M. S. (2001). *Sea-Level Changes in Bangladesh: The Last Ten Thousand Years*. Asiatic Society in Bangladesh, Asiatic Civil Military Press, Dhaka.

Islam, N. (1993). Rural housing in Bangladesh: An overview in search of new strategies. *Oriental Geographer*, 37, 47–59.

Islam, S. (2006). *Encyclopedia of Bengal*. Asiatic Society of Bangladesh, Asiatic Civil Military Press, Dhaka.

Islam, S. N. (2014). An analysis of the damages of Chakoria Sundarban mangrove wetlands and consequences on community livelihoods in southeast coast of Bangladesh. *International Journal of Environment and Sustainable Development*, 13(2), 153–171. https://doi.org/10.1504/IJESD.2014.060196.

Islam, S. N. (2016). Deltaic floodplains development and wetland ecosystems management in the Ganges–Brahmaputra–Meghna rivers delta in Bangladesh. *Sustainable Water Resources Management*, 2(3), 237–256. https://doi.org/10.1007/s40899-016-0047-6.

Islam, S. N. (2021). Floods, charland erosions and settlement displacement in the Ganges–Padma river basin. In Zaman, M., and Alam, M. (Eds.), *Living on the Edge*. Springer Geography, Springer, Cham. https://doi.org/10.1007/978-3-030-73592-0_125_18.

Islam, S. N., and Gnauck, A. (2007). Effects of salinity intrusion in mangrove wetlands ecosystems in the Sundarbans: An alternative approach for sustainable management. In Okruszko, Tomasz, Maltby, Edward, Szatylowicz, Jan, and Miroslaw-Swiatek, Dorota (Eds.), *Wetland: Monitoring and Management*. Taylor and Francis, Rotterdam, 315–322.

Islam, S. N., and Gnauck, A. (2007a). *Salinity Intrusion and Thresholds for the Protection of Mangrove Wetlands Ecosystems in the Sundarbans Region Modellierung und Simulation von Okosystemen*. Shaker-Verlag, Aachen, 187–201.

Islam, S. N., and Gnauck, A. (2008). Mangrove wetlands ecosystem in Ganges–Brahmaputra delta in Bangladesh. *Frontiers of Earth Science in China*, 2(4), 439–448.

Islam, S. N., and Gnauck, A. (2009). Variations in fluvial-deltaic and coastal reservoir deposited in tropical environments: The coastal mangrove wetland ecosystems in the Ganges delta: A case study on the Sundarbans in Bangladesh. In *AAPG HEDBERG Conference*, April 29-May 2, 2009 – Jakarta, Indonesia, 1–4.

Islam, S. N., and Gnauck, A. (2009a). Threats to the Sundarbans mangrove wetland ecosystems from transboundary water allocation in the Ganges basin: A preliminary problem analysis. *International Journal of Ecological Economics and Statistics (IJEES)*, 13(9), 64–78.

Islam, S. N., and Gnauck, A. (2011). Food security and ecosystem services under threat in the coastal region of Ganges delta in Bangladesh: Preliminary result analysis. In Gnauck, A. (Ed.), *Modelling and Simulation of Ecosystems*. Shaker Verlag, Aachen, 158–274.

Jansen, E. G. Dolman, A.J. Jerve, A.M. and Rahman (1994). *The country Boats of Bangladesh, Social and Economic Development and Decision-Making in Inland Water Transport*, University Press Limited, Dhaka, pp. 1–279.

Khan, S. R., and Islam, M. B. (2008). Holocene stratigraphy of the lower Ganges–Brahmaputra river delta in Bangladesh. *Frontiers of Earth Science in China*, 2(4), 393–399.

Lane, S., and Richards, K. S. (1997). Linking river channel form and process: Time, space and causality revisited. *Earth Surface Process and Landforms*, 22, 249–260.

Li, L., Lu, X., and Chen, Z. (2007). River channel changes during the last 50 years in the middle Yangzee River, the Jianli reach. *Geomorphology*, 85, 185–196.

Lijklema, L., Gelencser, P., Szilagyi, F., and Somlyody, L. (1986). Sediment and its interaction with water. In Somlyody, L., and van Straten, G. (Eds.), *Modeling and Management Shallow Lake Eutrophication*. Springer, Berlin, 156–182.

Miah, M. M. (1988). *Flood in Bangladesh: A Hydromorphologial Study of the 1987 Flood, Dhaka*. Academic Publishers, Dhaka.

Milliman, J. D., and Meade, R. H. (1983). World-wide delivery of river sediment to the oceans. *Journal of Geology*, 91, 1–21. https://doi.org/10.1086/628741.

Milliman, J.D., Broadus, J.M., and Gable, F. (1989). Environmental and economic implications of rising sea level and subsiding deltas: the Nile and Bengal examples. *Ambio*, 340–345.

Mills, W. B., Porcella, D. B., Ungs, M. J., Gherini, S. A., Summers, K. V., Lingfunk Mok, Rupp, G. L., Bowie, G. L., and Haith, D. A. (1985). *Water Quality Assessment: A Screening Procedure for Toxic and Conventional Pollutants. Part 1 and 2*. EPA-600/685-002a and b. US EPA, Athens, OH.

NEDECO (1967). Surveys of Inland Water and Ports. Hydrology and Morphology, Vol III, Part B Hydrological and morphological Phenomena, East Pakistan Inland Water Transport Authority, pp. 98.

Rahman, A., Mamunur, R., Dustegir, M., Karim, R., Haque, A., Nicholls, R. J., Darby, S. E., Nakagawa, H., Hussain, M., Dunn, F. E., and Akter, M. (2018). Recent sediment flux to the Ganges–Brahmaputra–Meghna delta system. *Science of the Total Environment*, 43(1), 1054–1064.

Rob, M. A. (1998). Changing morphology of the coastal region of Ganges delta. *Orient Geography*, 41(2), 49–64.

Sakar, M. H. (2008). Mophological response of the Brahmaputra–Padma–lower Meghna river system to the Assam earthquake of 1950. Unpublished PhD Thesis, School of Geography, University of Nottingham, Nottingham, 1–296.

Salarijazi, M., Abdolhosseini, M., Ghorbani, K., and Eslamian, S. (2016). Evaluation of quasi-maximum likelihood and smearing estimator to improve sediment rating curve estimation. *International Journal of Hydrology Science and Technology*, 6(4), 359–370.

Schulze, S. and Islam, S.N. (2012). Bangladesh (Country Profile) Encyclopedia of Global Warming & Climate, SAGE Publications, pp. 121–122.

Subramanian, V. (1987). Environmental geochemistry of Indian river basins - A review. *Journal of the Geological Society of India*, 29, 205–220.

Vanoni, V. A. (1977). *Sedimentation Engineering*. Manual No. 64. American Society of Civil Engineers, New York.

Vitousek, P. (1997). Human domination of earth's ecosystems. *Science*, 277, 494–490.

Water Resource Planning Organization (WARPO). (1998). *Ministry of Water Resources, National Water Management Plan*. Inception report, Vol. 1. Main Volume. WARPO, Dhaka.

Wischmeier, W. H., and Smith, D. D. (1978). *Predicting Rainfall and Erosion Losses – A Guide to Conservation Planning*. Agricultural Handbook No. 537. US Department of Agriculture, Washington, DC.

Zube, E. H. (1986). Landscape values: History, concepts and applications. In Smardon, F., and Fellman, J. P. (Eds.), *Foundation for Visual Projects*. Wiley, New York, 4–19.

16 Impact of Climate Change and Human Influence on the Sediment Yield of Indian Himalayan Rivers

Monica Sharma Shamurailatpam,
Arindan Mandal, and Alagappan Ramanathan

16.1 INTRODUCTION

Sediment transportation by the river system is an important indicator of the erosional processes and depositional activities that shape the morphology of the river basin (Dade and Friend, 1998; Church, 2006). Along with the sediment loads it carries essential dissolved nutrients to the riverine ecosystems (Reid et al., 2005; Julien, 2010; Apitz, 2012). Climatic influences such as temperature and rainfall are the factors that control the hydrological cycle and in turn affect the sediment loads in the river (Fujita, 2008; Zhu et al., 2008). The rivers in the Himalayas are amongst the highest contributors in terms of suspended sediment load, and it is approximated that over one-third of the total sediment inputs to the oceans. The Himalayan river also plays an essential role in the global sediment cycle and it exports nutrients and dissolved minerals downstream and is ultimately stored in the ocean (Krishnaswami and Singh, 2005; Krishna et al., 2016). It is estimated that the major rivers of the Himalayas, namely, the Ganges, Brahmaputra, and Indus, from the Tibetan region carried around 3 Gt year^{-1} or over 25% of the total sediment flux to the oceans (Figure 16.1) (Raymo and Ruddiman, 1992; Summerfield and Hulton, 1994). According to Subramanian and Ramanathan (1996), out of the global flux of 15×10^9 tonnes/yr (Milliman and Meade, 1983), it is estimated that as much as 90% of the sediment contribution is from the Himalayan rivers.

The high sediments load from these regions is responsible for its natural aspects such as steep slopes, highly erodible lithology, high elevations, and intensive rainfall (Galy and France-Lanord, 2001; Finlayson et al., 2002; Vance et al., 2003; Thiede et al., 2004; Burbank et al., 2012). The region is made of geologically weak rocks such as sandstone and grit conglomerates which are highly erodible. Moreover, natural hazards such as inundations, landslides, slope failures, and glacial lake outburst floods (GLOFs) further contribute to the large volumes of sediments. Further anthropogenic activities such as intensive agriculture, urbanization, and deforestation significantly altered the sediment load transportation in these river systems. Thus the magnitude of sediment transportation and erosion in the river pose serious concerns for the river discharge, reservoirs, and riverine ecosystem in the Himalayan region (Jiongxin, 2003; Walling, 2006; Kummu and Varis, 2007). Often the severity of the erosion in the riverine system is more likely to be increased with a reduction in vegetation area, where bare soils are mostly exposed (Walling, 1999; 2008; DeFries and Eshleman, 2004; Chakrapani, 2005). The process of sediment production, transportation, and depositional is a complex process involving wide variables and the complication increases in large catchment areas compared to smaller scales (Walling and Webb, 1996). Given the warming temperature and increased human actions, the natural flow of sediment

DOI: 10.1201/9781003473398-19

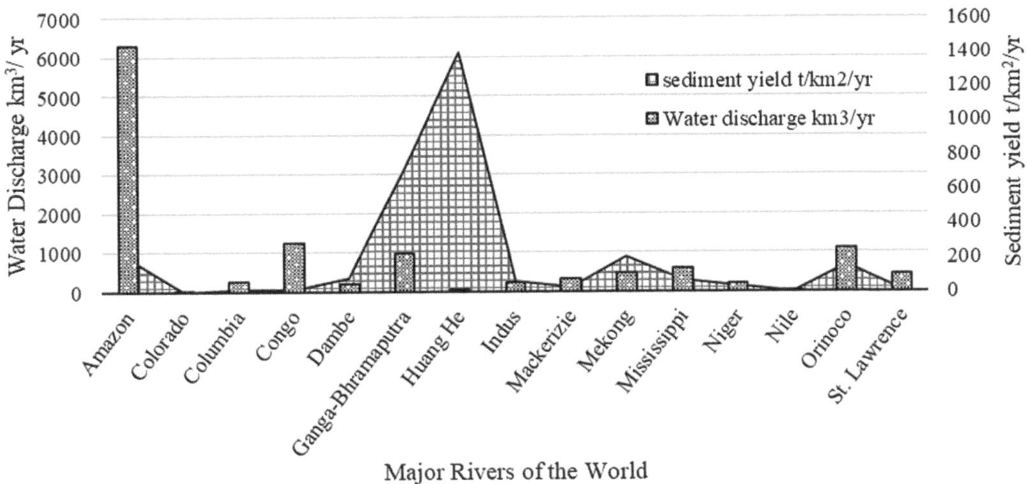

FIGURE 16.1 Suspended sediment yield (SSY) and water discharge of some major rivers of the world (adapted from McLennan, 1993).

transportation has been considerably affected (Walling and Fang, 2003; Walling, 2006; Darby et al., 2015). The glaciers in the Himalayas have been retreating in the last few years leading to an enhanced level of runoff and in turn intensified transporting of suspended load (Immerzeel et al., 2010). Studies have indicated that the hydrological cycles and sediment load from the glacierized basin can be altered significantly in the future because of climate change (Singh et al., 2007; Lu et al., 2011; Wulf et al., 2012). The melting of the glacier in the Himalayan region will affect the hydrology of the rivers (Singh et al., 1995; Haritashya et al., 2006; Kumar et al., 2018b), however, the impact may differ from catchment to catchment (Walling, 2008). For example, in the western Himalayas, the overall contribution of snow and glacial meltwater is significant (>50%), while in the eastern Himalayas, the annual discharge is lesser (~30% in Tsangpo River), and it becomes less significant in the central Himalayas (<20%) (Immerzeel, 2008; Bookhagen and Burbank, 2010). The summer monsoon rainfall, however, decreases in its strength along the east to the west trajectory of the Northern Himalayan ridge (Mukherjee et al., 2015), specifically in the western and central parts compared to the eastern part of the Himalayas (Basistha et al., 2008). The shifting monsoon trend of rainfall and the glaciation and deglaciation events that happened over the past and present also play a major factor in the variation in sedimentation rate both spatially and temporally in these regions (Arora et al., 2006; Bookhagen and Burbank, 2006; Mukherjee et al., 2015). For instance, the precipitation in the Ganges basin has been reduced by 11.25% from the 1950s compared to that of the 1990s (Agarwal Tushar, 2014).

Himalayan rivers are particularly vulnerable to climate change such as temperature and precipitation change as a vast proportion of it is comprised of snow and glacier covers, steep slopes, and high relief making it highly vulnerable to glacier and snow loss, surface erosion, landslides, and climatic disasters. Increased man-made infrastructures such as dams and reservoir construction, diversion of water resources, and increased deforestation led to severe degradation of the catchment area. It has already been reported in the literature (Walling and Fang, 2003; Darby et al., 2015; Fischer et al., 2017; Khan et al., 2018) that sediment has a tendency to be more sensitive to climate modifications and land covers conditions, therefore it is very important to analyse the sediment dynamics of the Himalayan rivers. This chapter aims to provide an improved understanding of erosion and sediment yield across the Himalayan river system from the western to eastern Himalayas and explore the different factors that are controlling sediment transportation in the region.

16.2 STUDY AREA

The Himalayan region is primarily comprised of the Indus, Ganges, and Brahmaputra River basins (Figure 16.2). The Indus basin has a total area of 1.12 million km² of which 39% of the basin area lies in India and other countries comprising Pakistan (47%), China (8%), and Afghanistan (6%) (FAO, 2016). The climatic condition in the Indus basin varies from subtropical dry and semi-arid to temperate sub-humid on the lowlands and high-altitude mountains in the north. The annual precipitation extends between 100 and 500 mm in the plains to a high of 2000 mm in highland gradients (FAO Aquastat, 2011). Snowfall at elevations greater than 2500 m provides for the largest part of the river runoff (Ojeh, 2006). The Ganges originates from the Himalayas and Tibetan Plateau and has a total area of 1.09 million km² flowing south through China (3%), India (79%), Nepal (14%), and, finally, joining with the Brahmaputra before entering Bangladesh (4%) (FAO Aquastat, 2011). Precipitation in the Ganges varies widely, in the upper Gangetic flood plain the average annual rainfall varies between 760–1020 mm, in the central part it receives around 1020–1520 mm, and in the mouth of the river is 1520–2540 mm (Palash et al., 2018). It receives precipitation mostly around July to October and gets only a lesser amount of rainfall occurs in December and January. The Brahmaputra originating from the Himalayan mountain range is an east-flowing river, the total basin area is 0.54 million km², of which half of the basin area lies in China (50%), followed by India (36%), Bhutan (7%), and Bangladesh (7%) (FAO Aquastat, 2011). The basin is severely influenced by the ISM with annual precipitation ranging between 1200–6000 mm (mean 2300 mm), except in the upper reaches of the basin which is mostly occupied by the Himalayan rain-shadow area (Nepal and Shrestha, 2015). Most of the rainfall (about >60%) is received during July–September.

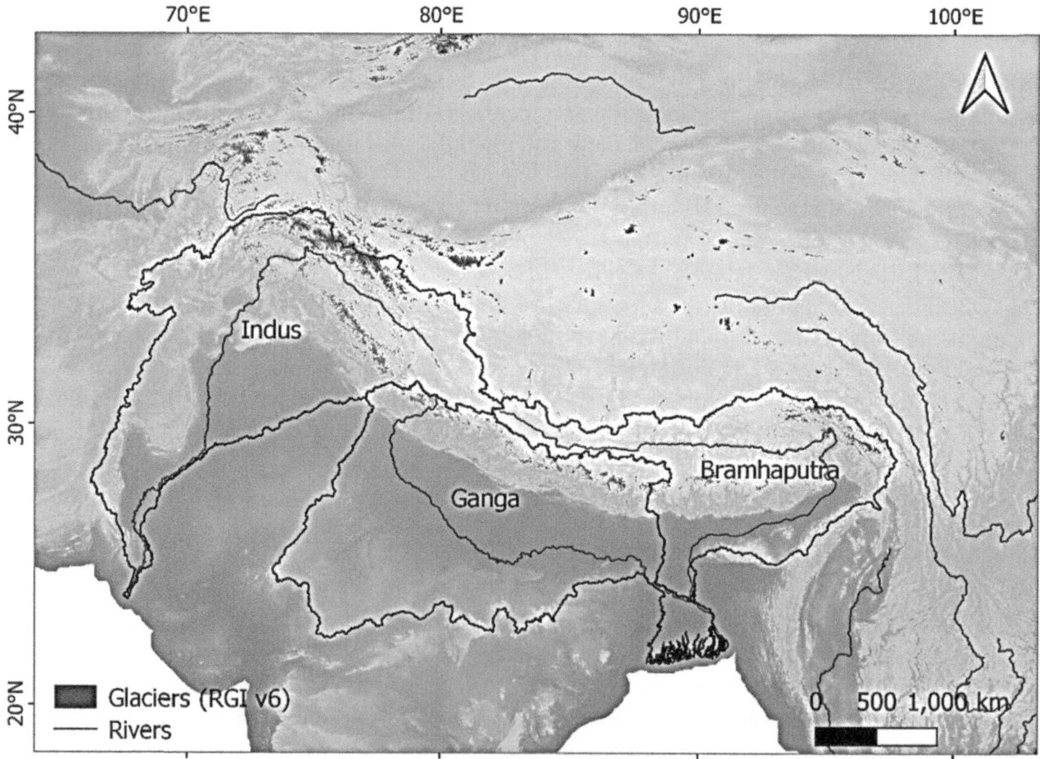

FIGURE 16.2 Map showing the Indus, Ganges, and Brahmaputra basins in the Indian Himalayan region. Dark patches showing the glacier cover from RGI v6.

16.3 ROLE OF SEDIMENT YIELD TO THE BIOGEOCHEMICAL CYCLES

Glaciers and ice sheets are considered a major source of various micro and macronutrients such as phosphorus, iron, silica, dissolved carbon, and nitrogen in the polar regions (Hawkings et al., 2014; 2015; 2016; 2017; Hood et al., 2015; Wadham et al., 2016). The proglacial and subglacial sites are active for both microbial and physical activities that could serve as a major source of nutrients downstream (Hodson et al., 2008; Achberger et al., 2017; Anesio et al., 2017). A recent study in the Himalayas also identifies the supraglacial environment as an active source of autochthonous organic matter (Nizam et al., 2020; Shamurailatpam et al., 2023) that can be potentially exported along with the cryoconite debris with the melting glacier surface. Such studies are still limited in the Indian Himalayan region and need extensive study to understand the nutrients associated with sediment debris and their consequences to the downstream ecosystem. Studies from the tributaries of the Tibetan Plateau indicate the export of a high concentration of dissolved inorganic carbon (30.7 mg/L), while the dissolved organic carbon concentration showed a low concentration (1.16 mg/L) due to the minimal productivity in such a cold environment (Qu et al., 2018). The labile dissolved organic carbon is potentially more bioavailable and can possibly influence the downstream carbon cycle (Anesio et al., 2009; Lawson et al., 2014; Hood et al., 2015; Qu et al., 2018). Dissolved nitrogen exports have also reportedly increased in this region due to discharge from agricultural and urban activities and hence can be expected to influence the regional nitrogen cycle in the future (Qu et al., 2018). The studies in the Ganges basin observed that the sediment concentration is especially high during the wet season compared to the post and pre-monsoon (Chakrapani and Saini, 2009). Considering the warming climate change, the Himalayan Rivers, because of their high relief, intensive water erosion, and intensive agricultural activities, have the possibility to increase exports of nutrients to the downriver and influence the biogeochemical cycles in the future. There remains uncertainty on the exports of sediment and associated nutrients from the glacier system that can change over time, due to changes in magnitude, timing, glacial area, vegetation coverage, and discharge. One recent study by Sarkar et al. (2023), has addressed that the glacial export of carbon and nitrogen fluxes associated with total suspended sediment (TSM) in the major river basin of the Ganges, Brahmaputra, and Mekong has been decreasing in the last decades because of change in precipitation patterns, low discharge flow, urbanisation, and water diversion projects. Therefore, changes in the flux of nutrients and sediments may also depend on spatial and temporal factors and can be highly diverse in the complex Himalayan cryosphere.

Most rivers in the Himalayas are perennial rivers as the runoff in these tributaries are dominated by glacial meltwater and precipitation. The glacierised basin in the Himalayan region provides a major source of sediments (Jain et al., 2003; Arora et al., 2015). The specific sediment yield (SSY) and drainage basin area of some of the major rivers in the Himalayan region stretch from the western to the eastern Himalayas during the summer (Figure 16.3; year of observations in Table 16.1). It is observed that SSY in the majority of the rivers in the Himalayan region is significantly higher than the mean global SSY (~150 tons km^{-2} yr^{-1}) (Jha et al., 1988). Average SSY in the Chandra basin has increased by two-fold since the observed period between 1978–1995 (Rao et al., 1997) to 2017 (Singh et al., 2020) (Figure 16.3). Such an increase in SSY could be due to the retreat and mass loss of glaciers in the Chandra basin as observed in previous literature (Tawde et al., 2017; Prakash and Nagarajan, 2018; Mandal et al., 2020) which resulted in considerably increased discharge and thus the sediment load and erosion rate in the Chandra basin (Singh et al., 2020). A study done by Singh and Ramanathan (2018) observed a higher yield of ~3000 tons km^{-2} yr^{-1} during the observed period of 2011–2014. Another study in the upstream catchment near the Gangotri Glacier reported an SSY of 4834 tons km^{-2} yr^{-1} (Haritashya et al., 2006; Wulf et al., 2012), while at the snout of the Gangotri Glacier, the SSY is much higher (~7663 tons km^{-2} yr^{-1}) (Singh et al., 2018). In the Central Himalayas, sediment transportation is influenced by the combined actions of steepness along with the strong ISM and physical erosion of the glacier (Kumar et al., 2016). The downstream basin of the Ganges at Alaknanda, Bhagirathi reported lowered SSY of 863 tons km^{-2} yr^{-1}, and 907 tons

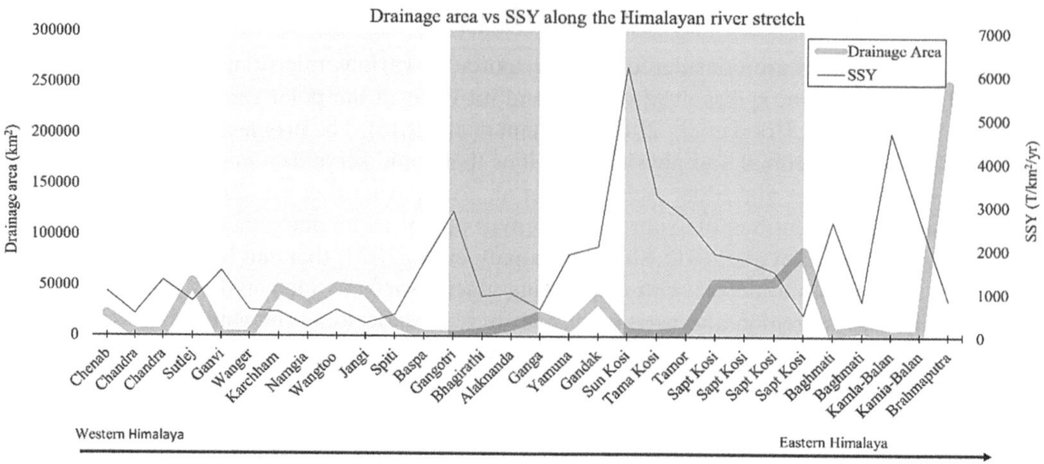

FIGURE 16.3 Figure showing the distribution of SSY and the drainage area in the Himalayan stretch from Western towards the Eastern Himalayas and reflects the summer only (Table 16.1).

km^{-2} yr^{-1} (Chakrapani and Saini, 2009), and 628 tons km^{-2} yr^{-1} at Rishikesh (Jha et al., 1988). The decrease in the sediment yield as it progresses downstream could be due to a decrease in the strength of glacial-induced physical erosional rate in the downstream area (Chakrapani and Saini, 2009).

The suspended load of the Ganges River is largely fine to very fine sands (Sinha and Friend, 1994) with the fine fractions essentially representing the wash load (Roy and Sinha, 2017). The increase in sediment loads, as it approaches the plain area, is because the tributaries pick up most of the load through catchment erosion and/or reworking small outlets (Sinha and Friend, 1994). Sediment load is observed to be at its peak in the wet/monsoon season mainly due to elevated discharge rate, heavy precipitation, and extensive agricultural land in the alluvial plain generated large bank erosion (Roy and Sinha, 2017). High SSY is often directly related to higher runoff (Figure 16.4). In the Western Himalayas, high magnitude sediment discharge is often related to peak events such as high rainfall, high snowfall, and tectonic events (Wasson, 2003; Bookhagen and Burbank, 2006; Wulf, 2011).

16.4 SEDIMENT LOAD RESPONSE DUE TO CLIMATIC CHANGE

Significant rises in temperature and precipitation in the Himalayan rivers have been confirmed by several studies (Bhutiyani et al., 2007, 2008; Zereini and Hötzl, 2008; Kumar et al., 2018b). This observation is consistent with the findings that the average temperature of India has risen by ~0.7°C in the period between 1901–2018 (Krishnan et al., 2020). The increase in temperature is, however, more sensitive in the Himalayas with an average rise of 1.5°C from 1982 to 2006 with the greatest increase of 1.7°C observed in the winter seasons and the least one during the summer (Shrestha et al., 2012). The study further suggested that the change in the temperature also varied spatially with the forest region of Brahmaputra that suffered the highest rate of warming of 2.0°C during the last 25 years. Another study by Shekhar et al., 2010 observed that the wintertime warming temperature has increased by ~2°C for the period 1984–2008 in the western Himalayas. In response to this increasing temperature, several authors have suggested the retreating of glaciers resulting in a drastic reduction of water supplies in several rivers in the long term (Bolch et al., 2012; Kääb et al., 2012) with an initial rise in meltwater volume. Some studies have claimed that the temperature rise may create positive feedback that may intensify the hydrological cycles and affect the hydrological regimes. However, this impact will vary temporally and spatially for a different region (Shekhar et al., 2010), and may benefit the water resources in some regions (Perry et al., 2009; Gornall et al., 2010). Numerous studies have indicated that climate change could have a substantial consequence

TABLE 16.1

Table Showing the Drainage Area, Runoff Data, Sediment Load, and Sediment Yield Observed in Different Studies in the Himalayan Rivers

River	Station	Observed period	Drainage area (km²)	Runoff (m/year)	Sediment load 10⁶ t /yr	SSY (T/km²/ yr)	Reference
Sutlej	Namgia	2005–2008	30950	0.06	6.9	223	(Wulf, 2011)
Sutlej	Jangi	2007	44738	0.13	13.5	302	(Wulf, 2011)
Sutlej	Karchham	2006–2007	46291	0.16	25.7	556	(Wulf, 2011)
Sutlej	Wangtoo	2004–2009	48316	0.20	29.7	615	(Wulf, 2011)
Sutlej	Kasol	1994–1996	53,768	-	43.9	816	(Jain et al., 2003)
Chandra	Ghousal	1978–1995	2,490	-	1.3	513	(Rao et al., 1997)
Chandra	Tandi	2017	2,440	1.9	3.1	1,285	(Singh et al., 2020; 2021)
Chenab	Akhnoor	1971–1995	21,808	-	22.4	1,029	(Rao et al., 1997)
Baspa	Sangla	2004–2008	989	1.14	1.7	1,717	(Wulf, 2011)
Spiti	Khab	2005–2008	12,477	0.26	6.2	499	(Wulf, 2011)
Wanger	Kafnu	2003	264	1.67	0.2	614	(Wulf, 2011)
Ganvi	Ganvi	2003	117	1.27	0.2	1,507	(Wulf, 2011)
Gangotri	Near snout	2000–2003	556.47	1.82	2.7	4,834	(Haritashya et al., 2006)
Bhagirathi	Maneri	2004	4,024	1.22	3.7	917	(Chakrapani and Saini, 2009)
Alaknanda	Srinagar	2004	10,237	1.70	10.2	995	(Chakrapani and Saini, 2009)
Yamuna	Tajewala	1983	9,572	1.10	18.1	1,889	(Chakrapani and Saini, 2009)
Ganges	Rishikesh	2004	20,600	1.15	12.9	628	(Jha et al., 1988)
Gandak	Triveni	1980–1989	37,845	1.53	78.5	2,074	(Sinha and Friend, 1994)
Sun Kosi	Dolalghat	1951–2007	4,842	0.27	30.0	6,200	(Sinha et al., 2019)
Tama Kosi	Busti	1951–2008	3,088	0.50	10.0	3,240	(Sinha et al., 2019)
Tamor	Mulghat	1951–2009	5,892	0.30	16.0	2,720	(Sinha et al., 2019)
Sapt Kosi	Chatara	1951–2010	52,730	0.02	101.0	1,915	(Sinha et al., 2019)
Sapt Kosi	Barahkshetra	1951–2011	52,735	0.02	93.9	1,780	(Sinha et al., 2019)
Sapt Kosi	Bizpur	1951–2012	54,089	0.02	81.1	1,500	(Sinha et al., 2019)
Sapt Kosi	Baltara	1951–2013	84,739	0.01	43.2	510	(Sinha et al., 2019)
Baghmati	Dheng	1951–2014	3,790	0.34	10.0	2,640	(Sinha et al., 2019)
Baghmati	Hayaghat	1980–1989	8,440	0.08	7.0	830	(Sinha and Friend, 1994)
Kamla–Balan	Jaynagar	1980–1989	2,131	0.46	10.0	4,690	(Sinha and Friend, 1994)
Kamla–Balan	Jhanjharpur	1980–1989	2,945	0.25	8.0	2,720	(Sinha and Friend, 1994)
Ganges	Hardinge Bridge	1980–1995	980,000	0.37	316	774	(Islam et al., 1999)
Brahmaputra	Pasighat	-	249,000	0.80	209.9	843	(Stewart et al., 2003)
Brahmaputra	Bahadurabad	1980–1996	640,000	0.95	721.0	1,126	(Islam et al., 1999)

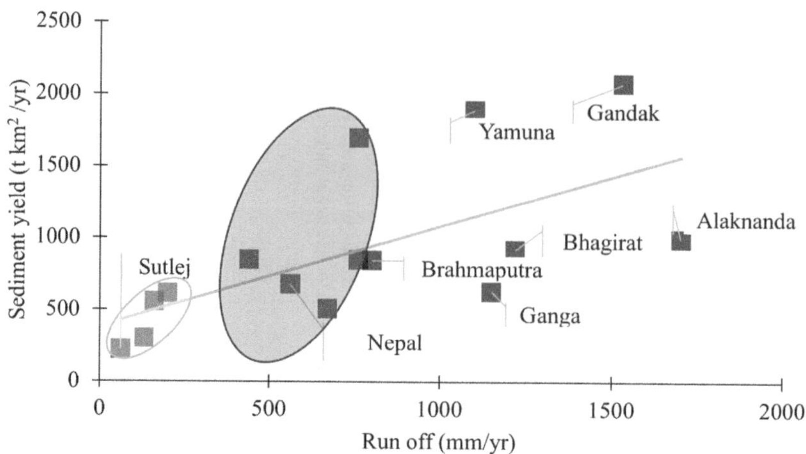

FIGURE 16.4 The plot indicates the link between SSY and runoff in some of the major rivers of the Himalayas (data from Table 16.1). Encircled data are observations for more than a year while the rest are only one seasonal observation.

on the hydrological dynamics (Nijssen et al., 2001; Menzel and Bürger, 2002; Sharif et al., 2010; Khan et al., 2018), soil denudation rate (Pruski and Nearing, 2002; Michael et al., 2005) and sediment transportation (Jiongxin, 2003; Syvitski et al., 2005; Khan et al., 2018).

The total sediment load from some of the rivers to the oceans has been reduced significantly in recent years in the Himalayas (Lu et al., 2011). For instance, the Ganges and Indus have been reported for substantial declines in both sediment load and water flow (Adel, 2001; Dai et al., 2008; Inam et al., 2008; Rahman et al., 2018) with major alterations in water flow under climate change and human influence (Dai et al., 2008; Milliman et al., 2008). Another key finding by Darby et al., 2015 projected that the sediment carrying capacity of the Ganges and Brahmaputra Rivers to the Bay of Bengal can potentially rise under the current anthropogenetic climate change, with an increase of 16% to 18% of sediment loads for the Ganges, and by 25% to 30% for the Brahmaputra rivers towards the mid-21st century. Towards the end of the 21st century, as liquid precipitation dominates over snowfall, the projected rise in sediment from the two basins is likely to reach between 34% and 37% for the Ganges and between 52% and 60% for the Brahmaputra (Darby et al., 2015). A similar finding by Fischer et al., 2017 also projected an increase of sediment loads by roughly 40% by the end of the current century to that of 1986–1991.

For the western Himalayas, the recent rise in air temperature is leading to change in the trend of snowfall in various parts, for instance, studies by Dimri et al. (2008), Shekhar et al. (2010), and Dimri and Dash (2012) have observed a reduction in snowfall in a warming climate over the various part of the Western Himalayas. As more precipitation is received as rainfall, induced by warming air temperature, this has been expected to be the cause of the glaciers melting observed in the last few decades (Mandal et al., 2016; Prakash and Nagarajan, 2018) exposing easily erodible paraglacial landscapes. The recent expansion of glacial lakes has been reported by several studies in the Himalayas (Prakash and Nagarajan, 2018; Kaushik et al., 2021; Zhang et al., 2021) due to changes in the temperature and precipitation patterns. Such changes due to climate change can trigger landslides, glaciation, erosion processes driven by glacial runoff, glacial lake outburst floods, and landslide lake outburst floods (LLOFs) enhancing the erosion intensity (Wulf et al., 2012).

16.5　HUMAN INFLUENCE

Sediment loads are significantly affected and more evident in areas where there are greater human interventions such as faulty agricultural practices, development of man-made infrastructures such as dams, canals, barrages over the river system, and deforestation. Cultivation in steep slopes due to shifting cultivation in the northeastern Himalayas has resulted in a land loss of roughly 30,000 km² (Narayana and Babu, 1983). In recent decades, the Himalayan region has experienced significant hydrogeological, social, and economic changes (Babel and Wahid, 2011; Surinaidu et al., 2020). The Himalayan mountains as a fragile ecosystem, being geologically unstable, and the region having an underdeveloped economy have been important concerns. The increase in population has placed great stress on these regions, leading to irrational land transformation, triggering environmental degradation, and disrupting the ecological balance through reducing groundwater recharge, increased runoff, and expanded sedimentation (Immerzeel et al., 2020). However, the relationship between anthropogenic activities and sediment loads is still poorly understood, especially in the Himalayas (Singh et al., 2010). In the Himalayan region, the majority of the inhabitants primarily rely on agricultural activities for their livelihoods (Rasul, 2010; Biemans et al., 2019). The Himalayan rivers provide for the livelihood of millions of inhabitants residing in the downstream region. It provides water supplies for use in irrigation, drinking, hydropower generation, and various other uses. An estimated irrigated area of around 144,900 km² in the Indus Basin, 156,300 km² in the Ganges Basin, and 6000 km² in the Brahmaputra Basin (Immerzeel et al., 2010) are directly sourced from this river water. The natural flow of the Ganges is highly limited and most of the surface runoff (80%) is in the wet season (June to September) (Chakrapani and Saini, 2009; Babel and Wahid, 2011; Kumar et al., 2018a). During the peak flow period, the bulk of the sediment and nutrients from the basin are exported with the flow thus providing higher loads to the rivers (Bhatt et al., 2018). A recent study by Singh and Singh, 2021, from the inter-montane cultivated catchment in the northwest Himalayas observed a high flow and spatial variation in sediment and nutrient loads and yield thereby potentially polluting the downstream. The study also proposed a dependency of the sediments and nutrients load with the discharge rate, with the highest sediment and nutrient loadings observed during the peak flow. The previous study also observed that highly disturbed agricultural and deforested basins generally have the greatest sediment fluxes, for instance, a study in the Central Himalayas estimated that agricultural and barren land had a denudation rate of 0.18 mm/yr compared to the natural forest which is significantly lower and varied between 0.02–0.04 mm/yr (Rawat and Rawat, 1994). This study indicated that poorly managed agricultural and deforested lands can increase the denudation rate by up to five to ten times despite the homogeneous bedrock with undulant terrain and cool temperate climate conditions in the Central Himalayas.

Another recent study by Khan et al., 2018 conclusively indicated that agricultural and barren land in the Ganges basin contributes to 87% of the total SSY with a total area of 77% whereas, natural forest contributes only 1% with a total area of 15% (Figure 16.5). The study also predicted a rise of 10–40% in sediment production by the middle of the century and a rise of ~35–79% by the year 2100, due to an increase in transport capacity projected by higher monsoon peak flow in the region.

The development of the dams, barrages, and diversion channels in the alluvial plains of the Himalayan region has considerably affected the river flows and sediment transportation (Bandyopadhyay, 1995). Throughout the past few decades, rivers from the Himalayan region have been dramatically changed in both water flow and sediment flux. Figure 16.6 shows the changing dynamics of the SSY in some of the major Himalayan rivers before the 1980s and 1990s. However, the magnitude and nature of sediment transport differ widely in the Himalayan regions depending on the degree of urbanization, uses of water resources, and climatic conditions (Subramanian and Ramanathan, 1996; Sarkar et al., 2023). The Ganges has been observed to decrease in sediment loads because of dams and abstractions, whereas increased erosion has been observed in the Nepal Mountains due to the removal of forested areas that have led to increased sediment loads on the Brahmaputra tributaries (Hossain, 1992; Sarkar et al., 2023). The construction of dams and extractions of both water and sediment (e.g., sand extractions) have disturbed the river flow regimes of the Ganges, especially down the Farakka Barrage (Mirza, 1998; Sarkar et al., 2023). A significant decrease in the rate of both water discharge (by 1.1%) and sediment suspended volume (by 6.2%) was observed in the Farakka gauging sites due to the intervention by the barrage

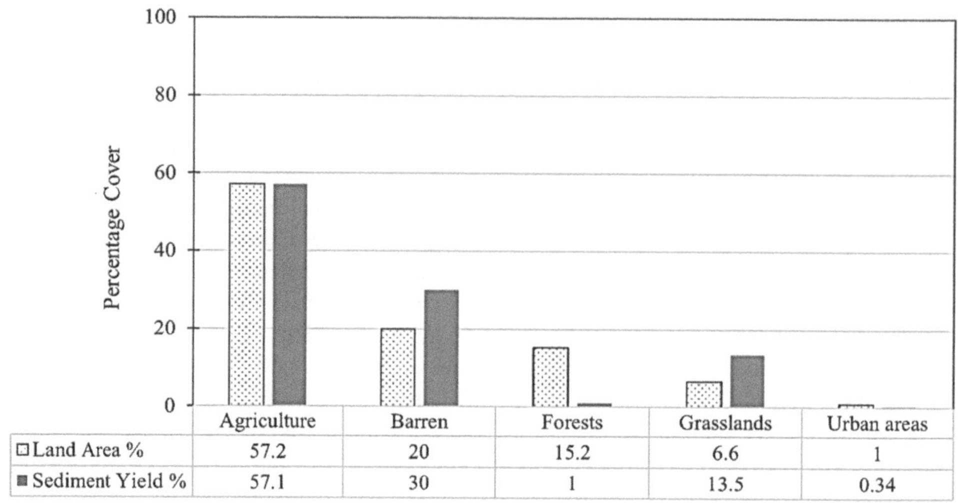

FIGURE 16.5 Histogram showing the land cover of the Ganges basin (Adapted from Khan et al. 2018).

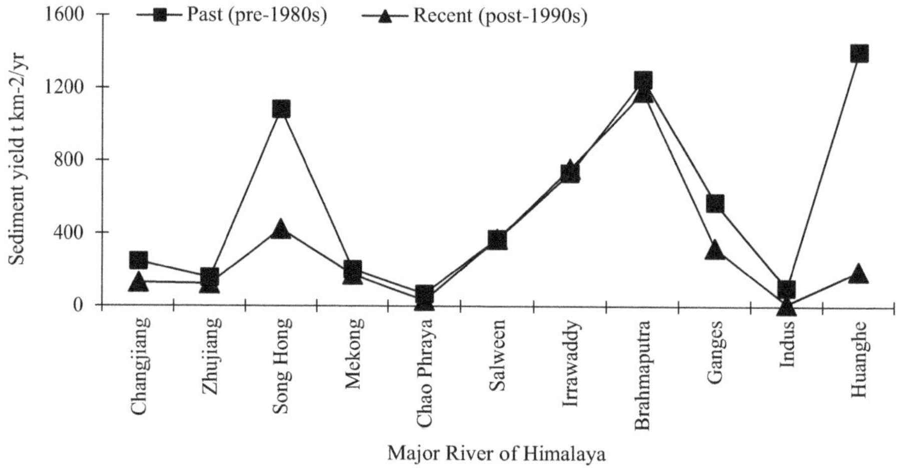

FIGURE 16.6 The graph shows the changing dynamics of the SSY in some of the major rivers of the Himalayas pre-1980s and post-1990s. (Adapted from (Lu et al., 2011) sources: (Milliman and Meade, 1983; Yang et al., 2006; Zhang et al., 2008; Thanh et al., 2005; Kummu and Varis, 2007; Furuichi et al., 2009; Islam et al., 1999; Inam et al., 2008; Wang et al., 2007).

downstream (Zakwan et al., 2018). Rivers such as the Brahmaputra and Irrawaddy do not show significant trends in both sediment transportation and river water flow (Islam et al., 1999; Furuichi et al., 2009) although some studies reported increasing water discharge in the Brahmaputra during the mid-1950s to 2000 (Milliman et al., 2008). However, until now little work has been done on the study of sediment dynamics on a basin scale, and there is no continuous measurement of suspended sediment transport of the rivers, thus creating a challenge in estimating the sediment load downstream and to the oceans. These led to large uncertainties in the sediments input estimation in the Ganges–Brahmaputra Delta mainly due to vagueness in climate estimation and unpredictable human impacts in the catchment in the future (Khan et al., 2018).

Table 16.2 shows the various estimated sediment loads of the Ganges–Brahmaputra Rivers from various published data. The annual sediment discharge in the Ganges–Brahmaputra reflects a wide

TABLE 16.2

Table Showing Estimated Sediment Load in Ganges and Brahmaputra Rivers from Published Data

Measurements of the sediment load of the Ganges River from various studies

Sediment load (10^6 t yr^{-1})	References	Sampling station	Period of measurement
1600	(Holeman, 1968)	Hardinge Bridge	1874–1879
485	(Coleman, 1969)	Hardinge Bridge	1958–1962
530	Bangladesh Water Development Board (Islam et al., 1999)	Hardinge Bridge	1966–1969
680	(Milliman and Meade, 1983)	Hardinge Bridge	1966–1967
350–600	(Hossain, 1992)	Hardinge Bridge	1980–1986
660	(Abbas and Subramanian, 1984)	Calcutta	1981
729	(Abbas and Subramanian, 1984)	Farakka	1981
403	(Singh, 1988)	Unknown	1984
375	Netherlands Engineering Consultants Ltd. In (Islam et al., 1999)	Unknown	1960
316	(Islam et al., 1999)	Hardinge Bridge	1979–1995
794	(Wasson, 2003)	Bengal Delta	unknown
262	(Rice, 2007)	Hardinge Bridge	2006
390–548	(Darby et al., 2015)	Hardinge Bridge	1981–2000
150–590	(Rahman et al., 2018)	Hardinge Bridge	1960–2008

Measurements of the sediment load of the Brahmaputra River from various studies

Sediment load (10^6 t yr^{-1})	References	Sampling Station	Period of measurement
800	(Holeman, 1968)	Bengal Delta	1874–1879
617	(Coleman, 1969)	Bahadurabad	1958–1962
540	Bangladesh Water Development Board (Islam et al., 1999)	Bahadurabad	1967–1969
1157	(Milliman and Meade, 1983)	Bahadurabad	1966–1967
402	(Goswami, 1985)	Pandu, Assam	1955–1979
710	(Subramanian, 1987)	Not mentioned	Not mentioned
750	Netherlands Engineering Consultants Ltd. In (Islam et al., 1999)	Bangladesh	1960
650	(Hossain, 1992)	Bahadurabad	1982–1988
721	(Islam et al., 1999)	Bahadurabad	1989–1994
590	(Wasson, 2003)	Bengal Delta	Unknown
387	(Rice, 2007)	Sirajganj	2006
595–672	(Darby et al., 2015)	Bahadurabad	1981–2000
135–615	(Rahman et al., 2018)	Bahadurabad	1960—2008

variation in short-term periods. For instance, the Ganges (Abbas and Subramanian, 1984) reported a total sediment of 729 Mt yr^{-1} while another study by Singh, 1988 around the same time reported 403 Mt yr^{-1}, similarly, Rice, 2007 reported a 262 Mt yr^{-1}, while a study by Wasson, 2003 reported an annual sediment load of 794 Mt yr^{-1}. Likewise in the Brahmaputra River, a study by (Islam et al., 1999) reported a sediment load of 540 Mt yr^{-1} during the study period of 1967–1969, while another study around the same period (1966–1967), (Milliman and Meade, 1983) estimated 1157 Mt yr^{-1}. Also in the recent period (Wasson, 2003; Darby et al., 2015) reported a sediment load of 590–672 Mt yr^1, while (Rice, 2007) reported 387 Mt yr^{-1}. Such large variations observed in the Ganges–Brahmaputra signify the challenge of measurement. Nevertheless, such large variations could be because of seasonal and inter-annual variations (Subramanian and Ramanathan, 1996) or due to extreme hydrological and weather events such as high monsoon rainfall, rapid snowmelt, or landslides (Thiede et al., 2004; Bookhagen et al., 2005; Decaulne et al., 2005; Burbank et al., 2012; Teng et al., 2020). On a long-term observation in the Ganges, it is noticed a net reduction from 90–60% of the sediment load to that reported earlier. For the Brahmaputra, the sediment load in the recent period has decreased by 23–83% from that reported by Holeman (1968) during the study period of 1874–1879 (Table 16.2). Therefore, to project the future changes in sediment loads, it is desirable to study for longer times to minimise the uncertainty and assumption related to environmental conditions.

16.6 CONCLUSIONS

The consequence of climate change will have a major impact on the precipitation, and glacier melt, which eventually affect the river runoff, sediment transportation, and sediment loads in the future. It can be concluded that the upper reaches of the Ganges and Brahmaputra are likely to increase in river flow and thereby sediment loads due to the projected increase in air temperature and precipitation while the Indus basin by an increase in glacier and snow melt. Human-induced climate change has a greater risk of altering the extent, occurrence, and destruction expenses of future extreme events in this region. There is likely evidence that increased runoff due to a warming climate, particularly from the agricultural and barren lands, can increase exports of sediments along with nutrients, which can alter the productivity of the riverine ecosystem and potentially increase the pollution downstream. However, changes in sediment load and associated nutrient flux can change over time and need extensive studies. In order to minimise the uncertainties in sediment base data measurement and projection, it is needed to measure and study the essential climate variabilities such as temperature change and precipitation on a long-term basis.

REFERENCES

Abbas, N., and Subramanian, V. (1984). Erosion and sediment transport in the Ganges river basin (India). *Journal of Hydrology*, 69, 173–182. https://doi.org/10.1016/0022-1694(84)90162-8.

Achberger, A. M., Michaud, A. B., Vick-Majors, T. J., Christner, B. C., Skidmore, M. L., Priscu, J. C., and Tranter, M. (2017). Microbiology of subglacial environments. In *Psychrophiles: From Biodiversity to Biotechnology*, 2nd ed. Springer, Berlin, 83–110. https://doi.org/10.1007/978-3-319-57057-0_5.

Adel, M. M. (2001). Effect on water resources from upstream water diversion in the Ganges basin. *Journal of Environmental Quality*, 30(2), 356–368. https://doi.org/10.2134/jeq2001.302356x.

Agarwal, T. (2014). The Ganges drainage basin: hydrological transitions due to anthropogenic water use. TRITA-LWR Degree Project 14:20

Anesio, A. M., Hodson, A. J., Fritz, A., Psenner, R., and Sattler, B. (2009). High microbial activity on glaciers: Importance to the global carbon cycle. *Global Change Biology*, 15, 955–960. https://doi.org/10.1111/j.1365-2486.2008.01758.x.

Anesio, A. M., Lutz, S., Chrismas, N. A. M., and Benning, L. G. (2017). The microbiome of glaciers and ice sheets. *NPJ Biofilms and Microbiomes*, 3, 10. https://doi.org/10.1038/s41522-017-0019-0.

Apitz, S. E. (2012). Conceptualizing the role of sediment in sustaining ecosystem services: Sediment-ecosystem regional assessment (SEcoRA). *Science of the Total Environment*, 415, 9–30.

Arora, M., Kumar, R., Malhotra, J., and Kumar, N. (2015). Characterization of suspended sediment in meltwater from Gangotri glacier. *International Journal of Engineering Research and Technology*, 3, 1–9.

Arora, M., Singh, P., Goel, N. K., and Singh, R. D. (2006). Spatial distribution and seasonal variability of rainfall in a mountainous basin in the Himalayan region. *Water Resources Management*, 20, 489–508. https://doi.org/10.1007/s11269-006-8773-4.

Babel, M. S., and Wahid, S. M. (2011). Hydrology, management and rising water vulnerability in the Ganges–Brahmaputra–Meghna River basin. *Water International*, 36, 340–356. https://doi.org/10.1080/02508060.2011.584152.

Bandyopadhyay, J. (1995). Sustainability of big dams in the Himalayas. *Economic & Political Weekly*, 30, 2367–2370.

Basistha, A., Arya, D. S., and Goel, N. K. (2008). Spatial distribution of rainfall in Indian Himalayas – A case study of Uttarakhand region. *Water Resources Management*, 22, 1325–1346. https://doi.org/10.1007/s11269-007-9228-2.

Bhatt, M. P., Hartmann, J., and Acevedo, M. F. (2018). Seasonal variations of biogeochemical matter export along the Langtang–Narayani river system in central Himalaya. *Geochimica et Cosmochimica Acta*, 238, 208–234. https://doi.org/10.1016/j.gca.2018.06.033.

Bhutiyani, M. R., Kale, V. S., and Pawar, N. J. (2007). Long-term trends in maximum, minimum and mean annual air temperatures across the Northwestern Himalaya during the twentieth century. *Climatic Change*, 85, 159–177. https://doi.org/10.1007/s10584-006-9196-1.

Bhutiyani, M. R., Kale, V. S., and Pawar, N. J. (2008). Changing streamflow patterns in the rivers of northwestern Himalaya: Implications of global warming in the 20th century. *Current Science*, 95, 618–626.

Biemans, H., Siderius, C., Lutz, A. F., Nepal, S., Ahmad, B., Hassan, T., Bloh, W. von, Wijngaard, R. R., Wester, P., Shrestha, A. B., and Immerzeel, W. W. (2019). Importance of snow and glacier meltwater for agriculture on the Indo-Gangetic Plain. *Nature Sustainability*, 2, 594–601. https://doi.org/10.1038/s41893-019-0305-3.

Bolch, T., Kulkarni, A., Kääb, A., Huggel, C., Paul, F., Cogley, J. G., Frey, H., Kargel, J. S., Fujita, K., Scheel, M., Bajracharya, S., and Stoffel, M. (2012). The state and fate of himalayan glaciers. *Science*, 336, 310–314. https://doi.org/10.1126/science.1215828.

Bookhagen, B., and Burbank, D. W. (2006). Topography, relief, and TRMM-derived rainfall variations along the Himalaya. *Geophysical Research Letters*, 33, L08405. https://doi.org/10.1029/2006GL026037.

Bookhagen, B., and Burbank, D. W. (2010). Toward a complete Himalayan hydrological budget: Spatiotemporal distribution of snowmelt and rainfall and their impact on river discharge. *Journal of Geophysical Research: Earth Surface*, 115, 3019. https://doi.org/10.1029/2009JF001426.

Bookhagen, B., Thiede, R. C., and Strecker, M. R. (2005). Abnormal monsoon years and their control on erosion and sediment flux in the high, arid northwest Himalaya. *Earth and Planetary Science Letters*, 231, 131–146. https://doi.org/10.1016/j.jpgl.2004.11.014.

Burbank, D. W., Bookhagen, B., Gabet, E. J., and Putkonen, J. (2012). Modern climate and erosion in the Himalaya. *Comptes Rendus - Geoscience*, 344, 610–626. https://doi.org/10.1016/j.crte.2012.10.010.

Chakrapani, G. J. (2005). Factors controlling variations in river sediment loads. *Current Science*, 88, 569–575.

Chakrapani, G. J., and Saini, R. K. (2009). Temporal and spatial variations in water discharge and sediment load in the Alaknanda and Bhagirathi Rivers in Himalaya, India. *Journal of Asian Earth Sciences*, 35, 545–553. https://doi.org/10.1016/j.jseaes.2009.04.002.

Church, M. (2006). Bed material transport and the morphology of alluvial river channels. *Annual Review of Earth and Planetary Sciences*, 34, 325–354.

Coleman, J. M. (1969). Brahmaputra river: Channel processes and sedimentation. *Sedimentary Geology*, 3, 129–239. https://doi.org/10.1016/0037-0738(69)90010-4.

Dade, W. B., and Friend, P. F. (1998). Grain-size, sediment-transport regime, and channel slope in alluvial rivers. *The Journal of Geology*, 106, 661–676.

Dai, S. B., Lu, X. X., Yang, S. L., and Cai, A. M. (2008). A preliminary estimate of human and natural contributions to the decline in sediment flux from the Yangtze River to the East China Sea. *Quaternary International*, 186, 43–54. https://doi.org/10.1016/j.quaint.2007.11.018.

Darby, S. E., Dunn, F. E., Nicholls, R. J., Rahman, M., and Riddy, L. (2015). A first look at the influence of anthropogenic climate change on the future delivery of fluvial sediment to the Ganges–Brahmaputra–Meghna delta. *Environmental Sciences: Processes and Impacts*, 17, 1587–1600. https://doi.org/10.1039/c5em00252d.

Decaulne, A., Sæmundsson, P., and Pétursson, O. (2005). Debris flow triggered by rapid snowmelt: A case study in the Gleidarhjalli area, northwestern Iceland. *Geografiska Annaler, Series A: Physical Geography*, 87, 487–500. https://doi.org/10.1111/j.0435-3676.2005.00273.x.

DeFries, R., and Eshleman, K. N. (2004). Land-use change and hydrologic processes: A major focus for the future. *Hydrological Processes*, 18, 2183–2186. https://doi.org/10.1002/hyp.5584.

Dimri, A. P., and Dash, S. K. (2012). Wintertime climatic trends in the western Himalayas. *Climatic Change*, 111, 775–800. https://doi.org/10.1007/s10584-011-0201-y.

Dimri, A. P., Kumar, A., Satyawali, P. K., and Ganju, A. (2008). Climatic variability of weather parameters over the western Himalayas: A case study. In Satyawali, P.K., and Ganju, A. (Eds.), *Proceedings of the National Snow Science Workshop*, 11–12. January 2008, Chandigarh, India. Chandigarh, Snow and Avalanche Study Establishment, 167–173.

FAO (2016). *Transboundary River Basin Overview – Kura Araks*. FAO Aquastat, Rome.

FAO Aquastat (2011). *Ganges-Brahmaputra-Meghna Basin*. Water Report 37 1–17. FAO Aquastat.

Finlayson, D. P., Montgomery, D. R., and Hallet, B. (2002). Spatial coincidence of rapid inferred erosion with young metamorphic massifs in the Himalayas. *Geology*, 30, 219–222.

Fischer, S., Pietroń, J., Bring, A., Thorslund, J., and Jarsjö, J. (2017). Present to future sediment transport of the Brahmaputra River: Reducing uncertainty in predictions and management. *Regional Environmental Change*, 17, 515–526. https://doi.org/10.1007/s10113-016-1039-7.

Fujita, K. (2008). Effect of precipitation seasonality on climatic sensitivity of glacier mass balance. *Earth and Planetary Science Letters*, 276, 14–19. https://doi.org/10.1016/j.jpgl.2008.08.028.

Furuichi, T., Win, Z., and Wasson, R. J. (2009). Discharge and suspended sediment transport in the Ayeyarwady River, Myanmar: Centennial and decadal changes. *Hydrological Processes*, 23, 1631–1641. https://doi.org/10.1002/hyp.7295.

Galy, A., and France-Lanord, C. (2001). Higher erosion rates in the Himalaya: Geochemical constraints on riverine fluxes. *Geology*, 29, 23–26.

Gornall, J., Betts, R., Burke, E., Clark, R., Camp, J., Willett, K., and Wiltshire, A. (2010). Implications of climate change for agricultural productivity in the early twenty-first century. *Philosophical Transactions of the Royal Society B: Biological Sciences*, 365, 2973–2989. https://doi.org/10.1098/rstb.2010.0158.

Goswami, D. C. (1985). Brahmaputra River, Assam, India: Physiography, basin denudation, and channel aggradation. *Water Resources Research*, 21, 959–978. https://doi.org/10.1029/WR021i007p00959.

Haritashya, U. K., Singh, P., Kumar, N., and Gupta, R. P. (2006). Suspended sediment from the Gangotri Glacier: Quantification, variability and associations with discharge and air temperature. *Journal of Hydrology*, 321, 116–130. https://doi.org/10.1016/j.jhydrol.2005.07.037.

Hawkings, J., Wadham, J., Tranter, M., Telling, J., Bagshaw, E., Beaton, A., Simmons, S. L., Chandler, D., Tedstone, A., and Nienow, P. (2016). The Greenland ice sheet as a hot spot of phosphorus weathering and export in the Arctic. *Global Biogeochemical Cycles*, 30, 191–210. https://doi.org/10.1002/2015GB005237.

Hawkings, J. R., Wadham, J. L., Benning, L. G., Hendry, K. R., Tranter, M., Tedstone, A., Nienow, P., and Raiswell, R. (2017). Ice sheets as a missing source of silica to the polar oceans. *Nature Communications*, 8, 14198. https://doi.org/10.1038/ncomms14198.

Hawkings, J. R., Wadham, J. L., Tranter, M., Lawson, E., Sole, A., Cowton, T., Tedstone, A. J., Bartholomew, I., Nienow, P., Chandler, D., and Telling, J. (2015). The effect of warming climate on nutrient and solute export from the Greenland Ice Sheet. *Geochemical Perspectives Letters*, 2000, 94–104. https://doi.org/10.7185/geochemlet.1510.

Hawkings, J. R., Wadham, J. L., Tranter, M., Raiswell, R., Benning, L. G., Statham, P. J., Tedstone, A., Nienow, P., Lee, K., and Telling, J. (2014). Ice sheets as a significant source of highly reactive nanoparticulate iron to the oceans. *Nature Communications*, 5, 1–8. https://doi.org/10.1038/ncomms4929.

Hodson, A., Anesio, A. M., Tranter, M., Fountain, A., Osborn, M., Priscu, J., Laybourn-Parry, J., and Sattler, B. (2008). Glacial ecosystems. *Ecological Monographs*, 78, 41–67. https://doi.org/10.1890/07-0187.1.

Holeman, J. N. (1968). The sediment yield of major rivers of the world. *Water Resources Research*, 4, 737–747. https://doi.org/10.1029/WR004i004p00737.

Hood, E., Battin, T. J., Fellman, J., O'neel, S., and Spencer, R. G. M. (2015). Storage and release of organic carbon from glaciers and ice sheets. *Nature Geoscience*, 8, 91–96. https://doi.org/10.1038/ngeo2331.

Hossain, M. M. (1992). Total sediment load in the lower Ganges and Jamuna. *Journal of the Institution of Engineers Bangladesh*, 20, 1–8.

Immerzeel, W. (2008). Historical trends and future predictions of climate variability in the Brahmaputra basin. *International Journal of Climatology: A Journal of the Royal Meteorological Society*, 28, 243–254.

Immerzeel, W. W., Beek, L. P. H. Van, and Bierkens, M. F. P. (2010). Climate change will affect the Asian water towers. *Science*, 328, 1382–1385. https://doi.org/10.1126/science.1183188.

Immerzeel, W. W., Lutz, A. F., Andrade, M., Bahl, A., Biemans, H., Bolch, T., Hyde, S., Brumby, S., Davies, B. J., Elmore, A. C., Emmer, A., Feng, M., Fernández, A., Haritashya, U., Kargel, J. S., Koppes, M., Kraaijenbrink, P. D. A., Kulkarni, A. V., Mayewski, P. A., Nepal, S., Pacheco, P., Painter, T. H., Pellicciotti, F., Rajaram, H., Rupper, S., Sinisalo, A., Shrestha, A. B., Viviroli, D., Wada, Y., Xiao, C., Yao, T., and Baillie, J. E. M. (2020). Importance and vulnerability of the world's water towers. *Nature*, 577, 364–369. https://doi.org/10.1038/s41586-019-1822-y.

Inam, A., Clift, P. D., Giosan, L., Tabrez, A. R., Tahir, M., Rabbani, M. M., and Danish, M. (2008). The geographic, geological and oceanographic setting of the Indus River. In: *Large Rivers: Geomorphology and Management*. John Wiley & Sons, Ltd, Chichester, UK, 333–346. https://doi.org/10.1002/9780470723722.ch16.

Islam, M. R., Begum, S. F., Yamaguchi, Y., and Ogawa, K. (1999). The Ganges and Brahmaputra rivers in Bangladesh: Basin denudation and sedimentation. *Hydrological Processes*, 13, 2907–2923. https://doi .org/10.1002/(SICI)1099-1085(19991215)13:17<2907::AID-HYP906>3.0.CO;2-E.

Jain, S. K., Singh, P., Saraf, A. K., and Seth, S. M. (2003). Estimation of sediment yield for a rain, snow and glacier fed river in the western Himalayan region. *Water Resources Management*, 17, 377–393. https:// doi.org/10.1023/A:1025804419958.

Jha, P. K., Subramanian, V., and Sitasawad, R. (1988). Chemical and sediment mass transfer in the Yamuna River - A tributary of the Ganges system. *Journal of Hydrology*, 104, 237–246. https://doi.org/10.1016 /0022-1694(88)90167-9.

Jiongxin, X. (2003). Sediment flux to the sea as influenced by changing human activities and precipitation: Example of the Yellow River, China. *Environmental Management*, 31, 328–341. https://doi.org/10.1007 /s00267-002-2828-y.

Julien, P. Y. (2010). *Erosion and Sedimentation*. Cambridge University Press, Cambridge.

Kääb, A., Berthier, E., Nuth, C., Gardelle, J., and Arnaud, Y. (2012). Contrasting patterns of early twenty-first-century glacier mass change in the Himalayas. *Nature*, 488, 495–498. https://doi.org/10.1038/ nature11324.

Kaushik, H., Ramanathan, A., Soheb, M., Shamurailatpam, M. S., Biswal, K., Mandal, A., and Singh, C. (2021). Climate change-induced high-altitude lake: Hydrochemistry and area changes of a moraine-dammed lake in Leh-Ladakh. *Acta Geophysica*, 69, 2377–2391. https://doi.org/10.1007/s11600-021 -00670-x.

Khan, S., Sinha, R., Whitehead, P., Sarkar, S., Jin, L., and Futter, M. N. (2018). Flows and sediment dynamics in the Ganga River under present and future climate scenarios. *Hydrological Sciences Journal*, 63, 763–782. https://doi.org/10.1080/02626667.2018.1447113.

Krishna, M. S., Prasad, M. H. K., Rao, D. B., Viswanadham, R., Sarma, V. V. S. S., and Reddy, N. P. C. (2016). Export of dissolved inorganic nutrients to the northern Indian Ocean from the Indian monsoonal rivers during discharge period. *Geochimica et Cosmochimica Acta*, 172, 430–443. https://doi.org/10.1016/j .gca.2015.10.013.

Krishnan, R., Sanjay, J., Gnanaseelan, C., Mujumdar, M., Kulkarni, A., and Chakraborty, S. (2020). *Assessment of Climate Change over the Indian Region: A Report of the Ministry of Earth Sciences (MOES), Government of India*, Government of India. https://doi.org/10.1007/978-981-15-4327-2.

Krishnaswami, S., and Singh, S. K. (2005). Chemical weathering in the river basins of the Himalaya, India. *Current Science*, 89, 841–849.

Kumar, A., Gokhale, A. A., Shukla, T., and Dobhal, D. P. (2016). Hydroclimatic influence on particle size distribution of suspended sediments evacuated from debris-covered Chorabari Glacier, upper Mandakini catchment, central Himalaya. *Geomorphology*, 265, 45–67.

Kumar, A., Verma, A., Gokhale, A. A., Bhambri, R., Misra, A., Sundriyal, S., Dobhal, D. P., and Kishore, N. (2018a). Hydrometeorological assessments and suspended sediment delivery from a central Himalayan glacier in the upper Ganga basin. *International Journal of Sediment Research*, 33, 493–509. https://doi .org/10.1016/j.ijsrc.2018.03.004.

Kumar, R., Kumar, R., Singh, S., Singh, A., Bhardwaj, A., Kumari, A., Randhawa, S. S., and Saha, A. (2018b). Dynamics of suspended sediment load with respect to summer discharge and temperatures in Shaune Garang glacierized catchment, Western Himalaya. *Acta Geophysica*, 66, 1109–1120. https://doi.org/10 .1007/s11600-018-0184-4.

Kummu, M., and Varis, O. (2007). Sediment-related impacts due to upstream reservoir trapping, the lower Mekong River. *Geomorphology*, 85, 275–293. https://doi.org/10.1016/j.geomorph.2006.03.024.

Lawson, E. C., Wadham, J. L., Tranter, M., Stibal, M., Lis, G. P., Butler, C. E. H., Laybourn-Parry, J., Nienow, P., Chandler, D., and Dewsbury, P. (2014). Greenland ice sheet exports labile organic carbon to the arctic oceans. *Biogeosciences*, 11, 4015–4028. https://doi.org/10.5194/bg-11-4015-2014.

Lu, X. X., Zhang, S. R., Xu, J. C., and Merz, J. (2011). The changing sediment loads of the Hindu Kush-Himalayan rivers: An overview. *IAHS-AISH Publication*, 349, 21–36.

Mandal, A., Ramanathan, A., Angchuk, T., Soheb, M., and Singh, V. B. (2016). Unsteady state of glaciers (Chhota Shigri and Hamtah) and climate in Lahaul and Spiti region, western Himalayas: A review of recent mass loss. *Environmental Earth Sciences*, 75, 1233. https://doi.org/10.1007/s12665-016-6023-5.

Mandal, A., Ramanathan, A., Azam, M. F., Angchuk, T., Soheb, M., Kumar, N., Pottakkal, J. G., Vatsal, S., Mishra, S., and Singh, V. B. (2020). Understanding the interrelationships among mass balance, meteorology, discharge and surface velocity on Chhota Shigri Glacier over 2002–2019 using in situ measurements. *Journal of Glaciology*, 66, 727–741. https://doi.org/10.1017/jog.2020.42.

McLennan, S. M. (1993). Weathering and global denudation on JSTOR. *The Journal of Geology*, 101, 295–303.

Menzel, L., and Bürger, G. (2002). Climate change scenarios and runoff response in the Mulde catchment (Southern Elbe, Germany). *Journal of Hydrology*, 267, 53–64. https://doi.org/10.1016/S0022 -1694(02)00139-7.

Michael, A., Schmidt, J., Enke, W., Deutschländer, T., and Malitz, G. (2005). Impact of expected increase in precipitation intensities on soil loss - Results of comparative model simulations. In *Catena*, 155–164. https://doi.org/10.1016/j.catena.2005.03.002.

Milliman, J. D., and Meade, R. H. (1983). World-wide delivery of river sediment to the oceans. *The Journal of Geology*, 91, 1–21.

Milliman, J. D., Farnsworth, K. L., Jones, P. D., Xu, K. H., and Smith, L. C. (2008). Climatic and anthropogenic factors affecting river discharge to the global ocean, 1951–2000. *Global and Planetary Change*, 62, 187–194. https://doi.org/10.1016/j.gloplacha.2008.03.001.

Mirza, M. M. Q. (1998). Diversion of the Ganges water at Farakka and its effects on salinity in Bangladesh. *Environmental Management*, 22, 711–722. https://doi.org/10.1007/s002679900141.

Mukherjee, S., Joshi, R., Prasad, R. C., Vishvakarma, S. C. R., and Kumar, K. (2015). Summer monsoon rainfall trends in the Indian Himalayan region. *Theoretical and Applied Climatology*, 121, 789–802. https://doi.org/10.1007/s00704-014-1273-1.

Narayana, D. V. V., and Babu, R. (1983). Estimation of soil erosion in India. *Journal of Irrigation and Drainage Engineering*, 109, 419–434. https://doi.org/10.1061/(asce)0733-9437(1983)109:4(419).

Nepal, S., and Shrestha, A. B. (2015). Impact of climate change on the hydrological regime of the Indus, Ganges and Brahmaputra river basins: A review of the literature. *International Journal of Water Resources Development*, 31, 201–218. https://doi.org/10.1080/07900627.2015.1030494.

Nijssen, B., O'donnell, G. M., Hamlet, A. F., and Lettenmaier, D. P. (2001). Hydrologic sensitivity of global rivers to climate change. *Climatic Change*, 50, 143–175. https://doi.org/10.1023/A:1010616428763.

Nizam, S., Sen, I. S., Vinoj, V., Galy, V., Selby, D., Azam, M. F., Pandey, S. K., Creaser, R. A., Agarwal, A. K., Singh, A. P., and Bizimis, M. (2020). Biomass-derived provenance dominates glacial surface organic carbon in the Western Himalaya. *Environmental Science and Technology*, 54, 8612–8621. https://doi .org/10.1021/acs.est.0c02710.

Ojeh, E. (2006). *Hydrology of the Indus Basin (Pakistan)*.

Palash, W., Jiang, Y., Akanda, A. S., Small, D. L., Nozari, A., and Islam, S. (2018). A streamflow and water level forecasting model for the Ganges, Brahmaputra, and Meghna rivers with requisite simplicity. *Journal of Hydrometeorology*, 19, 201–225. https://doi.org/10.1175/JHM-D-16-0202.1.

Perry, C., Steduto, P., Allen, R. G., and Burt, C. M. (2009). Increasing productivity in irrigated agriculture: Agronomic constraints and hydrological realities. *Agricultural Water Management*, 96, 1517–1524. https://doi.org/10.1016/j.agwat.2009.05.005.

Prakash, C., and Nagarajan, R. (2018). Glacial lake changes and outburst flood hazard in Chandra basin, North-Western Indian Himalaya. *Geomatics, Natural Hazards and Risk*, 9, 337–355. https://doi.org/10 .1080/19475705.2018.1445663.

Pruski, F. F., and Nearing, M. A. (2002). Climate-induced changes in erosion during the 21st century for eight U.S. locations. *Water Resources Research*, 38, 34-1–34-11. https://doi.org/10.1029/2001wr000493.

Qu, B., Sillanpää, M., Kang, S., Yan, F., Li, Z., Zhang, H., and Li, C. (2018). Export of dissolved carbonaceous and nitrogenous substances in rivers of the "Water Tower of Asia." *Journal of Environmental Sciences (China)*, 65, 53–61. https://doi.org/10.1016/j.jes.2017.04.001.

Rahman, M., Dustegir, M., Karim, R., Haque, A., Nicholls, R. J., Darby, S. E., Nakagawa, H., Hossain, M., Dunn, F. E., and Akter, M. (2018). Recent sediment flux to the Ganges–Brahmaputra–Meghna delta system. *Science of the Total Environment*, 643, 1054–1064. https://doi.org/10.1016/j.scitotenv.2018.06.147.

Rao, S. V. N., Rao, M. V, Ramasastri, K. S., and Singh, R. N. P. (1997). A study of sedimentation in Chenab basin in western Himalayas. *Hydrology Research*, 28, 201–216.

Rasul, G. (2010). The role of the Himalayan mountain systems in food security and agricultural sustainability in South Asia. *International Journal of Rural Management*, 6, 95–116. https://doi.org/10.1177 /097300521100600105.

Rawat, J. S., and Rawat, M. S. (1994). Accelerated erosion and denudation in the Nana Kosi watershed, Central Himalaya, India. Part I: Sediment load. *Mountain Research and Development*, 14, 25–38. https://doi.org /10.2307/3673736.

Raymo, M. E., and Ruddiman, W. F. (1992). Tectonic forcing of late Cenozoic climate. *Nature*, 359, 117–122. https://doi.org/10.1038/359117a0.

Reid, W. V, Mooney, H. A., Cropper, A., Capistrano, D., Carpenter, S. R., Chopra, K., Dasgupta, P., Dietz, T., Duraiappah, A. K., Hassan, R., et al. (2005). *Ecosystems and Human Well-Being-Synthesis: A Report of the Millennium Ecosystem Assessment*. Island Press, Washington, DC.

Rice, S. K. (2007). *Suspended Sediment Transport in the Ganges–Brahmaputra River System, Bangladesh.* Я т ы а т а т, Texas A&M University.

Roy, N. G., and Sinha, R. (2017). Linking hydrology and sediment dynamics of large alluvial rivers to landscape diversity in the Ganga dispersal system, India. *Earth Surface Processes and Landforms*, 42, 1078–1091. https://doi.org/10.1002/esp.4074.

Sarkar, S., Verma, S., Begum, M. S., Park, J. H., and Kumar, S. (2023). Sources, supply, and seasonality of total suspended matter and associated organic carbon and total nitrogen in three large Asian rivers—Ganges, Mekong, and Yellow. *Frontiers in Earth Science*, 11, 1067744. https://doi.org/10.3389/feart.2023.1067744.

Shamurailatpam, M. S., Telling, J., Wadham, J. L., Ramanathan, A. L., Yates, C. A., and Raju, N. J. (2023). Factors controlling the net ecosystem production of cryoconite on Western Himalayan glaciers. *Biogeochemistry*, 162, 201–220. https://doi.org/10.1007/s10533-022-00998-6.

Sharif, M., Burn, D., and Hussain, A. (2010). Climate change impacts on extreme flow measures in Satluj River Basin in India. In *World Environmental and Water Resources Congress 2010: Challenges of Change - Proceedings of the World Environmental and Water Resources Congress 2010*, 46–59. https://doi.org/10.1061/41114(371)7.

Shekhar, M. S., Chand, H., Kumar, S., Srinivasan, K., and Ganju, A. (2010). Climate-change studies in the western Himalaya. *Annals of Glaciology*, 51, 105–112. https://doi.org/10.3189/172756410791386508.

Shrestha, U. B., Gautam, S., and Bawa, K. S. (2012). Widespread climate change in the Himalayas and associated changes in local ecosystems. *PloS One*, 7, e36741. https://doi.org/10.1371/journal.pone.0036741.

Singh, A. T., Sharma, P., Sharma, C., Laluraj, C. M., Patel, L., Pratap, B., Oulkar, S., and Thamban, M. (2020). Water discharge and suspended sediment dynamics in the Chandra River, Western Himalaya. *Journal of Earth System Science*, 129, 206. https://doi.org/10.1007/s12040-020-01455-4.

Singh, A. T., Laluraj, C. M., Sharma, P., Redkar, B. L., Patel, L. K., Pratap, B., Oulkar, S., and Thamban, M. (2021). Hydrograph apportionment of the Chandra River draining from a semi-arid region of the Upper Indus Basin, western Himalaya. *Science of the Total Environment*, 780, 146500. https://doi.org/10.1016/j.scitotenv.2021.146500.

Singh, J., and Singh, O. (2021). Sediment and nutrient transfer from an inter-montane agricultural catchment in Himachal Himalayas of Northwestern India. *Journal of the Geological Society of India*, 97, 282–292. https://doi.org/10.1007/s12594-021-1679-1.

Singh, M., Singh, I. B., and Müller, G. (2007). Sediment characteristics and transportation dynamics of the Ganga River. *Geomorphology*, 86, 144–175. https://doi.org/10.1016/j.geomorph.2006.08.011.

Singh, O., Singh, P., Sarangi, A., Sharma, M. C., and Kumar, S. (2010). Anthropogenic impacts on the sediment flux in two alpine watersheds of the Lesser Himalayas. *Current Science*, 99, 608–618.

Singh, P., Ramasastri, K. S., Gergan, J. T., and Dobhal, D. P. (1995). Hydrological characteristics of the dokriani glacier in the Garhwal Himalayas. *Hydrological Sciences Journal*, 40, 243–257. https://doi.org/10.1080/02626669509491407.

Singh, S. K. (1988). Nature of chemical and sediment load in the Ganges River between Sone and Kosi. PhD thesis, Jawaharlal Nehru University, New Delhi.

Singh, V. B., and Ramanathan, A. (2018). Suspended sediment dynamics in the meltwater of Chhota Shigri glacier, Chandra basin, Lahaul-Spiti valley, India. *Journal of Mountain Science*, 15, 68–81. https://doi.org/10.1007/s11629-017-4554-1.

Singh, V. B., Ramanathan, A., Keshari, A. K., Ranjan, S., Kumar, N., and Yadav, S. K. (2018). Climatic influence on hydrogeochemistry of meltwater draining from Chhota Shigri Glacier, Himachal Pradesh, India. *Journal of Climate Change*, 4, 23–31. https://doi.org/10.3233/jcc-180003.

Sinha, R., and Friend, P. F. (1994). River systems and their sediment flux, Indo-Gangetic plains, Northern Bihar, India. *Sedimentology*, 41, 825–845. https://doi.org/10.1111/j.1365-3091.1994.tb01426.x.

Sinha, R., Gupta, A., Mishra, K., Tripathi, S., Nepal, S., Wahid, S. M., and Swarnkar, S. (2019). Basin-scale hydrology and sediment dynamics of the Kosi river in the Himalayan foreland. *Journal of Hydrology*, 570, 156–166. https://doi.org/10.1016/j.jhydrol.2018.12.051.

Stewart, R. J., Hallet, B., Zeitler, P. K., Malloy, M. A., Allen, C. M., and Trippett, D. (2003). Brahmaputra sediment flux dominated by highly localized rapid erosion from the easternmost Himalaya. *Geology*, 36, 711–714.

Subramanian, V. (1987). Environmental geochemistry of Indian river basins - A review. *Journal of the Geological Society of India*, 29, 205–220.

Subramanian, V., and Ramanathan, A. L. (1996). *Nature of Sediment Load in the Ganges-Brahmaputra River Systems in India*, pp. 151–168. https://doi.org/10.1007/978-94-015-8719-8_8.

Summerfield, M. A., and Hulton, N. J. (1994). Natural controls of fluvial denudation rates in major world drainage basins. *Journal of Geophysical Research: Solid Earth*, 99, 13871–13883. https://doi.org/10.1029/94jb00715.

Surinaidu, L., Amarasinghe, U., Maheswaran, R., and Nandan, M. J. (2020). Assessment of long-term hydrogeological changes and plausible solutions to manage hydrological extremes in the transnational Ganga river basin. *H2Open Journal*, 3, 457–480. https://doi.org/10.2166/h2oj.2020.049.

Syvitski, J. P. M., Kettner, A. J., Peckham, S. D., and Kao, S. J. (2005). Predicting the flux of sediment to the coastal zone: Application to the Lanyang Watershed, northern Taiwan. *Journal of Coastal Research*, 2005, 580–587. https://doi.org/10.2112/04-702A.1.

Tawde, S. A., Kulkarni, A. V, and Bala, G. (2017). An estimate of glacier mass balance for the Chandra basin, western Himalaya, for the period 1984–2012. *Annals of Glaciology*, 58, 99–109.

Teng, T.-Y., Huang, J.-C., Lee, T.-Y., Chen, Y.-C., Jan, M.-Y., and Liu, C.-C. (2020). Investigating sediment dynamics in a landslide-dominated catchment by modeling landslide area and fluvial sediment export. *Water*, 12, 2907.

Thanh, T. D., Yoshiki, S., Van, H. D., Cu, N. H., and Chien, D. D. (2005). Coastal erosion in Red River Delta: Current status and response [WWW Document]. *Mega-Deltas of Asia: Geological Evolution and Human Impact*. China Ocean Press, Beijing. https://www.researchgate.net/publication/258786038_Coastal_erosion_in_Red_River_Delta_current_status_and_response (accessed 3.22.21).

Thiede, R. C., Bookhagen, B., Arrowsmith, J. R., Sobel, E. R., and Strecker, M. R. (2004). Climatic control on rapid exhumation along the Southern Himalayan Front. *Earth and Planetary Science Letters*, 222, 791–806.

Vance, D., Bickle, M., Ivy-Ochs, S., and Kubik, P. W. (2003). Erosion and exhumation in the Himalaya from cosmogenic isotope inventories of river sediments. *Earth and Planetary Science Letters*, 206, 273–288.

Wadham, J. L., Hawkings, J., Telling, J., Chandler, D., Alcock, J., O'Donnell, E., Kaur, P., Bagshaw, E., Tranter, M., Tedstone, A., and Nienow, P. (2016). Sources, cycling and export of nitrogen on the Greenland Ice Sheet. *Biogeosciences*, 13, 6339–6352. https://doi.org/10.5194/bg-13-6339-2016.

Walling, D. E. (1999). Linking land use, erosion and sediment yields in river basins. *Hydrobiologia*, 410, 223–240. https://doi.org/10.1023/A:1003825813091.

Walling, D. E. (2006). Human impact on land-ocean sediment transfer by the world's rivers. *Geomorphology*, 79, 192–216. https://doi.org/10.1016/j.geomorph.2006.06.019.

Walling, D. E. (2008). The changing sediment loads of the world's rivers. In: *IAHS-AISH* Publication. *IAHS Publ*, 323–338. https://doi.org/10.2478/v10060-008-0001-x.

Walling, D. E., and Fang, D. (2003). Recent trends in the suspended sediment loads of the world's rivers. *Global and Planetary Change*, 39, 111–126. https://doi.org/10.1016/S0921-8181(03)00020-1.

Walling, D. E., and Webb, B. W. (1996). *Erosion and Sediment Yield: Global and Regional Perspectives*. In *Proceedings of IAHS*. IAHS Press, Publ. no. 236, 3–19.

Wang, H., Yang, Z., Saito, Y., Liu, J. P., Sun, X., and Wang, Y. (2007). Stepwise decreases of the Huanghe (Yellow River) sediment load (1950–2005): Impacts of climate change and human activities. *Global and Planetary Change*, 57, 331–354. https://doi.org/10.1016/j.gloplacha.2007.01.003.

Wasson, R. J. (2003). *A Sediment Budget for the Ganga–Brahmaputra Catchment*. Current Science, Bengaluru.

Wulf, H. (2011). *Seasonal Precipitation, River Discharge, and Sediment Flux in the Western Himalaya*. 120. Institutional Repository of the University of Potsdam.

Wulf, H., Bookhagen, B., and Barbara, S. (2012). Climatic and geologic controls on suspended sediment flux in the Sutlej River Valley, western Himalaya. *Hydrology and Earth System Sciences Discussions*, 9, 541. https://doi.org/10.5194/hessd-9-541-2012.

Yang, S. L., Li, M., Dai, S. B., Liu, Z., Zhang, J., and Ding, P. X. (2006). Drastic decrease in sediment supply from the Yangtze River and its challenge to coastal wetland management. *Geophysical Research Letters*, 33. https://doi.org/10.1029/2005GL025507.

Zakwan, M., Ahmad, Z., and Sharief, S. M. V. (2018). Magnitude-frequency analysis for suspended sediment transport in the Ganga River. *Journal of Hydrologic Engineering*, 23, 05018013. https://doi.org/10.1061/(asce)he.1943-5584.0001671.

Zereini, F., and Hötzl, H. (2008). Impact of climate change on water resources. *Climatic Changes and Water Resources in the Middle East and North Africa* 75–76. https://doi.org/10.1007/978-3-540-85047-2_7.

Zhang, M., Chen, F., Zhao, H., Wang, J., and Wang, N. (2021). Recent changes of glacial lakes in the high mountain Asia and its potential controlling factors analysis. *Remote Sensing*, 13, 3757. https://doi.org/10.3390/rs13183757.

Zhang, S., Lu, X. X., Higgitt, D. L., Chen, C. T. A., Han, J., and Sun, H. (2008). Recent changes of water discharge and sediment load in the Zhujiang (Pearl River) Basin, China. *Global and Planetary Change*, 60, 365–380. https://doi.org/10.1016/j.gloplacha.2007.04.003.

Zhu, Y. M., Lu, X. X., and Zhou, Y. (2008). Sediment flux sensitivity to climate change: A case study in the Longchuanjiang catchment of the upper Yangtze River, China. *Global and Planetary Change*, 60, 429–442. https://doi.org/10.1016/j.gloplacha.2007.05.001.

Part IV

East Asia

17 Impact of Large Dams on Local Climate Change

A Case Study on the Three Gorges Dam

Jihui Fan, Majid Galoie, Artemis Motamedi,
Mahdi Moudi and Saeid Eslamian

17.1 INTRODUCTION

According to the ICOLD (International Commission on Large Dams) definition, a dam with a height of 15 m or more from the lowest foundation to crest or a dam between 5 and 15 m that impounds more than 3 million m^3, is considered a large dam. It is obvious that any large dam can reduce the downstream flow and impound a large volume of water in its reservoir with a huge surface area, which increases the evaporation rate, especially in arid and semi-arid regions. Behind 58,713 large dams scattered all over the world, nearly 8000 km^3 of water is accumulated (https://www.icold-cigb.org/GB/world_register/general_synthesis.asp). In addition to the large dams, there are more than 16 million smaller dams worldwide with a reservoir area greater than 100 m^2, bringing the total reservoir area to 306,000 km^2 (Mulligan et al., 2020). Table 17.1 summarizes information about some famous large dams in the world.

Climate change is known as a change in global mean surface temperature, generally caused by the emission of greenhouse gases (CO_2, CH_4, N_2O) into the atmosphere (McCarthy et al., 2022). This phenomenon can be studied in two scales: local or global climate change. The impoundment of a large volume of water behind a dam can amplify the effects of local and even global climate change in several ways, which are briefly discussed below:

- The evaporation rate, and thus the relative humidity, in the dam increases when the surface area of the reservoir is increased, while the relative humidity decreases downstream of the reservoir when the river discharge is reduced or eliminated (Zeng et al., 2019). In addition, increasing aridity downstream of the dam leads to a decrease in vegetation cover and groundwater table, as well as an increase in temperature and drought development (Zhao et al., 2021; Zahraei et al., 2016).
- Decomposition of vegetation under high water pressure at the bottom of the reservoir and biochemical processes in the reservoir can release a large amount of CH_4 and N_2O into the atmosphere, the greenhouse gases responsible for climate change (Bastien et al., 2011). In addition, hydropower plants produce a large amount of CO_2 (from 1.5 g to 3747.8 g per kWh), which is one of the most important parameters for local and global climate change (Yadav et al., 2023).
- Retaining a huge amount of mass at a high level behind the dam can change the moment inertia of the earth or even tilt the Earth's axes (NASA, 2005; Tortajada et al., 2012). As a result, the Earth's rotation speed can be changed which may cause global climate change, season shifting (Wang et al., 2021), and intensive earthquakes (Gupta, 2002; NASA, 2005; Wang et al., 2022).

TABLE 17.1

List of Famous Large Dams with High Reservoir Capacity Globally

Dam Name	Location	Year	Reservoir Volume (km³)
Kariba Dam	Zambia and Zimbabwe	1959	180.6
Bratsk Dam	Russia	1964	169
Akosombo Dam	Ghana	1965	150
Daniel-Johnson Dam	Canada	1968	141.85
Guri Dam	Venezuela	1986	135
Aswan High Dam	Egypt and Sudan	1971	132
Grand Ethiopian Renaissance Dam	Ethiopia	2020	79
W. A. C. Bennett Dam	Canada	1967	74.3
Krasnoyarsk Dam	Russia	1967	73.3
Zeya Hydroelectric Station (ru)	Russia	1978	68.4
Robert-Bourassa generating station	Canada	1981	61.71
La Grande-3 generating station	Canada	1981	60.02
Ust-Ilimsk Dam	Russia	1977	59.3
Boguchany Dam	Russia	2012	58.2
Zhiguli Hydroelectric Station	Russia	1955	58
Cahora Bassa Dam	Mozambique	1974	55.8
Serra da Mesa Dam	Brazil	1998	54.4
Brisay generating station	Canada	1981	53.8
Bukhtarma Hydroelectric Power Plant	Kazakhstan	1967	53
Danjiangkou Dam	China	1962	51.6
Atatürk Dam	Turkey	1992	48.7
Irkutsk Dam	Russia	1956	46
Tucuruí Dam	Brazil	1984	45.54
Loma de la Lata Dam	Argentina	1973	43.5
Three Gorges Dam	China	2009	39.3
Hoover Dam	United States	1936	37.3
Roseires Dam	Sudan	1966	36.3
Vilyuy Dam (ru)	Russia	1967	35.9
Glen Canyon Dam	United States	1964	35.55
Lake Argyle Dam	Australia	1971	35

(https://en.wikipedia.org/wiki/List_of_reservoirs_by_volume)

The Three Gorges Dam (TGD) is a hydroelectric gravity dam constructed on the Yangtze River (Chang Jiang) west of the city of Yichang in Hubei Province, China (N 30° 45' 7.9416", E 111° 16' 1.92"). At the time of its completion in 2006, it was the largest dam structure in the world. The volume of the reservoir at a maximum water level of 185 m is almost 39.3 km³ and has a total area of 1045 km² (Three Gorges Dam, Wikipedia). The hydroelectric power plant of this dam generates an average of 95±20 TWh of electricity per year, depending on the annual rainfall in the river basin. Figure 17.1 shows an aerial view of the TGD in 2023.

There are 15 dams and reservoirs on the tributaries of the Yangtze River upstream of the TGD to a distance of nearly 500 km, which may affect the meteorological data. Figure 17.2 illustrates the location of all dam sites and the Yangtze River.

There are a number of scientific studies and literature on the impact of TGD on climate change, each of which has conducted a comprehensive analysis of meteorological observational data or

FIGURE 17.1   Aerial view of Three Gorges Dam in 2023 (source: www.xinhuanet.com).

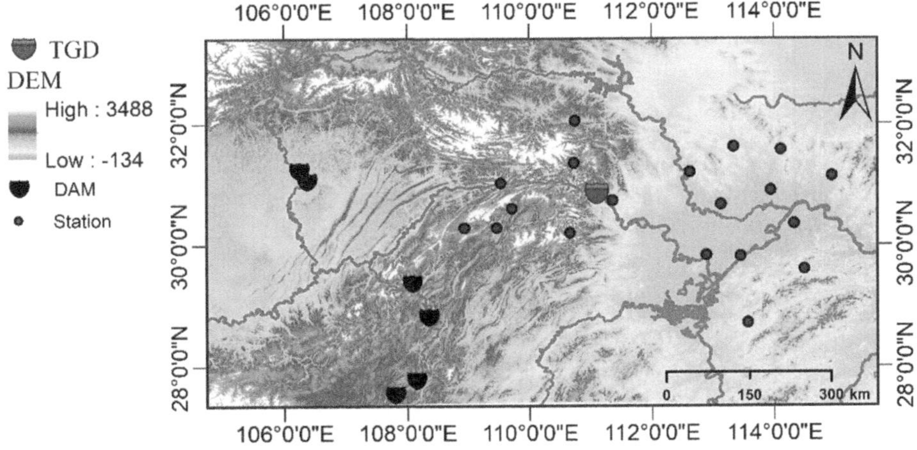

FIGURE 17.2   Location of TGD and the Yangtze River in China.

satellite maps and found significant trends and variations in precipitation (Zhao and Shepherd, 2012; Li et al., 2017, 2021; Lu et al., 2018), relative humidity (Wu et al., 2015; Zeng et al., 2019), air temperature (Song, 2017; Yu et al., 2021), downstream drought (Li et al., 2021; Cui et al., 2022), and even river water quality (Hou et al., 2022). In addition to the effects of TGD on hydrologic regime changes and climate data trends, analysis of long-term observed river water temperature in the Yangtze River has shown that TGD significantly affects water temperature in the lower reaches of the river (Cai et al., 2018).

This chapter briefly discusses the effects of TGD on the variation of trends in local meteorological data.

17.2   METEOROLOGICAL DATA

For a successive local climate change investigation, the meteorological data such as precipitation, relative humidity, temperature, wind speed, etc., should be collected from stations located upstream and downstream of the dam as long-term time series (e.g., 30–50 years). For this reason, all the stations scattered around the TGD basin up to a distance of almost 200 km were investigated and considered for preprocessing. Among them, 54 stations (34 stations located upstream and 20 downstream of TGD) have been selected for analysis. The criteria for this selection were the stations that contain long-term measured data series (1980–2018), have similar trends and correlations in the years prior to the dam construction (1980–2003), have minimum missing data, and those scattered uniformly throughout the basin.

17.3   IMPACT OF TGD ON METEOROLOGICAL DATA

17.3.1   Precipitation

The analysis of annual precipitation data showed that the effect of TGD on precipitation trends has been insignificant. Figure 17.3 represents the annual precipitation anomaly, upstream and downstream of the dam, before and after impounding in 2003. In statistical analysis, the time series anomaly chart is the simplest way to show the deviation of any values in the time series data from an accepted standard (e.g., the long-term average in meteorological data).

As can be seen in Figure 17.3, there are no noticeable trends in annual precipitation anomaly downstream of the dam but a slight increasing trend has occurred upstream of the dam after impounding. This is because of the fact that the relative humidity has increased significantly upstream of the dam after impounding.

The seasonal gridded mean precipitation maps were generated based on the inverse weighted interpolation (IDW) method in ArcGIS. The maps show that there are no significant variations in seasonal precipitation in the basin before and after impounding except for the summer. The analysis of the precipitation dataset for summer showed that these variations are dependent on the Indian monsoon rather than the TGD's impact. Indian monsoon winds in China occur from May to September which bring rain to the center and southern China. Since the spatial distribution of this rainfall is different from year to year, the gridded precipitation map for Summer is different in comparison with the other seasons.

17.3.2   Temperature

Figure 17.4 shows the annual temperature anomaly upstream and downstream of the dam before and after the impoundment in 2003. This figure shows an increasing trend with an almost constant

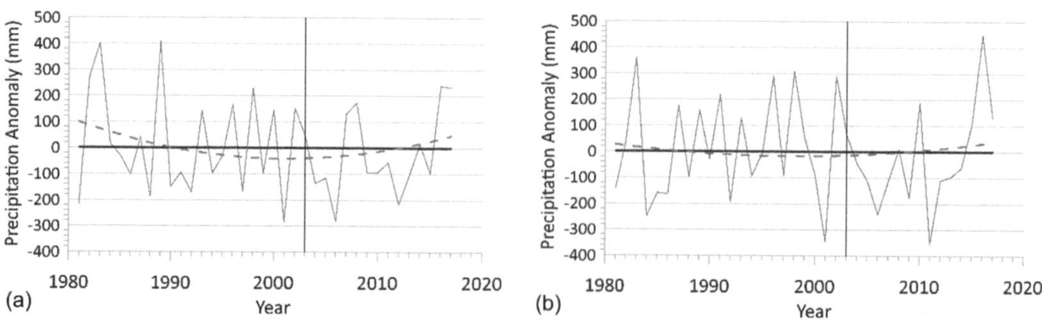

FIGURE 17.3   Variations in annual precipitation anomaly: (left) upstream of the dam and (right) downstream of the dam. The vertical line indicates the impounding period and the red curve shows the orthogonal trendline.

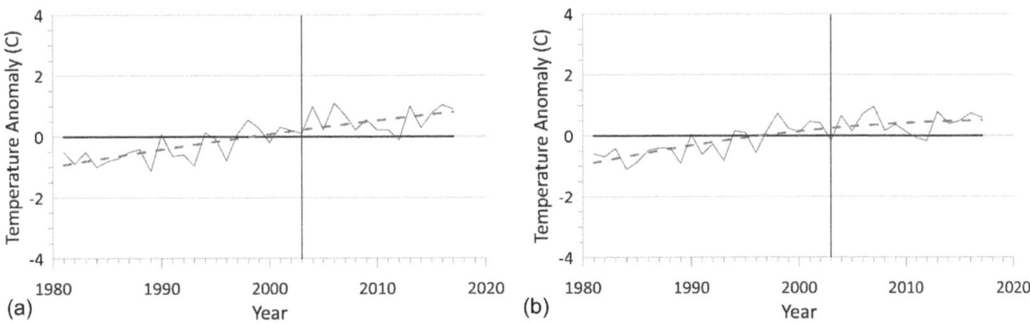

FIGURE 17.4 Variations in annual temperature anomaly: (left) upstream of the dam and (right) downstream of the dam. The vertical line shows the date of impoundment and the red curve shows the orthogonal trendline.

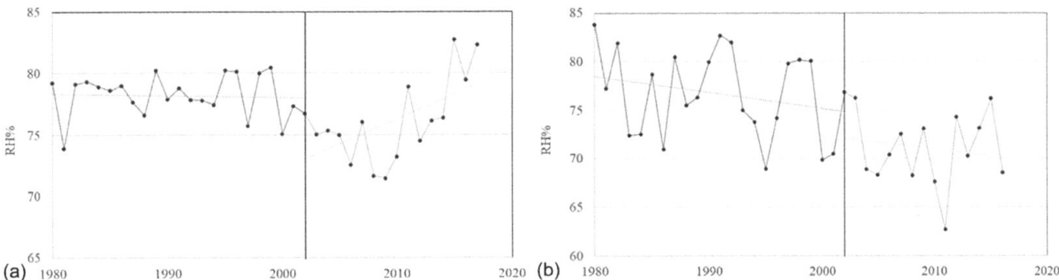

FIGURE 17.5 Variations in annual relative humidity anomaly: (left) upstream of the dam and (right) downstream of the dam. The vertical line indicates the impounding date and the red line shows the trendline.

rate throughout the basin, upstream and downstream of the TGD. This means that the surrounding area is affected by the impacts of global climate change.

The seasonal gridded mean temperature maps, before and after impounding were generated. It was seen that the impact of TGD on temperature was minor. Although the trends in annual precipitation have increased, climate change was not caused by the TGD, but by the effects of global climate change.

17.3.3 Relative Humidity (RH)

Figure 17.5 represents the variations in annual RH anomaly, upstream and downstream of the TGD. The variations in RH downstream of the TGD show a decreasing trend before and after impounding whereas upstream of TGD, the trend became positive after impounding. It is obviously due to the fact that after impounding, a huge volume of water with a vast surface was formed which has increased the evaporation rate upstream of the TGD.

17.4 CONCLUSIONS

With the increase in energy and water consumption in the world, the construction and operation of hydroelectric power plants, known as green and renewable energy, is increasing. There are many large dams around the world, each damming a huge amount of water and preventing the natural movement of water flow downstream. One of the major drawbacks of large dams is their negative impact on the ecology and the environment, which can take a long time to restore. In this chapter, the impact of the Three Gorges Dam on local climate change is investigated. The monthly data (1980–2019) of 54 stations was extracted. The analysis showed that although there were precipitation trends, the TGD had a minor effect on the variation of precipitation data. In addition, a shift in the spatial distribution of precipitation was observed, especially upstream of the dam, and the analysis showed that this shift was dependent on the wind speed and direction of the Indian monsoon. The data analysis also showed that the entire area has been affected by the impacts of global climate change, resulting in an increase in average temperature throughout the basin. In addition, the data analysis showed that the TGD had a strong influence on the increase in relative humidity upstream of the dam. Unlike upstream, relative humidity downstream of the TGD decreased after the impoundment in 2003.

ACKNOWLEDGMENTS

This research has been supported by the National Natural Science Foundation of China (Grant Nos. Nos. U2240216-1 and 41871072) and CAS President's International Fellowship Initiative; PIFI (Grant Nos. 2022VCC0004 and 2022VCC0005). The authors are grateful for all of their support.

REFERENCES

Bastien, J., Demarty, M., and Tremblay, A. (2011). CO_2 and CH_4 diffusive and degassing emissions from 2003 to 2009 at Eastmain 1 hydroelectric reservoir, Québec, Canada. *Inland Waters*, 1(2), 113–123.

Cai, H., Piccolroaz, S., Huang, J., Liu, Z., Liu, F., and Toffolon, M. (2018). Quantifying the impact of the Three Gorges Dam on the thermal dynamics of the Yangtze River. *Environmental Research Letters*, 13(5), 054016. https://doi.org/10.1088/1748-9326/aab9e0.

Cui, T., Chen, X., Zou, X., Zhang, Q., Li, S., and Zeng, H. (2022). State of the climate in the Three Gorges Region of the Yangtze River basin in 2020. *Atmospheric and Oceanic Science Letters*, 15(2), 100112. https://doi.org/10.1016/j.aosl.2021.100112.

Gupta, H. K. (2002). A review of recent studies of triggered earthquakes by artificial water reservoirs with special emphasis on earthquakes in Koyna, India. *Earth-Science Reviews*, 58(3–4), 279–310. https://doi.org/10.1016/S0012-8252(02)00063-6.

Hou, W., Wang, H., Zheng, Y., Wang, Y., Yang, D., and Meng, H. (2022). Seasonal variation characteristics of water quality in the Sunxi River Watershed, Three Gorges Reservoir area. *PeerJ*, 10, e14233. http://doi.org/10.7717/peerj.14233.

Li, Q., Liu, X., Zhong, Y., Wang, M., and Shi, M. (2021). Precipitation changes in the Three Gorges Reservoir area and the relationship with water level change. *Sensors*, 21(18), 6110. https://doi.org/10.3390/s21186110.

Li, X., Sha, J., and Wang, Z. L. (2021). Influence of the Three Gorges Reservoir on climate drought in the Yangtze River Basin. *Environmental Science and Pollution Research*, 28, 29755–29772. https://doi.org/10.1007/s11356-021-12704-4.

Li, Y., Zhou, W., Chen, X., Fang, D., and Zhang, Q. (2017). Influences of the Three Gorges Dam in China on precipitation over surrounding regions. *Journal of Meteorological Research*, 31(4), 767–773. https://doi.org/10.1007/s13351-017-6177-4.

Lü, M., Jiang, Y., Chen, X., Chen, J., Wu, S., and Liu, J. (2018). Spatiotemporal variations of extreme precipitation under a changing climate in the Three Gorges Reservoir area (TGRA). *Atmosphere*, 9(1), 24. https://doi.org/10.3390/atmos9010024.

McCarthy, J. J., Canziani, O. F., Leary, N. A., Dokken, D. J., and White, K. S. (Eds.) (2022). *Climate Change 2022: Impacts, Adaptation, and Vulnerability: Contribution of Working Group II to the Sixth Assessment Report of the Intergovernmental Panel on Climate Change (Vol. 6)*. Cambridge University Press, New York, 37–118. https://doi.org/10.1017/9781009325844.002.

Mulligan, M., van Soesbergen, A., and Sáenz, L. (2020). GOODD, a global dataset of more than 38,000 geo-referenced dams. *Scientific Data*, 7(1), 31. https://doi.org/10.6084/m9.figshare.10538486.

NASA (2005). NASA details earthquake effects on the earth. Accessed date Jan 10, 2005. http://www.nasa.gov/home/hqnews/2005/jan/HQ_05011_earthquake.html.

Song, Z., Liang, S., Feng, L., He, T., Song, X. P., and Zhang, L. (2017). Temperature changes in Three Gorges Reservoir area and linkage with Three Gorges project. *Journal of Geophysical Research: Atmospheres*, 122(9), 4866–4879. https://doi.org/10.1002/2016JD025978.

Tortajada, C., Altinbilek, D., and Biswas, A. K. (Eds.) (2012). *Impacts of Large Dams: A Global Assessment.* Water Resources Development and Management, Springer Science & Business Media, Berlin. 415. https://doi.org/10.1007/978-3-642-23571-9.

Wang, J., Guan, Y., Wu, L., Guan, X., Cai, W., Huang, J., Dong, W., and Zhang, B. (2021). Changing lengths of the four seasons by global warming. *Geophysical Research Letters*, 48(6), e2020. https://doi.org/10.1029/2020GL091753.

Wang, Y., Cao, Z., Pang, Z., Liu, Y., Tian, J., Li, J., Yin, L., Zheng, W., and Liu, S. (2022). Influence of Three Gorges Dam on earthquakes based on GRACE gravity field. *Open Geosciences*, 14(1), 453–461. https://doi.org/10.1515/geo-2022-0350.

Wu, M., Tang, W.C., Zhou, Y., Chen, S., Chen, G.W., Qin, Z.D., Wang, Y., Davis, P., and Sher, W. (2015). Modeling the impact of Three Gorges Dam on the cooling energy consumption of the reservoir cities. *Advanced Materials Research*, 1065, 3254–3259. https://doi.org/10.4028/www.scientific.net/AMR.1065-1069.3254.

Yadav, P. K., Kumar, A., and Jaiswal, S. (2023). A critical review of technologies for harnessing the power from flowing water using a hydrokinetic turbine to fulfill the energy need. *Energy Reports*, 9, 2102–2117. https://doi.org/10.1016/j.egyr.2023.01.033.

Yu, S., Sheng, L., Zhan, L., and Niluo, Z. (2021). Analysis of temperature variation characteristics in the Three Gorges Reservoir area after impoundment of the Three Gorges Dam. *Acta Ecologica Sinica*, 41(5), 384–389. https://doi.org/10.1016/j.chnaes.2020.06.003.

Zahraei, A., Eslamian, S., and Saadati, S. (2016). The effect of water extraction time from the river on the performance of off-stream reservoirs. *International Journal of Hydrology Science and Technology*, 6(3), 254–265.

Zeng, Y., Zhou, Z., Yan, Z., Teng, M., and Huang, C. (2019). Climate change and its attribution in Three Gorges Reservoir area, China. *Sustainability*, 11(24), 7206. https://doi.org/10.3390/su11247206.

Zhao, F., and Shepherd, M. (2012). Precipitation changes near Three Gorges Dam, China. Part I: A spatiotemporal validation analysis. *Journal of Hydrometeorology*, 13(2), 735–745. https://doi.org/10.1175/JHM-D-11-061.1.

Zhao, Y., Liu, S., and Shi, H. (2021). Impacts of dams and reservoirs on local climate change: A global perspective. *Environmental Research Letters*, 16(10), 104043. https://doi.org/10.1088/1748-9326/ac263c.

18 Historic Rainwater Harvesting System in Kampong Ayer

A Water Village in Brunei Darussalam

Shafi Noor Islam and Sandra Reinstädtler

18.1 INTRODUCTION

Negara Brunei Darussalam is a small country located on North Borneo Island. North Borneo includes the whole of the northern portion of the island of Borneo (HMSO, 1954). Brunei's 600-year-old monarchy and century-spanning history and cultural heritage have been steadfastly upheld to this day. The China Sea washes its western coast and the Sulu and Celebes Seas its eastern coasts. The official name for the country is Negara Brunei Darussalam which contains central mountains ranges from 4,000 to 6,000 feet in height, rising somewhat sharply from ranges of low hills nearer the coast. "Negara Brunei" means the state of Brunei and "Darussalam" means "Abode of peace". Its full name is Brunei Darussalam, is a compound of a Sanskrit name possibly meaning "sea people" or the name of a local tree that also gave its name to the entire island of Borneo, added to the name by Muslim Sultans in the 15th century.

These hills are traversed by valleys and occasional plains. The coastline is formed mainly of alluvial flats, with many creeks and swamps. Hills and valleys in most cases are covered with dense forests and many rivers (HMSO, 1954). The main harbor on the west coast is the island of Labuan, which lies to the north of Brunei Bay. Further north Jesselton, the capital of the Colony, has a good well-sheltered harbor for vessels of moderate size, which take away the bulk of the rubber produced on the west coast. At the most northern point of the colony is Marudu Bay, a former stronghold of Illanun pirates on its western shore, eleven miles from the entrance, is Kudat Harbour (HMSO, 1954).

The country consists of a flat coastal plain along the South China Sea in the western part of the country, with some hills further inland. Mountains rise in the eastern segment of Brunei, which is also the most undeveloped and inaccessible part of the Sultanate. An equatorial climate gives the area abundant rainfall, and most of the country remains heavily forested, with mangrove swamps along the coast. Most of the people live in the capital of Bandar Seri Begawan, along with its chief port of Muara, 25 mi (41 km) to the northeast. The other major town is Seria, the center of the oil and gas industry, in the western part of the country. Bandar includes great contrasts, between the Sultan's palace and the glittering Sultan Omar Ali Saifuddien Mosque and the world's largest stilt village, Kampung Ayer. This village, in existence for 400 years and providing housing for about 30,000 people, was declared a national monument in 1987 and is the most popular tourist attraction in the country.

The country has huge water resources but pure drinking water and water for irrigation are a problem for the whole nation. Historically, pure drinking water was only collected from rainfall and natural waterfalls. The most important drinking water source was only rainwater. Rain in Brunei occurs almost every day as it is located in the equatorial zone. Rainwater harvesting is a technology

DOI: 10.1201/9781003473398-22

that has been exercised worldwide for centuries. The technology has many advantages and disadvantages. Some of them are low-cost, easily maintained, sustainable, and simple construction. However, it also has numerous disadvantages. Because of climatic and geographical region considerations, there are issues of rainfall unpredictability, storage limitations, and vulnerable water quality. The applicability of rainwater harvesting is highly dependent on the climate condition, application of ability of technology, and economic condition of the country. This chapter's findings will provide an opportunity to cultivate a change in perception toward rainwater and wastewater among local Bruneians. Furthermore, the study offers a platform to establish a perspective for meeting demands with a sustainable alternative.

This chapter mainly focuses on the importance of harvesting rainwater in Kampong Ayer and in Brunei Darussalam as a whole. This is done by looking at the possible challenges faced by the government in order to maintain the Brunei River and Kedayan River in Kampong Ayer as the most potential water sources. Some issues can be derived such as ongoing threats due to infrastructural development, and frequent incidents and fires. The challenge is that due to the lack of large amounts of data concerning rainwater harvesting, management methods are very limited. As a result, the potential findings of the study could result in a lack of in depth analysis that is needed in order to satisfy the research aims and objectives.

18.2   THE GEOGRAPHICAL LOCATION AND PHYSICAL CHARACTERISTICS OF THE STUDY AREA

Negara Brunei Darussalam is located in South East Asia, with neighboring Malaysian states, Sabah and Sarawak, and the Indonesian state of Kalimantan. The country has four districts, namely Brunei–Muara, Tutong, Belait, and Temburong. In the accumulation of all four districts, the total land area of Brunei Darussalam is 5765 km^2, and a population of 405,900 (as of 2013) and an actual population of 455,858 (as of 2024) (World Population Review, 2024). As Brunei is located on the shoreline of Borneo, it has a coastline of approximately 161 km facing the South China Sea (Ibrahim, 2005; BBY, 2014, BBYB, 2015). Brunei Darussalam is situated within the Earth's middle location and carries the tropical and subtropical region's environment. Historically, the place of Borneo Island is the place of case research for geomorphic, anthropogenic, cultural, and language diversity, which makes the whole Borneo region a unique island in the South East Asian region. Brunei is also an incredibly diverse country derived from the country's historical links with the Hindu empire in the neighboring regions and modern-day Indonesia and Malaysia (BBY, 2014; BBYB, 2015).

The Kampong Ayer or Kampong "Aying" is located in Brunei Bay, especially on the meandering river banks of the Sungai Brunei. The geographical location of Brunei Bay extends approximately 40 km from east to west, while from south to north it spans around 23 km wide (Yunos, 2015). Kampong Ayer was a former capital city and commercial and administration center of Brunei. It was notorious for its powerful empire that spans throughout the Borneo Island and the Philippines archipelago. Due to this Kampong Ayer received an overwhelming number of visitors and traders from other regions. The majority of these traders came to Brunei through the coastal area to introduce Islam. During the reign of Sultan Bolkiah: the fifth Sultan of Brunei, Islam is willingly accepted by the Ruler and the local people as part of their faith and beliefs.

The climate of Brunei Darussalam is characterized by warm temperatures, high humidity, and high rainfall. Temperature shows little variation throughout the year. The monthly mean temperature of 28°C, high in the daytime, and the variation is 8°C to 12°C. The rainfall regime is about a bio model with the main wet season from November to January (northeast Monsoon) and a minor rainy season from May to October (southwest Monsoon). Between the monsoons, occasional drought may occur generally between late January to April. The mean annual rainfall ranges from 2200 mm in the coastal regions to about 400 mm in parts of the interior. The regional climate, temperature, rainfall, and landscape changes pattern is a potential factor of a region to manage and conserve nature, culture, and heritage (Tyler, 1996).

Brunei is a country in Southeast Asia, bordering the South China Sea and East Malaysia. Brunei Darussalam is located on the northwest coast of Borneo Island, near the equator, between approximately 114° 4'–115° 22' eastern longitude and 4°2'–5°3' northern latitude. The country is bounded along the southern and eastern sides by the Malaysian state of Sarawak. Sarawak has an enclave along the Limbang River that splits Brunei Darussalam into two separate parts (Brown, 1970, 132; Hassan, 1988; CDD, 1993, 2008; Ibrahim, 2005; Khatib et al., 2005).

The coastline faces the South China Sea from near the mouth of the Baram River in the southwest to the headland of Muara in the northeast. To the east of Muara is Brunei Bay, a large and protected shallow embayment. Brunei Darussalam consists of four districts: 1) the large Belait District in the southwest, 2) the Tutong District in the middle, 3) the Brunei Muara District that surrounds the capital, Bandar Seri Begawan, and 4) the separate Temburong District in the east (Figure 18.1) (CDD, 1993, 2008; Ibrahim, 2005; Khatib et al., 2005).

Its geographical coordinates are 4°30′N 114°40′E. Brunei shares a 266 km (165 mi) border with Malaysia and has a 161 km (100 mi) coastline. The area of the country is 2055 mi^2 (5770 km^2) (land: 5270 km^2; water: 500 km^2), the population of the country is 423,200 (in 2017) (Hassan, 1988; CDD, 1993, 2008; Ibrahim, 2005; Khatib et al., 2005; Omar, 2005). The main city and the capital of the country is Bandar Seri Begawan. Figure 18.1 shows the geographical and geomorphological overview of Brunei Darussalam. It also displays four district territorial boundaries and potential geographical features and landmarks in Brunei Darussalam. There are five major rivers dominating

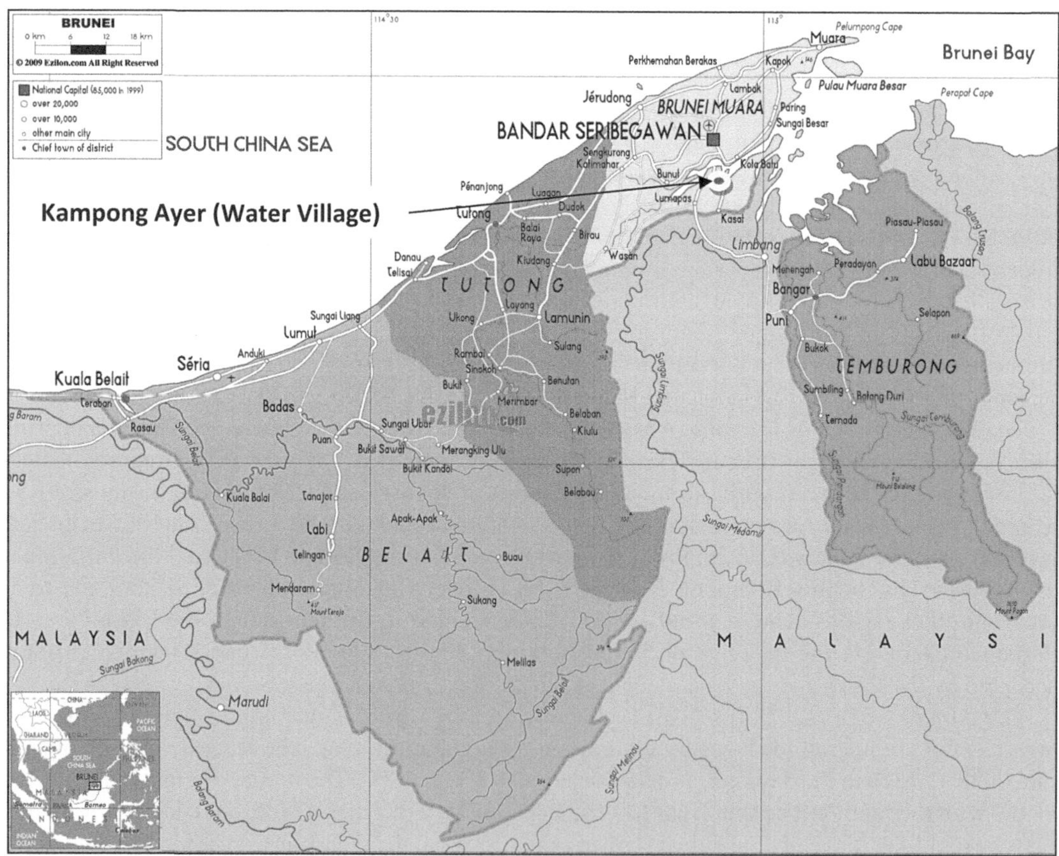

FIGURE 18.1 The geographical location and characteristics of Brunei Darussalam.

Source: Brunei physical map https://www.peta-hd.com/2019/03/peta-brunei-darussalam.html, accessed on 27th November 2019.

FIGURE 18.2   The view of Kampong Ayer (Water Village) Settlement pattern in Bandar Seri.

Source: Islam, S.N. 2019).

the floodplain morphology in Brunei Darussalam, which is situated within the three major deltaic regions Baram River Delta, Champion River Delta, and Meligan River Delta (UNESCO, 1972, 1999, 2004, 2008). The Baram River Delta covers most of the southwest corner floodplain in the Baram River catchment area and Champion River has covered the middle part as well as Tutong and Brunei Muara and the northern part of Kuala Belait district. The Temburong River has covered and connected the most part of Temburong district areas (Figure 18.1) (Hassan, 1988; CDD, 1993, 2008). Kampong Ayer water settlement is a unique element of the human landscape of Brunei Darussalam. It draws heavily on both historical and contemporary material to examine the impact of the changing social and economic development in Brunei on Kampong Ayer (Figure 18.2).

The historical geography of the early stage settlement is presented in this section. According to Yunos (2015), the current location of the water villages is suitable as it provides natural protection against the monsoon winds and pirate attacks in the South China Sea. Since the time of the British residency in 1906, Different motivational initiatives have always been made in 1980 and 2016 for the development of settlement and urbanization in the land area especially in the Kota Batu area. The resettlement of residents of Kampong Ayer on dryland, as well as provide them with modern amenities and facilities. Kampong Ayer, literally a "water village" but in reality a sprawling settlement of 42 villages of different sizes and shapes (Figures 18.2 and 18.3a, 18.3b) was and still is too many inhabitants, synonymous with Brunei itself.

Located on various sites in the vicinity of Brunei Bay over the years, Kampong Ayer has been central to the history and legends of Brunei Malaysia, and distinguished by its complex physical structure, diverse social characteristics, and unique traditions and social mores of its native people.

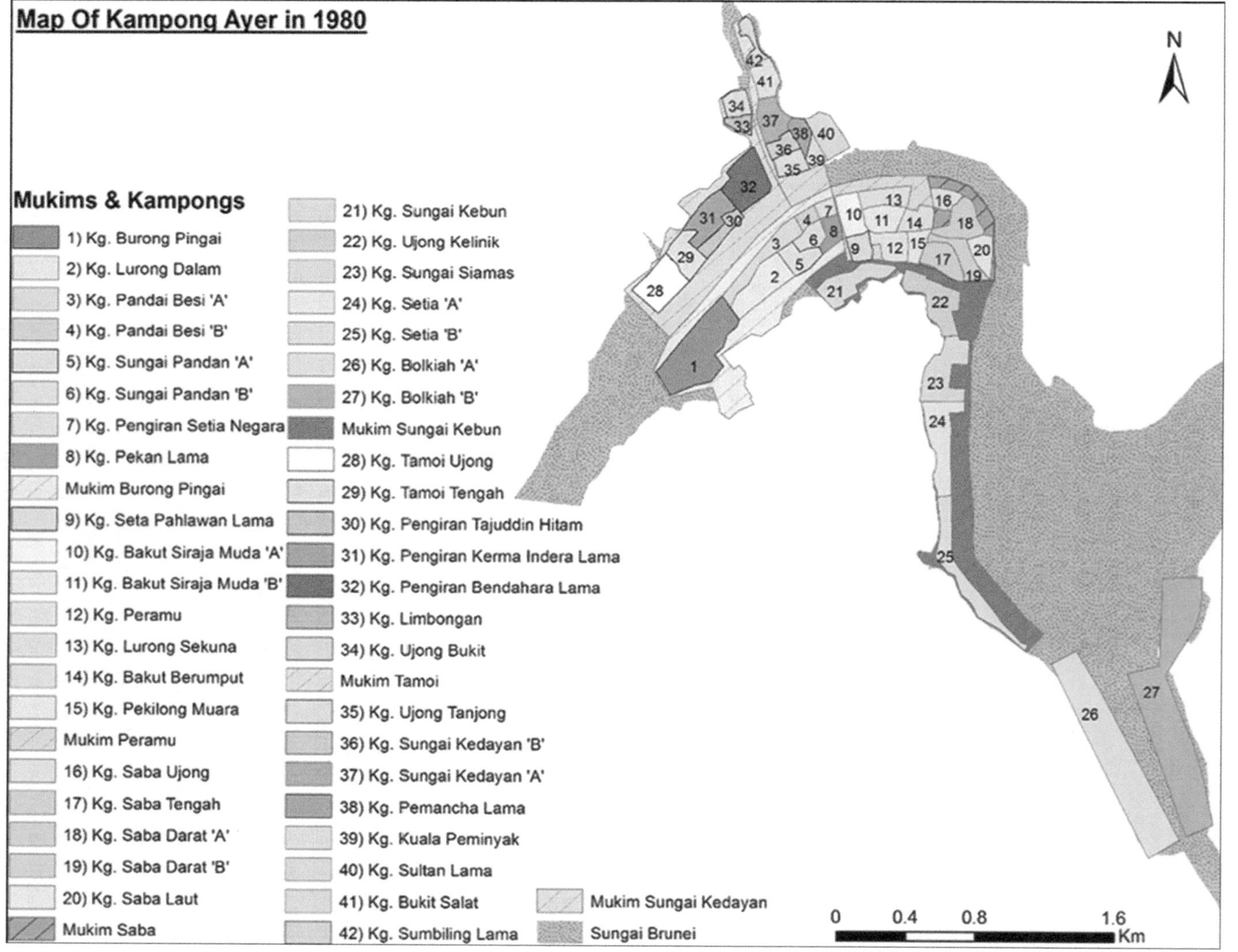

FIGURE 18.3A The Geographical locations and distribution of Villages over water in Kampong Ayer.

Sources: Islam, S.N, 2017.

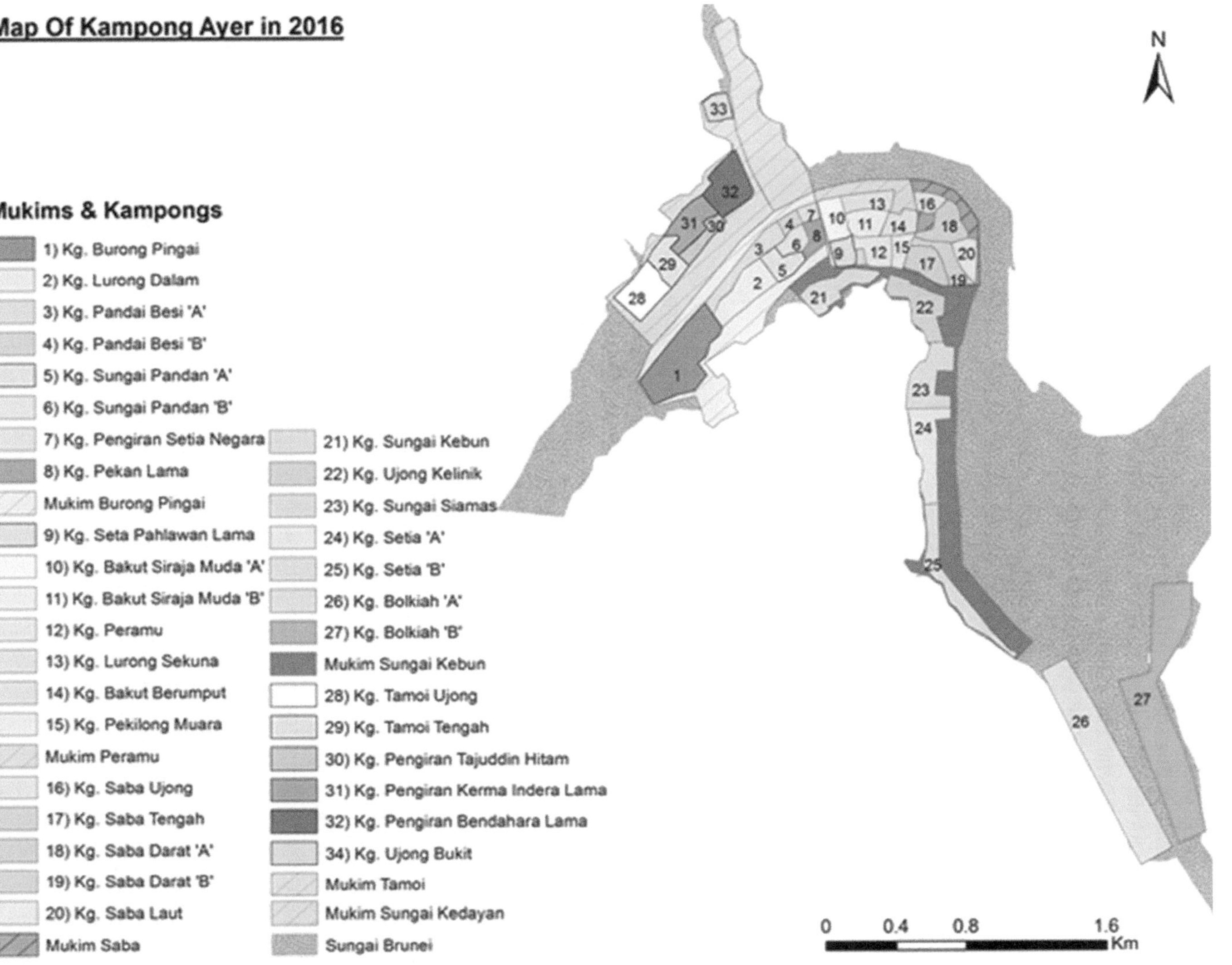

FIGURE 18.3B The Geographical locations and distribution of Villages over water in Kampong Ayer.

Sources: Islam, S.N, 2017.

18.3 AIM AND OBJECTIVES

The aim of this study is to outline the historical background of rainwater use and reuse scenario in Kampong Ayer and both the advantages and disadvantages of rainwater harvesting together with the ways to overcome the drawbacks and decide the applicability of the technology for water resources management in Brunei Darussalam.

The objectives of this study are to understand the unique characteristics and vulnerabilities of historical and modern uses and reuses of rainwater harvesting methods applied in Kampong Ayer (water village) in Bandar Seri Begawan, Brunei Darussalam. There are some specific objectives of this study, which have been recognized as follows:

- To understand the characteristics of rainwater harvesting and historical use and reuse patterns in Kampong Ayer in Bandar Seri Begawan.
- To investigate the water quality availability status and historical rainwater harvesting scenario of Kampong Ayer.
- To understand the water resources maintenance and management in Kampong Ayer, especially in the Brunei River basin area.
- The study seeks better maintenance and management of drinking and potable water resources sustainability in Kampong Ayer in Bandar Seri Begawan in Brunei Darussalam.

18.4 DATA AND METHODOLOGY

The study has been conducted based on primary and secondary data sources. The primary data on Brunei heritage information were investigated from field studies in Bandar Seri Begawan. A field investigation was arranged in the Brunei River Basin area on the Kampong Ayer complex, which is a unique culture and heritage site in Brunei Darussalam. Formal interviews with the inhabitants of the water village were one basis of the primary data set. A special discussion was arranged with the students of UBD, Brunei for primary information on Culture and Heritage, tangible and intangible cultural heritage in Brunei. Another field investigation with 58 students in the Kampong Ayer area and Mosques located areas in the Bandar Seri Begawan area in Brunei is also part of the investigations. Besides these field studies and interviews with different experts at the University of Brunei Darussalam, Brunei, the secondary data and information were collected from the UBD library and some other international organizations such as UNESCO, IUCN, and journal publications found through the internet. Besides this geomorphological, anthropological village-settled study information through informal discussion and with the inhabitants in the Kampong Ayer areas for benchmark data collection was conducted. A complete enumeration of the households covering the land holding pattern, tenancy, agricultural cropping system and marking, occupation, demographic character, literacy, etc., was collected in 2015. A number of standard approaches for information and data collection and analysis were used including a literature review of studies to get a basic understanding of culture and heritage and people's perception of cultural heritage in Brunei Darussalam. The secondary data inputs were obtained from different publications of government agencies, NGO reports, and research organizations in Brunei. Collected data and information were analyzed and visualized using EXCEL, and VISIO 32 software. These data and information were used to develop the management plan for the sustainability of cultural heritage conservation in Negara Brunei Darussalam.

18.5 PUBLIC PERCEPTION OF RAIN AND HARVESTING OF RAINWATER

Rainwater is a liquid in the form of droplets that have been considered from atmospheric water vapor and then precipitated that is, become heavy enough to fall under gravity. Rain is a major component of the water cycle and it is responsible for depositing most of the fresh water on the Earth's surface. In general perception of rain is the special system where water evaporates from the surface water and creates clouds, when clouds are heavy it comes back to the surface as rain. Rainwater is fresh and clean and it cools the land surface area. Rain is dominating so many different ecosystems on the surface and some ecosystem and ecosystem services are directly dependent on rain or rainwater because it provides favorable ecology and ecosystem as well as water source and supply for hydroelectric power plants and agricultural irrigation. It also maintains the water flows of river basins and reduces the salinity range in river water as well as maintaining the river water quality.

Rainwater harvesting is the gathering, accumulating, storing, and depositing of rainwater for use on-site, rather than allowing it to run off. Rainwater harvesting is also called roof water harvesting which involves the collection, storage, and distribution of rainwater from the roof for use inside and outside of the home or for agricultural or business purposes. Rainwater harvesting is the process of inception and concentration of runoff and its subsequent storage in the soil profile or in artificial reservoirs for crop production. On the other sense, after harvesting this water is used for agriculture and household cleaning, washing and bathing, and gardening purposes. Rainwater harvesting is a technology in which local surface runoff, rain, or surface water flow that happens when soil is infiltrated to its maximum capacity, is induced, collected, stored, and conserved (CTCN, n.d.). On the other hand, it can be said that rainwater is a process of collecting water from the rainfall, and later it is filtrated and gradually used for household, social, and cultural purposes. The collected harvesting water can be purified and used for drinking and potable use. It can also be used for daily social and cultural uses and other household uses such as dishwashing, laundry, watering garden plants, and for small-scale industries. These types of activities and rainwater use practices continued until the 1970s, very openly until the modern water supply and water pipeline construction was ready in Kampong Ayer.

In history, there are different types of methods and processes used for storage and purification and for production in industries, agriculture fields, and household activities. Rainwater harvesting processes and methods are very popularly used in water crisis regions such as dry regions, tropical and subtropical, arid, and semi-arid regions of the world. Rainwater provides an opportunity to stabilize agricultural landscapes and bring balance to ecology. The inhabitants of small island states popularly use rainwater for their agricultural production, gardening practices, household uses, bathing, cleaning, and fresh drinking water. In the Asia Pacific region, there are 17 countries that use rainwater harvesting as the main source of drinking water and household water for cleaning, bathing, and cooking purposes. Although these island countries are located inside the ocean or sea they suffer from a crisis of drinking water and fresh water. As seawater is highly saline and concentrated therefore the available large amounts of water is not suitable for use. Therefore, they have to wait for rainwater and different traditional collecting methods are used. In Southeast Asian countries terrace paddy cultivation is a very popular agricultural method in this region and the only water source is rainwater. The farmers of the terrace area of the Philippines, Indonesia, Malaysia, Thailand, Vietnam, Laos, Cambodia, China, Korea, Taiwan, Japan, and Borneo area are using rainwater for paddy cultivation and other gardening irrigation purposes. In Brunei, rainwater was the most important and only source of freshwater for drinking, cooking, cleaning, and washing at the community level as well as in Kampong Ayer (water village) almost for 100 years. The alternative fresh drinking water was available from five waterfalls in the Subok Valley area near Brunei town (such as Subok 1, Subok 2, Subok 3, Subok 4, and Subok 5, and in Bandar Sungai Pangea). The other most important drinking water source was located in Kampong Tamoei of Kampong Ayer it is the Haji Tarrek Kandong well. For a long time, this Kandong or water well was frequently used for water collection, the Haji Tarrek Kandong location is still visible however, it is closed and not in use anymore. The Subok Valley water sources were used very frequently by traditional boatmen of Kampong Ayer and they collected this freshwater through a bamboo pipe or Bamboo canalize method from uphill to the down of the hill and transferred to water barrels (drums) in traditional boats or water taxies are standing on the Sungai Brunei (Brunei River). Later the boatmen distributed the water barrels from house to house or family to family. This fresh drinking water distribution was done voluntarily and in some cases, householders were paid a minimum amount of money. There was a drinking water well that was also frequently used which is Sri Lampai Kandong (well) it was located in the Sungai Mattan whose new name is Sungai Pintu Malin today. All these collected fresh water from different water sources was used for fresh drinking water and water carried by traditional Bruneian boats and these small entrepreneurs were selling water or supplying water through boats in every house in Kampong Ayer. This was practiced until the 1970s and into the 1980s. The water collectors or the group of water collectors and entrepreneurs group were collecting water from Subok and Subok Valley's waterfalls in five locations through a bamboo canal system water went down to where water collectors collected water in plastic drums or barrels and the water drums were carried by water boat/water taxi in the Brunei river and the Kedayan river and supplying to the homes in water villages in the Kampong Ayer. This was the general scenario in the Kampong Ayer.

18.6 HISTORICAL BACKGROUND OF WATER SOURCES, WATER SUPPLY, AND SANITATION IN KAMPONG AYER

According to the water services annual report in 1995, the water sources in Brunei depend solely on rainwater collected in various basins found in the country, and groundwater is not one of the sources like in other countries due to the presence of high iron content in the water, high abstraction coast incurred and low amount output. The demand for potable water supply in Brunei–Muara District is currently fulfilled by three dams Benutan Dam in Tutong District and Mengkubau and Tasek Dam located in Brunei–Muara District. Two dams named Kargu and Ulu Tutong are still under construction to meet the increasing public consumption.

Bandar Seri Begawan was made accessible to untreated pipe water in the year 1948. The only water source available during that period of time actually came from one of the Sungai Brunei's tributaries, Tasek. The treated water only came about in 1953 after the Tasek treatment works were completed. In 1964 Tasek Dam was developed and allowed people in Bandar Seri Begawan to enjoy the potable water service at a rate of 10,000 cu.m/day. However, Tasek Dam is not the only source of water and it serves to augment the supply to Bandar Seri Begawan from the main source at Sungai Tutong. The Sungai Tutong can be regarded as the most significant water source in Brunei since it was commissioned in the year 1973. The Layong treatment plant was constructed in the year 1973 with a capacity of 41,000 cu.m/day (stage 1) and upgraded by means of extension to about 77,000 cu.m/day (stage 2) reaching its ultimate capacity of 109,000 cu.m/day in the year 1984 (stage 3) to support not merely Brunei–Muara but Tutong district's water requirement. The Bukit Barun treatment plant was also introduced to make sure the demand was met. The treated water is subsequently pumped and transported to three terminal reservoirs which include Tutong Town.

Terminal, Bukit Barun, and Sungai Burong. Sungai Burong terminal reservoir located on the eastern part of the Tutong is responsible for the supply of water to the Brunei–Muara district through the Gadong booster pumping station constructed in 1979 (stage 2). The Brenutan Dam, a so-called earth dam was completed in 1989 to supplement and ensure the demand for water supply is met. Apart from that the purpose of Benutan Dam is to avoid salt intrusion that is possible during dry periods with deficit water abstraction.

Rozan Yunos (2015) wrote in his article regarding the Mengkubau Dam and water treatment plant located in Brunei–Muara District can be considered relatively young in providing water services to people who live in Kota Batu and Muara as it was only introduced in the 21st century. An increasing number of terminal reservoirs were also constructed such as Lambak and Kiarong and more to ensure adequate among to be supplied to the public (Yunos 2015).

With regards to the location of the research area in Brunei–Muara District, the sources of water supply services are believed to be provided by Tasek Dam and water treatment plant, the laying water treatment plant, the Bukit Brunei treatment plant, and the Benutan Dam.

18.7 HISTORICAL BACKGROUND OF RAINWATER HARVESTING IN KAMPONG AYER (WATER VILLAGE)

Kampong Ayer is literally a "water village" but in reality a sprawling settlement of 42 villages located not only on the river but also on the river itself with an area of over 10 km^2 (Yunos, 2015; Ghazali, 2000), which is home to about around 28,000 in 1981 and reduced to 13,000 in 2011 (KBPP, 2016). It has been located on various sites within the vicinity of Brunei Bay over many years (Aziz, 1996) and has been central to the history and legends of Brunei Malaya, and is distinguished by its complex physical structure, diverse social characteristics and the unique traditions and social mores of its people (Aziz, 1996).

Kampong Ayer is culturally and historically the most famous, as well as the largest water settlement in the Southeast Asian region and the world (Aziz, 1996; Yunos, 2015; Ghazali, 2000). Thus it is one of the Brunei heritages that needs to be conserved for the future generation. Its beauty lies it is the uniqueness of "floating" above seawater, appearing as an island when viewed from the air. However, the people in Kampong Ayer have faced a number of problems since long ago. One of the problems is water quality issues which getting serious each day. Although the local people have resided in Kampong Ayer for more than a century, they are still facing problems from water pollution, especially solid waste that deteriorates the environmental quality (Aqilah, et al., 2013).

Also, the major problem of water pollution can be linked to the people living in Kampong Ayer and the continuous development of the area. Kampong Ayer has been experiencing worse pollution due to the increase of waste coming from the capital city and the negative practice of people dumping rubbish into the river. A number of studies were carried out for water quality in Kampong Ayer but the three studies were of utmost importance A study made by Goh (1991) found that every family in Kampong Ayer produced an average of 24.23 kg of trash each week. A cleanup project was conducted on the river yielded around 1,700 kg of rubbish mostly comprised of wood, plastic, and solid waste.

The simplest method to harvest rainwater is using storage tanks, either on-ground or in-ground (AA, 2017). There are different size ranges of tanks available. The best catchment area, the area on which the rainfall or water runoff is initially captured, is the rooftops of buildings. Provided they are of large size and enough to harvest daily water requirements. This method of using storage tanks and rooftops is better known as rooftop rainwater harvesting. It is the least expensive method and is very effective. And, with proper implementation, the groundwater level of the land can be augmented.

18.8 SMART PLANT FOR RAINWATER USE WITHIN ETHICS AND CULTURAL PRACTICES

The demand for water and the climatic condition, the uncontrolled disposal in the environment of municipality, industrial, and agricultural liquid, solid, and gaseous waters constitutes one of the most serious threats to the sustainability of human civilization by contaminating the water, land, and air through global warming (Ferdaush et al., 2015; Beecher, 2001). With a continuously increasing population, academic and economic growth, and industrial development, treatment, and safe disposal of wastewater is essential to preserve public health and reduce intolerable levels of environment degradation in the dry and water crisis regions (Reinstädtler, 2011, 2013, 2017; Reinstädtler et al., 2017).

In addition, adequate wastewater treatment and management is also required to prevent contamination of water bodies to preserve the sources of clean water. Effective rainwater and wastewater management planning is well established in developed dryland and water crisis areas but is still limited and there are some limitations to water reuse and rainwater harvesting and management systems. At present the requirements for a sustainable rainwater harvesting and treatment system have

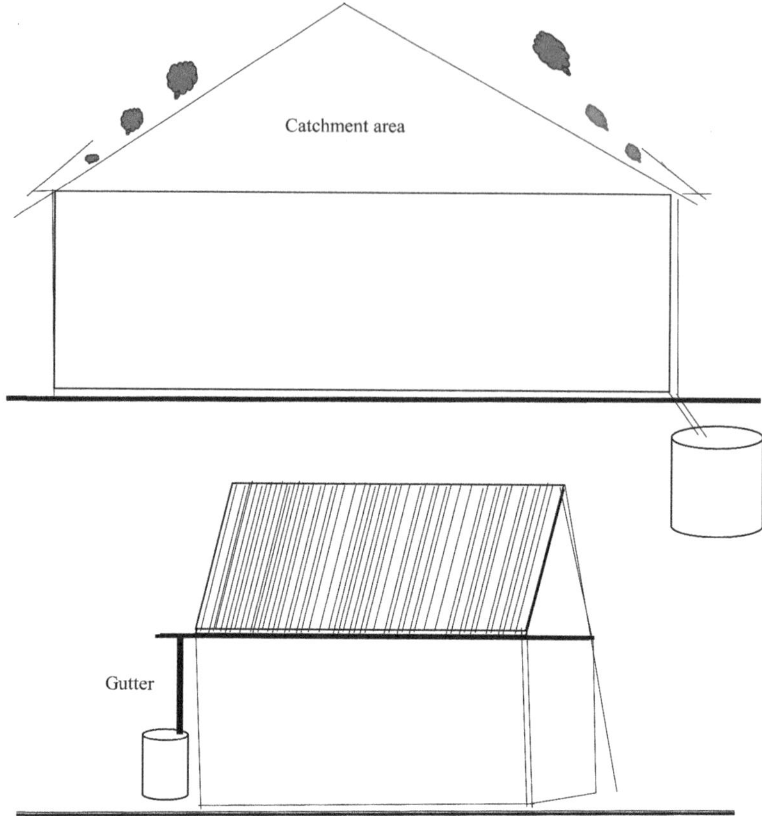

FIGURE 18.4   Rooftop rainwater harvesting methods in Kampong Ayer.

Source: Islam, S.N. 2020- this method are popularly used before 1984.

been presented, there are several options one can choose from in order to find the most appropriate technology for different climatic and particular environmental regions. This chapter will discuss a traditional model of microscale level rainwater harvesting and the use of the plant at the community level for the freshwater crisis and dryland regions.

Before the 1980s rainwater harvesting methods were popularly used in Kampong Ayer as rainwater, waterfall, and only one water reservoir were sources of fresh water in the Brunei Muara district. All three sources are used for fresh and drinking water in the water villages of Kampong Ayer. The traditional methods were used in every house in the water villages of Kampong Ayer this is the rooftop method for the collection of rainwater. In this method, the house rooftops are commonly used to store water in a drainage system or pipeline connected with plastic drams with water stored at the plastic drams. Later this stored water was purified for household use and use for drinking purposes. The method was popularly used until the 1980s. Gradually the level of household use of rainwater has decreased.

The reasons for this are the modern water supply pipeline was constructed in 1983–1984, most portions of Kampong Ayer have been constructed and the supply line has also connected the supply network from house to house. The traditional indigenous rainwater harvesting methods were used and practiced for more than a thousand years in the water villages in the Kampong Ayer of Bandar Seri Begawan in Brunei Darussalam. These traditional water harvesting methods were used not only in Brunei Darussalam but also in the whole Asia Pacific region and over 325 other water villages worldwide. However, the majority of water villages are located in the Asia Pacific region.

The figure (Figure 18.4) above shows the basic components of rooftop rainwater harvesting consist of catchment (roof), transportation (gutter and pipes), first flush (outlet tap), reservoir (storage tanks), and filter. The rainfalls or rainwater on the roof as the catchment area and collected by gutters, which then transports them through the pipes toward the tanks. The tanks then store the water which is filtered before being used.

These were the traditional water collection and rainwater harvesting methods that had been used for thousands of years. The community people of Kampong Ayer were practicing these freshwater collection and harvesting methods within the community's engagement and participation in Kampong Ayer until the 1980s, just before the modern water pipeline construction in Kampong Ayer.

18.9 THE MODERN WATER SUPPLY AND MANAGEMENT SYSTEM IN KAMPONG AYER

Kampong Ayer has remained an important settlement in the history of Brunei Malays. The water village, with more than a thousand houses, is still home to about 30,000 people. These houses are built on stilts, just as it was done centuries ago (Aziz, 1996) However, the government today has made Kampong Ayer almost self-sufficient by providing it with modern conveniences such as piped water and electricity together with such community facilities like schools, shops, police station, clinic and transport and fire services (Aziz, 1996). Besides, most of the houses in the water village

FIGURE 18.5 The traditional and renovated in between the pathway and water supply pipe line.

Source: Islam, S.N. 2019.

FIGURE 18.6 The fresh and drinking water supply and electricity gas supply system in Kampong Ayer (Water Village).

Source: Islam, S.N, 2019.

have all the luxuries that characterize modern living with modern water supply, electricity supply, and gas supply available. The government is giving subsidies to the inhabitants of Kampong Ayer. The change from traditional building styles and materials is seen in the fact that as much as 66% of the houses in the water village are supported by concrete piles at regular intervals. With the use of concrete piles, the water settlement has now even encroached further into the deeper waterways. Consequently, some of the water passage channels have been occupied by housing extensions and additional houses. These water passage channels also acted as natural fire breaks and in some cases as natural breaks between clusters of houses. An assessment of the building conditions carried out by the Kampong Ayer Improvement Consortium (1994), showed that a large number of houses, almost 55%, were found to be in fair condition. In fact, about 95% of the houses have been identified as being in poor conditions were found to be located on the side of the river (Aziz, 1996). As much as 97% of the houses are residential units, while there are as many as 63 small commercial establishments consisting mainly of retail shops and a few eating houses that are located within the residential premises (Aziz, 1996).

The shape and size of Kampong Ayer is gradually reshaping and being reconstructed in different stages. More than 12 water villages already shifted from Kampong Ayer to the land areas in different parts of Brunei Darussalam. The figure 18.6 is displaying the entry and the connectivity from land area to the Kampong Ayer with Wooden connectivity pathway (Wooden Bridge) and water supply connectivity. The Kampong Ayer has some potential in becoming a UNESCO World Heritage site and it is the landmark of Brunei Darussalam. Recently, Kampong Ayer has been considered a potential place for education and research fields.

Figure 18.6 demonstrates the water and gas supply pipe line which has been constructed after 1983. The constructed wooden bridge and the pipe lines (Water and Gases and sanitary pipelines) were following the same direction and it is on the river basin. The river basin is a joint basin connected from Kedayan River and Brunei River joining point is the place of water villages (Kampong Ayer). In Figure 18.7, a group of students from university of Brunei Darussalam were investigating the tangible and intangible cultural heritage in Kampong Ayer. Figure 18.8 displays scenario of low tide time in the Kampong Ayer where a group of students are observing the water quality tide and high tide low tide condition and water biodiversity in the water villages in 2019. The research group also investigating the popularity of rainwater harvesting method and using pattern and the benefit of rainwater practices in Kampong Ayer.

Figure 18.8 demonstrates the renovated part of the Kampong Ayer where the modern building has been reshaped and the re designed the architecture.

FIGURE 18.7   A Group of Students investigating the water supply System in Kampong Ayer.

Source: Islam, S.N. 2019.

FIGURE 18.8   The renovated part of Kampong Ayer and the water supply and management system.

Source: Islam, S.N, 2019.

The traditional Malay architectural pattern is changing where the rainwater harvesting system is not attached to this building over water in the water villages. For rainwater harvesting, even the traditional method could be introduced with new modern buildings. As for rainwater harvesting the following issues should be considered in any region of the world:

a) The site location where the amount of rainfall received by the region for Kampong Ayer is already recognized as a suitable region for more than a thousand years.

b) The size of the catchment area (roof size).

c) The rainwater storage capacity – the areas with lower available precipitation may require larger tanks to provide more storage capacity. It would be increased but an increase will increase equipment costs.

d) Roof pitch and type – the roof and pitch material influence the amount of water that can be harvested. The lower-pitched roofs tend to lose less water than steeply-pitched roofs, roof texture could facilitate runoff better than textured roofs.

e) Water rates – the areas with more expensive water rates will make rainwater harvesting more economically viable.

f) Permits – rainwater harvesting permits may be required for some countries but in Brunei traditionally it is not a barrier from the Government as it was practiced in Kampong Ayer from ancient times.

g) Operation and maintenance – rainwater harvesting systems require regular operation and maintenance, monitoring tank level, cleaning system, monitoring for leaks and maintaining treatment system, and finally testing water quality.

h) Harvesting potential – the amount of rainfall that can be collected from a rooftop or other hard surfaces can be estimated using a rainwater harvesting tool. Monthly rainwater harvest can be calculated through the following formula:

Monthly Rainwater Collected (gal) = Catchment Area (roof size) (Square feet) × Monthly Rainfall (Inches) × Conversion Factor × Collection Factor

The above formula can be used to size a tank and the conversion factor is a factor of 0.62 used to convert inches of rain that fall onto the roof area and monthly harvesting potential.

18.10 DISCUSSION

Almost 1,000 world natural and cultural sites are inscribed in the World Heritage list through UNESCO and most of the sites are facing climatic and anthropogenic disturbances (Islam et al., 2016). The global climate change impacts and anthropogenic influences on cultural properties all over the world are more or less similar (UNESCO, 1972, 1999, 2008). The heritage sites in Brunei Darussalam are also facing a similar global problem. It is a new threat to the protection and conservation of cultural and natural heritage sites. Based on the present threatened situation the approach model has been proposed for the future implementation and development of cultural heritage sites in Brunei Darussalam.

Based on the present scenario on the water resources in Kampong Ayer are displaying that inhabitants from Mukim Saba Darat B and a few from Mukim Tamoi who happened to be the village headman perceived their back then was much better as compared to what they have now. In the past, the turbidity of the water in Kampong Ayer was quite low, and unpolluted. This was further highlighted since the majority of the people in Kampong Ayer were mostly in the water ecosystem. As an example it can be taken of people in Kampong replying on the river ecosystem would include fishing activities, trawling net catch, or even operators such as water taxis. The water was also used for traditional healing, in which they used the river for swimming.

Presently, the perceptions of the people in Kampong Ayer have changed due to several factors that affect the water quality in Kampong Ayer. Nowadays, the turbidity of the water is quite high above 2 to 3 m deep (Aqilah, et al., 2013). The river channel also getting smaller due to construction activities in a particular area adjacent to the Sungai Kebun bridge construction near Mukim Saba Darat. This could be one of the factors that affect the perception of the people on the water quality. They also persisted that the water quality has changed and due to that, the reliance of the people of Kampong Ayer on the river ecosystem has reduced in numbers. As quoted from one participant during the interview stated "Before it was so easy to catch fish, prawns from the river, but now, it is quite difficult to even see the fish, the size of the prawns were also reduced in quantity.". Rainwater and river water are the most important issues in nature and natural resource effectiveness and management. Most of the river water and wetlands are very complex regarding their hydrologic, biological, and geomorphological functioning and dynamics (Reinstädtler, 2011, 2013, 2017). The rainwater and watershed of river basin areas have been utilized by farmers, fishermen, and other stakeholders in terms of drainage and annual banning of stock farming for hundreds or even thousands of years. Rainwater and river water are used as the driving force of ecosystem services, biodiversity conservation, and community livelihood sustainability (Islam et al., 2014; Ferdaush et al., 2015).

18.11 FUTURE APPLICABILITY OF RAINWATER HARVESTING IN KAMPONG AYER (WATER VILLAGE)

In contrast to the world population without access to clean water supply, Brunei Darussalam's achievement in the supply of clean piped water to the urban population is recognized and ranked 5th out of 48 countries in the Asia Pacific region by the Asian Development Bank in the year 2016 (Hong, 2016). The Government commitment can be witnessed from the well-oriented water distribution network available in place to provide water services as shown before in this paper. Apart from that, their dedication to ensuring the demand for potable water service is adequate to meet public consumption can be witnessed from the history of water services.

Sanitation in Brunei also raked with a score of 90 rewarded by the Asian Development Bank (Hong, 2016). It cannot be seen in the urban areas in Brunei but something is neglected in Kampong

Ayer. Therefore the present status of the water supply and sanitation system in the untouched part of Kampong Ayer is really not a good condition. Especially the sanitation system is still practicing the traditional open sanitation system when the waste is going to the river directly. Which is really a big threat to the river water quality and water ecosystem as well as water biodiversity in water villages. The old generation of the Kampong Ayer inhabitants still expect that they will use the traditional rainwater for many social and cultural activities. There is a strong religious faith and beliefs and values from a cultural point of view on rainwater use and harvesting. Religiously the inhabitants of Kampong Ayer believed that rainwater is holy water and it can be used for any social, cultural, and religious purposes. The rainwater was used in special cooking and making special cakes and food items where they used rainwater as fresh water and holy water. Which can make the food delicious and at the same time believe that it is fresh food as they used the rain holy water in their food items. Uninterruptedly the inhabitants of Kampong Ayer were using the rainwater for bathing and cleaning their hair and they believed that rainwater was making their hair smooth and their skin feeling soft and smooth. Therefore still rainwater could be used as they were used before or in ancient times. For the future use and rainwater sustainability concept for harvesting and use, Islam (2014) proposed a model that could be applied for rainwater harvesting and use and water resources management and ensure the sustainability of Kampong Ayer rainwater resources.

The proposed model (Figure 18.9) could be implemented in Kampong Ayer (water villages) as well as other parts and locations in Brunei and the whole of Borneo Island. For example, the

FIGURE 18.9   Model of water resources management and rainwater harvesting in Kampong Ayer.

Source: after Islam et al., 2014.

Kampong Ayer water village catchment area is a hotspot of sensitive water biodiversity, and mangrove biodiversity, and a national cultural identity and landmark in Bandar Seri Begawan in Brunei. A large number of inhabitants of Kampong Ayer (water village) directly depend on the natural and cultural properties and natural resources and they can apply this model for the sustainable use of rainwater in Kampong Ayer (BBYB, 2015; BBY, 2014).

18.12 CONCLUSIONS

Globally and regionally rainwater and river water resources are potentially used for human life and socio-cultural economic and industrial development for human wellbeing. Water is a driving force for community, social, and economic development all over the world. The world's geographical features do not consider the matter of water distribution, but climate issues are a matter of productivity and development. Climate change and tropical and subtropical climates are potential factors for quality water distribution, availability, uses, and future development issues. They have the ability to focus tremendous energy and to generate significant creative and economic betterment.

In general, the country's rainwater natural resources as well as the ecosystems are degrading due to anthropogenic influences and natural calamities. The rain freshwater is always playing a potential role in the community economy and local ecology. The hydrological cycle of any potential area is losing its balance. The rainwater harvesting and maintenance in Kampong Ayer and the surrounding are not properly maintained. The area needs an adequate interdisciplinary policy and strategies and political will to improve it for sustainable management and protection of traditional rainwater harvesting technology. The traditional and modern methods for the generation of transformation knowledge to achieve and maintain natural resources, agro-farming system development, fishing development, and, in general, the wise use of rainwater as natural resources could protect the micro scale level biodiversity in the freshwater ecosystem and river water salinity reduction could project the local ecology.

The climate change impacts as well as drought are a new threat to the control, management, and conservation of rainwater resources in tropical and subtropical climatic regions in the world as well in Borneo Island and in Brunei Darussalam. In general, the country's rainwater resources are reducing due to anthropogenic influences, uses, and practices of modern application of engineering tools in water resources. The scenario in Brunei also indicates similar decreasing characters and it is the result of mismanagement and unprofessional planning for rainwater uses, storage, purification, and conservation in Kampong Ayers (water villages) in Brunei Darussalam.

Therefore, Kampong Ayer needs an adequate interdisciplinary policy and framework strategies as well as the political will to implement it for sustainable uses, management practices, and protection of rainwater resources in Brunei. The traditional methods for the generation of transformation knowledge to achieve and maintain rainwater resources, and natural properties such as fishing development and in general wise use of rainwater resources that could protect the local water ecology, and biodiversity in the tropical region.

The following recommendations could be followed in implementing the management stages. Further, the issues are also strongly recommended for implementation in the Kampong Ayer –water village in Brunei Darussalam particularly for better management of rainwater harvesting and uses.

- To increase community awareness concerning rainwater harvesting, its benefits and cost-effectiveness.
- Provide information to the global rural and urban communities concerning rainwater harvesting, availability, and use benefits from social, cultural, and religious perspectives.
- Ensure community involvement in maintaining and protecting traditional tools for rainwater harvesting methods in Kampong Ayer.
- Technology and awareness knowledge transfer to the urban and rural communities for sustainable use of rainwater for protecting human health and keeping traditional cultural beliefs.

- The need to continue environmental and cultural discourse and dialogues for understanding the importance and inevitability of capacity-building skills to protect the landscapes, land use, and rainwater and their appropriate use within drought-hazardous conditions in Brunei Darussalam.
- To develop and train local communities, local governance, stakeholders, NGOs, national policymakers, and planners directly or indirectly involved in water resources conservation and management planning activities.
- Therefore, an integrated rainwater harvesting, conservation, management, policy guideline framework, and strategic environmental planning (SEA) is urgently needed for sustainable Kampong Ayer rainwater management and quality maintenance and protection within the tropical climatic conditions.

BIBLIOGRAPHY

AloAqua (AA). (2017). Innovative technologies for sustainable water use. Rainwater harvesting advantages-disadvantages. http://www.aloaqua.co.nz/blogs/news/rainwater-harvesting-advantages-disadvantages.

Aqilah, M. S., Qasrina, N. H. H. O., Zahidah, N. B., and Nurazri, M. A. (2013). *Water Quality in Kampong Ayer. Report on Geomatics Module in 2013 at the Geography and Environmental Studies.* University of Brunei Darussalam, Brunei Darussalam.

Aziz, A. F. K. (1996). Housing Kampong Ayer Residents in Brunei Darussalam. In *28th International Geographical Congress*, The Netherlands.

Beecher (2001). *The Ethics of Water Privatization in Navigating Rough Water: Ethical Issues in the Water Industry.* American Water Works Association, Denver, CO, p. 247.

Borneo Bulletin Yearbook (BBY) (2014). *Brunei 2035 Development, Education and Health-Seeds of Success.* Bandar Seri Begawan, Brunei.

Borneo Bulletin Year Book (BBYB) (2015). *The Guide to Brunei Darussalam 2015.* Brunei Press Sdn Bhd, Bandar Seri Begawan.

Brown, D. E. (1970). *Brunei: The Structure and History of a Bornean Malay Sultanate.* Brunei Museum, Brunei.

Climate Technology Centre & Network (CTCN) (n.d.). *Rainwater Harvesting.* UNFCCC, United Nations Industrial Development Organization (UNIDO). UN City, Copenhagen. http://www.ctc-n.org/technologies/rainwater-harvesting.

Curriculum Development Department (CDD) (1993). *History for Brunei Darussalam (Secondary 3).* EPB Pan Pacific, Singapore.

Curriculum Development Department (CDD) (2008). *History for Brunei Darussalam: Sharing our Past.* (Secondary 1 & 2), EPB Pan Pacific, Singapore.

Edson, G. (2004). Heritage, pride or passion, product or service? *International Journal of Heritage Studies*, 10(4), 333–348.

Ferdaush, J., Islam, S. N., Reinstädtler, S., and Eslamian, S. (2015). Ethical and cultural dimensions of water reuse in global aspects. In Saeid Eslamian (Ed.), *Urban Water Reuse Handbook.* Taylor and Francis Group, CRC Press, New York, 285–295.

Ghazali, A. (2000). *Kampong Ayer, the Water Village.* Heritage of Brunei Darussalam, Ministry of Home Affairs, Bandar Seri Begawan.

Goh, K. C. (1991). *Garbage Production and Disposal in Kampong Ayer and Environmental Implications.* Working paper no. 7. Faculty of Arts and Social Sciences, Universiti Brunei Darussalam, Bandar Seri Begawan.

Hassan, A. B. H. (1988). *Elementary Geography for Brunei Darussalam.* Federal Publications Ltd., Singapore.

Her Majesty's Stationery Office (HMSO) (1954). Colonial Reports North Borneo, 1953, Government Printing Department North Borneo, London.

Hong, J. (2016). ADB ranks Brunei fifth in Asia-Pacific for water security. *The Brunei Times.* http://www.bt.com.bn/news-national/2016/31/adb-ranks-brunei-fifth-asia-pacific-water-security.

Ibrahim, H. A. L. H. (2005). Culture and counter cultural forces in Contemporary Brunei Darussalam. In Thumboo, E. (Ed.), *Cultures in ASEAN and the 21st Century.* UniPress, Singapore, 22–34.

Islam, S. N. (2014). Biosphere reserve, heritage identity and ecosystem services of Sundarbans Transnational site between Bangladesh and India. In *Conference Proceeding of World Heritage Alumni Conference*, 22–25 October 2014, 224–244. http://www.IAWHA.com/.

Islam, S. N., Karim, R., Islam, A. N., and Eslamian, S. (2014). Wetland hydrology. In *Handbook of Engineering Hydrology: Fundamentals and Applications*. Taylor & Francis Group, CRC Press, New York, 581–605.

Islam, S. N., Reinstädtler, S., Aparecida de Sa´Xavier, M., and Gnauck, A. (2017). Food security and nutrition policy. In Eslamian, S., and Eslamian, F. A. (Eds.), *Handbook of Drought and Water Scarcity. Vol. 1: Principles of Drought and Water Scarcity*. CRC Press, Taylor & Francis, New York, 26.

Islam, S. N., Reinstädtler, S., and Eslamian, S. (2016). Water reuse sustainability in cold climate regions. In Eslamian, S. (Ed.), *Urban Water Reuse Handbook*. Taylor and Francis, CRC Group, New York, 875–886.

KBPP (2016). Population and Housing Census Update Final Report 2016 (PDF). www.deps.gov.bn. Department of Statistics. February 2018. Retrieved 15 November 2021 under: https://deps.mofe.gov.bn/DEPD%20 Documents%20Library/DOS/KBPP/finalreport2016/KBPP_2016.pdf)

Kershaw, R. (2001). Brunei. Malay, monarchical, micro-state. In Funston, J. (Ed.), *Government and Politics in Southeast Asia*. Seng Lee Press Pte Ltd., Singapore, 1–35.

Khatib, M. A. B. P., and Sirat, D. P. H. A. (2005). Some aspect of the cultural trends for tomorrow's Brunei-A strategy. In Thumboo, E. (Ed.), *Cultures in ASEAN and the 21st Century*. UniPress, Singapore, 35–49.

LAPORAN AKHIRKEMASKINIBANCIPENDUDUK DAN PERUMAHAN - (KBPP) (2016). Population and Housing Census Update Final Report 2016 (PDF). www.deps.gov.bn. Department of Statistics. February 2018. Retrieved 15 November 2021 under: https://deps.mofe.gov.bn/DEPD%20Documents %20Library/DOS/KBPP/finalreport2016/KBPP_2016.pdf

Mahmud, R. (2017). *The Borneo Bulletin*, 16 October 2017. http://borneobulletin.com.bn/new-bridge-a-boon -to-business-commuters/.

Morote, A.-F., Hernández, M., and Eslamian, S. (2020). Rainwater harvesting in urban areas of developed countries. The state of the art (1980–2017). *International Journal of Hydrology Science and Technology*, 10(5), 448–470.

Omar, D. P. H. M. (2005). The making of a national culture: Brunei's experience. In Thumboo, E. (Ed.), *Cultures in ASEAN and the 21st Century*. UniPress, Singapore, 5–21.

Omar, M. (1981). *Archaeological Excavations in Protohistoric Brunei*. Meseum Brunei, The Government Printing Department, Brunei.

Peter, C. (1998). *Nature, Western Attitudes since Ancient Times*. University of California Press Ltd., London.

Reinstädtler, S. (2011). Sustaining landscapes – Landscape units for climate adaptive regional planning. In *Conference Presentation and Proceeding of Abstracts of the International World Heritage Studies (WHS) – Alumni – Conference "World Heritage and Sustainable Development,"* 16–19.06.2011, IAWHP (International Association for World Heritage Professionals), Cottbus, 1–107. http://www.iawhp .com/wp-content/uploads/2012/01/IAWHP2011_Book-Abtracts.pdf.

Reinstädtler, S. (2013). Sustaining landscapes – Landscape units for climate adaptive regional planning. In *Conference Proceeding of the International World Heritage Studies (WHS) – Alumni – Conference "World Heritage and Sustainable Development"*, 16–19.06.2011, IAWHP (International Association for World Heritage Professionals), Cottbus, 45–65. http://www.iawhp.com/wp-content/uploads/2013/05 /World-Heritage-and-Sustainable-Development-_-IAWHP-eV_2011.pdf.

Reinstädtler, S. (2017). Why spatially determining of SDGs matters for landscapes in the German Lusatia Region. Oral Presentation in form of a TED-Talk as a Landscape Talk at 20.12.17. Forum Contribution at the "Global Landscape Forum 2017, Bonn from 19–20.12.2017, organized by CIFOR, Bonn.

Reinstädtler, S., Islam, S. N., and Eslamian, S. (2017). Drought management for landscape and rural security. In Eslamian, S., and Eslamian, F. (Eds.), *Handbook of Drought and Water Scarcity, Vol. 3: Management of Drought and Water Scarcity*. Taylor and Francis, CRC Press, New York, 195–234.

The Brunei Museums (2005). *The Brunei Museums – 40th Anniversary of the Brunei Museums Department (1965–2005)*. Perletakan Oscar Sdn Bhd, Malaysia.

Tyler, R. (1996). *Brunei Darussalam-the Making of a Modern Nation*. Nutmeg Publishing, Bandar Seri Begawan.

United Nations Educational Scientific and Cultural Organization (UNESCO) (1972). *Convention Concerning the Protection of the World Cultural and Natural Heritage*. World Heritage Centre, Paris.

United Nations Educational Scientific and Cultural Organization (UNESCO) (1999). *Report 1999/3*. World Heritage Centre, Paris.

United Nations Educational Scientific and Cultural Organization (UNESCO) (2004). *Brunei Darussalam UNESCO Country Programming Document 2013–2016*. Organized by UNESCO Jakarta, Bandar Seri Begawan.

United Nations Educational Scientific and Cultural Organization (UNESCO) (2008). *Policy Document on the Impacts of Climate Change on World Heritage Properties*. World Heritage Centre, Paris.

World Population Review (2024). Brunei Population 2024 (Live). Accessed 28.02.2024 under https://worldpo pulationreview.com/countries/brunei-population.

Yogerst, J. (1994). *The Golden Legacy-Brunei Darussalam*. Syabas, Bandar Seri Begawan.

Yunos, R. (2015). *Our Brunei Heritage-A Collection of Brunei Historical Accounts*. Perpustakaan Dewan Bahasa dan Pustaka Brunei, Bandar Seri Begawan.

Yunos, R. (2015). Kampong Ayer-past and present. *Brunei Daily Resources*. http://bruneiresources.blogspot .com/2015/12/kampong-ayer-past-and-present.html.

Yunos, R. (2017). *The Golden Islamic Heritage of Brunei Darussalam-A Collection of Selected Historical Articles*. UNISSA Press, Gadong.

19 Role of Wetlands in Flood Prevention in Brunei Darussalam
A Case Analysis of Tutong District

Shafi Noor Islam, Azlyana Syahirah binti Haji Ibrahim, and Sandra Reinstädtler

19.1 INTRODUCTION

Brunei Darussalam is a small country located in Southeast Asia, divided into four administrative regions, the districts of Brunei–Muara, Tutong, Kuala Belait, and Temburong (Azffri et.al., 2022). The small country of Brunei Darussalam, officially the Nation of Brunei, the Abode of Peace, is situated at 40 30' N, 1140 40' E on the northwestern edge of the island of Borneo (King and Knudsen, 2021; Azffri et.al., 2022). Apart from its coastline with the South China Sea, it is completely surrounded by the state of Sarawak, Malaysia; and it is separated into two parts by the Sarawak district of Limbang. It is the only sovereign state completely on the island of Borneo (King and Knudsen, 2021; Azffri et.al., 2022). The country has enormous levels of both natural and water resources (NIDM, 2014). Rainfall is an essential phenomenon in Brunei Darussalam. Brunei's saline seawater and significant rivers play an important role in protecting the natural landscapes and developing and fertilizing the land (King and Knudsen, 2021; Azffri et al., 2022). Tutong District is located in the middle of the country and has proximity to the South China Sea to the west. Since the embankment of the 21st century, the spread of urbanization has been rapidly steady throughout countries as it is supported by the interconnectedness of globalization. On the other hand, wetlands are no less important than urbanization as it is believed to bring benefits to the ecosystem and society. Wetlands are defined as:

> …areas of marsh, fen, peatland or water, whether natural or artificial, permanent or temporary, with water that is static or flowing, fresh, brackish or salt, including areas of marine water the depth of which at low tide does not exceed six metres

> (RCSG, 2016)

In spite of the importance of wetlands in recent times there is still a negative impression of wetlands as a wasteland due to the perception that it does not provide roles to others. Additionally, wetlands have triggered the growth of mosquitoes which creates tension with society (Rey et al., 2012). This is due to the harmonization of wetlands and water which is a suitable place for mosquito breeding, fortunately, this growth contributes as a food source for other animal species which potentially alleviates the phenomena of endangered habitats. Many wetland scientists and flood management experts have decided and declared that wetlands have been recognized as valuable resources and it is the place which can play a potential role to mitigate floods in the floodplain areas. Besides this many other studies that have been observed their benefits are actively playing a role in environment, economy, and society, as well as in the ecosystem as a whole. One such benefit is flood prevention. To what extent wetlands can be a protection against this natural disaster, this study is willing to examine the mechanism of wetlands in flood prevention by referring to previous studies on wetlands flood prevention. Moreover, it will investigate the function through case analysis in Brunei Darussalam to demonstrate the justification of wetland ability in the country.

19.2 LITERATURE REVIEW

With the awareness of the ability of wetlands in flood prevention, several studies have been conducted by a case which principally studied flood-prone areas. Some potential case studies have been reviewed on global aspects as well on Brunei case. According to Faisal, Kabir and Nishat (1999), Dhaka, the capital city of Bangladesh, is surrounded by a network of rivers. The water level in all rivers rise in June to September due to rainfall and transboundary flow in which makes the city vulnerable to flooding. In 1988, the city was hugely hit by a disastrous flood, causing enormous damages to life, property and income. Due to this, the structural measures for flood damage mitigation was being looked into. However, when the city was hit by another catastrophic flood, it was proved that, the non-structural measures contributed significantly to flood damage reduction. The non-structural measures taken were flood for casting and warning retention ponds and the drainage network; land use planning, introducing flood insurance and other activities practiced by various groups such as emergency services' flood shelter; flood proofing and recovery and reconstruction (Faisal et al., 1999). The Mississippi River has been one of the most prone areas in the United States since 1993 since it impacts many aspects, especially agriculture which is predominantly land use in the vicinity of water thus flood control is necessary as an act in response to overcome the issue (Galoie et al., 2012). The issue of flood control, however, is a question of expertise and experiences as well as would require a large amount of money considering the area needing to be managed. However, merely letting floods happen is not a wise course of action as it would cause damage to the land that would cost billions of dollars. In this case, a realistic alternative would be the restoration of wetlands as a long-term solution, as well as providing greater benefits to the environment, ecosystem, and society. A study concluded that to control the overall flow of the Mississippi River to overcome flooding, it would require as much as 3 million acres of wetland (Hey and Philippi, 1995). Even though, the perception of wetlands as a hindrance to the economy, structuring drainage to hold the water has been proven unsuccessful throughout history and uniformly worsens the rate of flooding. Although, like Mississippi, several countries have experienced wetland deterioration which can be caused by many factors, nonetheless, the cause of wetland degradation in one place may differ, for instance, the Mississippi River which has been mentioned as mainly human factors as a reason to the degradation of wetland which resulted in negative consequences. Natural factors have been seen as the factors of degradation of the wetlands in Sanjiang Plain, China. The linear regression analysis has specified and observed a reduction to the mean annual rainfall rate of 1.586 mm in the past 45 years has been observed (Zhou et al., 2009). Although the reduction may not be seen as significant, in the long-term perspective it may be responsible for the deterioration of wetland conditions. This is also supported by other natural factors such as temperature, where the issue of climate change has been an issue for decades, most of the time the thinning of the ozone layer traps pollution and carbon dioxide for longer which causes warming. China, an extremely urbanized and industrialized country, is not an exception. Although it is undeniable how human factors may also play a role in wetland degradation in Sanjiang Plain, China, as the agriculture activities have become diverse throughout the year to accommodate the people and the economy, as a result, the inputs of water irrigation are increased and trigger wetland degradation. It shows how crucial the presence of hydrological aspects are in the maintenance of wetland ecology. In general, we can understand how wetlands are linked with the hydrological cycle where water flows are stored, the evaporation process which lessens the runoff that may cause a flood, however, there are studies that might not see wetlands in the same perspective. Wetlands are seen as neither increasing nor decreasing the flow variability nor are wetlands seen as fuel to the increase in flow variability. A total of 19 studies on headwater wetlands were reviewed. It was shown that 10 out of 19 studies conducted in the United States of America and Europe observed an increase in flow variability, while 5 observed neither increased nor decreased flow variability. The remaining observed decreasing flow variability (Bullock and Acreman, 2003).

The Tutong River rises in the south of Brunei District on the border with Malaysia and has a length of approximately 137 km with an estimated catchment area of 1300 km^2 while the river entrance is located in the central part of Brunei's coastline at the South China Sea. Despite the relatively large catchment area, there are several events where flooding occurs that significantly affect the surrounding area, especially the local community settlements and public infrastructures. The effectiveness of wetlands as flood management can be questioned from the claimed studies which invite more detailed studies on the role of wetlands in flood prevention. Although the studies were conducted in various countries, it should be noted that the geographical locations were located in the northern and western parts of the map. In addition, it should be noted the different types of wetlands, climate, soil, areas of wetlands, and any relevant aspects which may cause different results in different studies. In the above literature review, several studies on wetlands are observed from the United States, China, and European countries which gives motivation to the study of wetlands as flood prevention in Brunei Darussalam.

19.3 METHODOLOGY

The study is conducted from September 2021, which was when Brunei is experiencing the second wave of the COVID-19 pandemic which restrains field inspection within the area studied. However, another alternative is conducted accordingly. Some secondary and primary data have been collected through literature reviews during the COVID-19 period and after COVID-19. Some data have been collected through a field trip in April 2022. Our team member Sandra Reinstädtler, who is based in Germany, arranged a field trip to Brunei and a field observation was arranged for 2 days in the Tutong District wetlands areas and coastal wetlands, and Tutong River catchment areas on the 19th and 20th of April 2022. There was direct observation and discussion with local people and professionals. The present status of the wetlands and flooding issue in Tutong town and the catchment area of Tutong River was investigated. Some discussion was done on the field level and based on the status of positions of Tutong River where – next to the Nature-Based Solutions (NBS) – some alternative and engineering concepts have been proposed for the mitigation and flood prevention issues.

19.3.1 The Geographical Location and Characteristics of the Study Area

Brunei Darussalam is located on the northwest coast of Borneo Island, an island that also belongs to Indonesia and Malaysia. Apart from its coastline on the South China Sea, it is surrounded by the Malaysian state of Sarawak. Brunei shares a 266 km (165 mi) (CFE-DM, 2018) border with Malaysia, and Sarawak has an enclave, the district of Limbang, along the Limbang River that splits Brunei Darussalam into two separate parts (Figure 19.1) (Brown, 1970, 132; Hassan, 1988; CDD, 1993, 2008; Ibrahim, 2005; Khatib et al., 2005). Thus, Brunei Darussalam is the one state located entirely on the island of Borneo. Situated on the equatorial line between the southern and northern hemispheres, Brunei is placed at latitudes and longitudes of approximately 4°30'N and 114°40'E, respectively. In addition, the country has a tropical equatorial climate with high temperatures high humidity, and episodic monsoon rains with an average precipitation amount of about 2909 mm per year (NIDM, 2014). Annual rainfall varies from 2,500 mm on the coast to 7,500 mm in the interior of the country (Gupta, 2010). The average annual temperature is 26.1 °C (79.0 °F), with the AprilMay average of 24.7 °C (76.5 °F) and the OctoberDecember average of 23.8 °C (74.8 °F) (NIDM, 2014). Brunei has an equatorial climate characterized by uniformly high temperature, high

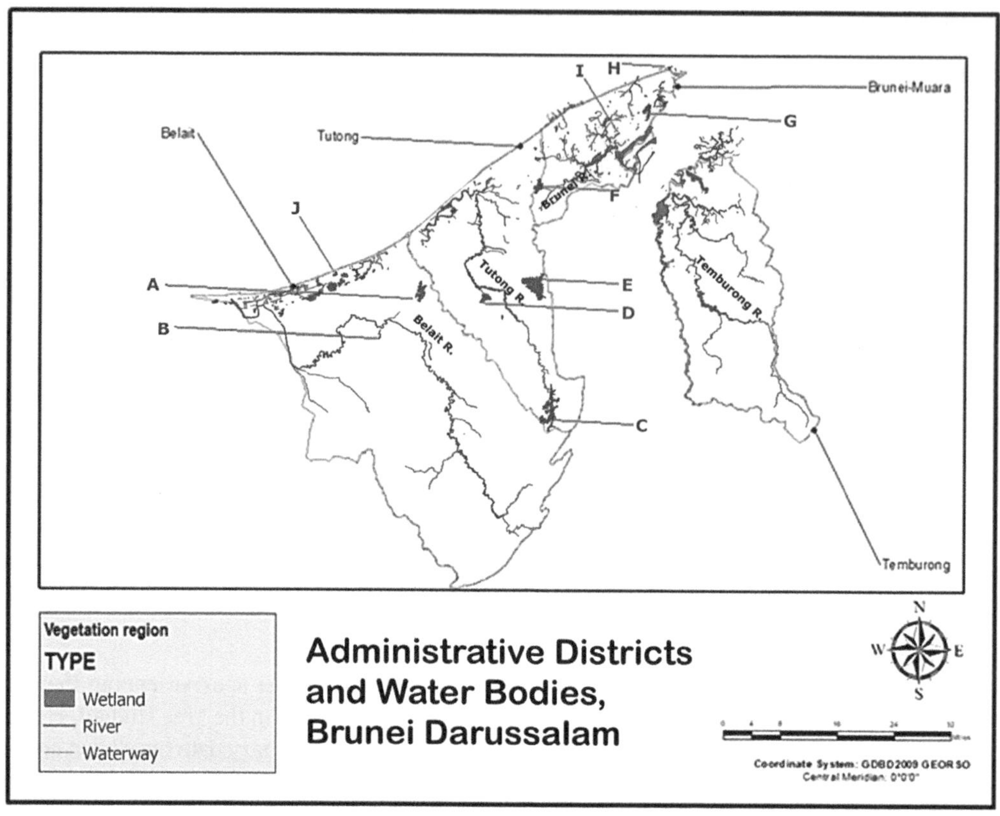

FIGURE 19.1 Observation area: Shows the study area for this paper which is located in Tutong district, the map highlights the water bodies being identified next to the interacting rivers in the entire Brunei Darussalam: A. Andulau Reservoir; B. Lalak Lake; C. Belaban Ulu Tutong Golden Jubilee Reservoir; D. Tasek Merimbun; E. Benutan Reservoir; F.Imang Reservoir; G. Mengkubau Reservoir; H. Api-Api; I. Tasek Lama Dam; J.Anduki Recreational Park (slightly changed on basis of map source: Azlyana, 2021).

humidity, and heavy rainfall. Temperatures range from 23–32 degrees Celsius, while annual rainfall varies from 2,500 mm on the coast to 7,500 mm in the interior (Gupta, 2010).

Wetlands in Brunei Darussalam have become an increasing discussion among organizations and scholars as the advantages of wetlands are tremendous. Countries that experience hot and wet climates have encouraged the development of land classifications, initially identified by the Forestry Department (n.d.) there are two major forest classifications in Brunei Darussalam which occur in different environments; a forest that is found in low-lying and seasonal land and a forest that is commonly found in high and dry land consisted different types of forest. The wetland in the country is found on low-lying and seasonal land which consists of mangrove, freshwater, and peat swamp forests.

The area is comprised of approximately 51,500 people (Department of Economics Planning and Statistics, 2020). One of the district landmarks, and the largest lake in the district, Tasek Merimbun Heritage Park, is which is considered to be a protected forest and is located near the wetlands.

The study area of wetlands as flood prevention will be the wetlands in Tutong district which consists of three types of wetlands, freshwater swamps, mangroves, and peat swamp forests. Tutong District is seen as a suitable area for the study due to the fact that the area is not highly urbanized and is less likely to interrupt the analysis of wetlands as flood prevention as well as providing diverse types of wetlands within moderate areas covered. The study will observe two areas in Tutong District which are in the north and the center of the district where the formation of wetlands is commonly found. The northern part of Tutong district consists of the following areas:

a) Wetland adjacent to Pekan Tutong up to Kuala Abang
b) Kampung Penabai
c) Kampung Kiudang
d) Kampong Bukit Sibut
e) Kampung Serambangun
f) Parts of Kampung Bukit Beruang

The center of Tutong district consists of the following areas:

a) Lamunin
b) Kampung Rambai
c) Benutan
d) Belaban
e) Panchong
f) Tasek Merimbun Heritage Park and nearby

19.3.2 Aims and Objectives

The aim is to examine the wetlands within Brunei Darussalam and the view on wetlands in attenuation of flood with the inclusion of relevant aspects to provide a detailed study on whether wetlands have prevented floods in the country.

The objectives of this paper are as follows:

i. To utilize Geographical Information System (GIS) software to facilitate analyzing the spatial area and data presentation.
ii. Examination of other studies on the wetland in Brunei Darussalam is also necessary to conduct a comparison as well as to fortify more data for the paper.

19.3.3 Data collection

The method of the paper is quantitatively conducted where data is gathered and analyzed into specific results in determining the role of wetlands in the case of flood. As the study will be entirely conducted virtually, spatial data can be collected through Google Earth which offers diverse tools for measuring distance and elevation, and observing changes in land cover and land use through the time span provided. In order to extract data from Google Earth into another GIS software, namely ArcGIS 10.5.1, any production of polypoint, polyline, or polygon will need to be converted from a KML file.

Additionally, data may not be sufficient to be dependent on GIS software thus references to other studies on the targeted wetland for our study are necessary as a supplementary in justifying the study in conducting the case analysis. Data obtained is in the form of journal articles, research papers, and peer-reviewed academic sources mainly available in Google Scholar.

19.4 WETLAND AS FLOOD PREVENTION IN TUTONG DISTRICT

Tutong District has been categorized as a low-lying area thus it is known to be a flood flood-prone area, where water from precipitation subsides in the low-lying area and is hence accumulated. The condition of long-term rainfall in Tutong considerably occurs every year which often happens in the northeast monsoon during late October and the southwest monsoon which happens during May while at the same time is supported by the Inter-Tropical Convergence Zone (ITCZ).

In the study of wetlands as flood prevention, elevation is one of the aspects other than climatic aspects that have always been investigated to find out the low and high points of the area. The low

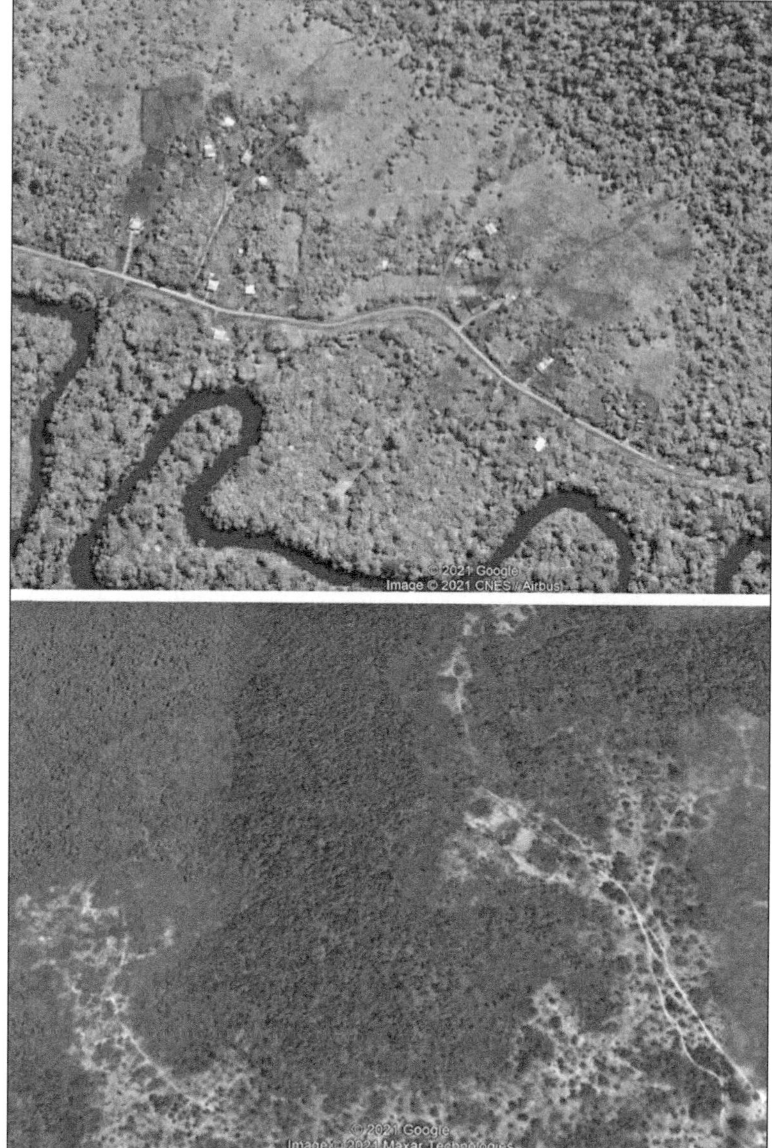

FIGURE 19.2 Excerpt of a wetland area: On the upper picture shows one of the wetland areas in the center of the district concurrently transportation link is also constructed for the purpose of accessibility of the residence and others while on the down picture shows combination of white sand with peat swamp in the northern district.

point can determine the area in which high concentrations of precipitation and runoff are stored. On the other hand, high elevation points can be observed to have less water intake as water is mainly runoff to the ground or to the lower elevation point. The Google Earth examination found a difference in the elevation of the northern and central parts of wetlands in the district that shows the northern part tends to have lower elevation points with an average of 14 m for mangroves and 16 m for peat swamp forest. In contrast, there is no mangrove forest detected in the major center wetland but peat swamp forest in the area shows an average elevation of 38 m and 23 m for freshwater swamp forest.

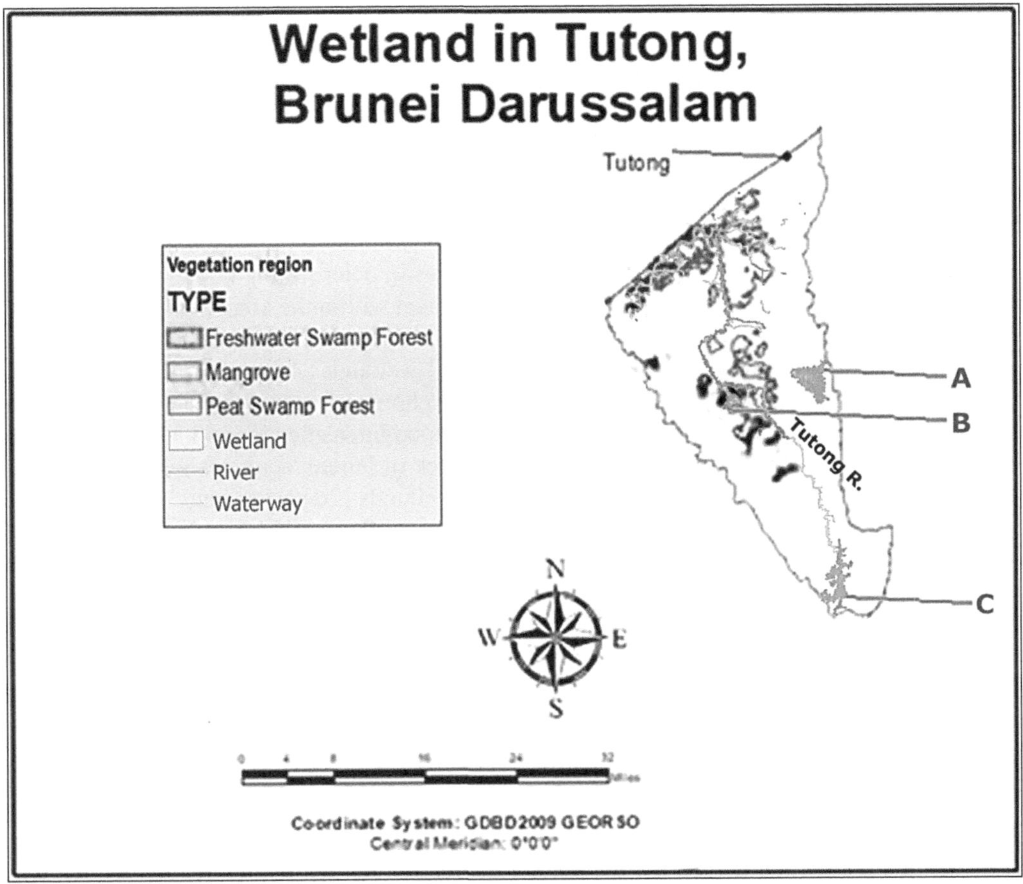

FIGURE 19.3 Determination of wetland areas: Shows the formation of wetland in Tutong district where two major area of wetland can be identified and three main water bodies next to Tutong River interact: A. Benutan Reservoir; B. Tasek Merimbun; C. Belaban Ulu Tutong Golden Jubilee Reservoir (slightly changed on basis of map source: Azlyana, 2021).

Although there has not yet been research published on the degradation of wetlands in the Tutong District and all of the swamp forest found along the Tutong River are considered to be intact (Forestry department, n.d.), wetland areas have been affected by human activities, such as construction, especially the construction of transportation within wetland areas (Figure 19.2.). With the construction of transportation links, residential houses were established along the road which resulted in land clearing.

19.5 DISCUSSIONS

This study is divided into two areas of case study within Tutong District in Brunei Darussalam which are wetlands in the northern part, and wetlands in the central part, as shown in Figure 19.3. The study has delivered dissimilar outcomes in terms of flood intensification. Due to the monsoon season and the low-lying areas in Tutong District, it has been experiencing floods despite the formation of wetlands in the district, however, wetlands located in the northern part of the district are less likely to experience severe flood compared to the central wetland in the district which observed through yearly flood and the recent occurrence of flood in 2021 where it severely affected Kampung Benutan, Ukong, Kuala Ungar, Bangkuru, and Layong (Noor, 2021). Through remote sensing inspection, the severe flood in 2021 appeared to be in the proximity of wetlands in the center of the district which is mostly a remote area, although transportation links and dispersed housing settlements can be identified. Due to land clearing and the construction of transportation links, the degradation of wetlands can be identified with the existence of patches and sparse growth of wetlands in the proximity of rivers as housing appears along the Tutong River.

Wetlands are undeniably able to mitigate flood occurrence from the observation to the northern part of the wetlands. In the area even though the elevation of the area is much lower than the central part of wetlands, it shows the ability to control the flood intensity at which flood levels tend to be low. This is due to the fact that some factors play roles in determining flood intensification and occurrence such as wetland coverage and flood management within the area. Figure 19.3 shows that wetlands in the northern part are more pristine even with the proximity to a town which is highly compact with human activities compared to the central wetlands. Adjacent to the Pekan Tutong or the Tutong Town, mangroves remained intact with no human activities found thus even though Pekan Tutong is precisely located close to the river, the flood intensification is not severe, especially with drainage system support. Contrarily, despite the lack of human activities within proximity to the central wetland in the district, the condition of the wetlands is deteriorating. Where swamp forests are inconsistent and unorganized thus water storage will be less effective in some areas which results in more severe flooding and high water levels despite being higher in elevation compared to the wetlands in the northern part of Tutong District.

The Tutong River is considered a large body of water with many distributaries, however, this study will focus on key areas that are wetland areas, prone to flooding, and experienced floods over a period of time. Figure 19.2 displays the main body of the Tutong River basin area and floodplain areas, where the depressed areas and wetlands areas are seen very clearly. The upstream and downstream areas are very critical geomorphological features. On such geomorphological characteristics condition to mitigate floods within natural processes is not so easy. The Tutong River catchment area is a flood-prone area in Brunei which is very frequently flooded every year in the monsoon period, the potential wetlands areas can be used for storage of flood water and monsoon rainwater and gradually these wetlands areas can store huge quantities of water, and within this process floods damages would be reduced and as a whole floods could be prevented and be mitigated in the Tutong River catchment area in Tutong District, Brunei Darussalam.

19.6 CONCLUSIONS

This research has studied about the role of wetlands in flood prevention by taking into consideration previous studies in the United States and China as a reference to find the role of wetlands in flood control where studies show a positive correlation. This research adopted the statement into the case study at the local level in the Tutong District of Brunei Darussalam by investigating two areas of major wetland formation that are in the northern part and the center of the district. Based on the study wetlands undeniably play a role in flood prevention, however, with the low-lying area, the district is considered to be a flood-prone area where floods occur every year. Nonetheless, wetlands have aided in the lowering of flood intensification and the northern wetlands have been slightly affected by floods compared to the center, due to less disruption and a better draining system despite lower elevation. The formation of wetlands in the center of the district has been inconsistent thus resulting in some areas in the proximity of the river being highly affected by floods. Therefore, there is still a need to have awareness of the importance of ensuring that wetlands are in good condition as it benefits all aspects by restoring and limiting the disruption to the wetland.

CONFLICTS OF INTEREST

The authors declare that the research was conducted in the absence of any financial or commercial relationships and therefore without any potential conflict of interest.

REFERENCES

Azffri, S. L., Azaman, A., Sukri, R. S., Md Jaafar, S., Ibrahim, M. F., Schirmer, M., & Godeke, S. H. (2022).. Soil and groundwater investigation for sustainable agricultural development: A case study from Brunei Darussalam (Sustainability 2022, 14, 1388). MDPI. https://doi.org/10.3390/su14031388.

Brown, D. E. (1970). Brunei the structure and history of a Borneoan Malay Sultanate. Brunei Museum, The Government Printing Department, Bandar Seri Begawan, Brunei Darussalam.

Bullock, A., and Acreman, M. (2003). The role of wetlands in the hydrological cycle. *Hydrology and Earth System Sciences*, 7(3), 358–289.

CDD – (Curriculum Development Department). (1993). History for Brunei Darussalam (Secondary3). EPB Pan Pacific, Singapore, 1–144.

CDD – (Curriculum Development Department). (2008). History for Brunei Darussalam: Sharing ourpast (Secondary 1 & 2). EPB Pan Pacific, Singapore, 1–108.

CFE-DM. (2018). Indonesia disaster management reference handbook center for excellence in disaster management and humanitarian assistance (CFE-DM).

Department of Economic Planning and Statistics. (2020). *Population*. Ministry of Finance and Economy, Brunie Darussalam.

Forestry Department. (n.d.). *4th National Report: Convention of Biological Diversity*. Government of Brunei Darussalam.

Forestry Department. (n.d.). *Forest Types*. Ministry of Primary Resources and Tourism.

Galoie, M., Zenz, G., Eslamian, S., and Motamedi, A. (2012). Numerical simulation of flood due to dam-break flow using an implicit method. *International Journal of Hydrology Science and Technology*, 2(2), 117–137.

Gupta, S. (2010). Synthesis report on ten ASIAN countries disaster risks assessment, ASEAN disaster risk management initiative by UNISDR and World Bank, Retrieved August 2, 2020, from https://www.preventionweb.net/tiles/18872_ascan.pdf.

Hassan, A. B. H. (1988). *Elementary Geography for Brunei Darussalam*. Federal Publications (s). Ltd, Singapore, 1–51.

Hey, D. L., and Philippi, N. S. (1995). Commentary flood reduction through wetland restoration: The upper Mississippi river basin as a case history. *Restoration Ecology*, 3(1), 4–17.

Ibrahim, H. A. L. H. (2005). Culture and counter cultural forces in contemporary Brunei Darussalam. In Thumboo, E. (Ed.), *Cultures in ASEAN and the 21st Century*. Uni Press, Singapore, 22.

Khatib, M. A. B. P., and Sirat, D. P. H. A. (2005). Some aspect of the cultural trends for tomorrow's Brunei-Astrategy. In Thumboo, E. (Ed.), *Cultures in ASEAN and the 21st Century*. Uni Press, Singapore, 35–49.

King, V. T., & Knudsen, M. (2021). The Iban of Temburong: Migration, Adaptation and identity in Brunei Darussalam, Working Paper 64, Institute of Asian Studies (IAS), Universiti Brunei Darussalam, Brunei Darussalam.

NIDM - (National Institute for Disaster Management). (2014). Country profile: Brunei Darussalam. New Delhi. https://nidm.gov.in/easindia2014/err/pdf/country_profile/brunei_darussalam.pdf.

Noor, A. (2021). Armed forces assist in Tutong flood relief efforts. *Borneo Bulletin*, September 5. https://borneobulletin.com.bn/armed-forces-assist-in-tutong-flood-relief-efforts-2/.

RCSG. (2016). *An Introduction to the Convention on Wetlands*, Ramsar Convention Secretariat, Gland.

Rey, J. R., Walton, W. E., Wolfe, R. J., Connelly, C. R., O'Connell, S. M., Berg, J., ... Laderman, A. D. (2012). North American wetlands and mosquito control. *International Journal of Environmental Research and Public Health*, 9(1), 4537–4605. https://doi.org/10.3390/ijerph9124537.

Zhou, D., Gong, H., Wang, Y., Khan, S., and Zhao, K. (2009). Driving forces for the marsh wetland degradation in the Honghe national nature reserve in Sanjiang Plain, Northeast China. *Environ Model Assess*, 14(1), 101–111. https://doi.org/10.1007/s10666-007-9135-1.

Part V

The Middle East

20 Projected Changes in Climate Extremes Using a High-Resolution Regional Climate Model over Türkiye

Selahattin İncecik, Yurdanur S. Ünal, Merve Açar,
Ferat Çağlar, Asude Hanedar, Ceren Ballı Gözen, and
Erdem Görgün

20.1 INTRODUCTION

Recent IPCC reports indicated that global warming will lead to the strengthening of extreme weather events, becoming more frequent and more severe around the world and in particular Europe and the Mediterranean regions [1]. Specifically, the Mediterranean is one of the most sensitive regions to climate change [2, 3]. The observed trends and the climate model projections indicate a strong vulnerability to changes in hydrological regimes and consequent threats to water availability in the Mediterranean region [4]. Therefore, the most critical point in regional climate change is the increase in extreme events. Increasing frequency and intensity of extreme weather events have serious consequences for society-economies, environments, and ecosystems at regional and local scales. Thus, it is essential to identify the future change in extremes and this information is vital to assess the planning approaches for mitigation measures, implementation of adaptation, and managing the impact of climate change. Specifically, sectors such as agriculture, energy, water resources, and transportation are very sensitive to extreme weather events [5]. For instance, growing season temperature and precipitation extremes have significant influences on agricultural yield while urban life and its infrastructure will need adaptation to more extreme weather conditions [6–8]. The impacts of climate change and extreme weather events are crucial in Türkiye since Türkiye (36° to 42°N; 26° to 45°E) has distinct climate zones. The climate regions are shaped by seas surrounding the country and complex topographical structures with marked mountain ranges at the east, south, and west Anatolia [9]. For example, the annual average precipitation is 653.6 mm and it varies from region to region. For instance, central Anatolia receives the minimum precipitation of about 300 mm in mostly winter months while the Black Sea coasts in the northeast receive precipitation throughout the year with annual totals of more than 2000 mm.

Extreme indices have been extensively analyzed in numerous studies [1, 5, 10–19]. However, ESMs do not resolve the topography, land use, and cloud features at regional scales because of their coarse resolution [20]. Specifically, in the case of complex topography, statistical or dynamical downscaling techniques are used to address the scale differences between coarse-resolution ESM outputs and the observations. In dynamic downscaling, high-resolution information can be derived using regional climate models for a particular domain using ESM outputs as initial and boundary conditions. The regional climate downscaling techniques are increasingly being utilized to produce regional climate information for impact and adaptation studies since dynamic downscaling generally improves ESM outputs and produces realistic results [11, 21].

DOI: 10.1201/9781003473398-25

Analysis of the extreme climate indices based on daily temperature and precipitation in Europe and the Mediterranean indicates that cold nights showed negative trends during the winter and summer while the frost day index increased throughout the year in the Eastern Mediterranean [22]. Several studies such as Klein Tank et al. (2002) and Ramos et al. (2011) concluded that all temperature indices increased specifically in the last quarter of the 21st century, particularly, increasing trends of consecutive dry days in the Eastern Mediterranean area [23–25].

Several studies have investigated future changes in regional climate projected for Türkiye under different scenarios of greenhouse gas emissions [19, 26–33]. For example, Önol and Semazzi (2009) studied the potential role of global warming in modulating the future climate over the eastern Mediterranean region using the Regional Climate Model (RegCM3) [28]. Moreover, Önol and Unal (2014) investigated the expected climate change under the SRES A2 emission scenario over climate zones of Türkiye for the 21st century by using the regional climate model forced by the NASA Finite Volume GCM [34]. Sen et al. (2013) evaluated the regional climate model simulations for Türkiye for the 21st century [19]. Bozkurt and Sen (2011) found that the simulations reveal that the eastern Mediterranean Sea has the biggest potential to affect the precipitation in the Anatolian peninsula [35]. Bozkurt et al. (2012) assessed the capability of the ECHAM5, CCSM3, and HadCM3 models in simulating the climatology of the region for the eastern Mediterranean–Black Sea region using RegCM3 [36]. Bozkurt and Sen (2013) presented climate change projections based on the A2 scenario with the regional climate model forced by ECHAM5 based on 27 km resolution over Türkiye [33]. These results exhibit a coherent change in extreme climate events throughout Türkiye. In particular, there is an increase in warm extreme events and a decrease in cold extreme events, owing to the warmer climate. The most recent of these results indicates the change in precipitation patterns with an increase along the Black Sea coast and a decrease in the southern parts of the country through the end of the century.

The data used in this study were obtained from the open-source data of the project "Climate Change Impact on Water Resources in Turkey" completed under the contract of the Ministry of Agricultural and Forestry. The aim of the project was to create a high-quality database of daily temperature and precipitation and their extremes up to 2100 under RCP4.5 and RCP8.5 scenarios in order to be used in impact studies on different sectors since it is expected that the sectoral challenges will occur in drinking water, agriculture, food, tourism, industry, transport, and hydroelectric energy sectors over Türkiye in coming decades. In the project, besides the climate projections, hydrological, hydrogeological, and hydraulic projections were also estimated in 25 river basins over all of Türkiye for the 2015–2100 projection period [37].

Regional responses to the climate change might have a profound effect. Especially extreme weather events are expected to generate significant risks for current infrastructure in water supply, drainage, energy, and public transport. In order to prevent future risks from the negative effects of extreme weather events climate models are used in the prediction of extreme climate indices. In recent decades, climate modeling at regional and global scales has provided an opportunity to utilize Earth system models for projections of extreme temperature and precipitation indices [12, 38].

The goal of the present work is to investigate the ability of the regional climate model RegM4.3 forced by HadGEM2-ES, MPI-ESM-MR, and CNRM-CM5.1 Earth System Models to simulate the present-day climatology at 10 km horizontal resolution over Türkiye. In this work, the extreme climate indices for the reference period of 1971–2000 using the climate simulations carried out by a high-resolution regional climate model (RegCM4.3) are evaluated. The ability of the RegCM4.3 is analyzed and these results will help us further use the output of the regional climate model for climate change impact assessments. Due to the high resolution of the simulations, it is very likely that this study is expected to be significant for climate change impact and adaptation studies in Türkiye.

In the presented study, the impact of climate change is analyzed on a range of extreme indices focusing on temperature and precipitation. The nine indices of climate extremes selected from the list of core climate extreme indices recommended by the WMO are used in the study and listed in Table 20.1. Each index describes a wide range of concerns about the potential impacts of climate change. For example,

TABLE 20.1
Annual Average of the Climate Indices Produced from Observations and Model Results in the Reference Period

Index	Obs.	RCM-HadGEM	RCM-MPI	RCM-CNRM
FD0 (day)	79.01	79.27	81.90	78.92
SU25 (day)	110.98	108.71	111.07	106.21
TX35 (day)	10.37	12.25	14.68	16.34
RX5 (mm)	69.67	63.60	67.72	60.32
RX1 (mm)	34.71	34.17	36.06	31.04
R10 (day)	16.03	13.49	14.11	13.27
R25 (day)	2.54	2.02	2.21	1.69
CDD (day)	74.07	82.19	75.62	68.81
CWD (day)	16.31	20.06	19.46	21.15

FD and CDD are important for agricultural production. High values of TX35 and SU25 can have serious impacts on human health. The number of days with daily precipitation larger than 25 mm and 10 mm are important factors for hydrological applications. In the second part of the study, the projections of nine extreme climate indices are evaluated for three-time slices with 30-year periods (2015–2040, 2041–2070, and 2071–2100) under RCP4.5 and RCP8.5 emission scenarios.

This chapter is organized as follows. In Section 20.2 the models are described, along with the data and methodology used in the study. In Section 20.3, the ability of the models is examined to reproduce the extreme climate indices. In Section 20.4, the projections of climate extremes up to 2100 were presented. Concluding remarks are presented in Section 20.5.

20.2 METHODOLOGY AND DATA

20.2.1 STUDY AREA

The area of interest in the presented study is Türkiye, which lies in Asia and partly in Europe in a unique geographic position. As shown in Figure 20.1, the country is located in the southeastern part of Europe and Asia. This large area of 780,000 km^2 and surrounded by the Black Sea on the north side, the Aegean Sea on the west side, and the Mediterranean on the south side, and the great differences in altitude lead to the formation of various climate types.

In Türkiye, the average altitude is about 1100 m and gradually increases from central Anatolia to the east, south, and west. For instance, in Eastern Anatolia, the Taurus Mountains at the South and East Black Sea Mountains are characterized by mountain peaks up to 3000 m and valleys. The highest peak (5137 m) is located in the East Anatolia region. The major differences in topography affect the distribution of stations measuring atmospheric variables.

20.2.2 REGIONAL CLIMATE MODELING

In this study, the Earth System Models (HadGEM2-ES, MPI-ESM-MR, and CNRM-CM5.1) from the Fifth Phase of Coupled Model Intercomparison Project (CMIP5) archive, coupled with the regional climate model (RegCM4.3) is used. A model grid with a horizontal resolution of 10×10 km is selected to conduct the simulation experiments.

The three Earth System Models were selected from CMIP5 based on their performance and availability. For the dynamical downscaling, RegCM4.3 was chosen with a target resolution of

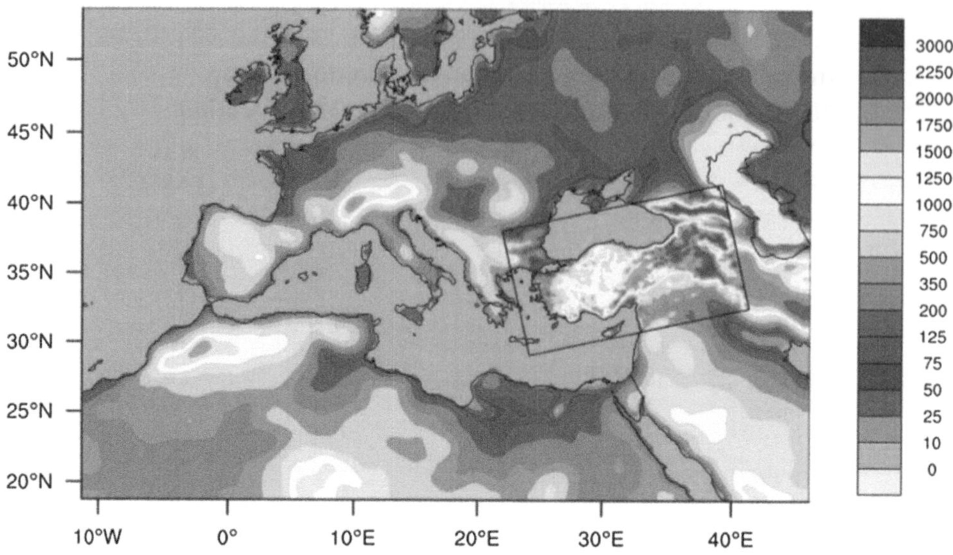

FIGURE 20.1　2003 The model domain and the orography covering all parts of Türkiye.

10 km. Regional climate simulations are first carried out over the Med-CORDEX domain with a resolution of 50 km, and then the 10 km simulations are forced by these low-resolution simulations. The span of the model domain is 39.49–42.00° N, 17.00–35.75° E. The model domain and the orography covering all parts of Türkiye are shown in Figure 20.1. The RegCM4.3 is based on the MM5 hydrostatic mesoscale meteorological model of PSU/NCAR and is continuously developed by the International Centre for Theoretical Physics, ICTP [39]. In this chapter, 23 vertical sigma levels are used of which 5 levels are within the atmospheric boundary layer from the surface. The radiative transfer package of the NCAR Community Climate Model v.3. Kiehl et al. (1996), the mass flux cumulus cloud scheme of Grell (1993), and the turbulence and boundary scheme by Holtslag et al. (1999) are used in the regional climate model set up [40, 41, 42]. The land-surface processes are incorporated via the Biosphere–Atmosphere Transfer Scheme (BATS) which employs 20 vegetation types [43]. The United States Geological Survey (USGS) and Global Land Cover Characteristics (GLCC) data sets are used for the model topography and land use.

Climate change projections used in this study, are based on two of the four representative concentration pathways (RCPs) identified by their radiative forcing in 2100 relative to 1750 [1]. We focus on only RCP4.5 and RCP8.5 scenarios, which represent radiative forcing peaks at 4.5 W/m² (RCP4.5) and 8.5 W/m² (RCP8.5) by 2100, respectively. RCP4.5 is a scenario that stabilizes radiative forcing at 4.5 watts per meter squared in the year 2100 without ever exceeding that value and assumes the imposition of emissions mitigation policies while the RCP8.5 scenario corresponds to the pathway with the highest greenhouse gas emissions. GHG emissions at the RCP8.5 scenario continue to rise as a result of the high fossil fuel usage of the energy sector [15]. In RCP8.5 CO_2-eq. emissions more than double by 2050 and increase by three-fold to about 120 Gt CO_2-eq. by 2100.

20.2.3　Extreme Climate Indices

Extreme weather events always have risk to human society. To analyze the impact of climate change on a range of threshold and duration based extreme indices which focus on temperature and precipitation, the nine indices with three frost days (FD0), summer days (SU25), and hot days (TX35) for temperature and six maximum 1-day precipitation (RX1), maximum 5-day precipitation (RX5), number of heavy precipitation days (R10), number of very heavy precipitation days (R25), consecutive dry days (CDD), and consecutive wet days (CWD) for precipitation from the list of core climate extreme indices recommended by the WMO are employed. Each index describes a wide range of concerns about the potential impacts of climate change. The threshold indices (FD0, SU25, TX35, RX1, RX5, R10, and R25) count the number of days when a fixed temperature or precipitation is exceeded. The duration indices (CDD, CWD) describe the length of wet or dry spells.

The indices were calculated from high-resolution regional climate simulations driven by ESMs and the projected changes are analyzed in the three time slices, 2015–2040, 2041–2070, and 2071–2100.

20.2.4 DATA

In this study, daily maximum and minimum temperatures (°C), and daily total precipitation (mm) data are obtained from 276 meteorological stations collected by the Turkish State Meteorological Service (TSMS) network for the period of 1971 to 2000. A quality control procedure is performed by the TSMS.

Meteorological stations are used in the validation phase of the model simulations only if daily temperature and precipitation data are available for at least 80% of the reference period (1971–2000) as suggested in Prein and Gobiet, (2016) [17]. The spatial coverage of meteorological stations presents a significant variation throughout the country due to the complex topography described above. For example, about 76% of the meteorological stations in the network are located at below-average altitudes (1100 m), while the rest are located at higher altitudes (e.g., Eastern Anatolia, the Taurus Mountains in the south, and the West Black Sea Mountains, which are characterized by mountain peaks and valleys). Western parts of Türkiye exhibit higher station densities while stations with data scarcities are concentrated over South Eastern Anatolia and mountainous areas. The changes in the topography greatly influence the variation of climatic variables and their extremes. Therefore, in order to increase the certainty of the model simulations, a large number of meteorological stations is needed over the high-elevation topography.

The data sets provided from irregularly spaced observations are problematic for interpolation methods in regions of complex terrain [44]. Therefore, the gridded data sets provide a better representation of the climate parameters even at the distant observation stations in addition to minimizing the impact of data quality at individual stations due to averaging [45, 46]. Additionally, the high-resolution gridded surface climate dataset is of great value for the validation phase of the high-resolution climate models. There are various methods applied to move observation data to grid points. Among them, interpolation is the most commonly used method [47]. However, in order to achieve correct results with this interpolation process, the topography must also be taken into consideration. A good example is the PRISM (Parameter-Elevation Relationships on Independent Slopes Model) model which was developed by the PRISM Climate Group at Oregon State University and takes into account the complexity of the region used for the interpolation of observations [48].

To validate model performance, the surface observations are used from 276 meteorological stations over Türkiye and then the daily temperature and precipitation observations are interpolated onto regular grids which coincide with the regional climate model's grids by using the PRISM approach. The interpolation scheme uses the weights depending on the differences between the targeted grid point and the station in terms of distance, elevation, coastal proximity, topographic facets, and topographical position. In the interpolation process, only observations by the TSMS are used. Therefore, the interpolated values along the country borders are based on the internal network data. The model PRISM gridding algorithm has been used by a number of studies for gridding data sets of climate indices [49, 50, 51].

20.2.5 BIAS CORRECTION

The typical resolutions of the regional climate models range between 10 km to 50 km depending on the resolution of the forcing reanalysis products or general circulation models. However, even finer resolution regional climate models are subjected to considerable biases especially over complex topography when comparing the simulated climate at present with the corresponding observations. Particularly, the small-scale distributions of surface meteorological variables such as daily precipitation amounts, and daily maximum and minimum temperatures are highly affected by model resolution and parameterization schemes selected. The accuracy of regional climate model simulations shows seasonal and regional dependences. Biases in products of ESMs and regional climate models lead many researchers to avoid the direct use of climate model outputs for climate impact studies which are based on threshold values. If the region of the study has complex topography and land-sea contrast, adjustments of the simulations are required to reproduce local climate characteristics and reduce the systematic deviations of the model from observations. Therefore, several bias correction methods ranging from simple scaling to sophisticated distribution mapping have been proposed in literature within the last decade. The biases in the climate model outputs are adjusted by using statistical properties obtained from observations for the same period and it is assumed persistence of the climate model error over time.

In this chapter, the extreme indices are calculated after bias correction is applied to the regional climate model outputs. The simplest method of delta change is selected to correct the mean bias of the regional climate model.

The mean temperatures of the RegCM driven by HadGEM2-ES, MPI-ESM-MR, and CNRM-CM5 (RCM-HadGem, RCM-MPI, and RCM-CNRM) are 3.0°C, 2.9°C and 4.2°C colder than the observations over Türkiye, respectively. The largest temperature biases of all model simulations are obtained for the spring season, and they range between –3.6°C and –5.1°C. For temperature bias correction, first monthly mean values of the model simulations are subtracted from the simulations, and then the monthly mean values of the gridded observations are added to the simulations. Similarly, for future simulations, the monthly means of the reference period simulations are subtracted from them, and the monthly means of the observations for the reference period are added.

The precipitation biases are found generally within ± 15% for the RCM-HadGEM simulations between –5 and +10 by RCM-MPI and around ± 20% by RCM-CNRM. All RegCM simulations produce more precipitation for the winter and spring seasons. Precipitation is particularly higher in the Southeast Taurus Mountains, Küre Mountains, Canik Mountains, and Eastern Black Sea Mountains, while it is 10–40% less than normal in the lee side of the mountain ranges. On the other hand, the regional climate simulations of RCM-HadGEM and RCM-CNRM are drier than the observations during the mid-summer over the eastern part of Türkiye and wetter in autumn season while the RCM-MPI produces drier conditions in summer for all over Türkiye and in autumn for the southern latitudes of Türkiye. The models have a positive bias in regions with high topography. Here, the problem of representation of the mountainous regions of the meteorological observation network should not be ignored. The overestimation of precipitation in mountainous areas should not be considered directly as a model error.

For precipitation correction, the direct method, sometimes referred to as the delta change method was applied [52, 53, 18, 54]. The direct method scales the simulated precipitation by multiplying it by the ratio of observed and simulated precipitation. Therefore, correction (rescaling) factors are calculated for each month separately to converge the annual variability of the model to the observations. For the reference period, every month's correction factors are computed as a ratio of the mean observation to the mean of the model simulation at each grid point. Similarly, the bias-corrected future simulations for precipitation at each grid point are derived by multiplying the simulations with the correction factor obtained at the same grid point for the reference period.

20.3 PERFORMANCE OF THE MODELS

20.3.1 Overall Performance of the Simulations

The nine climate indices are calculated by using the daily gridded meteorological observations and the bias-corrected model simulations. The performance of the regional climate model driven by different earth system models in reproducing the mean of the climate indices over Türkiye during the 1971–2000 reference period is quantitatively assessed. The list of indices is categorized into two main classes: absolute-threshold indices (e.g., SU25, FOD, TX35, R10, and R25) and duration indices (e.g., CDD, CWD). Indices have been calculated on each grid for each coupled model and time-averaged over the study period.

The mean values of the FD0, SU25, and TX35 and the RX1, RX5, CWD, CDD, R25, and R10 for observations and simulations of the RCM-HadGEM, RCM-MPI, and RCM-CNRM for the reference period are listed in Table 20.1 which shows the overall performance of the each coupled model system in simulating the extreme climate indices.

The temperature extreme indices are generally well reproduced by the three simulations in the baseline period, whereas the precipitation extreme indices of RCM-MPI reproduce slightly better results than the others.

Moreover, the models predict approximately one-week expansion of the CWD, over Türkiye. A wide disagreement is found between the models and observations for the CDD index. For example, RCM-CNRM underestimates the CDD index except in the Eastern Anatolia region, while RCM-HadGEM overestimates the CDD index throughout Türkiye and RCM-MPI only overestimates the CDD in the inland regions of Türkiye. The RCM-MPI and RCM-CNRM are generally more skillful in simulating CDD. RCM-HadGEM presents less skillful compared to the other RCM simulations. Compared with regional climate model solutions based on HadGEM2-ES, MPI-ESM-MR, and CNRM-CM5.1 models show that RCM-MPI gives more accurate results, particularly for extreme climate indices focusing on precipitation. It is also more compatible with observations of RCM-HadGEM and RCM-CNRM models for frost days only.

The results indicated that in general, models can properly reproduce the RX1 and RX5 indices with the exception of RCM-CNRM that overestimated over Türkiye. Among the precipitation indices, RX1 which defines the highest precipitation amount in a year, is better represented by the models. Whereas the RX5 index the maximum consecutive 5-day precipitation is slightly underestimated by 7.6%. However, the heavy rain indices R10 and R25 are highly underestimated by 14.5% and 33%, respectively. For temperature, SU25 and FD0 indices are also better reproduced by the regional climate model driven by these three ESMs than the TX35 which is the number of days with maximum temperature higher than 35°C. As seen in Table 20.1, the RCM poorly predicts extremely hot days (TX35). However, it is more difficult to simulate the CDD and CWD indices than the other precipitation indices used in the study. The CWD index which is directly related to precipitation patterns is overestimated by about 25% by each model. The CWD is overestimated by 25%, while predicted CDD values have a significant variability compared to the reference period.

20.3.2 Spatial Variations of the Simulations

Moreover, in order to understand how model successes, change in Türkiye, the spatial variations of mean biases in simulating the threshold indices for each model are examined. Spatial differences in the indices between the observations and the model simulations are given in Figure 20.2 and Figure 20.3 for indices based on temperature and precipitation, respectively. Units of the first two indices in Figure 20.3, RX1 and RX5 are given in millimeters, the others are in days. Table 20.2 presents the comparison metrics of Root Mean Square Error (RMSE), Correlation Coefficient (CC), and Mean Bias Errors (MBE) for each index simulated by RCM driven by three ESMs.

The number of frost days (FD0) difference between the observations and simulations is about ±10 days, implying that two models (RCM-HadGEM and RCM-CNRM) are reliable to simulate future FD0 data while RCM-MPI slightly under-predicts the FD0. In general, FD0 on the west and coastal regions of Türkiye presents negative biases while central and eastern Anatolia shows positive biases. On the other hand, the distribution of the biases for FD0 based on RCM-MPI illustrates slightly different spreading on the central and eastern Anatolia regions where the temperature probability distribution function of the model simulations might be positively skewed compared to the

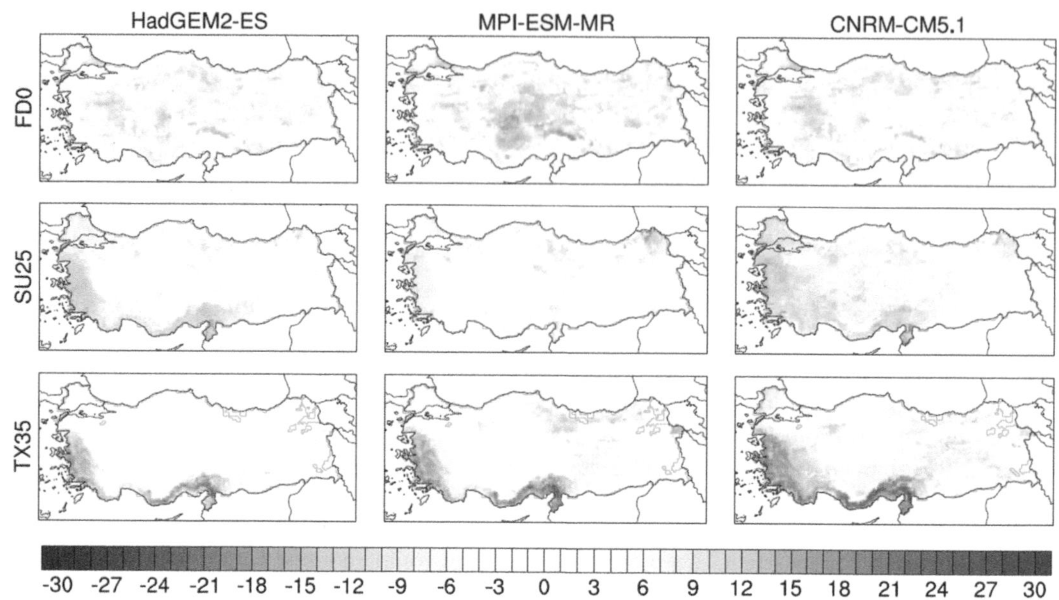

FIGURE 20.2   Spatial variations of the TX35, SU25, and FD0 indices between observations and model simulations.

observations. If the general performance of the model simulations is evaluated, all show similar FD0 distributions over Türkiye with almost perfect correlation coefficients. The highest RMSE is obtained for RCM-MPI with 8.47 days because of the cold tendency of RCM-MPI over the inland regions. However, variabilities of the RMSE and the MBE might be attributed to the complex topography, similar to findings by Dong et al. (2015) in China [55].

For SU25, a large negative bias is found in the Aegean and Mediterranean coastal areas in all RCM simulations with values greater than 6 days. However, RCM-CNRM shows the largest negative biases greater than 10 days in the western part of Türkiye. The correlations are higher than 0.99 for all simulations and indicate that the distribution of the SU25 values throughout Türkiye is well simulated. The RMSE values are in the range of 4.7–6.7. The mean biases are negative for RCM-HadGEM and RCM-CNRM. On the other hand, the extreme temperatures index of TX35 reveals higher positive biases, up to 30 days for all simulations, especially over the Aegean and Mediterranean coastlines. RX5 exceeds a month over the Aegean region, eastern Mediterranean, and southeastern Anatolia in the reference period. The correlation coefficients are slightly smaller than the SU25 but still higher than 0.96 since the general distribution of the index throughout Türkiye is well simulated. However, the RMSEs are similar to the errors obtained for SU25. It indicates that the number of days for TX35 is in close range of the observations.

Changes in precipitation extremes are spatially more complex and exhibit a less widespread spatial coverage than the temperature indices. This may be caused by the irregular topography in the country. RCM-MPI gives less biased results in particular for extreme climate indices focusing on precipitation. Both simulations RCM-HadGEM and RCM-MPI predict the RX1 and RX5 indices in a similar way throughout Türkiye, while there is an underestimation of the extreme 1-day and 5-day precipitation on the western part of Türkiye and overestimation of the eastern high elevation regions. On the other hand, the RCM-CNRM model mostly underestimates almost the entire region of Türkiye. The average biases for RX1 for all simulations range in ±18 mm while for RX5, the biases are almost twice of these values. The correlation coefficients are higher than 0.81 and indicate that the extreme precipitation pattern over Türkiye is well captured by all models. The RMSEs for RX5 and RX1 change in the range of 13.18–15.65 mm and 8.57–5.79 mm, respectively.

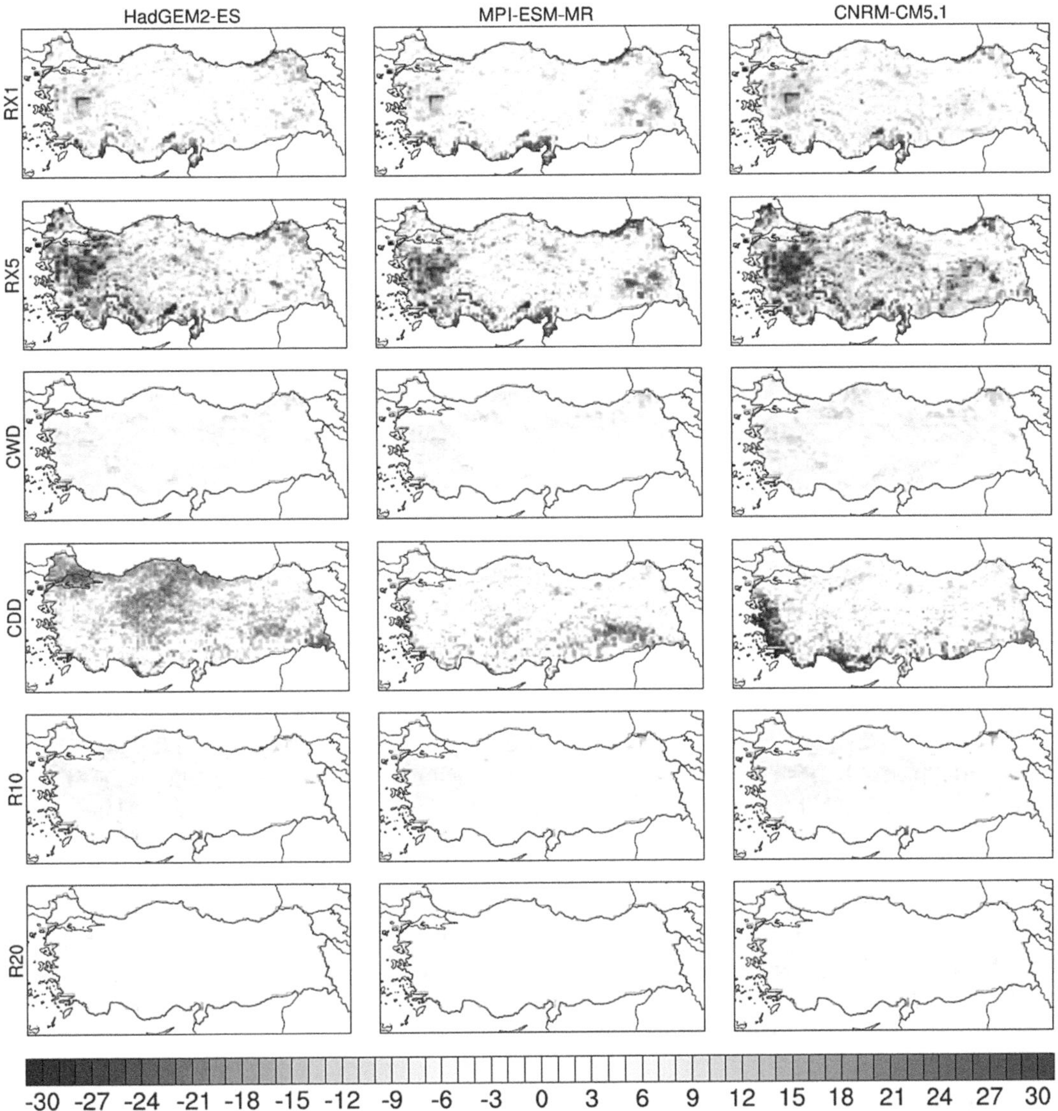

FIGURE 20.3   Spatial variations of the RX1, RX5, CWD, CDD, R10, and R20 indices between observations and model simulations.

All models show lower estimates than the observation for the extremes except RCM-MPI for RX1. Specifically, regional climate simulation driven by MPI-ESM-MR is better at reproducing the RX5 index. This coupling gives less biased results in particular for extreme climate indices focusing on precipitation extremes.

Simulated R10 and R20 show largely underestimation in Türkiye while overestimation in limited coastal areas at the Aegean and Mediterranean. However, the extreme precipitation patterns are well captured since the correlations are quite high (0.90–0.95). The negative biases for days of precipitation larger than 10 mm are concentrated on the western part of Türkiye in all of the RCM simulations and the largest discrepancies are in the order of 10 days. However, the RMSEs are around 3–4 days. The MBEs are mostly negative and up to 3 days on R10 and less than a day for RX25.

All models predict the consecutive wet days (CWD) more than the observations for the reference period throughout Türkiye with the high values at high elevations. Whereas for the CDD extreme

TABLE 20.2

Comparison Metrics of RMSE, Correlation Coefficient, and MBE for the Spatial Distribution of the Climate Indices between Observations and Model Results in the Reference Period

	RCM-HadGEM			RCM-MPI			RCM-CNRM		
Index	RMSE	CC	MBE	RMSE	CC	MBE	RMSE	CC	MBE
FD0 (day)	4.74	1	0.26	8.47	0.99	2.89	7.54	0.99	−0.09
SU25 (day)	5.41	0.99	−2.27	4.2	0.99	0.1	6.74	0.99	−4.77
TX35 (day)	4.74	0.98	1.88	6.34	0.97	4.31	7.88	0.96	5.97
RX5 (mm)	14.28	0.87	−6.07	13.18	0.88	−1.95	15.65	0.88	−9.35
RX1 (mm)	8.57	0.81	−0.55	8.58	0.82	1.34	8.79	0.83	−3.67
R10 (day)	3.93	0.93	−2.55	3.12	0.95	−1.92	4.23	0.93	−2.76
R25 (day)	1.31	0.9	−0.51	1.07	0.93	−0.32	1.46	0.9	−0.84
CDD (day)	12.26	0.88	8.12	8.56	0.92	1.56	11.24	0.85	−5.26
CWD (day)	4.32	0.61	3.76	3.82	0.64	3.15	5.38	0.64	4.84

index, a wide disagreement was found between the models and observations. For example, the distributions of the biases of RCM-CNRM and RCM-HadGEM for CDD are quite the opposite in Türkiye. RCM-CNRM underestimates the CDD index except in the eastern Anatolia region, while HadGEM overestimates the CDD index. On the other hand, RCM-MPI biases remain relatively small with positive biases in the central and southern Anatolia regions. The correlation coefficients for the CWD are quite low which implies the consecutive wet days are not well represented by the models. On the other hand, even though the biases are higher for the CDD, the general pattern of the consecutive dry days over Türkiye produced by RCMs is similar to the observations.

In conclusion, the number of frost days and summer days is better represented by the models than the CDD and TX35. For precipitation, the maximum 1-day precipitation is better represented than the other indices based on precipitation such as Rx5, Rx10, R10, R25, and CWD. The overall results indicated that, in general, models can properly reproduce the selected climate indices, with the exception of RCM-CNRM, which overestimates it over Türkiye.

20.4 PROJECTIONS

Temperature and precipitation extreme indices were calculated for future periods of 2015–2040, 2041–2070, and 2071-2100 under RCP4.5 and RCP8.5 scenarios. The indices are compared to the observed 1971–2000 period. Table 20.3 shows the averaged indices for the 30-year periods over Türkiye and the signs next to the numbers represent the trends within the periods. Results showed widespread significant changes in warm indices consistent with warming, for instance, increases in warm extremes and decreases in cold extremes over Türkiye (Tables 20.3 a, b, c). Changes in precipitation extremes are spatially more complex and exhibit a less widespread spatial coverage of the temperature indices.

20.4.1 FD0

The occurrence of frost days (when Tmin<0°C) is expected mainly in the winter. However, due to the various climate regions that occur in Türkiye, frost days may occur in the spring and fall seasons. The RCM driven by three ESMs predicts decreases in FD0 at the end of the projection period (2071–2100). According to the simulations, 78.9–81.9 days can be considered frost days

TABLE 20.3

30-year Averages of the Climate Indices Calculated for Reference Period and Future Periods under RCP4.5 and RCP8.5 Scenarios. a) RCM-HadGEM, b) RCM-MPI, and c) RCM-CNRM

a	Observation 1971-2000	RCP4.5			RCP8.5		
		2015-2040	2041-2070	2071-2100	2015-2040	2041-2070	2071-2100
FD0 (day)	79	61 ↓	56 ↓	51	61	52 ↓	35 ↓
SU25 (day)	111	132 ↑	141 ↑	146 ↑	131	149 ↑	166 ↑
TX35 (day)	10	24 ↑	34 ↑	38 ↑	26 ↑	43 ↑	66 ↑
RX5 (mm)	70	69	69	69	68	68	70
RX1 (mm)	35	36	36	36	36	36	37
R10 (day)	16 ↓	13 ↓	13	13	13	13 ↓	13
R25 (day)	3 ↓	2 ↓	2	2	2	2	2
CDD (day)	74	89	90	94	86 ↓	92	95
CWD (day)	16	20	19	19	20	19 ↓	19

b	Observation 1971-2000	RCP4.5			RCP8.5		
		2015-2040	2041-2070	2071-2100	2015-2040	2041-2070	2071-2100
FD0 (day)	79	72 ↓	65	61	74 ↓	61	49 ↓
SU25 (day)	111	122	128	131	123	138 ↑	157 ↑
TX35 (day)	10	20 ↑	22	29	22 ↑	32 ↑	51 ↑
RX5 (mm)	70	71	71	71	72	72	73
RX1 (mm)	35	38	38 ↓	38	39	39 ↓	40
R10 (day)	16 ↓	13	14	13	14	14 ↓	13
R25 (day)	3 ↓	2	2	2	2	2 ↓	2
CDD (day)	74	77 ↑	79	80	75	81 ↑	85
CWD (day)	16	18 ↓	19	19	19	18	17

c	Observation 1971-2000	RCP 4.5			RCP 8.5		
		2015-2040	2041-2070	2071-2100	2015-2040	2041-2070	2071-2100
FD0 (day)	79	70	65	56	72	57 ↓	45 ↓
SU25 (day)	111	114	122	130	124	136	150
TX35 (day)	10	21	27	30	24	35 ↑	48 ↑
RX5 (mm)	70	64	64	64	62	63	65
RX1 (mm)	35	33	33	34	32 ↑	33	35
R10 (day)	16 ↓	14	14	14	13	13	13
R25 (day)	3 ↓	2	2	2	2 ↑	2	2
CDD (day)	74	69	69	71	70	73	72
CWD (day)	16	22	21	21	21	20	21

in Türkiye, while the annual average of the reference period is 79.1 days. RCM-HadGEM and RCM-CNRM better predict the FD0 index in the reference period. RCM-HadGEM projects a significant decrease in FD0 days for both RCP4.5 and RCP8.5 scenarios. For instance, at the end of the century, the annual number of frost days is projected to be 52 days for RCP4.5 and 35 days for RCP8.5, respectively. It decreased to 34% and 56% for both scenarios at the end of the century, respectively. The simulations by the RCM-MPI also present the decreasing values of the annual average of FD0 days for both scenarios. However, decreasing ratios on the FD0 days are less than RCM-HadGEM. At the end of the century, the numbers of FD0 days are 61 and 49 days for RCP4.5 and RCP8.5, respectively. Whereas, RCM-CNRM simulations of the FD0 index present lower values of 56 and 45 days, respectively for both scenarios. These results seem to be closer to RCM-HadGEM. The decrease in the number of frost days every 30 years is expanding from the coastline to the inland of Türkiye. The western mountain range of the Aegean Sea lies perpendicular to the coastline. Therefore, the decrease of the frost days extends gradually to the interior of central Anatolia more than the other shorelines. The RCP8.5 scenario indicates more decrease, especially at high-altitude regions where a high number of frost days are experienced, the region stretches from the Middle Taurus Mountains to the Tahtalı Mountains and decreases to around 70–80 days. A similar situation exists in the northern parts of the Euphrates Tigris Basin. The reflections of the minimum temperature decrease are also seen in the amount of snowfall in the same regions. Snow volume results of model simulations indicate that the character of the precipitation in these regions often turns from snow to rain.

20.4.2 SU25

The simulations performed with all three models predict an increase in the 3rd projection period for the SU25, summer day index. The highest increases were in the RCM-HadGEM with 34% and 49% for RCP 4.5 and RCP 8.5, respectively, followed by RCM-MPI with 18% and 41%. As seen in Section 20.3, skillful results were found by the RCM-HadGEM and RCM-MPI models in the validation process of the SU25. This increases the significance of these two model results in terms of the projections of the SU25. For example, the annual number of summer days is projected to exceed 149 days for RCP4.5 and 166 days for RCP8.5 by the end of the century (2071–2100), respectively. Although the RCM-CNRM expectations for an increase in average temperatures are between the other models, the expected rise in extreme temperatures is smaller than the RCM-HadGEM and RCM-MPI. The decrease in the frequency of the cold end values of the probability distribution function of the temperatures is less than the other models, and the decrease in the warm end is less than the other models. It might be related to the higher precipitation prediction of RCM-CNRM. However, all of the models estimated for SU25 are similar around the southern Aegean coastline, Mediterranean coastline, and southeastern Anatolia and increase to almost 200 days for RCP8.5. These results indicate that the days with maximum temperatures higher than 25°C are more than five months almost over the entire Türkiye area at the end of the century according to RCP8.5.

20.4.3 TX35

The simulation results for TX35 suggest that the projections by all models used in this study present increasing rates over all the future periods compared to a historical period. The highest increases relative to the reference period have been found by the RCM-HadGEM simulations with 34 and 66 days for RCP4.5 and RCP8.5 scenarios, respectively. The RCM-CNRM and RCM-MPI followed by RCM-HadGEM model which has more skill in the reference period than the other models used in the study. TX35 increases are more pronounced over the regions near the Aegean Sea and the Mediterranean Sea and in the southeastern part of Türkiye in which the number of TX35 days exceeds three months.

20.4.4 RX1 AND RX5

Evidently, both the annual maximum 1-day precipitation (RX1) and the annual maximum of wettest consecutive 5-day precipitation (RX5) indices by the near future (2015–2040) are not likely to change much in Türkiye for the RCP4.5 scenario (Table 20.3). The RCP4.5 and RCP8.5 results of the RCM-HadGEM and RCM-MPI models largely overlap with the reference period values, and there are no major differences between the two scenarios. The RX1 index changes show similarities in all three projection periods, except for the Black Sea coastline. Simulated maximum 1-day total precipitations are projected to increase gradually for RCM-HadGEM and RCP-MPI relative to the reference period (1971–2000). However, RCM-CNRM which has less skill compared to RCM-HadGEM and RCM-MPI and underpredicts the precipitation-based indices in the reference period indicates a decrease throughout the projection period. The 30-year averages of the maximum 1-day precipitation for RCM-HadGEM and RCM-MPI vary between 20 and 40 mm inland regions and between 40–100 mm in all coastal regions. An increase in the amount of RX1 is expected in all models in the Marmara Region, the Aegean Sea, the Black Sea coastline, and the Euphrates--Tigris Basin. The highest increases are predicted by RCM-MPI in both scenarios. Extreme temperature increases in the region may lead to convective instability along with atmospheric moisture transport and then an increase in extreme precipitation.

RX5 overall changes in Türkiye show a similar spatial pattern with slight increases along the coastal strip of the Black Sea and Mediterranean Sea for RCP4.5 and RCP8.5 scenarios. RCM-MPI estimates larger increases of RX5 over southeastern Anatolia (Euphrates and Tigris river basins) through the end of the projection period for the RCP8.5 scenario. It is predicted that the maximum five-day precipitation intensifies in the eastern Black Sea, Mediterranean, and especially in the southeastern Anatolia regions in the RCP8.5 scenario. All models foresee that the long-term average of maximum 5-day precipitation totals in the Central Anatolia Region will be between 20–40 mm and 100–200 mm along the Black Sea and Mediterranean coastline.

RX1 simulated by RCM-HadGEM is expected to increase in Türkiye by 2.8% and 5.7% for the RCP4.5 and RCP8.5 scenarios at the end of the century. RX1 of RCM-MPI increases much more than RCM-HadGEM. For instance, the RCM-MPI simulations present an increase of 6.6% and 14.3% for RCP4.5 and RCP8.5 at the end of the century. However, RCM-CNRM project decreases. This is strongly related to the underestimation of the RX5 day by the RCM-CNRM. RCM-HadGEM shows a decrease in the RX5 index by 2.8 % for RCP4.5 at the end of the projection period. The simulations by the RCM-HadGEM model under RCP8.5 do not change relative to the reference period.

However, by the end of the century (2071–2100) the annual percentage of RX1 is likely to increase by about 5.7% for RCP8.5 while RX5 is not expected to change. The projected changes for these two indices are similar in space with a slightly wider spread for RX1.

20.4.5 R10 AND R25

The RCM-HadGEM and RCM-MPI model simulations indicate a likely decrease in the number of heavy precipitation days (R10) compared to the reference period by each future time slice for both RCP4.5 and RCP8.5 emission scenarios. Similar to the RX1 and RX5 indices, the simulated heavy precipitation days, R10 are projected to decrease in Türkiye by RCM-CNRM model. Very heavy precipitation days (R25) are also likely to decrease in Türkiye by RCM-CNRM model under both the RCP4.5 and RCP8.5 emission scenarios. The annual average of R25 days is projected to decrease by 1 day in comparison to the reference period for each future time slice by RCM-HadGEM, RCM-MPI, and RCM-CNRM models. The value of this index is between 18–28 days in the Eastern Black Sea coastal regions, and around 12–14 days in the region of the eastern Mediterranean and around Lake Van. While there is no significant difference between the RCP4.5 and RCP8.5 scenarios for the RCM-HadGEM, the RCM-MPI predicts that the regions with more than 25 mm of precipitation will expand towards the inland edges of these areas for the RCP8.5 scenario.

20.4.6 CDD AND CWD

The Consecutive Dry Days (CDD) are calculated from the simulated daily precipitation amounts for three-time slices. The changes in the number of maximum length of dry days, CDD are more remarkable than the maximum length of wet days, CWD. For instance, the simulations of CDD by RCM-HadGEM and RCM-MPI models suggest that the annual number of CDD is projected to increase for both scenarios while CDD of RCM-CNRM, in each future time slice, is below the reference period value for both emission scenarios. The skill of the RCM-CNRM and RCM-HadGEM in producing CDD index for the reference period is relatively poor compared to RCM-MPI as seen in Table 20.1. The RCM-HadGEM overestimates the average value of CDD over Türkiye while the RCM-CNRM underestimates. However, the distribution of the CDD throughout Türkiye is slightly better predicted by the RCM-HadGEM compared to the RCM-CNRM even though the intensity is positively biased.

The rate of increase in CDD is about 20% for RCM-HadGEM and 4% for RCM-MPI at the first projection period under RCP4.5. These rates rise for RCP8.5. Regional climate simulations driven by RCM-HadGEM and RCM-MPI estimate the expansion of dry consecutive days at the end of the century about 27%, 8% for RCP4.5, and 28%, 15% for the RCP8.5 scenario, respectively. Expected changes are larger for 2071–2100 than for 2015–2040 in Türkiye. On the other hand, CDD is projected to decrease by about 6.7% for RCM-CNRM from 2015 to 2040 under RCP4.5 and to decrease by about 4% at the end of the projection period under RCP4.5.

Even though the changes in the averaged CDD over Türkiye are small, the regional impacts might be large especially over the regions with longer CDD in the reference period. 5–30 days of increases in the CDD are seen in the eastern Mediterranean, central Anatolia the southern part of the Eastern Taurus Mountains, and the Euphrates-Tigris river basin. An increase of 10 days is expected in the CDD on the Aegean coastal regions extending towards central Anatolia. The CDD index values reach 140–160 days at the end of the century on the southeastern Anatolia, and 30–70 days on the Black Sea coast. Although the RCM-CNRM predicts precipitation distribution similar to the other models, it has a tendency to produce more precipitation under the scenarios. Unlike the others, the RCM-CNRM expects a partial decrease in consecutive dry days in southeast Anatolia for the RCP8.5. Thus, projected increases suggest the extension of the dry period which causes more severe drought, especially by the end of the 21st century in some places of Türkiye.

There is no significant change in the number of CWD averaged over Türkiye from 2015 to 2100 for both emission scenarios.

20.5 CONCLUSIONS

In this study, the projected change in nine extreme climate indices in the 21st century was investigated under scenarios RCP4.5 and RCP8.5, respectively, by analyzing the simulated results of a regional climate model, RegCM4.3 coupled with the three Earth System Models in the CMIP5 archive, MPI-ESM-MR, HadGEM2-ES, and CNRM-CM5.1. The reference period is taken from 1971 until 2000. In this work, extreme climate indices simulations are derived from a regional climate model coupled with the three ESMs from the CMIP5 archive. Validation of the indices is carried out by comparing the annual averages from the three model simulations and observations. The projections are shown in three time slices, 2015–2040, 2041–2070, and 2071–2100.

The main findings can be summarized as follows:

- The model simulations are closer to observations in terms of MBE and RMSE for most indices and models over Türkiye. Compared with observations, models perform well and they are able to capture the main features of the spatial distribution of indices during 1971–2000.
- RCM-MPI gives more accurate results in particular for extreme climate indices focusing on precipitation.

- Both RCM-HadGEM and RCM-MPI models predict the RX1 and RX5 indices in a similar way throughout Türkiye, while the RCM-CNRM model underestimates in the Eastern Anatolia region where the highest altitudes.
- Moreover, all of the models predict the CWD overestimate throughout Türkiye. Whereas for the CDD index, wide disagreement was found between the models and observations. For example, RCM-CNRM underestimates the CDD index except in the Eastern Anatolia region, while RCM-HadGEM overestimates the CDD index throughout Türkiye and RCM-MPI only overestimates the CDD in the inner regions of Türkiye.
- The projections of the models used in the study showed a widespread significant change in climate extremes consistent with warming. For instance, future projections of warm temperature extremes continue to increase and cold temperature extremes continue to decline in the models for Türkiye.
- All warm indices (SU25, TX35, and CDD) indicate increasing rates for Türkiye. A notable decline in the number of FD0 days is expected in the study while the number of summer days (SU25) and very hot days (TX35) is expected to increase.
- In Türkiye, for each period, CDD is projected to increase, resulting in more severe drought, especially by the end of the 21st century in some places.
- There was an insignificant decrease in heavy and very heavy rainy days for the future time slices under both the RCP4.5 and RCP8.5 scenarios.

In summary, this work is essential for water resources research and development strategies of Türkiye to adapt to climate change in many sectors such as agriculture and food, industry, energy, and tourism.

ACKNOWLEDGMENT

Republic of Türkiye, Ministry of Agriculture and Forestry, Water Management General Directorate, offers a great contribution to the sustainable management of water resources against the effects of climate change on water resources with the projects they conduct. The project completed in 2016 and entitled *"Climate Change Impacts on Water Resources"*, which inspired this study, is one of them. The authors are grateful to the General Directorate for realizing such an important project and for providing the open-source data.

REFERENCES

1. IPCC (2013). *Intergovernmental Panel on Climate Change*. AR5-Fifth Assessment Report. Cambridge: Cambridge University Press.
2. Giorgi, F. (2006). Climate change hot-spots. *Geophysical Research Letters*, 33, L08707.
3. Vasileiou, P. K., Philippopoulos, I., Yiannikopoulou, C., Adamantiadou, C., and Deligiorgi, D. (2013). Statistical assessment of extreme climate indices of air temperature over eastern Greek Island and Cyprus. In *Proceedings of the 13th International Conference on Environmental Science and Technology*, Athens, Greece, 5–7 September.
4. Ludwig, R., Roson, R., Zografos, C., and Kallis, G. (2011). Towards an inter-disciplinary research agenda on climate change, water and security in Southern Europe and neighboring countries. *Environmental Science and Policy*, 14, 794–803.
5. Yun, K., Heo, K., Chu, J., Ha, K., Lee, E., Choi, Y., and Kitoh, A. (2012). Changes in Climate classification and extreme climate indices from a high-resolution future projection in Korea. *Asia-Pacific Journal of Atmospheric Sciences*, 48(3), 213–226.
6. Hay, J. E., Easterling, D., Ebi, K. L., Kitoh, A., and Parry, M. (2016). Introduction to the special issue: Observed and projected changes in weather and climate extremes. *Weather and Climate Extremes*, 11, 1–3.
7. European Commission (2009). Directive 2009/29/EC of the European Parliament and of the Council of 23 April 2009 amending Directive 2003/87/EC so as to improve and extend the greenhouse gas emission allowance trading scheme of the Community, European Parliament.

8. Eslamian, S. (Ed.) (2014). *Handbook of Engineering Hydrology, Vol. 2: Modeling, Climate Change and Variability*. Taylor and Francis, CRC Group, New York.

9. Deniz, A., Toros, H., and Incecik, S. (2011). Spatial variations of climate indices in Turkey. *International Journal of Climatology*, 31(3), 394–403.

10. Kiktev, D., Sexton, D. M. H., Alexander, L., and Folland, C. K. (2003). Comparison of modeled and observed trends in indices of daily climate extremes. *Journal of Climate*, 16, 3560–3571.

11. Farzaneh, M. R., Eslamian, S. S., Samadi, Z., and Akbarpour, A. (2012). An appropriate general circulation model (GCM) to investigate climate change impact. *International Journal of Hydrology Science and Technology*, 2(1), 34–47.

12. Kiktev, D. B., Caesar, J., and Alexander, L. (2009). Temperature and precipitation extremes in the second half of the twentieth century from numerical modeling results and observational data. *Atmospheric and Oceanic Physics*, 45(3), 284–293.

13. Beniston, M., and Stephenson, D. B. (2004). Extreme climatic events and their evolution under changing climatic conditions. *Global and Planetary Change*, 44, 1–9.

14. Alexander, L. V., Zhang, X., Peterson, T. C., Caesar, J., Gleason, B., Klein Tank, A. M. G., Haylock, M., Trewin, B., Rahimzadeh, F., Tagipor, A., Rupa Kumar, K., Revadekar, J., Griffiths, G., Vincent, L., Stephenson, D. B., Burn, J., Aguilar, E., Brunet, M., Taylor, M., New, M., Zhai, P., Rusticucci, M., and Vazquez-Aguirre, J. L. (2006). Global observed changes in daily climate extremes of temperature and precipitation. *Journal of Geophysical Research*, 111, 1–22.

15. Riahi, K., Rao, S., Krey, V., Cho, C., Chirkov, V., Fischer, G., Kindermann, G., Nakicenovic, N., and Rafaj, P. (2011). RCP 8.5-A scenario of comparatively high greenhouse gas emissions. *Climatic Change*, 109, 33–57.

16. Efthymiadis, D., Goodess, C. M., and Jones, P. D. (2011). Trends in Mediterranean gridded temperature extremes and large-scale circulation influences. *Natural Hazards and Earth System Sciences*, 11, 2199–2214.

17. Prein, A. F., and Gobiet, A. (2016). Impacts of uncertainties in European gridded precipitation observations on regional climate analysis. *International Journal of Climatology*, 37(1), 305–327.

18. Wetterhall, F., Pappenberger, F., He, Y., Freer, J., and Cloke, H. L. (2012). Conditioning model output statistics of regional climate model precipitation on circulation patterns. *Nonlinear Processes in Geophysics*, 19(6), 623–633.

19. Sen, O. L., Bozkurt, D., Göktürk, O. M., Dündar, B., and Altürk, B. (2013). Climate change and its possible impacts in Turkey. In *National Overflow Symposium*, Istanbul.

20. Hassan, M., Penfei, D., Iqbal, W., Can, W., Wei, F., and Ba, W. (2014). Temperature and precipitation climatology assessment over South Asia using the regional climate model (RegCM4.3): An evaluation of the model performance. *Journal of Earth Science and Climatic Change*, 5, 214.

21. Ruti, P. M., Marullo, S., D'Ortensio, F., and Tremant, M. (2008). Comparison of analyzed and measured wind speeds in the perspective of oceanic simulations over the Mediterranean basin: Analyses, QuikSCAT and buoy data. *Journal of Marine Systems*, 70, 33–48.

22. Kostopoulou, E., and Jones, P. D. (2005). Assessment of climate extremes in the Eastern Mediterranean. *Meteorology and Atmospheric Physics*, 89, 69–85.

23. Klein Tank, A. M. G., Wijgaard, J. B., Können, G. P., Böhm, R., Demaree, G., Gocheva, A., Mileta, M., Pashiardis, S., Hejkrlik, L., Kern-Hansen, C., Heino, R., Bessemoulin, P., Müller-Westermeier, G., Tzanakou, M., Szalai, S., Palsdottır, T., Fitzgerald, D., Rubin, S., Capaldo, M., Maugeri, M., Leitass, A., Bukantis, A., Aberfeld, R., Van Engelen, A. F. V., Forland, E., Mietus, M., Coelho, F., Mares, C., Razuvaev, V., Nieplova, E., Cegnar, T., Antonio Lopez, J., Dahlström, B., Moberg, A., Kirchhoffer, W., Ceylan, A., Pachaliuk, O., Alexander, L. V., and Petrovic, P. (2002). Daily dataset of 20th-century surface air temperature and precipitaton series for the European Climate Assessment. *International Journal of Climatology*, 22, 1441–1453.

24. Ramos, A. M., Ramos, R., Sousa P., Trigo, R. M, Janeira, M., and Prior, V. (2011). Cloud to ground lightning activity over Portugal and its association with circulation weather types. *Atmospheric Research*, 101, 84–101.

25. Oikonomou, C., Flocas, H. A., Hatzaki, M., Asimakopoulos, D. N., and Giannakopoulos, C. (2008). Future changes in the occurrence of extreme precipitation events in Eastern Mediterranean. *Global Nest Journal*, 10(2), 255–262.

26. Önol, B., Semazzi, F. H. M., Unal, Y., and Dalfes, H. N. (2006). Regional climatic impacts of global warming over the eastern mediterranean. In *International Conference on Climate Change and the Middle East: Past, Present and Future*. November 2016. Istanbul Technical University, Turkey.

27. Unal, Y. S., Önol, B., Menteş, S., Borhan, Y., Kahraman, A., and Ural, D. (2006). *Assessment of Global Climate Change Impact on Turkey by Regional Climate Model, TUJJB-TUMEHAP, 2006–2010.* Ministry of National Defence General Command of Mapping, Ankara.

28. Önol, B., and Semazzi, F. (2009). Regionalization of climate change simulations over the eastern mediterranean. *Journal of Climate*, 22(8), 1944–1961.

29. Bozkurt, D., Turuncoglu, U., Sen, O. L., Önol, B., and Dalfes, B. H. N. (2011). Downscaled simulations of the ECHAM5, CCSM3 and HadCM3 global models for the Eastern Mediterranean-Black Sea Region: Evaluation of the reference period. *Climate Dynamics*, 39, 207–225.

30. Demir, İ. (2011). Bölgesel İklim Modeli Projeksiyonları: RCHAM5-B1. In *5th Atmospheric Science Symposium* (s. 153–160). Istanbul Technical University, Istanbul.

31. Önol, B. (2012). Impacts of coastal topography on climate: High-resolution simulation with a regional climate model. *Climate Research*, 52, 159–174.

32. Unal, Y. S., Deniz, A., Toros, H., and Incecik, S. (2012). Temporal and spatial patterns of precipitation variability for annual, wet, and dry seasons in Turkey. *International Journal of Climatology*, 32(3), 392–405.

33. Bozkurt, D., and Sen, O. L. (2013). Climate change impacts in the Euphrates–Tigris Basin based on different model and scenario simulations. *Journal of Hydrology*, 480, 149–161.

34. Önol, B., and Unal, Y. (2014). Assessment of climate change simulations over climate zones of Turkey. *Regional Environmental Change*, 14(5), 1921–1935.

35. Bozkurt, D., and Sen, O. L. (2011). Precipitation in the Anatolian Peninsula: Sensitivity to increased SSTs in the surrounding seas. *Climate Dynamics*, 36(3–4), 711–726.

36. Bozkurt, D., Turuncoglu, U., Sen, O. L., Onol, B., and Dalfes, H. N. (2012). Downscaled simulations of the ECHAM5, CCSM3 and HadCM3 global models for the eastern Mediterranean–Black Sea region: Evaluation of the reference period. *Climate Dynamics*, 39(1–2), 207–225.

37. Republic of Turkey Ministry of Agricultural and Forestry (MAF) (2016). *Climate Change Impacts on Water Resources Project-Final Report*, Ankara, Turkey.

38. Tebaldi, C., Hayhoe, K., Arblaster, J. M., and Meehl, G. A. (2006). Going to the extremes an intercomparison of model-simulated historical and future changes in extreme events. Climatic Change, 79, 185–211.

39. Giorgi, F., and Bates, G. T. (1989). The climatological skill of a regional model over complex terrain. *Monthly Weather Review*, 117(11), 2325–2347.

40. Kiehl, J. T., Hack, J. J., Bonan, G. B., Boville, B. A., Briegleb, B. P., Williamson, D. L., and Rasch, P. J. (1996). *Description of the NCAR Community Climate Model* (CCM3) (No. NCAR/TN-420+STR). University Corporation for Atmospheric Research, Boulder, CO.

41. Grell, G. A. (1993). Prognostic evaluation of assumptions used by cumulus parameterizations. Monthly Weather Review, 121(3), 764–787.

42. Holtslag, A. A. M., van Meijgaard, E., and de Rooy W. C. (1995). A comparison of boundary layer diffusion schemes in unstable conditions over land. *Boundary-Layer Meteorology*, 76, 69–95.

43. Dickinson, R. E., Henderson-Sellers, A., and Kennedy, P. J. (1993). *Biosphere Atmosphere Transfer Scheme (BATS) Version le as Coupled to the NCAR Community Climate Model.* NCAR Tech. Note. National Center For Atmospheric Research, Boulder, CO, 72.

44. Abatzoglou, J. T. (2011). Development of gridded surface meteorological data for ecological applications and modeling. *International Journal of Climatology*, 33(1), 121–131.

45. Donat, M. G., Alexander, L. V., Yang, H., Durre, I., Vose, R., Dunn, R. J. H., Willett, K. M., Aguilar, E., Brunet, M., Caesar, J., Hewitson, B., Jack, C., Klein Tank, A. M. G., Kruger, A. C., Marengo, J., Peterson, T. C., Renom, M., Oria Rojas, C., Rusticucci, M., Salinger, J., Elrayah, A. S., Sekele, S. S., Srivastava, A. K., Trewin, B., Villarroel, C., Vincent, L. A., Zhai, P., Zhang, X., and Kitching, S. (2012). Updated analyses of temperature and precipitation extreme indices since the beginning of the twentieth century: The HadEX2 dataset. *Journal of Geophysical Research: Atmospheres*, 118, 2098–2118.

46. Van Den Besselaar, E. J. M., Haylock, M. R., Van Der Schrier, G., and Klein Tank, A. M. G. (2011). A European daily high-resolution observational gridded data set of sea level pressure. Journal of Geophysical Research, 116, D11110.

47. Sluiter, R. (2009). Interpolation methods for climate data. Literature review. Intern Rapport: IR 2009–04, De Bilt.

48. Daly, C., Halbleib, M., Smith, J. I., Gibson, W. P., Doggett, M. K., Taylor, G. H., Curtis, J., and Pasteris, P. P. (2008). Physiographically sensitive mapping of climatological temperature and precipitation across the conterminous United States. International Journal of Climatology, 28(15), 2031–2064.

49. Daly, C., Slater, M. E., Roberti, J. A., Laseter, S. H., and Swift, L. W. (2017). High-resolution precipitation mapping in a mountainous watershed: Ground truth for evaluating uncertainty in a national precipitation dataset. *International Journal of Climatology*, 37(S1), 124–137.

50. Taylor, G. H., Daly, C., Gibson, W. P., and Sibul-Weisberg, J. (1997). Digital and map products produced using PRISM. In *Proceedings of 10th AMS Conference on Applied Climatology*, American Meteorological Society, Reno, NV, 20–24 October, 2 pp.

51. Daly, C., Neilson, R. P., and Philips, D. L. (1994). A statistical-topographic model for mapping climatological precipitation over mountain terrain. *Journal of Applied Meteorology*, 33(2), 140–158.

52. Lehner, B., Döll, P., Alcamo, J., Henrichs, T., and Kaspar, F. (2006). Estimating the impact of global change on flood and drought risks in Europe: A continental, integrated analysis. *Climatic Change*, 75, 273–299.

53. Lenderink, G., van Ulden, A., van den Hurk, B., and Keller, F. (2007). A study on combining global and regional climate model results for generating climate scenarios of temperature and precipitation for the Netherlands. *Climate Dynamics*, 29, 157–176.

54. Ballı, C. (2014). Bias correction of precipitation simulated by regional climate model with different configurations over Turkey. M.Sc. Thesis, Istanbul Technical University, Institute of Science and Technology, Turkey.

55. Dong, S., Xu, Y., Zhou, B., and Shi, Y. (2015). Assessment of indices of temperature extremes simulated by multiple CMIP5 models over China. Advances in Atmospheric Sciences, 32, 1077–1091.

21 Time Series Analysis for Watershed Meteorological Data

A Case Study on Temperature Prediction

Saeid Eslamian, Faezeh Jannesary, Mousa Maleki, and Fatemeh Jannesary

21.1 INTRODUCTION

Generally, Iran is classified as one of the most arid and semi-arid climates, with a lack of water resources and inadequate distribution of precipitation have urged optimizing water resources management (Nowrozi Moghadami, 2012). Prediction of meteorological parameters and the investigation of their trends in water resources play a major role in water resources management. The stochastic nature of meteorological and hydrological data requires the use of time series analysis for data assessment purposes. Time series is defined as a set of data with equal and regular intervals. The purpose of time series analysis is to predict the desired data, obtain a model of changes, and identify future trends (Khazaei and Mirzaei, 2014). The basic property of a time series is that its observations are correlated with each other. Most of the standard random-based statistical methods are not applicable for this purpose, therefore, other relatively accurate methods are required. Some of these methods are descriptive and are given more emphasis (Zhou et al., 2018). Many researchers considered the trend components, periodic changes, seasonal changes, and irregular changes to describe the behavior of time series. Based on the principles of statistics and probability, weather models are of special importance and have unlimited applications. The use of statistical models to reconstruct past values and regenerate future data values is called time series analysis. Many time series models are based on stationarity values (Shabani et al., 2012).

What is specific about this study is the number of primary tests used for the meteorological parameters, and the watershed under study would be one of the most problematic basins considered in the world and Iran. The periods studied in the case study were monthly, seasonal, and daily time scales. The periods used were chosen for the following reasons: (i) the more data, the less error, and the intervals chosen were used for the accuracy of the results. (ii) In using time series, the higher the frequency of data is, the more accurate will be the final model (Jannesary, 2021).

In this chapter, the aim is to demonstrate the primary statistical tests and time series analysis of the meteorological data records particularly watershed temperature.

The temperature data is assumed stochastic and the effect of man-made and human factors on the data was assumed negligible.

DOI: 10.1201/9781003473398-26

21.1.1 Previous Investigators

Helmy et al. (2020) studied the seasonal autoregressive integrated moving average (SARIMA) model for the different statistical samples of precipitation to predict drought using the standardized precipitation index (SPI). The time scale used was 12 months with seven synoptic stations and a recording duration of 20 years. First, the stationarity and seasonality of the data were investigated and the best model was obtained using the criteria of Akaike and Schwartz. The results showed that the SARIMA model was accurate for all seasons, the residual graphs were close to the normal distribution, and the prediction performance was satisfactory (Helmi et al., 2020). Kocsi et al. (2020) studied the climate data in western Hungary and examined the breakpoints in the annual, seasonal, and monthly precipitation time series, without looking at the breakpoints. A significant decreasing trend of 0.2 to 0.7 mm per year was observed with the Mann–Kendall test. The results showed that the seasonal precipitation reduction in the autumn ranges from 0.15 to 0.38 mm per year. It was concluded that signals of changes in precipitation related to climate change could be detected in a long series (Kocsi, 2020). Polwiang (2020) studied dengue fever and its association with climatic factors in Bangkok, Thailand from 2003 to 2017. The time series model was obtained seasonally, and the results represented that the amount of rainfall and humidity affect the transmission of dengue disease and it was strongly recommended that more variables should be taken into account to increase the accuracy.

The model accuracy measures Mean Absolute Error (MAE), Mean Absolute Scaled Error (MASE), Accuracy Percent, Root Mean Squared Error (RMSE), Mean Absolute Percent Error (MAPE), and Akaike Information Criterion (AIC) are based on the time series data that is used to generate the model. The MAE, RMSE, and MAPE measures were also used for the validation analysis (Polwiang, 2020). Kalamaras et al. (2017) examined the mean, maximum, and minimum temperatures on Crete Island, Greece, using the time series method. The data used were on a daily scale from January 1973 to December 2010. The behavior of the time series models (daily average, maximum, and minimum temperatures) with the local climate was almost the same (Kalamaras et al., 2017). Asfaw et al. (2018) observed precipitation and temperature data in Northern Ethiopia, and then the data were analyzed using the Mann–Kendall trend test, and the existence of an increasing trend of mean and low temperature was significant at the level of 5% and 10%. Hence, it has been recommended that the nature of decreasing and erratic rainfall and the trend of rising temperatures should be taken into account for the strategies designed in the agricultural sector (Asfaw et al., 2018). Latifo et al. (2020) studied 20 meteorological stations in the eastern, western, and northern basins of India, China, and Afghanistan. The variables included the maximum, minimum, and average temperature and the result was that the temperature increased between 0.26 and 0.13°C in a decade, which leads to a decrease in polar ice if the decrement trend is followed by precipitation and an increment in temperature, which in turn the endangers water resources conditions (Latif et al., 2020).

The purpose of time series analysis is to predict the desired data, get a pattern of changes, and identify their trends (Dudangeh et al., 2012). Time series analysis includes the temporal structure recognition, identifying stationary and non-stationary data, examining the effect of external variables, examining the effect of data correlation structure and its autocorrelation, and identifying the natural seasonality of the data and the immediate presentation of forecasts in the next phase, but not in large quantities. In hydrology, there is no guarantee that forecasting will be accurately predicted and estimated.

The static time series model includes (i) autoregressive processes (AR(p)), (ii) Moving average processes MA(q), (iii) autoregressive moving average ARMA(p,q), and (iv) In the non-stationary time series that includes the autoregressive moving average integrated average ARIMA model, which will lead to solving the instability in the mean and the instability in the variance.

21.2 MATERIALS AND METHODS

Time series models include autoregressive (AR), moving average (MA), and autoregressive moving average (ARMA) models, and these models are used in hydrological processes such as annual flow, precipitation, and temperature. Typically, these models can preserve past statistics, including the mean, variance, and covariance. Parametric methods differ significantly from their non-parametric alternatives. Parametric methods require assumptions about the marginal probability distribution of the variables and the dependence structure of the spatial and temporal covariance, while non-parametric methods rely on preserving the empirical structure of the observed variables. Methods based on nonparametric data have become popular in recent decades.

Time series analysis includes a time domain approach and a frequency domain approach. In the time domain approach, time functions such as auto-correlation functions (ACF) and partial auto-correlation functions (PACF) are used to describe the characteristics of a time series process and their evolution is shown through different time delay ratios. In the frequency domain method, an attempt is made to use a spectral function to study how it works (Little, 2013).

The most important applications of time series analysis are in basin management, droughts and floods, and climate change. One of the most vital climatic elements is temperature, which has always been the focus of attention for climatologists and meteorologists due to its importance. The Earth's temperature has increased by 0.6°C between 1986 and 2000, which has caused many changes in snowfall, rainfall, storms, and floods (Alsharif et al., 2019) and according to NASA data, from 2000 to 2021, the temperature has been increased by 0.46°C and this trend is expected to increase in the following decades. For instance, Iran is one of the most arid and semi-arid countries, which has a critical shortage of water resources and inadequate water distribution that has made it necessary to plan for the optimal use of water resources (Alsharif et al., 2019).

Eslamian (2014) classified the normal steps in the analysis of an abnormal and non-stationary time series as follows:

(i) test data for normality and transform the data to a normal form, (ii) test for trends and changes and remove them if necessary, and (iii) if there is a seasonal pattern in the mean or variance, remove it and relate the steps normality to the model creation of a time series that includes the model identification, parameter estimation, and model testing and validation.

To convert to the normal state, it is assumed that most of the time series models that were in the normal test are rejected, which can be converted to the normal state by using a variety of parametric transformations. Common transformations include the Box–Cox transformation and detrending between them (Eslamian, 2014).

21.2.1 MODEL SELECTION

Eslamian (2014) states that in model selection, ACF and PACF are often used to estimate the appropriate model, and the model is often selected as the best model by examining the minimum variance of the residuals, and the residuals should follow the normal distribution. The residuals must have ACF and PACF values close to zero and be independent of other variables used in the model. Non-normal residuals can be shown as an increasing trend in ACF and PACF, which requires a higher-order model. A common method for modeling is to use criteria to select the best model based on the number of parameters involved. The AIC criterion and the Schwartz criterion (SBIC), which is also known as the Bayesian information criterion, and the Hanan–Quinn criterion (HQIC) are also used (Eslamian, 2014) (Table 21.1).

TABLE 21.1

Criteria for Estimating the Best Model

Number of data	Criterion
Up to 200	(AIC)
200–100	(HQIC)
Under 100	(SBIC)

21.2.2 The Time Series Models

21.2.2.1 Auto-regressive Model (AR)

This model is one of the most common stochastic models that perform regression on the sentences and is performed on the past values of Z_t. In addition, it is used in static-non-stationary time series. However, this model is generally used for static time series.

$$AR(p) \rightarrow y_t = \mu + \varnothing_1\left(y_{\tau-1}-\mu\right) + \varnothing_2\left(y_{\tau-2}-\mu\right) + \ldots + \varepsilon_\tau \tag{21.1}$$

μ: mean

Φ: Autoregressive coefficients and parameters

ε_τ: Error sentence (normal distribution)

21.2.2.2 Moving Average Model (MA)

The general form is as follows:

$$Z_t = a_t + \theta_{1a_{t-1}} + \theta_{2a_{t-2}} + \ldots + \theta_{qa_{t-q}} \tag{21.2}$$

θ: Coefficients and parameters of the moving model

Mixed models: models that have auto-correlation and moving average conditions.

21.2.2.3 Moving Average Autoregressive Models (ARMA)

By combining the autoregressive model with order p and the moving average model with order q, the ARMA model with order (p, q) is created and the general structure is as follows.

$$Z_t = \varnothing_1 Z_{t-1} + \varnothing_2 Z_{t-2} + \ldots + \varnothing_p Z_{t-p} + a_t - \theta_{1a_{t-1}} - \theta_{2a_{t-2}} + \ldots - \theta_{qa_{t-q}} \tag{21.3}$$

$$Z_t = \sum_{j=1}^{p} \varphi_j Z_{t=j} + \varepsilon_t - \sum_{j=1}^{q} \theta_j \varepsilon_{t=j}$$

21.2.2.4 Cumulative Moving Average Autoregressive Models (ARIMA)

The ARIMA model has been developed for non-stationary series and the ARIMA model is created (Modares and Eslaminan, 2015).

21.2.2.5 Model Assumption and Initial Modeling

Autocorrelation (ACF) and partial auto-correlation (PACF) plots are used for initial modeling and model assumption.

ACF

$$\rho_k = \frac{Y_k}{Y_0} = \frac{Cov\left(Z_t, Z_{t-k}\right)}{\sqrt{Var\left(Z_t\right)}\sqrt{Var\left(Z_{t+k}\right)}} \tag{21.4}$$

$$\hat{\rho}_k = \frac{\hat{Y}_k}{\hat{Y}_0} = \frac{\sum_{t=1}^{n=k}\left(Z_t - \bar{Z}\right)\left(Z_{t+k} - \bar{Z}\right)}{\sum_{t=1}^{n}\left(Z_t - \bar{Z}\right)^2} \tag{21.5}$$

PACF

$$\varnothing_{kk} = \left(Z_t, Z_{t+k}, Z_{t+1}, \ldots Z_{t+k-1}\right) \tag{21.6}$$

Z_t: Past values

To obtain the best model, Akaik, Schwartz, and Hannan-Quinn criteria are used. Then, based on the obtained model, the other tests are estimated.

21.2.3 STEPS TAKEN FOR MODELING

21.2.3.1 Check Outlier Data

The outlier data include Grubzobek's test and normality test of univariate methods and using a box plot. In this study, the normal method was used (Nohegar et al., 2016). The normality test is from the univariate methods for identifying outlier data:

$$Y = Y_m \pm 3S_t \tag{21.7}$$

21.2.3.2 Trend Analysis

Climate trends can vary over time, for example, the global temperature is increasing and predictions of the expected future trends could be obtained from the general models of climate change. The trend could be removed by non-parametric filtering, but a parametric form of the trend is often desired so that the trend can be extracted for prediction and simulation purposes (Ghodoosi et al., 2014). In the data set, the mean and variance are often influenced by the annual cycle, and there is a periodic structure in the data, where the mean and variance are not constant. The seasonal mean in each season is eliminated by subtracting it from the sample mean, and in the same way, the seasonality in the variance will also be eliminated (Ghodoosi et al., 2014; Ajamzadeh et al., 2017).

Detrending is performed through the various methods of average detrending, variance trend targeting, mean and variance detrending, variance shift detrending, and mean and variance shift detrending.

21.2.3.3 Stationary Analysis

The concept of stationary is that data does not change over time. That is, the process is in statistical equilibrium, and the mean and variance of the series remain constant over time (Shabani et al., 2012). The stationarity state is the non-dependence of the data to time, and the basic and main assumption for using statistical methods is random, so if this test is rejected, the modeling will be complicated as the factors that affect the instability must be included in the modeling and analysis. There are several methods to verify the stationarity of the data, which can be referred to as the GLS-DF (Dickey–Fuller Generalized Least Squares) unit root test, ADF (Augmented Dickey–Fuller) unit root test (Kazroni et al., 2014). A data series is stationary if there is no regular change in its mean and variance and the periodic changes are eliminated. Non-stationary time series can be transformed into stationary time series by differentiating or stabilizing the variance (Gardfarstatisticszi and Saberi, 2017).

The following methods are used to make data stationary:

If the variance is not constant, the Box–Cox transformation is used.
If the average is not constant, the difference is used.

The variance should first be fixed and then the mean.

21.2.3.3.1 ADF Unit Root Test
ADF unit root test is based on below:

$$\nabla d_{ij} = a_{jt} + \alpha_{jt}t + \beta_{ij}d_{ijt-1} + \sum_{s=1}^{Pij} \delta_{ijs}\nabla d_{ij,t-s} + \vartheta_{ijt} \tag{21.8}$$

$\Delta d_{ij,t-s}$: Difference
P_i: Lag
a_{jt}: y-intercept
t: Time parameter

21.2.3.3.2 Unit Root Test GLS-D
This test is one of the tests that solve the problem of the sensitivity of the test to the sample size and the low power of the ADF and KPSS (Kwiatkowski–Phillips–Schmidt–Shin) tests.

$$\nabla d_{ij,t}^{d} = \beta_{ij}d_{ij,t}^{d} + \sum_{s=1}^{Pij} \delta_{ijs}\nabla d_{ij,t-s}^{d} + \vartheta_{ijt} \tag{21.9}$$

$$d_{ij,t}^{d} = d_{ij,t} - \hat{\beta}_0 - \hat{\beta}_1 t \tag{21.10}$$

where β_1 and β_0 are obtained by regressing d_{ij} on z_{ij}.

21.2.3.4 Homogeneity Check
Data should be homogenized for long-term climate analysis, especially climate change studies. The meaning of data homogeneity is that there is no data caused by human activity or problems related to the observation or changes for the measurement tools of weather elements. After data quality control, the data can be included in the analysis or removed. In addition, the analysis of homogeneity in time series is necessary for hydrological modeling, water resource management, and climate change studies. It is important to identify the sudden change points in the weather and climate because these natural and man-made changes can significantly change the results of weather and climate trends, their variability, and their analysis. These changes, especially in long-term records, are often unavoidable and can cause data inconsistencies. Therefore, it is necessary to correct and homogenize the data for climate research and other applications. The results of climatic studies highly depend on the continuity of the data, that is, on the absence of a point of change (Rafati and Karimi, 2017). To use the statistical tests in a time series, it is necessary that the data must be homogeneous. There are different methods to check homogeneity, the most common is SNH (Standard Normality Test).

21.2.3.4.1 Standard Normal Homogeneity
Normal homogeneity test is usually more sensitive to the first and the last lag.

This test performs the average of the first observation to the nth observation with the average of $n-k$ data (k is the interval).

$$T(k) = kz_1^{-2} + (n-k)z_2^{-2} \tag{21.11}$$

$(k = 1, 2, \dots n)$

$$\bar{z_1} = \frac{1}{k} \sum_{i=1}^{k} \left(Y_i - \bar{Y} \right) \Big/ s \tag{21.12}$$

$$\bar{z_2} = \frac{1}{k-1} \sum_{i=k+1}^{k} \left(Y_i - \bar{Y} \right) \Big/ s \tag{21.13}$$

$$T_0 = max T_k \left(1 \le k \le n \right) \tag{21.14}$$

where z_1 and z_2 are the statistical parameters $T_{(k)}$. k is the recording years, Y is observational data, and S is the standard deviation.

The statistical value of T_0 in the standard normal test is as follows:

Statistically significant values of T_0 are given in Table 21.2. According to the normal homogeneity standard test, if the T_0 statistic exceeds the critical values, the null hypothesis is rejected.

A table of critical values of 1% T_0 in the natural standard homogenous test is obtained as a reference and at the 5% level in the same way.

21.2.3.5 Check for Normality

To use statistical methods, it is necessary to check the normality of the data. Different methods are used to check the normality of the data (De Gois et al., 2020; Choudhury and Urena, 2020). The most common test is the Kolmogorov-Smirnov test (Modares, 2012).

$$D = MAX|FN\left(x \right) - F)\left(x \right)| \tag{21.15}$$

FN is a statistic based on the deviations of the sample distribution function from the hypothesized function.

After data modeling, it should also be noted that the changes that have been made in the data, including trends, stationary, and homogeneity are completed (Semakosh et al., 2017). After predicting the data, other tests need to be done to check the correctness of the model.

21.2.3.6 Test the Correctness of the Model

This test is used to check the correctness of the model.

21.2.3.6.1 Pert–Manto test

Ljung Box Statistic (LBQ) is used to test the assumption of auto-correlation of data up to lag k, and the null hypothesis is that the auto-correlation of lags is equal to zero.

$$H_0 = \rho_1 = \rho_2 = \ldots = \rho_k = 0 \tag{21.16}$$

TABLE 21.2

T_0 Value

			T_0 value			
Number	n =20	n =30	n =40	n =50	n =70	n =100
99	9/56	10/45	11/1	11/38	11/89	12/32
95	6/95	7/65	8/10	8/45	8/8	9/15

(Kazemzadeh and Malekian, 2018)

And LBQ, which is also called modified Q, is as follows:

$$Q(LBQ) = n(n+2)\sum_{h=1}^{k} n - h^{-1}\rho_h^2 \qquad (21.17)$$

Where n is the number of observations and m is the number of estimated parameters, and the null hypothesis is rejected if the value of the statistic Q is greater than the value of the table. In the obtained model for all delays, the P value is greater than five percent, which confirms the hypothesis of uncorrelated residuals.

21.2.3.7 Estimate Error

For the best model, should check the error rate of the data using factors including MS, MSE (Mean Squared Error), SSR (Regression Sum of Squares), SE (Standard Error), and RMSE (Root Mean Square Error) factors. (Semakosh et al., 2017)

As an example, the checked error in the form of MSE, and the error percentage, which is as follows:

$$MSE = \frac{\Sigma(\hat{}_t - y_t)^2}{N} \qquad (21.18)$$

$$\frac{\text{actual data-predicted data}}{\text{actual data}} = (21.19)$$

21.3 CASE STUDY

The study area is located in the Zayandehrud Basin and the mean elevation is about 1466 m above sea level, the total area is 41,500 square km², most of the basin has an annual precipitation of less than 150 mm, and the annual evapotranspiration of this basin is about 1500 mm per year. The Zayandehrud Basin includes 12 synoptic stations (Figure 21.1). In this study, the Mobarakeh synoptic station was used as a case study reference station. The data gathered by this station is from September 24, 1993, to December 30, 2019. The meteorological parameter of the maximum temperature in the daily time interval had an increasing trend. This station has a seasonal trend in the monthly period.

According to the past droughts and climate changes, it seems that climate change had an impact on this meteorological parameter and has caused an increment in temperature due to the increasing trend.

21.3.1 MODELING STEPS

1- First, auto-correlation (ACF) and partial auto-correlation (PACF) are checked in the review program and a correlogram is drawn.
2- The unit root is checked through the ADF test and the GLS-D unit root test to determine the order of the difference and then modeling begins.
3- For modeling in the Eviews program, the best model is selected based on the coefficient of determination (R²) and based on the number of data, the Akaik, Schwartz, or Hannan-Quinn criteria and the estimation has been done.
4- Finally, modeling was implemented in Minitab software and the graphs are output.
5- Several data were used to validate the model based on the real data.
6- After the error validation, the models are checked and the best model is fitted.
7- Using the Kolmogorov–Smirnov test, for checking the normality of the residuals in the Minitab software.

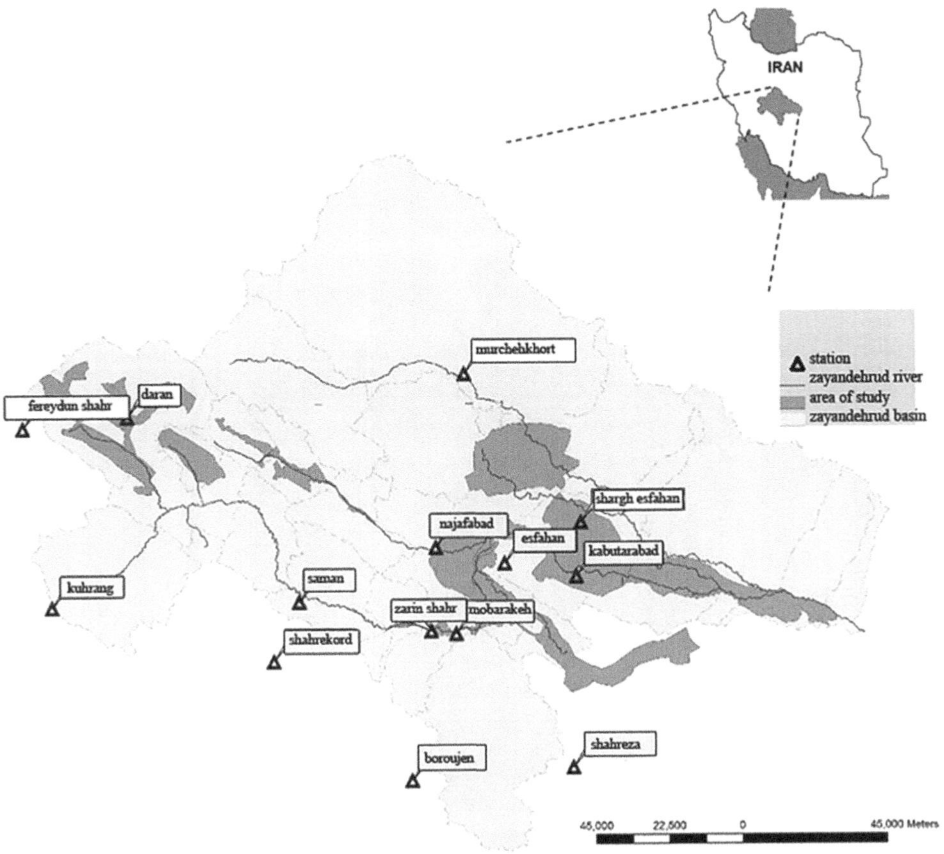

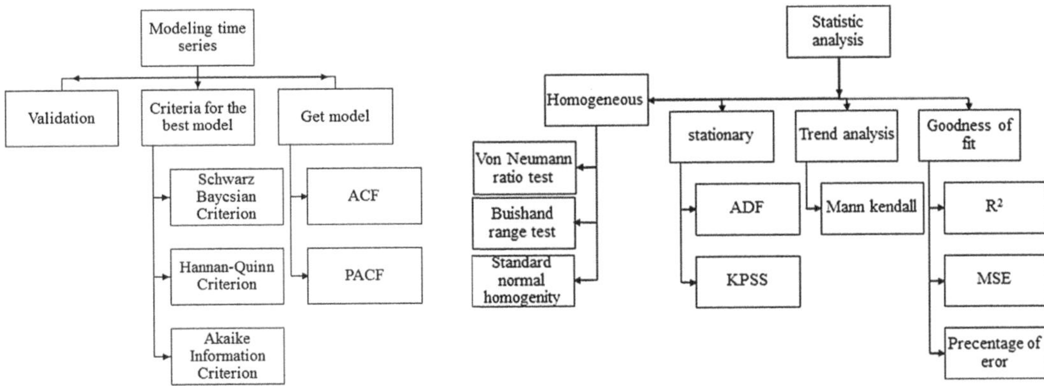

FIGURE 21.1   Selected stations in the Zayandehrud Basin, Iran.

21.3.2 SOFTWARE USED

(i) The trend and homogeneity were checked using the formulas coded in MATLAB.

(ii) The Minitab software is a software with different capabilities in the field of statistics, but in the case of time series, it has some limitations in the ranking of models. It also has Kolmogorov–Smirnov and Pratt–Manto tests and the error values.

FIGURE 21.2   Methodology flow chart.

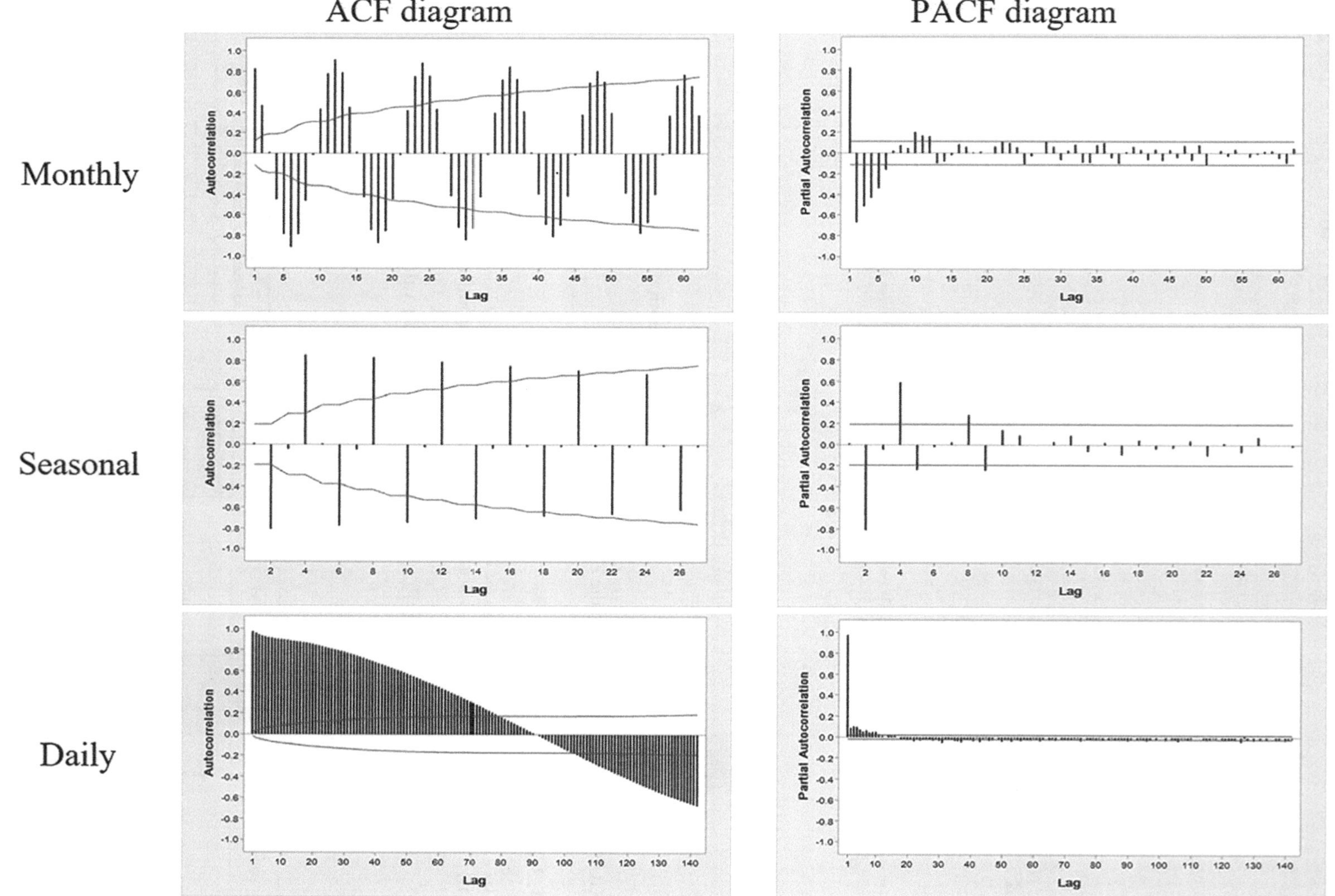

FIGURE 21.3 Data auto-correlation maximum temperature of the Mobarakeh station.

(iii) Eviews is a software for time series with extensive capabilities where the value of the ranks of modeling parameters is not limited like Minitab software. It also has the Watson R^2 and modified R^2 camera tests and Akaik, Schwartz, and Hannan-Quinn criteria.

Among the stations, the diagrams of Mobarakeh station for the maximum temperature meteorological parameter are given as a reference.

Data stationery is determined by drawing auto-correlation and partial auto-correlation diagrams. If the statistically significant lag is greater than 0.95, it shows that the data is not stationary and another method is to use the auto-correlation graph and the partial autocorrelations graph to check the seasonality of the data. If a trend was observed in the specific intervals in the auto-correlation and partial auto-correlation plots, the data appeared to be seasonal.

From Figure 21.3, it is obvious that data in monthly intervals have a seasonal feature.

To check the correctness of the model used, the auto-correlation and partial auto-correlation of the residuals are examined. In such a way, no correlation should be observed until intervals 3 to 4, if it is observed, the model obtained seems to have to change.

In general, by examining the auto-correlation and partial auto-correlation plots, the correct model can be identified. After modeling, histograms and time series plots were drawn with the predicted values, and the upper and lower limits of the predicted values were considered with a significance level of 5%.

It looks like there is the skewness in monthly one, but it looks that way because the outliers are not removed, and the reason for not removing these data is that the data is observational data and cannot be removed. For validation data in a daily time interval, 5% of the data and in monthly time intervals 10% of the data, and in seasonal time intervals 20% of the data were considered. The predicted data were compiled in Table 21.3.

The maximum temperature data in the daily time interval had an increasing trend, which was eliminated for modeling by the time series method. The models were rated as normal in Table 21.4 and the R^2 values in Table 21.6 were very good and had a high amount.

Modeling was performed based on the criteria and review of partial autocorrelation diagrams and the results were inserted in Table 21.5

The observational and predicted data had a good determination coefficient that is given in Table 21.6 also the model was examined in terms of criteria and the results are presented in Table 21.6.

The number of predicted data and the average error of the predicted data are given in Table 21.7.

TABLE 21.3

Predicted Values of Maximum Temperature (°C) of Mobarakeh Station

Seasonal		Monthly		Daily	
n	t_{max}	n	t_{max}	n	t_{max}
1	40.15	1	31.61	1	13.09
2	39.01	2	23.38	2	13.18
3	30.72	3	20.16	3	13.29
4	25.76	4	18.11	4	13.36
5	37.84	5	20.67	5	13.43
6	37.79	6	24.10	6	13.49
7	30.16	7	28.39	7	13.55
		8	34.36	8	13.60
		9	39.36	9	13.66

Table: n refers to the number of predicted data.

TABLE 21.4

Check the Normality and Accuracy of the Model From Two Methods (Durbin–Watson and Pert–Manto) and Error Percentage of all Stations

Station & Period		Normality t_{max} k	j	S	Station & Period		Model Accuracy t_{max} d	l	Error% t_{max}	Num. t_{max}
Mobarakeh	Monthly	>0.15			Mobarakeh	Monthly		×	−8	9
	Seasonal	>0.15				Seasonal	×	×	−9.5	7
	Daily			1.3		Daily			4.7	10
				−0.4						

Table: k =the Kolmogorov–Smirnov method. J= Jark–Bera method. S=a statistical method in which the first parameter means kurtosis and the second parameter means skewness. d- Durbin–Watson method. l: method of Pert–Manto.

TABLE 21.5

Models Obtained From Other Stations

Fitted Model of t_{max}	Period	Station
SARIMA(2,1,3)(1,0,2)	Monthly	Mobarakeh
ARIMA(4,0,3)	Seasonal	
ARIMA(3,0,2)	Daily	

TABLE 21.6

Criteria Obtained by Other Stations

Station	Criteria	Monthly	Seasonal	daily
Mobarakeh	R^2	0.83	0.85	0.89
	AIC	5.3	4.9	5.1
	SBIC	5.3	5	5.1
	HQIC	5.3	4.9	5.1
	MSE	0.1	0.21	0.02
	RMSE	5.01	2.62	3.12

TABLE 21.7

Number of Predicted Values and Error of Stations in the Maximum Temperature Parameter

Station Name	Period	Error (%)	Num.
Mobarakeh	Monthly	−8	9
Mobarakeh	Seasonal	−9.5	7
Mobarakeh	Daily	5.8	9

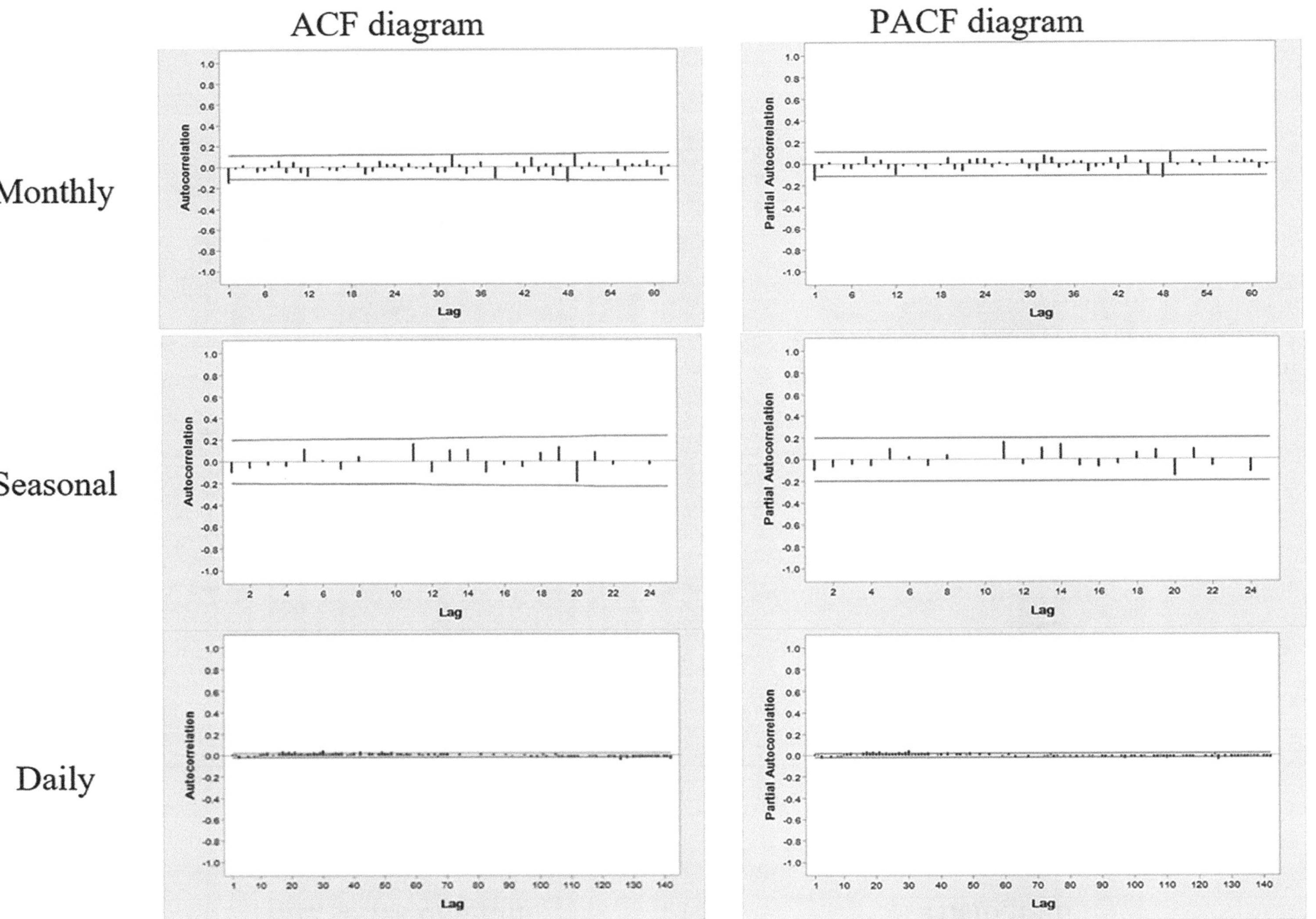

FIGURE 21.4 Auto-correlation of residuals maximum temperature of Mobarakeh station.

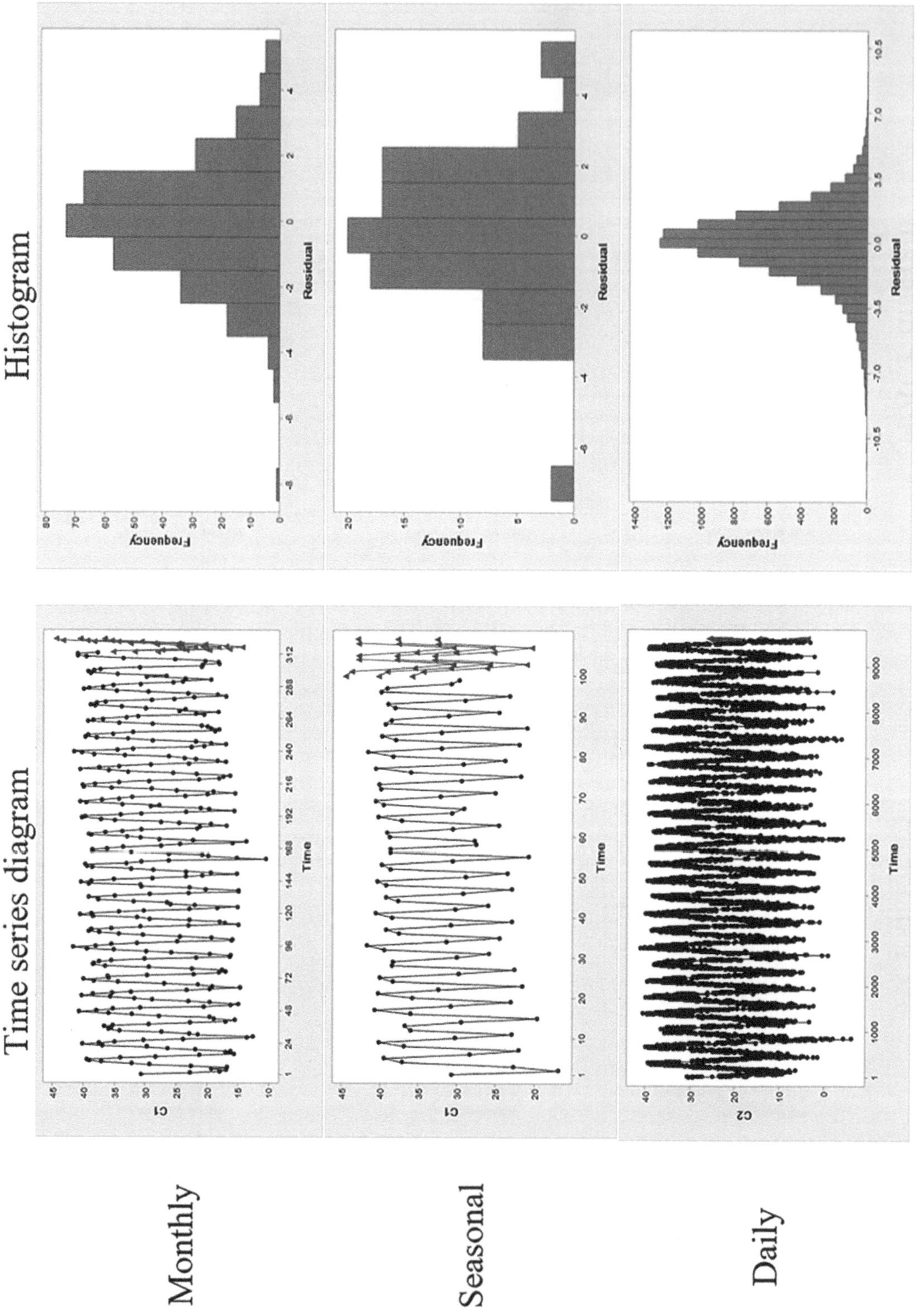

FIGURE 21.5 Correct examination of the maximum temperature model of Mobarakeh station.

21.4 CONCLUSIONS

The determination of the trend was verified by the non-parametric Mann–Kendall test and had an increasing linear trend in the daily interval. The method that was used in this study was the Box–Jenkins method. In such a way, that no correlation should be observed until intervals 3 to 4, if it is observed, the obtained model seems to change and the Akaike, Schwartz, and Hannan-Quinn criteria were used, then the normality of the data was verified and for its evaluation, the Durbin–Watson and Pert–Manto tests were used.

The maximum temperature interval (as a meteorological parameter) was evaluated in monthly, seasonal, and daily time intervals, and as the number of data decreases, the amount of estimation error increases. Box–Cox conversion was not used because the number obtained was beyond the desired range. For validation, R^2, MSE, and the percentage error are used. R^2 values were above 0.7 and the percentage error was below 20%. In terms of normality, all of the stations were similar (in terms of tests used at different periods). All of the data were analyzed at, monthly and seasonal periods using the Kolmogorov–Smirnov and Jarque–Bera tests. The models obtained in the intervals were the following: for monthly intervals was SARIMA and for daily and seasonal intervals was ARIMA.

To assess other meteorological parameters such as wind speed, minimum temperature, maximum and minimum humidity, and sunshine in time intervals of daily, 7-day, 15-day, monthly, and seasonal time frames and in the meteorological parameter of precipitation in monthly, seasonal, and 6-month time frames (Jannesary, 2021).

It is recommended: (i) the effect of climate change on the prediction of the results could be considered, (ii) wavelet transform in time series analysis could be used, and (iii) use and compare the ARCH and GARCH time series models.

BIBLIOGRAPHY

Ajamzadeh, A., Mollaeinia, M. R., and Kandahari, Q. (2017). Comparison of some artificial intelligence methods in predicting daily time series of minimum, maximum, and rainfall temperature of Tangab dam station located in Fars province. *Geospace*, 17, 205–228.

Alsharif, M. H., Younes, M. K., and Kim, J. (2019). Time series ARIMA model for prediction of daily and monthly average global solar radiation: Tcastud Seo. *Symmetry*, 11, 240–257.

Aqalpour, P., and Nadi, M. (2018). Assessing the accuracy of the SARIMA model in modeling and long-term forecast of average monthly temperature in different climates of Iran. *Climatological Research*, 9, 113–126.

Asadi Zarch, M. A., Malekinezhad, H., Mobin, M. H., Dastorani, M. T., and Kousari, M. R. (2011). Drought monitoring by reconnaissance drought index (RDI) in Iran. *Water Resources Management*, 25, 3485–3504.

Asfaw, A., Simane, B., Hassen, A., and Bantider, A. (2018). Variability and time series trend analysis of rainfall and temperature in northcentral Ethiopia: A case study in Woleka sub-basin. *Weather and Climate Extremes*, 19, 29–41.

Batty, M. (2008). The size, scale, and shape of cities. *Science*, 319, 769–771.

Choudhury, A., and Urena, E. (2020). Forecasting hourly emergency department arrival using time series analysis. *British Journal of Healthcare Management*, 26, 34–43.

De Gois, G., De Oliveira-Júnior, J. F., Da Silva Junior, C. A., Sobral, B. S., De Bodas Terassi, P. M., and Junior, A. H. S. L. (2020). Statistical normality and homogeneity of a 71-year rainfall dataset for the state of Rio de Janeiro Brazil. *Theoretical and Applied Climatology*, 141, 1573–1591.

Di Persio, L., and Frigo, M. (2016). Gibbs sampling approach to regime-switching analysis of financial time series. *Journal of Computational and Applied Mathematics*, 300, 43–55.

Dudangeh, A., Abedi Koopai, J., and Gohari, J. (2012). Application of time series models to determine the trend of future climatic parameters to manage water resources. *Water and Soil Science*, 16, 59–74.

Eslamian, S. (2014). *Handbook of Engineering Hydrology: Environmental Hydrology and Water Management.* CRC Press, New York.

Gerdfaramarzi, S. S., Saberi, A., and Gheisouri, M. (2017). Determining the best time series model in forecasting annual rainfall of selected stations in West Azerbaijan province. *Applied Research in Geographical Sciences*, 17, 87–105.

Ghodoosi, M., Morid, S., and Delavar, M. (2014). Comparison of detrending methods in temperature and precipitation time series. *Journal of Agricultural Meteorology*, 1(2), 32–45.

Hadizadeh, R., Eslamian, S., and Chinipardaz, R. (2013). Investigation of long-memory properties in streamflow time series in Gamasiab River, Iran. *International Journal of Hydrology Science and Technology*, 3(4), 319–350.

Helmi, M., Bakhtiari, B., and Ghaderi, K. (2020). Modeling and forecasting of meteorological drought using the SARIMA time series model in different climatic samples of Iran. *Iranian Journal of Irrigation and Drainage*, 14, 1079–1090.

Jain, G., and Mallick, B. (2017). A study of time series models ARIMA and ETS. *International Journal of Modern Education and Computer Sciences*, 4, 57–63.

Jannesary, F. (2021). Prediction of meteorological parameters of Zayandehrud basin using time series. Master Thesis, Isfahan University of Technology, Isfahan.

Kalamaras, N., Philippopoulos, K., Deligiorgi, D., Tzanis, C. G., and Karvounis, G. (2017). Multifractal scaling properties of daily air temperature time series. *Chaos, Solitons and Fractals*, 98, 38–43.

Kazemzadeh, M., and Malekian, A. (2018). Homogeneity analysis of streamflow records in arid and semi-arid regions of northwestern Iran. *Journal of Arid Land*, 10, 493–506.

Kazroni, A., Asgharpour, H., and Rezaei, K. (2014). Comparison of the degree of convergence of the price level of goods and services between the provinces of Iran. *Economic Research*, 49, 599–620.

Khatar, B., and Bahmani, A. (2015). Predicted temperature of deep soil layers using time series models. *Journal of Soil Research*, 29(2), 199–210.

Khazaei, M., and Mirzaei, M. (2014). Prediction of climatic variables using time series analysis of Zohreh watershed. *Applied Research in Geographical Sciences*, 34, 233–250.

Kocsi, T., Kovács-Székely, I., and Anda, A. (2020). Homogeneity tests and non-parametric analyses of tendencies in precipitation time series in Keszthely, Western Hungary. *Theoretical and Applied Climatology*, 139, 849–859.

Latif, Y., Yaoming, M., Yaseen, M., Muhammad, S., and Atif Wazir, M. (2020). Spatial analysis of temperature time series over the Upper Indus Basin (UIB) Pakistan. *Theoretical and Applied Climatology*, 139, 741–758.

Little, T. D. (2013). *The Oxford Handbook of Quantitative Methods in Psychology*. Oxford University Press, Oxford.

Masoumpour Samakosh, J., Jalilian, A., and Yari, A. (2017). Analysis of seasonal rainfall time series in Iran. *Natural Geography Research*, 49, 475–457.

Modares, R. (2012). Prediction of hydrological drought using time series analysis in one of the tributaries of Zayandehrud River. Master's Thesis, Faculty of Natural Resources, Isfahan University of Technology, Isfahan.

Modares, R., and Eslamian, S. (2015). Time series modeling of Zayandehrud river flow. *Iranian Journal of Science and Technology*, 30, 570–567.

Nohegar, A., Kazemi, M., Ahmadi, J., Gholami, H., and Mahdavi, R. (2016). Using univariate and multivariate methods to detect outliers in sediment fingerprinting method, case study: Tange Bostanak Watershed. *Soil Conservation and Watershed Research Institute*, 9, 412–398.

Nowrozi Moghadami Tayol, M. (2012). KPSS test for the mean of spatial point processes. Master Thesis, School of Mathematics, Shahid Beheshti University. Tehran.

Nury, A. H., Hasan, K., and Alam, M. J. B. (2017). Comparative study of wavelet-ARIMA and wavelet-ANN models for temperature time series data in northeastern Bangladesh. *Journal of King Saud University*, 29, 47–61.

Polwiang, S. (2020). The time series seasonal patterns of dengue fever and associated weather variables in Bangkok. *BMC Infectious Diseases*, 20, 1–10.

Qin, R. X., Xiao, C., Zhu, Y., Li, J., Yang, J., Gu, S., Xia, J., Su, B., Liu, Q., and Woodward, A. (2017). The interactive effects between high temperature and air pollution on mortality: A time-series analysis in Hefei, China. *Science of the Total Environment*, 575, 1530–1537.

Rafati, S., and Karimi, M. (2017). Investigating the homogeneity of climate data and the trend of temperature change. *Journal of Earth and Space Physics*, 44, 199–214.

Shabani, B., Mousavi Baighi, M., Jabari Nougabi, M., and Ghahreman, B. (2012). Modeling and forecasting the maximum and minimum monthly temperature of Mashhad Plain using time series models. *Water and Soil Journal*, 27(5), 906–896.

Soltani, S., Modarres, R., and Eslamian, S. (2007). The use of time series modeling for the determination of rainfall climates of Iran. *International Journal of Climatology*, 27, 819–829.

Zhang, J., Zhao, Z., Xue, Y., Chen, Z., Ma, X., and Zhou, Q. (2017). Time series analysis. *Journal of Physics and Chemistry of Solids*, 4, 269–285.

Zhongda, T., Shujiang, L., Yanhong, W., and Yi, S. J. C. (2017). A prediction method based on wavelet transform and multiple models fusion for chaotic time series. *Chaos, Solitons and Fractals*, 98, 158–172.

Zhou, Z., Wang, L., Lin, A., Zhang, M., and Niu, Z. (2018). Innovative trend analysis of solar radiation in China during. *Renewable Energy*, 119, 1962–2015.

22 Irrigation with Superabsorbent Polymer under Drought Stress Condition in a Temperate Arid Region

Davoud Khodadadi Dehkordi and Saeid Eslamian

22.1 INTRODUCTION

Agriculture has been negatively influenced by low obtainable water because of climate change, making water stress conditions for economically crucial plants such as legumes (Ma et al., 2017). Many countries have inadequate water supplies to meet their current urban, environmental, and agricultural needs (Valipour et al., 2015). In the face of the increased water scarcity, population and water demands continue to grow (Postel et al., 1996; Bouwer, 2002). The challenge is to grow enough food for 2 billion more people over the next 50 years while supplying growing urban and environmental needs for water (Gordon et al., 2005; Gupta and Deshpande, 2004). Some analysts have estimated that 60% of the added food required will come from irrigation (Plusquellec, 2002). Raising food production to support this larger world population requires sustaining the improved performance of irrigation (Oster and Wichelns, 2003; Sarwar and Perry, 2002; Ward and Velazquez, 2008). Water deficit is known as the most important limiting factor in agricultural products, especially in the arid and semi-arid regions (Khodadadi Dehkordi, 2016). As Iran is a major region consisting of arid and semi-arid areas with limited water resources, in case the minimum plant water use is not maintained, the plant would experience drought stress and the products would suffer irreparable losses (Lafitte, 2002; Boroomand Nasab et al., 2010; Hooshmand et al., 2019; Neissi et al., 2019a; Neissi et al., 2019b; Neissi et al., 2020). Iran is extremely exposed to the negative influences of climate change. One way for optimal use of the water resources and their preservation is the use of superabsorbent polymers (SAPs) which not only provide conditions for improved product quality but also result in increased water consumption efficiency in the arid and semi-arid areas (Fazeli Rostampour et al., 2011). SAPs could absorb and store water up to several times their weight. Due to the drying up of the environment, the water retained in the superabsorbent polymer is gradually discharged and thus the soil would remain moist for a long time without needing further irrigation. This property has great importance in confronting water shortage and reducing the harmful effects of drought stress in plantations (Haghighi et al., 2014; Wu et al., 2008). SAPs cause water retention in the soil and reduce the number of irrigation frequencies up to 50% (Nazarli et al., 2010). SAPs have been established as a soil conditioner to reduce soil water loss and increase crop yield. They are hydrophilic networks that can absorb and retain 1000 times more water or aqueous solutions than their original size and weight (Sojka and Entry, 2000). Thus, the application of SAPs to soil may increase water-holding capacities and nutrient utilization efficiency (Lentz and Sojka, 1994; Lentz et al., 1998) and reduce water loss (Al-Omran and Al-Harbi, 1997). SAPs are used in soil to create a water reserve, near the rhizosphere zone (roots) and benefit agriculture (Zohuriaan-Mehr and Kabirim, 2008; Han et al., 2010). Because of the water resource crisis, water-saving

DOI: 10.1201/9781003473398-27

agriculture is essential for the sustainable development of human societies. Furthermore, droughts are predicted to be increasingly severe due to climate change (Gornall et al., 2010). SAPs are moderately cross-linked 3D hydrophilic network polymers that can absorb and conserve considerable amounts of aqueous fluids even under certain heat or pressure. Because of the unique properties superior to the conventional absorbents, SAPs have found potential application in many fields such as agriculture (Karadag et al., 2000; Liu et al., 2006; Puoci et al., 2008), hygiene products (Kosemund et al., 2009), wastewater treatment (Kasgoz and Durmus, 2008; Kasgoz et al., 2008; Wang et al., 2008), sealing material (Vogt et al., 2005), and drug delivery system (Sadeghi and Hosseinzadeh, 2008). Currently, the further extension of application domains of SAPs is limited because the practically available SAPs are mainly petroleum-based synthetic polymers with high production costs and poor environmental properties (Kiatkamjornwong et al., 2002). Hence, the development of multi-component SAPs derived from natural polymer and eco-friendly additives become the subject of great interest due to their unique commercial and environmental advantages (Kurita, 2001), and such materials have also been honored as the material families "in greening the 21st-century materials world" (Ray and Bousmina, 2005). Thus far, many natural polymers such as starch (Lanthong et al., 2006; Li et al., 2007), cellulose (Suo et al., 2007), chitosan (Mahdavinia et al., 2004; Zhang et al., 2007), guar gum (Wang and Wang, 2009), and gelatin (Pourjavadi et al., 2007) have been utilized for fabricating multicomponent SAPs, and it was concluded that the composition and preparation technologies of SAPs are the dominant factors affecting the properties of SAPs (Wang and Wang, 2010). Many kinds of materials have been used for preparing SAPs. In addition, most of the traditional water-absorbing materials are acrylic acid and acrylamide-based products. They have poor degradability. About 90% of SAPs are used in disposable products and most of them are disposed of by landfills or by incineration (Kiatkamjornwong et al., 2002). In addition, there will be an environmental problem with SAPs (Zhang et al., 2007, 2006). Meanwhile, it has a low absorption rate under the high concentration of electrolytes, undesirable water-keeping capacity, and high cost (Wang and Liu, 2004). In order to solve those problems, considerable attention has been paid to the naturally available resources such as polysaccharides and inorganic clay minerals (Ray and Bousmina, 2005; Li et al., 2007). It has good commercial and environmental value with the advantages of low-cost, renewable, and biodegradable polysaccharides for deriving SAPs (Pourjavadi and Mahdavinia, 2006; Yoshimura et al., 2005). Recently, a series of new SAPs characterized by eco-friendliness and biodegradability were made from some natural materials, such as starch, cellulose, chitosan (Lanthong et al., 2006; Peng et al., 2008; Farag and Al-Afaleq, 2002; Wu et al., 2008), which were used to react through radical graft polymerization with vinyl monomers and crosslinking agent (Ma et al., 2011). Teimouri and Sharifan (2013) evaluated the effect of two monovalent salts (NaCl and KCl) in different concentrations on the hydrate and dehydrate of some SAPs including Aquasorb, Stockosorb, Clophony, and A 200. The results showed that A 200 and Clophony had the most hydrate and dehydrate, respectively. SAPs minimize micronutrients from washing out to water tables and cause more efficient water consumption, reduction in irrigation costs and intervals by 50%, water stress and mechanical damages to transplants during transferring; providing plants with eventful moisture and nutrients (Abedi Koupai and Mesforoush, 2009) and improving plant viability, seed germination, ventilation, and root development. SAPs can increase the water-holding capacity of light soils and keep this capability for about 2 to 4 years (Khoram-Del, 1997). SAPs were introduced to the markets in the early 1960s, by the American Company of Union Carbide (Dexter and Miyamoto, 1959). The product absorbed water, 30 times as much as its weight, but did not last long and was sold to greenhouse retail markets. Soon it was determined that the product was unsuccessful in the market because of its low swell (high cost per unit of water held) and short life (Joao et al., 2007). In fact, materials having the capacity to absorb water 20 times more than their weights are considered superabsorbent polymers (Abedi Koupai and Sohrab, 2006). Hydrogels are 3-D networks of SAPs swelling in the aquatic environment. Because of their cross bonds, they tend to hold a part of the solvent in their structure instead of dissolving. Their performance is dependent on chemical properties such as molecular weight, and formation condition along with the chemical composition of soil's solution or irrigation water (Abedi Koupai and Asadkazemi, 2006; Abedi Koupai et al., 2008). Hydrophilic polymers are of three types, natural (polysaccharide derivatives), semi-artificial (cellulosic primitive derivatives), and artificial (Mikkelsen, 1999). Artificial polymers are used more than natural ones because has more stability against environmental breaking down (Peterson, 2002). SAPs do not threaten human life and the environment (Boatright et al., 1997; Shooshtarian et al., 2011). A famous SAP in Iran is Super AB A 200, made by Rahab Resin, licensed under the Polymer and Petrochemical Institute of Iran. This superabsorbent is a tripolymer of acrylamide, acrylic acid, and acrylate potassium that is shown in Figure 22.1 (Khodadadi Dehkordi et al., 2013).

Many studies have indicated that SAPs cause improvement in plant growth by increasing water holding capacity in soil (Boatright et al., 1997; De Varennes and Queda, 2005; Karimi et al., 2008) and delaying the duration to the wilting point in drought stress (Gehring and Lewis, 1980). Water-conserving by gels creates a buffered environment, which is effective in short-term drought tensions and can reduce the losses amount in the establishment phase of some plant species (Johnson and Leah, 1990). Li et al. (2014) evaluated two types of SAPs, Jaguar C (JC) and Jaguar S (JS), which were applied at 200 kg ha^{-1} by bulk and spraying treatments in a field trial to investigate their effects on winter wheat growth, the soil's physical properties, and microbial abundance and activity. It was found that the addition of SAPs promoted the formation of macro soil aggregates (particle size>0.25 mm) and soil bacterial abundance under winter wheat cultivation. SAPs also significantly increased the soil water content (SWC) and soil maximum hygroscopic moisture in the booting and filling stages but had no effects on the soil available water-holding capacity compared with the control in the filling stage. The effects of SAPs depended on the application strategy, as only the bulk JC treatment improved the wheat yield, soil microbial biomass carbon, and soil microbial respiration. The results showed that the application of SAPs did not lead to detectable adverse effects on the soil microbial community and might even enhance soil microbial activity. Fazeli-Rostampoor et al. (2011) reported that drought stress and applying SAP (Taravat A 200) had a significant effect on corn grain yield and water use efficiency (WUE). In this study, three different depths of irrigation were considered as the main treatment I_1, I_2, and I_3 as 100, 70, and 40% of the water requirement of plants, respectively, and different levels of SAP were used as secondary treatment including S_0, S_1, S_2, and S_3, equal to 0 (control), 35, 75, and 105 kg ha^{-1}, respectively. Most corn grain yield and WUE were related to I_1 and S_3 treatments and the least of them were related to I_3 and S_0 treatments. Zangooei-Nasab et al. (2012) reported that applying SAP (Stockosorb) had a significant effect on plant height, dry weight of aerial organs, root dry weight, and root length of the Saxaul plant. In this study, it was considered three different irrigation intervals including I_1, I_2, and I_3 as daily, 3-day, and 5-day, respectively, and different levels of SAP including S_0, S_1, S_2, and S_3, equal to 0 (control), 0.1, 0.2, 0.3, and 0.4 weight percent, respectively. The most impact of SAP was related to 0.4% treatment had not any significant difference with 0.3% treatment and the most effect of irrigation interval was related to three-day treatment. Rahmani et al. (2010) evaluated the effects of different levels of water deficiency stress and SAP (Super AB A 200) on yield, antioxidant enzyme activity, and cell membrane stability in mustard. The treatments were five levels of water deficiency stress including irrigation after 80 mm evaporation from a class A evaporation container, cut irrigation from the booting stage, flowering stage, slicing stage, and grain-filling stage. Besides, three levels

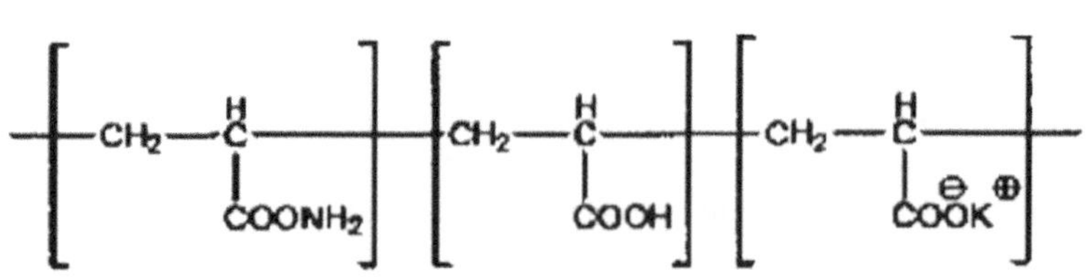

FIGURE 22.1   Chemical structure of Super AB A 200 SAP.

of SAP included the amount of 0, 5, and 7 weight percentages. The results showed that the effects of water deficiency stress and SAP were significant. The mutual impact of water deficiency stress and SAP was not significant for any treatments. Normal irrigation and use of 7% SAP with a mean of 5.803 t h^{-1} and 25.78 t h^{-1} showed the most seed and biological performance, respectively. Cut irrigation from the booting stage without applying SAP with a mean of 206.1 mg.protein^{-1} and 5231 mg.protein^{-1} showed the most activity of catalase and superoxide dismutase and with a mean of 2683 Ls cm^{-1} showed the least cell membrane stability, respectively. Banej Shafie et al. (2012) evaluated the effects of SAP (Manufactured by Flowergel Co., of the Netherlands) application and irrigation interval on the growth of pistachio seedlings. In this study, the treatments were three levels of SAP (0, 50, and 100 gr), irrigation amount in two levels (5, and 10 l), and three levels of irrigation interval (10, 20, and 30 days). The results showed that using 50 gr SAP halved the irrigation amount (5 l) in 10-day period intervals. Using 100 gr SAP with 10 l water, the increased period of irrigation from 10 to 20 days. Also, the SAP application increased the height growth and diameter at collar growth compared with the controls. It was concluded that SAP could diminish the water consumption and irrigation frequency by 50%. Khodadadi Dehkordi (2020) assessed the impacts of the varied irrigation regimes on the yield, irrigation water productivity (IWP), and crop water productivity (CWP) of *Arachis hypogaea* L. under pressurized irrigation in South West of Iran (Hamidiyeh, Khuzestan province, Iran). The test was performed in a split-plot pattern. The main treatments were three irrigation frequencies (Fs) based on the varied levels of cumulative pan evaporation (CPE) amounts (F_1: 35 mm; F_2: 60 mm; F_3: 85 mm of CPE). The sub-treatments were seven irrigation water levels (Is) (I_1=60%, I_2=85%, I_3=100%, I_4=135% (of actual evapotranspiration (ET_a)), I_5=PRZD$_{60}$, I_6=PRZD$_{85}$, and I_7=PRZD$_{100}$ treatments) with four replications. The partial root-zone drying (PRZD) treatments received 60, 85, and 100% of the actual evapotranspiration (ET_a), respectively, from replacing laterals. Crops were planted on June 1, 2017, and June 1, 2018. The consequences indicated that the F_2I_4 treatment in both years showed the greatest peanut yield (PY). Besides, the PRZD method for avoiding greater PY loss was a proper and favorable option compared with traditional deficit irrigation (DI). However, PY declined by an average of 11.1 and 3.4% as compared to the F_2I_4 treatment in 2017 and 2018, respectively, but the F_2I_7 treatment got about 16.3 and 16.1% lower water value than the F_2I_4 treatment. PRZD treatments (I_5 and I_6) have more effective amended PY than deficit irrigation treatments (I_1 and I_2) in both growing seasons whereas these treatments got the same amount of moisture for all Fs (F_1, F_2, and F_3). In the end, crop water productivity and IWP amounts also declined with rising irrigation frequency.

The scientific name for the navy bean is *Phaseolus vulgaris* L. which belongs to the Fabaceae plant family. It is an annual plant in the family of beans and in addition to its effect on soil fertility, its remains could be well stored in silos. It is also a good source of protein. Navy bean is a valuable plant rich in protein, carbohydrates, fat, minerals, and vitamins. It could be used both for human and animal consumption it is also used in the pharmaceutical industry (Shukla and Dixit, 1996). This plant is one of the popular productions in Khuzestan province, Iran, and is cultivated by farmers mostly. Amiri Deh Ahmadi et al. (2010) in research concluded that the water stress at the flowering stage of the chickpea plant reduced the grain yield (GY), relative growth rate, plant growth rate, pure photosynthesis rate, and increased the leaf area. Allahyari et al. (2013) reported that the SAPs had a significant effect on the increase of the biologic yield, the number of pods in the chickpea plant, and the 100-grain weight in comparison with the control treatment. Rajabi et al. (2012) reported that the effects of salicylic acid and superabsorbent and their reciprocal effects on the number of pods in the plant, the 1000-grain weight, the GY, and the biologic yield of the chickpea plant are significant. Abhari et al. (2017) through investigating the effect of superabsorbent on the yield and the yield components of the chickpea plant in drought stress conditions, reported that the use of superabsorbent to achieve a desirable economic performance is the best possible way. Teimouri et al. (2013) through investigating the effects of drought stress and the superabsorbent on the physiologic properties of the dry bean plant reported that the drought stress and superabsorbent both influenced the bean physiologic properties. So that they reported the highest rates of leaf area and the relative water content values belonged to the application of superabsorbent. Also, the use of SAPs caused increased biological yield, number of pods in a plant, and 100-grain weight in comparison with the control treatment. Abbaslou et al. (2014) reported that drought stress had a significant effect on the reduction of leaf area index (LAI), photosynthesis rate, stomatal conduction, and total chlorophyll of a chickpea plant. In arid and semi-arid regions, usually identified by increased evapotranspiration rate, extensive soil salinity, and restricted water supply, water use efficiency needs to be increased (Lo Bianco et al., 2012). This condition needs more accurate control levels of irrigation water (IW) to be examined which could store more water without losing production (Ro, 2001). Therefore, it is crucial to apply the deficit irrigation to achieve a better ratio of IW amount to optimum yield. In previous studies, the irrigation method was based on surface irrigation; however, few have assessed the effects of superabsorbent polymer and DI in drip irrigation systems (Khodadadi Dehkordi, 2016). This chapter aims to evaluate the superabsorbent polymer effect on WUE and grain yield of navy beans in drip irrigation systems under drought-stress conditions.

22.2 MATERIALS AND METHODS

22.2.1 EXPERIMENTAL DETAILS

The experimental farm was conducted (31°48′30′ ′N and 48°46′15′ ′E, and an elevation of 11 m) in Hamidiyeh, Khuzestan province, Iran. Hamidiyeh has a hot and relatively arid climate and has a hot summer and a Mediterranean winter. The average annual precipitation is about 210 ml and the average temperature is about 5 in the winter and over 50°C in the summer. Table 22.1 shows the physical and chemical properties of the tested soil and Table 22.2 shows the chemical properties of the IW.

The navy bean cultivar used in this study was named *Dorsa*. The growing of this cultivar, characterized by a hot and dry climate, is only related to irrigation. Because the value of precipitation is not enough to provide the necessary water supply for irrigation. The grains were hand sown on the rows with 3 m long and 2 m apart on June 22, 2017–2018 (two growing seasons). Before sowing, a drip irrigation system was located at each row with 25 cm drippers apart. The value of water used in every irrigation event has been recorded by a flow meter. The amount of 30 mm for IW was given after sowing and after the appearance of crops, they were thinned to stable one plant every 25 cm in the rows (12 crops for row) and, in front of each dripper an individual navy bean crop was located.

TABLE 22.1

The Physical and Chemical Properties of the Soil

Soluble potassium (mg kg⁻¹)	Soluble phosphorus (mg kg⁻¹)	Bulk density (g cm⁻³)	pH	Total nitrogen (g kg⁻¹)	Soil organic matter (g kg⁻¹)	EC (dS.m⁻¹)	Soil texture	Size of the soil particles (%)		
								Sand	Silt	Clay
353	50.6	1.59	7.4	3.3	45.7	6.85	Loam	34	44	22

TABLE 22.2

The Chemical Properties of the IW

EC (dS.m⁻¹)	Na (meq.l⁻¹)	Ca (meq.l⁻¹)	Mg (meq.l⁻¹)	SAR
1.1	5.5	4.4	2.5	2.96

After the appearance of plants, to optimize IW usage and decrease water use, Super-AB-A-200 polymer (Table 22.3) was combined with the soil at a depth of 20 cm near the crop. This polymer is a granular type and is produced by Rahab Resin Co. with product license holding of Iran Polymer and Petrochemical Institute (Rahab Resin Co, 2016). This hydrophilic polymer is a tripolymer of acrylamide, acrylic acid, and acrylate potassium.

The experimental plan was conducted as a randomized complete block design with four replicates (Figure 22.2).

For approving the efficacy of superabsorbent polymer, only a few irrigation treatments with different levels of Super-AB-A-200 polymer were considered: treatment I_1S_0, including 100% ETc demand as irrigation amount and with 0.0 g SAP (I_1:100% – S_0:0.0 g SAP); Treatment I_2S_1, including 75% ETc demand as irrigation amount and with 7.0 g Super-AB-A-200 polymer for plant (I_2:75% – S_1:7 g SAP); Treatment I_3S_2, including 55% ETc demand as irrigation amount and with 14.0 g Super-AB-A-200 polymer for plant (I_3:55% – S_2:14 g SAP).

TABLE 22.3
The Properties of Super-AB-A-200 polymer

Characteristics	Super-AB-A-200 polymer
Shape	granular
Density	1.4–1.5 (gr.cm^{-3})
Size of particles	50–150 (µm)
Maximum stability in soil	7 (year)
Practical capacity of water uptake	220 (g.g^{-1})

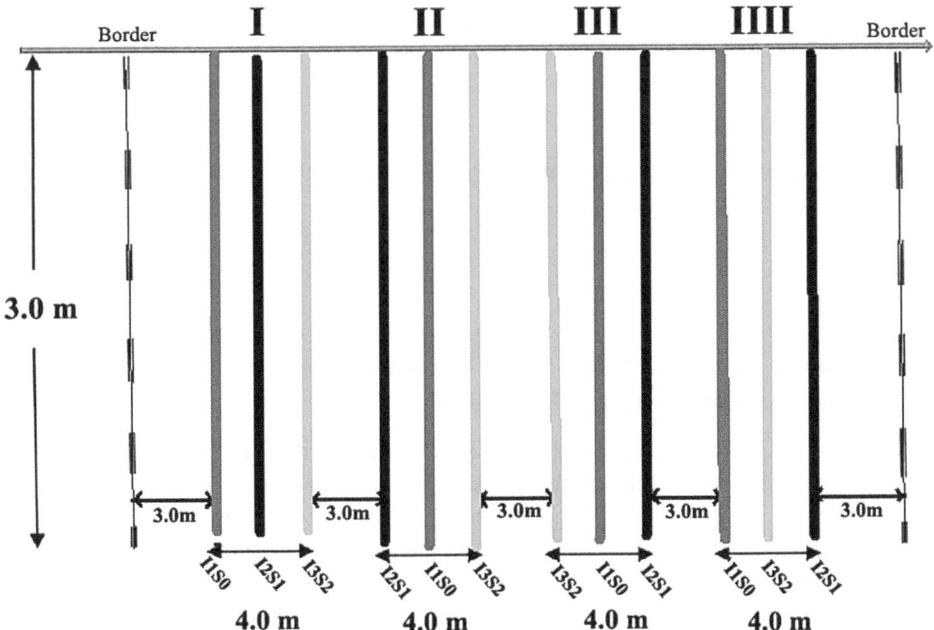

FIGURE 22.2 Layout of the plan (without scale). I_1S_0, I_2S_1, and I_3S_2 present navy bean rows with 0.0, 7.0, and 14.0 g of Super-AB-A-200 polymer in the soil and with IW treatments related to 100%, 75% and 55% ET$_c$ demand, respectively.

22.2.2 Assessment Tasks

The mean daily reference crop evapotranspiration (ET_0) was estimated using the Penman–Monteith method (Allen, 1998) and the meteorological data obtained by the Ahvaz weather station that was near the farm. For determination of the crop potential evapotranspiration (ET_c) or crop water requirement were used ET_0 multiplied by a crop coefficient (K_c) (Allen, 1998). The applied irrigation water (AIW) in each irrigation during the vegetative stage to provide 100% ET_c demand was calculated using the following equation (Satriani et al., 2018; Allen, 1998):

$$AIW = \left(\frac{SWD + ETc}{Ea} \right) \tag{22.1}$$

where
    AIW: Applied irrigation water (mm),
    Ea: Application efficiency (%), considered by 90% for drip irrigation,
    ETc: Crop evapotranspiration (mm day^{-1}),
    SWD: Soil water deficiency one day ago (mm day^{-1}).

The soil water content was monitored using the multiplexing TDR system (HandiTrase Soilmoisture Meter w/FCT Probe, model 6085K3, USA). TDR system was remotely regulated by a computer. In the center of the block rows, twelve 40 cm-long TDR waveguides were fixed vertically in the 5–45 cm depths of the soil surface. The soil surface layer is open to soil water (SW) variations, therefore, the TDR waveguides were fixed after the first 5 cm of soil especially due to perfect insulation (Satriani et al., 2018). For measuring the GY of the navy bean plant, the pods of 12 crops were removed from each row during harvest and after abandoning the crops to dry on the farm which occurred on October 13, 2017–2018. Irrigation WUE was determined as GY divided by applied IW all along the entire growing season. For determining the agricultural water productivity (AWP), GY was divided by the entire water consumption (IW plus precipitations) (Abhari et al., 2017; Molden et al., 2010). The results presented in the Figures and Tables were the mean values of the two growing seasons.

22.2.4 Statistical Analysis

The data analysis was performed using SPSS 22.0 software. Variance analysis (ANOVA) was used to check the variations between the treatments used. P < 0.05 was considered for determining statistically significant levels

22.3 RESULTS AND DISCUSSION

22.3.1 Variations of Soil Moisture and ET_0 under Varied Water and Superabsorbent Treatments

Readily available water (RAW) for navy bean plants that were cultivated on loam soil and with the highest root depth of almost 80 cm was 51.48 mm before moisture stress (Allen, 1998). Thus, due to this amount, the navy bean plant was irrigated when the SWD became higher or equal to the readily available water. The initial SW was supposed to be at field capacity (FC) because extensive irrigation occurred directly after sowing (IW depth was 30 mm equal to 200 m^3 ha^{-1}). This conveyed that soil water deficiency was 0.0 at the start of the growing cycle and this presumption was approved by the SW assessment. According to Equation (22.1), the total value of AIW was 1015 mm (related to 100% ETc demand), 761.25 mm (related to 75% ETc demand), 558.25 mm (related to 55% ETc demand), for I_1S_0, I_2S_1, and I_3S_2 treatments, respectively. Table 22.4 shows the monthly time series of rainfall, temperatures, and IW amount for the *Dorsa* cultivar.

TABLE 22.4

Daily Mean Estimated ET$_0$, Precipitation, Daily Mean Temperature, and IW Amount For Navy Bean Plant

Month	Precipitation (mm)	The mean of daily temperature (°C)			The mean of daily estimated ET$_0$ (mm day^{-1})	Irrigation 100% of ETc demand (mm)
		Min.	Max.	Average		
July	0	27.9	50.9	40.3	11.0	273
August	0	27.9	49.4	39.1	10.7	325
September	0	22.1	47.6	34.8	9.7	327
October	0	14.4	42.9	28.2	7.1	90

Figure 22.3 indicates the cumulative ET$_0$ and cumulative IW consumed for the navy bean plant in the whole treatment. All along the vegetative stage and particularly after appearance, daily ET$_c$ rose significantly up to nearly 80 days after sowing.

Soil water content is directly dependent on the ET procedure and the trends of SWC related to the whole treatments are indicated in Figure 22.4. Allen (1998) showed that the optimum conditions of SW are the actual field situations in well-administered fields. In this experiment, despite the water supply being lower, the SW was not significantly decreased in the treatments under DI. In correspondence with the reports of Abhari et al. (2017); Dabhi et al. (2013); Parvathy et al. (2014); Satriani et al. (2018) and Khodadadi Dehkordi (2016), the same trend could be because of the fact that the superabsorbent polymers take up water and then gradually release it back to the soil to balance the lower water reservoir. Other studies have verified that the superabsorbent polymer application in water deficit situations has reduced the adverse influence of DI (Satriani et al., 2018; Islam et al., 2011; Khodadadi Dehkordi, 2016; Sayyari and Ghanbari, 2012). Besides, Khodadadi Dehkordi (2016) and Fallahi et al. (2015) reported that the application of superabsorbent polymer could increase the plants' irrigation intervals. In addition, Wang and Wang (2010) and Khodadadi Dehkordi (2016) have shown that the water evaporation rate was decreased in sandy soil with adding superabsorbent polymer compared to the soil without SAP. Khodadadi Dehkordi (2016); Dorraji et al. (2010) and Akhter et al. (2004) reported similar effects of superabsorbent polymer on the water-holding properties of loamy and sandy soils.

22.3.2 Navy Bean Grain Yield

The navy bean plants should be irrigated all along the growing season and was considered reasonable to investigate the influence of superabsorbent polymer and IW on GY. According to the results, the highest amount of GY was achieved when the plant was irrigated with the value of IW by 100 and 75% of ETc demand, while a lower amount was achieved when the plant was irrigated with the value of IW by 55% of ET$_c$ demand. Besides, the soil treated with superabsorbent polymer, in the mild DI treatment (I$_2$S$_1$), made it possible to obtain the same GY as the complete irrigation treatment (I$_1$S$_0$) and there was no significant difference between them. This could related to the fact that superabsorbent polymer is able to retain water and nutrients and release them in water-stress situations, rebuilding the optimum conditions of SW appropriate for crop growth. This result is confirmed by Khodadadi Dehkordi (2016); Abhari et al. (2017) and Satriani et al. (2018). Figure 22.5 indicates a significant second-degree polynomial relation between navy bean GY and seasonal IW applied.

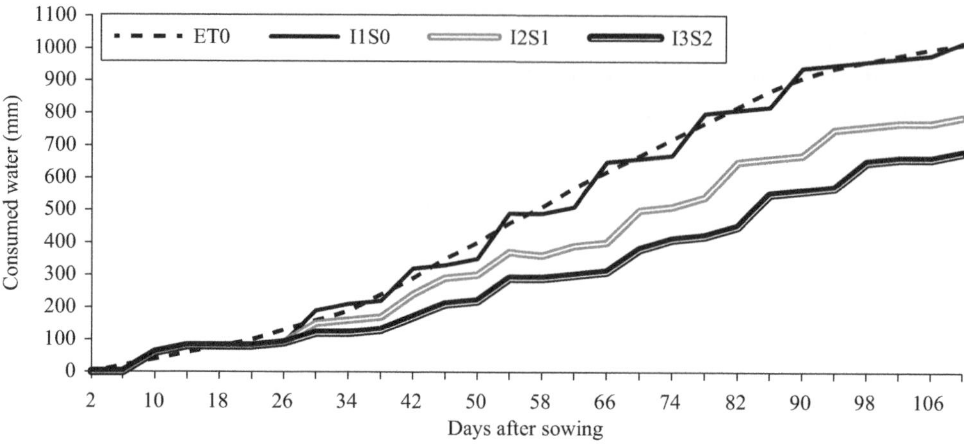

FIGURE 22.3 Cumulative ET_0 and cumulative IW consumed for navy bean plant in the whole treatments all along the vegetative stage.

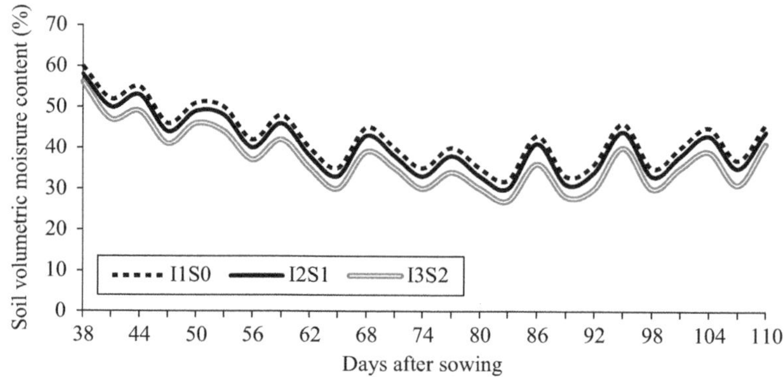

FIGURE 22.4 Soil volumetric water content in navy bean rows with varied superabsorbent polymer and water values.

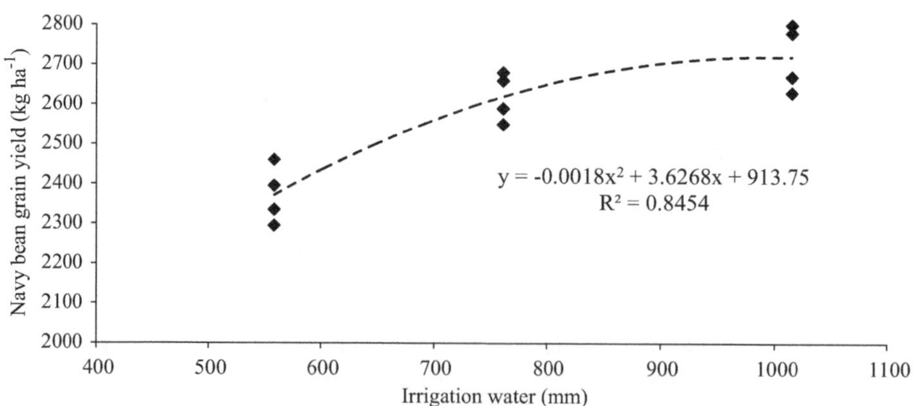

FIGURE 22.5 The relation between navy bean GY and seasonal IW applied.

TABLE 22.5

Navy Bean GY, Irrigation, and Precipitation Values, WUE, and AWP data kg (ha mm^{-1})

Irrigation and superabsorbent polymer treatments	Navy bean GY (kg. ha^{-1})	Irrigation + precipitation (mm)	WUE	AWP
I_1S_0	2750a	1015+0	2.71c	2.71c
I_2S_1	2680a	761.25+0	3.52b	3.52b
I_3S_2	2395b	558.25+0	4.29a	4.29a

This relation is curvilinear and indicates that navy bean GY rises with the IW amount, and when a particular level is achieved, it does not increase anymore. Research by Khodadadi Dehkordi (2015, 2016) approved these consequences. He investigated the application of superabsorbent polymer on corn yield under the DI conditions and reported that by enhancing the intensity of drought stress, corn GY decreased, however, there was no significant variation between 100 and 75% of ET_c demand treatments for the corn crop, with 0.0 and 45.0 g.m^{-2} of superabsorbent polymer respectively. On the contrary, the minimum GY was achieved in the treatments with 50% of ET_c demand for the crops. The results shown in Table 22.5 indicate that the maximum amounts of IWUE and AWP with significantly varied averages (P < 0.05) were achieved in the I_2S_1 and I_3S_2 treatments compared to the control (I_1S_0 treatment).

Khodadadi Dehkordi (2016) and Zhang et al. (2008) reported that IWUE was conversely related to the total IW requirement all along the growing cycle, and for higher IWUE, the water would be decreased under the IW demand for the maximum GY. Thus, the mix of mild DI and superabsorbent polymer (I_2S_1), that leads to an increase in the AWP, was an acceptable policy to optimize navy bean GY and water utilization. These achieved consequences were approved by the researches on maize, sorghum, cotton, peanut, bean, and chickpea (Khodadadi Dehkordi, 2015, 2016; Abhari et al., 2017; Hazrati et al., 2017; Timouri et al., 2013; Mazen et al., 2015; Najafinezhad et al., 2014; Satriani et al., 2018).

22.4 CONCLUSIONS

The findings in this study indicated that using a superabsorbent polymer for retaining IW conserved the GY and at the same time decreased the IW value. Applying Super-AB-A200 polymer for navy bean plants in drought stress conditions conserved a suitable SW capability. The maximum amounts of navy bean GY under the varied levels of superabsorbent polymer and IW values were achieved when the navy bean plant was irrigated with the IW value by 100 and 75% of ET_c demand, but the mix of DI and superabsorbent polymer led to the highest AWP. Thus, the use of Super-AB-A200 polymer in the irrigated plants should be promoted, also based on the achieved consequences and as shown in the text, approved by other researchers on the varied plants. This study has a strong interest in the hot and dry regions where the aridity situations in the summer season permit planning the test for investigating the effect of the SAP on the varied cultivations in the experimental farms. It is suggested that the new experiments for evaluating the optimum value of applied superabsorbent polymer in the soil, the SAP type, and its levels depending on the factors like the plant and soil type are needed.

ABBREVIATIONS

AIW: Applied irrigation water;
AWP: Agricultural water productivity;

DI: Deficit irrigation;
ET_c: potential crop evapotranspiration;
ET_0: Reference crop evapotranspiration;
E_a: Application efficiency;
FC: Field capacity;
GY: Grain yield;
IW: Irrigation water;
IWUE: Irrigation water use efficiency;
K_c: Crop coefficient;
LAI: Leaf area index;
RAW: Readily available water;
SAP: Superabsorbent polymer;
SWD: Soil water deficiency;
SWC: Soil water content;
SW: Soil water;
WUE: Water use efficiency.

REFERENCES

Abbaslou, L., Kazemeini, S. A. R., Edalat, M., and Dadkhodai, A. (2014). Effect of drought stress and planting manner on the some physiologic and biologic characteristics of two *Cicer arietinum* L. cultivar. *Journal of Crop Improvement*, 16(4), 933–943. (In Persian)

Abedi Koupai, J., and Asadkazemi, J. (2006). Effects of a hydrophilic polymer on the field performance of an ornamental plant (*Cupressus arizonica*) under reduced irrigation regimes. *Iranian Polymer Journal*, 15, 715–725.

Abedi Koupai, J., Eslamian, S. S., and Asad Kazemi, J. (2008). Enhancing the available water content in unsaturated soil zone using hydrogel, to improve plant growth indices. *Ecohydrology and Hydrobiology*, 8(1), 3–11.

Abedi Koupai, J., and Mesforoush, M. (2009). Evaluation of superabsorbent polymer application on yield, water and fertilizer use efficiency in cucumber (*Cucumis sativus*). *Journal of Irrigation and Drainage*, 2, 100–111. (In Persian)

Abedi Koupai, J., and Sohrab, F. (2006). Effect evaluation of superabsorbent application on water retention capacity and water potential in three soil textures. *Journal Science Technology Polymer*, 17, 163–173. (In Persian)

Abhari, A., Azizi, A., and Hareth Abadi, B. (2017). The effect of superabsorbent on pea yield under drought stress of the season end. *Journal of Crop Production*, 10(1), 191–202. (In Persian)

Akhter, J., Mahmood, K., Malik, K. A., Mardan, M., Ahmad, M., and Iqbal, M. M. (2004). Effects of hydrogel amendment on water storage of sandy loam and loam soils and seedling growth of barley wheat and chickpea. *Plant, Soil and Environment*, 50(10), 463–469.

Allahyari, S., Golchin, A., and Vaezi, A. R. (2013). Study on effect of super absorbent polymer application on yield and yield components of two chickpea cultivars under rainfed conditions. *Journal of Plant Production*, 20(1), 125–140. (In Persian)

Allen, R. G. (1998). *Crop Evapotranspiration: Guidelines for Computing Crop Water Requirements*. FAO Irrigation and Drainage, Paper 56, Rome, Italy.

Al-Omran, A. M., and Al-Harbi, A. R. (1997). Improvement of sandy soils with soil conditioners. In Wallace, A., and Terry, R. E. (Eds.), *Handbook of Soil Conditioners: Substances that Enhance the Physical Properties of Soil*. Marcel Dekker, Inc., New York, 363–384.

Amiri Deh Ahmadi, S. R., Parsa, M., Nezami, A., and Ganjeali, A. (2010). The effects of drought stress at different phenological stages on growth indices of chickpea (*Cicer arietinum* L.) in greenhouse conditions. *Iranian Journal of Pulses Research*, 1(2), 69–84. (In Persian)

Banej Shafie, A., Eshaghi-Rad, J., Alijanpour, A., and Pato, M. (2012). Effects of super-absorbent application and irrigation period on the growth of pistachio seedlings (*Pistacia atlantica*), (case study: Dr. Javanshir nursery, Piranshahr). *Iranian Journal of Forest*, 4, 101–112.

Boatright, J. L., Balint, D. E., Mackay, W. A., and Zajicek, J. M. (1997). Incorporation of a hydrophilic polymer into annual landscape beds. *Journal of Environmental Horticulture*, 15, 37–40.

Boroomand Nasab, S., Albaji, M., and Naseri, A. A. (2010). Investigation of different irrigation systems based on the parametric evaluation approach in Boneh Basht plain, Iran. *African Journal of Agricultural Research*, 5(5), 372–379.

Bouwer, H. (2002). Integrated water management for the 21st century: Problems and solutions. *Journal of Irrigation and Drainage Engineering*, 128, 193–202.

Dabhi, R., Bhatt, N., and Pandit, B. (2013). Superabsorbent polymers an innovative water saving technique for optimizing crop yield. *International Journal of Innovative Research in Science, Engineering and Technology*, 2, 5333–5340.

De Varennes, A. D., and Queda, C. (2005). Application of an insoluble polyacrylate polymer to copper-contaminated soil enhances plant growth and soil quality. *Soil Use and Management*, 21, 410–414.

Dexter, S. T., and Miyamoto, T. (1959). Acceleration of water uptake and germination of sugar beet seed balls by surface coatings of hydrophilic colloids. *Agronomy Journal*, 51, 388–389.

Dorraji, S. S., Golchin, A., and Ahmadi, S. (2010). The effects of hydrophilic polymer and soil salinity on corn growth in sandy and loamy soils. *Clean Soil Air Water*, 38, 584–591.

Fallahi, H. R., Taherpour-Kalantari, R., Aghhavani-Shajari, M., and Soltanzadeh, M. G. (2015). Effect of super absorbent polymer and irrigation deficit on water use efficiency, growth and yield of cotton. *Notulae Scientia Biologicae*, 7, 338–344.

Farag, S., and Al-Afaleq, E. I. (2002). Preparation and characterization of saponified delignified cellulose polyacrylonitrile-graft copolymer. *Carbohydrate Polymers*, 48, 1–5.

Fazeli Rostampour, M., Theghat al-Islami, M. G., and Mousavi, S. G. R. (2011). Effect of water stress and polymer (Superjazeb A200) on yield and water use efficiency of corn (*Zea mays* L.) in Birjand region. *Environmental Stresses in Crop Sciences*, 4(1), 11–19. (In Persian)

Gordon, L. J., Steffen, W., Jonsson, B. F., Folke, C., Falkenmark, M., and Johannessen, A. (2005). Human modification of global water vapor flows from the land surface. *Proceedings of the National Academy of Sciences of the United States of America*, 102, 7612–7617.

Gornall, J., Betts, R., Burke, E., Clark, R., Camp, J., Willett, K., and Wiltshire, A. (2010). Implications of climate change for agricultural productivity in the early twenty-first century. *Philosophical Transactions of the Royal Society B: Biological Sciences*, 365, 2973–2989.

Gehring, J. M., and Lewis, A. J. (1980). Effect of hydrogel on wilting and moisture stress of bedding plants. *Journal of the American Society for Horticultural Science*, 105, 511–513.

Gupta, S. K., and Deshpande, R. D. (2004). Water for India in 2050: First-order assessment of available options. *Current Science*, 86, 1216–1224.

Haghighi, M., Mozafarian, M., and Afifi-pour, Z. (2014). Impact evaluation of superabsorbent polymer and different levels of deficit irrigation on growth and some qualitative and quantitative properties of *Lycopersicum esculentum* L. *Journal of Horticulture Science*, 28(1), 125–133. (In Persian)

Han, Y. G., Yang, P. L., Luo, Y. P., Ren, S. M., Zhang, L. X., and Xu, L. (2010). Porosity change model for watered super absorbent polymer-treated soil. *Environmental Earth Sciences*, 61, 1197–1205.

Hazrati, S., Tahmasebi-Sarvestani, S., Mokhtassi-Bidgoli, A., Modarres-Sanavya, S. A. M., Mohammadi, H., and Nicola, S. (2017). Effects of zeolite and water stress on growth yield and chemical compositions of *Aloe vera* L. *Agricultural Water Management*, 181, 66–72.

Hooshmand, M., Albaji, M., and Zadeh Ansari, N. A. (2019). The effect of deficit irrigation on yield and yield components of greenhouse tomato (Solanum lycopersicum) in hydroponic culture in Ahvaz region, Iran. *Scientia Horticulturae*, 254, 84–90.

Islam, M. R., Xue, X., Mao, S., Ren, C., Eneji, A. E., and Hu, Y. (2011). Effects of water saving superabsorbent polymer on antioxidant enzyme activities and lipid peroxidation in oat (*Avena sativa* L.) under drought stress. *Journal of the Science of Food and Agriculture*, 91, 680–686.

Joao, C. M., Bordado, P., and Gomez, J. F. (2007). New technologies for effective forest fire fighting. *International Journal of Environmental Studies*, 64, 243–251.

Johnson, M. S., and Leah, A. (1990). Effects of superabsorbent polyacrylamides on efficiency of water use by crop seedlings. *Journal of the Science of Food and Agriculture*, 52, 431–434.

Khodadadi Dehkordi, D. (2015). Evaluation of superabsorbent application on corn yield under deficit irrigation. *International Journal of Agricultural and Biological Engineering*, 9(7), 806–810.

Khodadadi Dehkordi, D. (2016). The effects of superabsorbent polymers on soils and plants. *Pertanika Journal of Tropical Agricultural Science*, 39(3), 267–298.

Khodadadi Dehkordi, D. (2020). The effect of different irrigation treatments on yield and water productivity of *Arachis hypogaea* L. under semi-arid conditions in Iran. *Irrigation and Drainage*, 69(4), 646–657.

Khodadadi Dehkordi, D., Kashkuli, H. A., Naderi, A., and Shamsnia, S. A. (2013). Evaluation of deficit irrigation and superabsorbent hydrogel on some growth factors of SCKaroun701 corn in the climate of Khuzestan. *Advances in Environmental Biology*, 7, 527–534.

Karadagˇ, E., Saraydin, D., Caldiran, Y., and Guven, O. (2000). Swelling studies of copolymeric acrylamide/crotonic acid hydrogels as carriers for agricultural uses. *Polymers for Advanced Technologies*, 11, 59–68.

Karimi, A., Noushadi, M., and Ahmad-zadeh, M. (2008). Effect of water superabsorbent (Igita) amendment material on water soil, plant growth and irrigation intervals. *Journal of Science and Technology of Agriculture and Natural Resources*, 46, 403–414. (In Persian)

Kasgoz, H., and Durmus, A. (2008). Dye removal by a novel hydrogel-clay nanocomposite with enhanced swelling properties. *Polymers for Advanced Technologies*, 19, 838–845.

Kasgoz, H., Durmus, A., and Kasgoz, A. (2008). Enhanced swelling and adsorption properties of AAm-AMPSNa/clay hydrogel nanocomposites for heavy metal ion removal. *Polymers for Advanced Technologies*, 19, 213–220.

Khoram-Del, N. (1997). Effects of PR3005A as a Superabsorbent polymer on some physical properties of soils. M.A. Dissertation, University of Tarbiat Modaress, Department of Agriculture, Tehran. (In Persian)

Kiatkamjornwong, S., Mongkolsawat, K., and Sonsuk, M. (2002). Synthesis and property characterization of cassava starch grafted poly [acrylamide-co-(maleic acid)] superabsorbent via c-irradiation. *Polymer*, 43, 3915–3924.

Kosemund, K., Schlatter, H., Ochsenhirt, J. L., Krause, E. L., Marsman, D. S., and Erasala, G. N. (2009). Safety evaluation of superabsorbent baby diapers. *Regulatory Toxicology and Pharmacology*, 53, 81–89.

Kurita, K. (2001). Controlled functionalization of the polysaccharide chitin. *Progress in Polymer Science*, 26, 1921–1971.

Lafitte, R. (2002). Relationship between leaf relative water content during reproductive stage water deficit and grain formation in rice. *Field Crops Research*, 76(2–3), 165–174.

Lanthong, P., Nuisin, R., and Kiatkamjornwong, S. (2006). Graft copolymerization, characterization, and degradation of cassava starch-g-acrylamide/itaconic acid superabsorbents. *Carbohydrate Polymers*, 66, 229–245.

Lentz, R. D., and Sojka, R. E. (1994). Field result using polyacrylamide to manage furrow erosion and infiltration. *Soil Science*, 158, 274–282.

Lentz, R. D., Sojka, R. E., and Robbins, C. W. (1998). Reducing phosphorus losses from surface irrigated fields: Emerging polyacrylamide technology. *Journal of Environmental Quality*, 27, 305–312.

Liu, M. Z., Liang, R., Zhan, F. L., Liu, Z., and Niu, A. Z. (2006). Synthesis of a slow-release and superabsorbent nitrogen fertilizer and its properties. *Polymers for Advanced Technologies*, 47, 430–438.

Li, A., Zhang, J. P., and Wang, A. Q. (2007). Utilization of starch and clay for the preparation of superabsorbent composite. *Bioresource Technology*, 98, 327–333.

Li, X., He, J. Z., Hughes, J. M., Liu, Y. R., and Zheng, Y. M. (2014). Effects of super-absorbent polymers on a soil–wheat (Triticum aestivum L.) system in the field. *Applied Soil Ecology*, 73, 58–63.

Lo Bianco, R., Talluto, G., and Farina, V. (2012). Effects of partial root zone drying and rootstock vigor on dry matter partitioning of apple trees (*Malus domestica* cvar Pink Lady). *Journal of Agricultural Science*, 150, 75–86.

Ma, L., Ahuja, L. R., Islam, A., Trout, T. J., Saseendran, S. A., and Malone, R. W. (2017). Modeling yield and biomass responses of maize cultivars to climate change under full and deficit irrigation. *Agricultural Water Management*, 180, 88–98.

Ma, Z., Li, Q., Yue, Q., Gao, B., Xu, X., and Zhong, Q. (2011). Synthesis and characterization of a novel superabsorbent based on wheat straw. *Bio Resource Technology*, 102, 2853–2858.

Mahdavinia, G. R., Zohuriaan-Mehr, M. J., and Pourjavadi, A. (2004). Modified chitosan III, superabsorbency, salt- and pH-sensitivity of smart ampholytic hydrogels from chitosan-g-PAN. *Polymers for Advanced Technologies*, 15, 173–180. (In Persian)

Mazen, A. M., Radwan, D. E. M., and Ahmed, A. F. (2015). Growth responses of maize plants cultivated in sandy soil amended by different superabsorbent hydrogels. *Journal of Plant Nutrition*, 38, 325–337.

Mikkelsen, R. L. (1999). Using hydrophilic polymers to control nutrient release. *Nutrient Cycling in Agroecosystems*, 38, 53–59.

Molden, D., Oweis, T., Steduto, P., Bindraban, P., Hanjra, M. A., and Kijne, J. (2010). Improving agricultural water productivity: Between optimism and caution. *Agricultural Water Management*, 97, 528–535.

Najafinezhad, H., Tahmasebi Sarvestani, Z., Modarres Sanavy, S. A. M., and Naghavi, H. (2014). Evaluation of yield and some physiological changes in corn and sorghum under irrigation regimes and application of barley residue, zeolite and superabsorbent polymer. *Archives of Agronomy and Soil Science*, 61, 891–906.

Nazarli, H., Zardashti, M. R., Darvishzadeh, R., and Najafi, S. (2010). The effect of water stress and polymer on water use efficiency, yield and several morphological traits of sunflower under greenhouse conditions. *Notulae Scientia Biologicae*, 2(4), 53–58.

Neissi, L., Tishehzan, P., and Albaji, M. (2019a). Chemical assessment of surface water quality in upstream and downstream of Jare Dam, Khuzestan, Iran. *Environmental Earth Sciences*, 78(3), 83.

Neissi, L., Albaji, M., and Boroomand Nasab, S. (2019b). Site selection of different irrigation systems using an analytical hierarchy process integrated with GIS in a semi-arid region. *Water Resources Management*, 33(14), 4955–4967.

Neissi, L., Albaji, M., and Boroomand Nasab, S. (2020). Combination of GIS and AHP for site selection of pressurized irrigation systems in the Izeh plain, Iran. *Agricultural Water Management*, 231(2020), 106004.

Oster, J. D., and Wichelns, D. (2003). Economic and agronomic strategies to achieve sustainable irrigation. *Irrigation Science*, 22, 107–120.

Parvathy, P. C., Jyothi, A. N., John, K. S., and Sreekumar, J. (2014). Cassava starch based superabsorbent polymer as soil conditioner: Impact on soil physico-chemical and biological properties and plant growth. *Clean Soil Air Water*, 42(11), 1610–1617.

Peng, G., Xu, S. M., Peng, Y., Wang, J., and Zheng, L. (2008). A new amphoteric superabsorbent hydrogel based on sodium starch sulfate. *Bioresource Technology*, 99, 444–447.

Peterson, D. (2002). Hydrophilic polymers and uses in landscape. *Horticultural Science* 75, 10–16.

Plusquellec, H. (2002). Is the daunting challenge of irrigation achievable? *Irrigation and Drainage*, 51, 185–198.

Postel, S. L., Daily, G. C., and Ehrlich, P. R. (1996). Human appropriation of renewable fresh water. *Science*, 271, 785–788.

Pourjavadi, A., Hosseinzadeh, H., and Sadeghi, M. (2007). Synthesis, characterization and swelling behavior of gelatin-g-poly (sodium acrylate)/kaolin superabsorbent hydrogel composites. *Journal of Composite Materials*, 41, 2057–2069.

Pourjavadi, A., and Mahdavinia, G. R. (2006). Superabsorbency, pH-sensitivity and swelling kinetics of partially hydrolyzed chitosan-g-poly (acrylamide) hydrogels. *Turkish Journal of Chemistry*, 30, 595–608.

Puoci, F., Iemma, F., Spizzirri, U. G., Cirillo, G., Curcio, M., and Picci, N. (2008). Polymer in agriculture: A review. *American Journal of Agricultural and Biological Sciences*, 3, 299–314.

Rahab Resin Co. (2016). Rahab Resin Company. Tehran, Iran. http://www.bizearch.com/company/Rahab _Resin_Co_280864.htm.

Rahmani, M., Habibi, D., Shiranirad, A. H., Daneshian, J., Valadabadi, S. A. R., Mashhadi-Akbar Boujar, M., and Khalatbari, A. H. (2010). The effect of superabsorbent polymer on yield, antioxidant enzymes (catalase and superoxide dismutase) activity and cell membrane stability in mustard under water deficiency stress condition. *Plant and Ecosystem*, 6, 19–38. (In Persian)

Rajabi, L., Sajedi, N. V., and Roshandel, M. (2012). Yield and yield components reaction of rainfed pea to salicylic acid and superabsorbent polymer. *Journal of Crop Production Research*, 4(4), 343–353. (In Persian)

Ray, S. S., and Bousmina, M. (2005). Biodegradable polymers and their layered silicate nanocomposites: In greening the 21st century materials world. *Progress in Materials Science*, 50, 962–1079.

Ro, H. M. (2001). Water use of young "Fuji" apple trees at three soil moisture regimes in drainage lysimeters. *Agricultural Water Management*, 50, 185–196.

Sadeghi, M., and Hosseinzadeh, H. J. (2008). Synthesis of starch-poly (sodium acrylate-co-acrylamide) superabsorbent hydrogel with salt and pH-responsiveness properties as a drug delivery system. *Journal of Bioactive and Compatible Polymers*, 23, 381–404.

Sarwar, A., and Perry, C. (2002). Increasing water productivity through deficit irrigation: Evidence from the Indus plains of Pakistan. *Irrigation and Drainage*, 51, 87–92.

Satriani, A., Catalano, M., and Scalcione, E. (2018). The role of superabsorbent hydrogel in bean crop cultivation under deficit irrigation conditions: A case-study in Southern Italy. *Agricultural Water Management*, 195, 114–119.

Sayyari, M., and Ghanbari, F. (2012). Effects of super absorbent polymer A200 on the growth, yield and some physiological responses in sweet pepper (*Capsicum annuum* L.) under various irrigation regimes. *Journal of Agriculture and Food Research*, 1, 1–11.

Shooshtarian, S., Abedi Koupai, J., and Tehranifar, A. (2011). Evaluation of application of superabsorbent polymers in green space of arid and semi-arid regions with emphasis on Iran. *Journal of Biodiversity and Ecological Sciences*, 1, 258–269.

Shukla, S. K., and Dixit, R. S. (1996). Nutrient and plant population management in summer green gram (*Phaseolus radiates*). *Indian Journal of Agronomy*, 41(1), 78–83.

Sojka, R. E., and Entry, J. A. (2000). Influence of polyacrylamide application to soil on movement of microorganisms in runoff water. *Environmental Pollution*, 108, 405–412.

Suo, A. L., Qian, J. M., Yao, Y., and Zhang, W. G. (2007). Synthesis and properties of carboxymethyl cellulose-graft-poly (acrylic acid-co-acrylamide) as a novel cellulose-based superabsorbent. *Journal of Applied Polymer Science*, 103, 1382–1388.

Teimouri, A., Shiroui, H., and Mohamadi Babazeidi, H. (2013). Evaluation of drought stress and superabsorbent effects on physiologic properties of fava bean varieties. In *The First International Conference of Planning, Conservation, Environmental Protection and Sustainable Development*, Hamedan, Iran. (In Persian)

Teimouri, F., and Sharifan, H. (2013). The evaluation of monovalent salts effect on superabsorbent hydrogels hydrate. In *The First National Conference in Drainage and Sustainable Agriculture*, Tehran, Iran. (In Persian)

Valipour, M., Gholami Sefidkouhi, M. A., and Eslamian, S. (2015). Surface irrigation simulation models: A review. *International Journal of Hydrology Science and Technology*, 5(1), 51–70.

Wang, L., Zhang, J. P., and Wang, A. Q. (2008). Removal of methylene blue from aqueous solution using chitosan-g-poly (acrylic acid)/montmorillonite superadsorbent nanocomposite. *Colloid Surface A*, 322, 47–53.

Wang, W., and Wang, A. (2010). Preparation, swelling and water retention properties of crosslinked superabsorbent hydrogels based on guar gum. *Advanced Materials Research*, 96, 117–182.

Wang, W. B., and Wang, A. Q. (2009). Preparation, characterization and properties of superabsorbent nanocomposites based on natural guar gum and modified rectorite. *Carbohydrate Polymers*, 77, 891–897.

Wang, Z., and Liu, Z. (2004). Properties of super water absorbent polymers and their applications in agriculture. *Chinese Journal of Soil Science*, 35, 352–356.

Ward, F. A., and Velazquez, M. P. (2008). Water conservation in irrigation can increase water use. *Proceedings of the National Academy of Sciences*, 105, 18215–18220.

Wu, L., Liu, M., and Liang, R. (2008). Preparation and properties of a double-coated slow-release NPK compound fertilizer with superabsorbent and water retention. *Bioresource Technology*, 99(3), 547–554.

Yoshimura, T., Uchikoshi, I., Yoshiura, Y., and Fujioka, R. (2005). Synthesis and characterization of novel biodegradable superabsorbent hydrogels based on chitin and succinic anhydride. *Carbohydrate Polymers*, 61, 322–326.

Zangooei-Nasab, S., Imami, H., Astaraei, A. R., and Yari, A. R. (2012). The effects of stockosorb hydrogel and irrigation on growth and establishment of saxaul plant. In *The First National Conference on Farm Water Management*, Karaj, Iran. (In Persian)

Zhang, J., Li, A., and Wang, A. (2006). Study on superabsorbent composite. VI. Preparation, characterization and swelling behaviors of starch phosphate-graft-acrylamide/attapulgite superabsorbent composite. *Carbohydrate Polymers*, 65, 150–158.

Zhang, J. P., Wang, Q., and Wang, A. Q. (2007). Synthesis and characterization of chitosan-g-poly (acrylic acid)/attapulgite superabsorbent composites. *Carbohydrate Polymers*, 68, 367–374.

Zhang, X., Chen, S., Sun, H., Pei, D., and Wang, Y. (2008). Dry matter, harvest index, grain yield and water use efficiency as affected by water supply in winter wheat. *Irrigation Science*, 27(1), 1–10.

Zohuriaan-Mehr, M. J., and Kabiri, K. (2008). Superabsorbent polymer materials: A review. *Iranian Polymer Journal*, 17, 451–477.

23 Wind Drift and Evaporation Losses from Sprinkler Irrigation in a Semi-Arid Climate

Abolfazl Nasseri, Ahmad Lotfi-Sadigh,
Saeid Eslamian, and Zekiye Şengül

23.1 INTRODUCTION

23.1.1 THE IMPORTANCE OF WIND DRIFT AND EVAPORATION LOSSES

Sprinkler irrigation is the distribution of water in the form of droplets on the ground surface. Droplets are made by the pressurized water flow through small orifices. Part of the drops are evaporated or drift from the farm due to meteorological and/or operational factors before reaching the ground surface. The loss of water in this manner is called wind drift and evaporation losses (WDEL). In sprinkler irrigation systems, WDEL is considered to be a major source of water loss. WDEL affects water application efficiency and distribution uniformity in sprinkler irrigation in the farm [8]. Therefore, the design and application of sprinkler irrigation systems depend on WDEL during irrigation.

23.1.2 DEFINITION OF COMPONENTS OF WDEL

WDEL consists of two separate components wind drift (WD) and evaporation loss (EL). In other words, WDEL can be written as:

$$WDEL = WD + EL \tag{23.1}$$

Wind drift component can be defined as the fraction of the volume of irrigation water that is carried outside the field by the wind, while evaporation loss can be defined as the fraction of the total sprinkler water converted into a vapor of the atmosphere [15]. Sprinkler water components comprising reached water to the ground surface and WDEL are shown in Figure 23.1.

23.1.3 EFFECTIVE FACTORS ON WDEL

To investigate the effective factors in WDEL, various experiments have been conducted across the globe (Table 23.1). Some important results and findings are presented below. Findings showed that meteorological variables, operational factors; and soil, crop, and water properties affect WDEL. Christiansen (1942) reported that WDEL ranges were from 10% to 42% for afternoon experiments, while was 4% for early morning experiments in California. He found that WDEL significantly correlated with the evaporative demand of the atmosphere while did not find any correlation between WDEL and meteorological variables [4]. Frost and Schwalen (1955) reported that WDEL varied directly with air temperature (T), wind speed (WS), and operational pressures (P); and varied inversely with relative humidity (RH) and nozzle diameter (D) [10]. They found that WDEL ranged

DOI: 10.1201/9781003473398-28

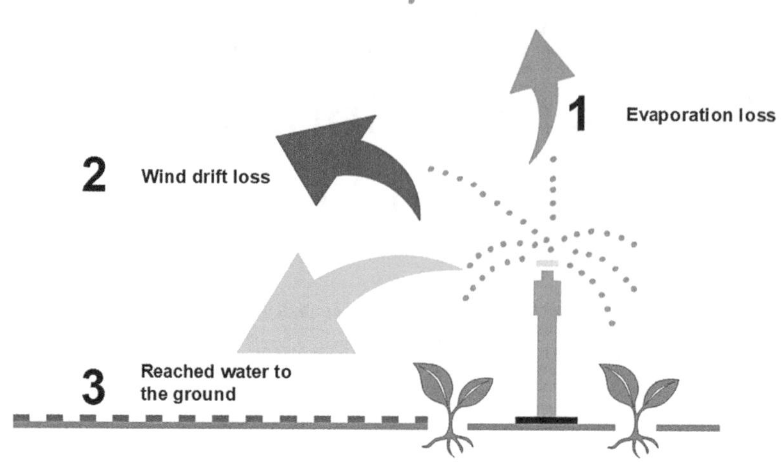

FIGURE 23.1 Sprinkler water components.

TABLE 23.1

Factors Affecting Wind Drift and Evaporation Losses (WDEL) from Sprinkler Irrigation.

| Meteorological variables | | | | | | | Operational factors | | | | | Soil, crop and water properties | | | | |
Air temperature	Relative humidity	Wind speed	Vapor pressure deficit	Pan evaporation	Cloud cover	Solar radiation	Droplet size	Nozzle diameter	Nozzle height	Operational pressures	Application rate	Crop	Crop height	Soil moisture retention	Water temperature	References
●	●	●								●	●					[10]
		●	●		●		●				●	●	●	●	●	[9]
	●															[12]
●	●			●							●					[11]
			●													[5]
						●										[23]
	●															[2]
●	●	●								●						[3]
		●	●													[6]
●	●	●														[24]
●			●													[25]
		●	●													[30]
●	●	●	●	●		●										[17]
		●	●													[27]
		●														[19]
		●						●								[1]
		●														[14]
	●	●							●					●		[31]

from 35% to 45% for high temperature and low humidity conditions in Arizona. Frost (1963) separated effective factors on WDEL into the meteorological, operational, and soil–water–crop factors [9]. The meteorological factors were vapor pressure deficit, wind velocity, and cloud cover (Eslamian and Amiri, 2011). While, operational factors were droplet size and application rate; and soil–water–crop factors comprised crop, crop height; soil moisture retention; and water temperature. Kraus (1966) found that WDEL varied inversely with relative humidity, WDEL was 3.4% for RH of 78.4%, and was 17% for RH of 37% [12]. Sternberg (1967) presented WDEL results from day and night irrigation systems. He reported that about 60% of WDEL belongs to the wind drift component [26]. Hermsmeier (1973) found that air temperature, relative humidity, and application rate were factors affecting WDEL in California. He found that WDEL ranged from 0 to 50% and could be modeled by pan evaporation and application rate [11]. Clark and Finley (1975) showed that evaporation loss depends on vapor pressure deficit. While air temperature and operating pressure had a minor effect on evaporation losses [5]. Seginer and Kostrinsky (1975) reported that the correlation between WDEL and solar radiation was significant. While the correlation between WDEL and wind speed was not significant [23]. Ali (1977) reported that WDEL ranged from 15% to 52% for different spray nozzles. He concluded that relative humidity and nozzle diameter affected significantly WDEL [2]. Arshad Ali and Barefoot (1981) found that WDEL had a direct relation with air temperature, wind speed, and operating pressure, while had inversely with relative humidity. The found WDEL ranges from 0 to 48% for a full-circle sprinkler head in Oklahoma [3]. Dylla and Shull (1983) developed a liner relation describing WDEL as a function of vapor pressure deficit and wind speed for Minnesota climate conditions [6]. Spurgeon et al. (1983) found that WDEL had a direct relation with air temperature and wind speed; and had an inverse relation with relative humidity [24]. Steiner et al. (1983) found that WDEL is a function of a vapor pressure deficit and temperature and WDEL was 39% for a center pivot system in Kansas [25]. According to the report of Yazar (1984), wind speed played a major role in both components of evaporation loss and wind drift. In addition, the vapor pressure deficit was also effective in evaporation loss. The report showed that ranges in evaporation loss were from 1.5 to 16.8% and wind drift ranged from 1.5 to 15% for an impact sprinkler at the University of Nebraska [30]. In addition to the above-mentioned findings, due to the importance of WDEL, extensive research on WDEL from sprinkler irrigation systems has also been carried out in recent years [8]. Recently, Manke et al. (2019) reported that WDEL averaged 12% for the full-length lateral-move systems [1,14,18,20-22,27].

Reducing WDEL increases irrigation efficiency and uniformity of distribution in sprinkler irrigation systems. Therefore, since 1942 extensive research has been conducted worldwide on the evaluation of WDEL. The reported results were sometimes different. The main reason for these findings is the difference in climatic and application conditions of sprinkler systems. Therefore, it seems that there is a need to investigate the factors affecting WDEL from a sprinkler in a semi-arid environment. The objective of the present study is to evaluate field-measured wind drift and evaporation losses from a sprinkler under different operating treatments in order to develop relationships among WDEL and affecting meteorological and operational factors under semi-arid conditions by multiple regression.

23.2 CASE STUDY: WDEL MODELING BY MULTIPLE REGRESSION TECHNIQUE

23.2.1 EXPERIMENT SITE

A case study was carried out at a field located in the Agricultural Faculty of Tabriz University, Iran. The location of the experiment field was 38° 05′ N, 46° 17′E, and 1360 m above sea level (Figure 23.2). The experiments were conducted during the irrigation season from June to September. The soil of the experiment field is classified as loamy, mixed, mesic, and *Typic calcixerept*. The field soil's physical properties are presented in Table 23.2.

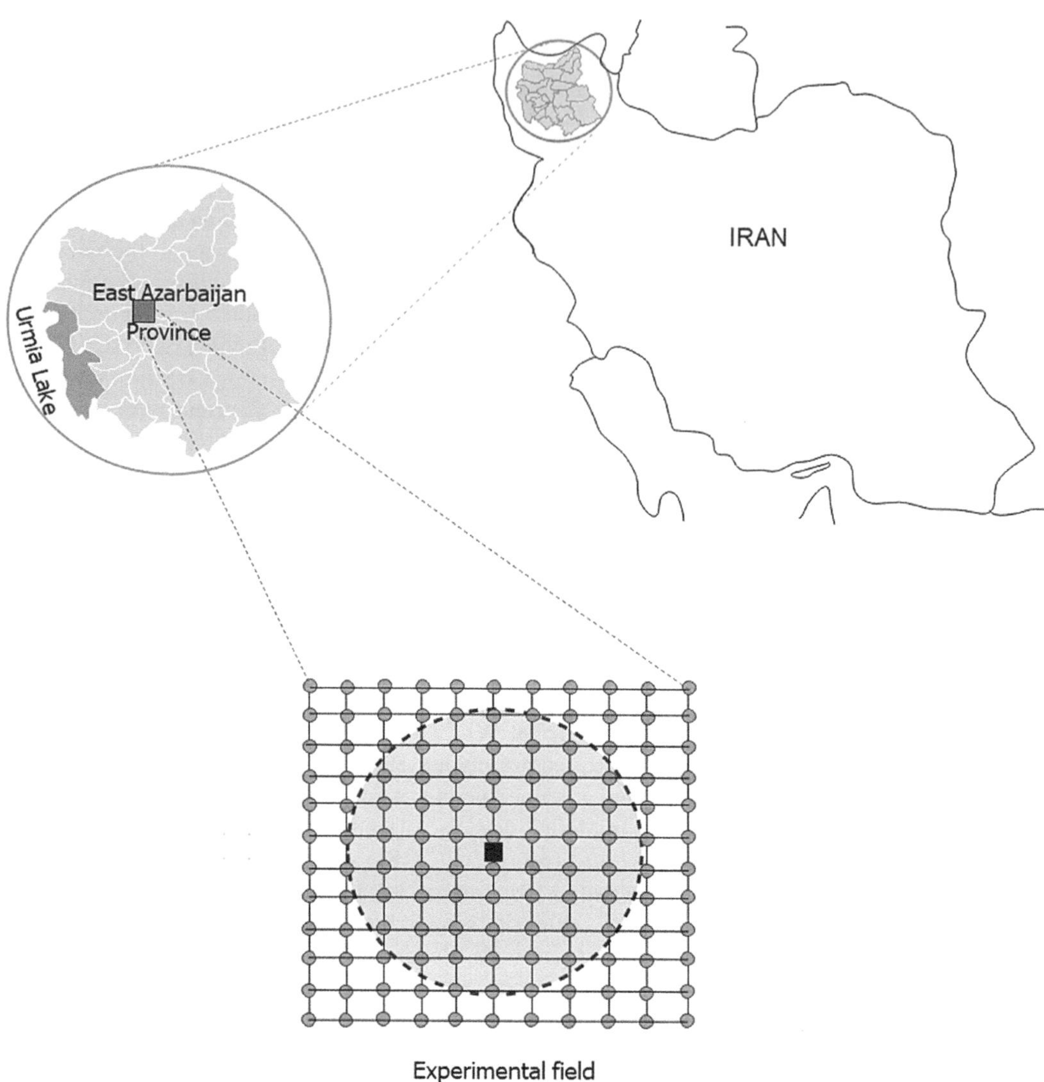

FIGURE 23.2 Experiment field location.

TABLE 23.2
Field Soil Physical Properties

Depth	Sand	Silt	Clay	Texture	Field capacity	Wilting point	Bulk density
(cm)		(%)			vol %		g cm^{-3}
0-25	72	24	4	Sandy loam	13.5	4.6	1.58
25-40	60	30	10	Sandy loam	18.5	8.0	1.57
40-65	66	30	4	Sandy loam	15.0	4.7	1.58
65-90	80	16	4	Loamy sand	12.0	4.5	1.61

The experiment site had a cold semi-arid climate with an annual rainfall of 233 mm, relative humidity of 52%, air temperature of 13.6°C (max. of 19.7°C and min. of 7.6°C), and average temperature of, with a wind speed of 5.3 m s^{-1}.

23.2.2 Experiment Setup

A single sprinkler was placed in the center of the main grid with dimensions of 30 m × 30 m with 100 subplots with each plot size of 3 m × 3 m (Figure 23.3). Sprinklers were type MZ-30 with 28° angle with three nozzle sizes of 11/64, 13/64, and 15/64 in; and a riser diameter of 1.5 in and height of 1 m. The riser was connected to the sub-main pipe by a pipe with a diameter of 2 in and 60 m long (Figure 23.3). The flow meter and pressure gauges were checked and calibrated before and after experiments. To collect water reached to ground surface from the sprinkler, 121 plastic catch cans with a space of 3 m × 3 m were placed in the field. The can's volumes were recorded after each experiment and were applied to estimate WDEL. A temporary meteorological station was

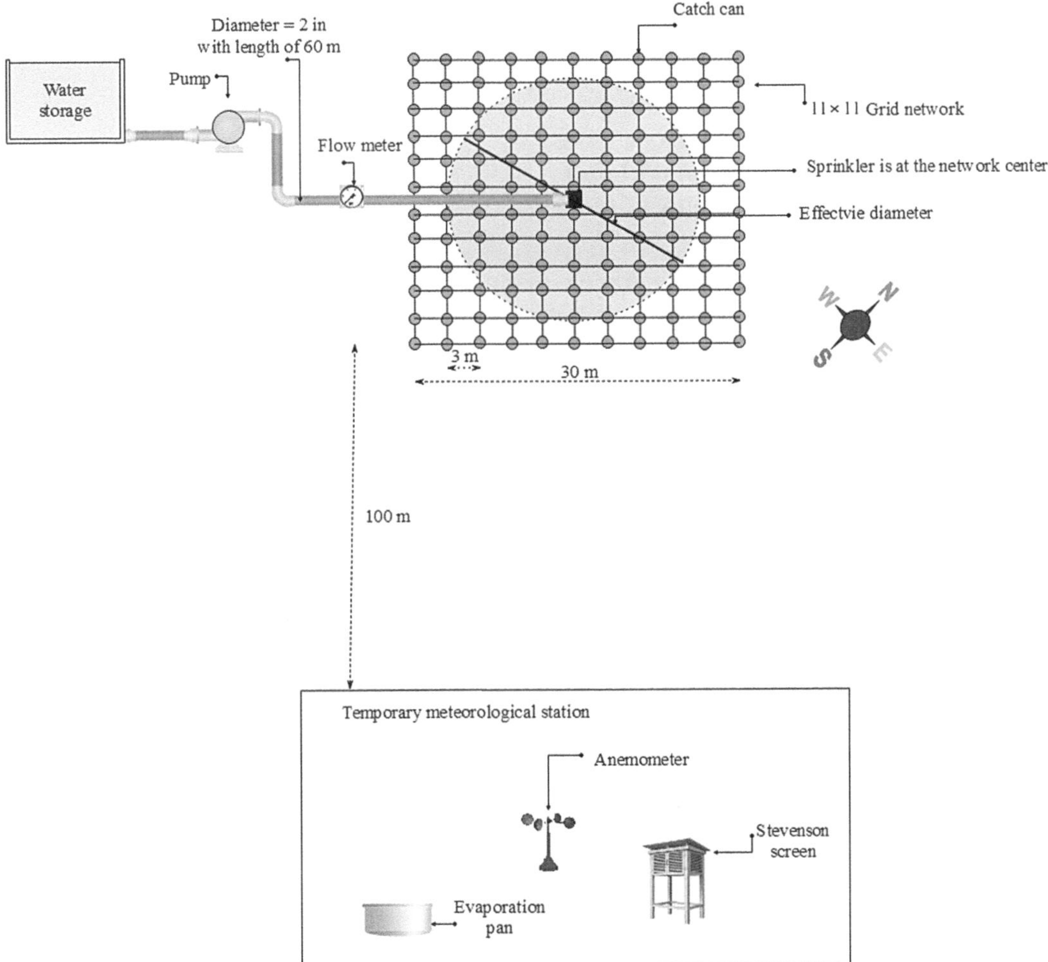

FIGURE 23.3 Schematic diagram of the experiment setup.

considered to measure air temperature, relative humidity, wind speed, pan evaporation, and daily sunshine duration on the south of the experiment site (Figure 23.3).

The vapor pressure deficit was estimated using the following relation [16]:

$$\left(es - ea\right) = 0.611 \times \exp\left(\frac{17.27 \times T}{237.3 + T}\right) \times \left(1 - \frac{RH}{100}\right) \tag{23.2}$$

in which
T= Dry bulb temperature (°C),
RH= Relative humidity (%).

23.2.3　WDEL Determination

WDEL is part of the water coming out of the sprinklers that does not reach the ground surface. Therefore, it either drifts by the wind to another location and/or evaporates into the air. The following relation based on the principle of conservation mass was applied to determine WDEL [12,13]:

$$WDEL\left(\%\right) = \left(1 - \frac{VWG\left(m^3\right)}{VWS\left(m^3\right)}\right) \times 100 \tag{23.3}$$

where
WDEL = Wind drift and evaporation losses (%),
VWS = Emitted water from sprinkler (m³),
VWG = Reached water to the ground (m³).

23.2.4　Statistics to Evaluate the Accuracy of the Models

The following statistics were applied to evaluate the accuracy of the models and to compare the results of WDEL experiments with the multiple regression models. The statistics are mean squared errors (MSE), Mallows' statistic (Cp), and determination coefficient (R^2).

$$R^2 = \frac{\left(\sum_{j=1}^{n}\left[WDEL1 - \overline{WDEL1}\right]\left[WDEL2 - \overline{WDEL2}\right]\right)^2}{\sum_{j=1}^{n}\left[WDEL1 - \overline{WDEL1}\right]^2 \sum_{j=1}^{n}\left[WDEL2 - \overline{WDEL2}\right]^2} \tag{23.4}$$

where
$\overline{WDEL1}$ = Measured wind drift and evaporation losses from a sprinkler (%),
$\overline{WDEL1}$ = Mean value for measured wind drift and evaporation losses from a sprinkler (%),
$\overline{WDEL2}$ = Predicted wind drift and evaporation losses from a sprinkler (%),
$\overline{WDEL2}$= Mean value for measured wind drift and evaporation losses from a sprinkler (%),
n= Number of measurements.

$$MSE = \frac{1}{n} \times \sum_{j=1}^{n}\left[WDEL1 - WDEL2\right]^2 \tag{23.5}$$

$$C_p = \frac{SSE}{\tilde{A}^2} + 2p - n \tag{23.6}$$

in which
SSE = Error sum of squares for the fitted model,

σ^2 = Pure error variance,
P = Number of predictor variables, and
N = Number of measurements.

23.3 RESULTS AND DISCUSSIONS

23.3.1 Correlation between WDEL and Effective Variables

WDEL was determined by the principle of conservation. The WDEL from 27 experiments with different nozzle sizes and meteorological variables is presented in Table 23.3. The losses ranged from 23.1% (at an air temperature of 27.0°C and relative humidity of 38.0%, at a wind speed of 1.1 m s^{-1} with a vapor pressure deficit of 2.2 kpa, at solar radiation of 10.7 mm day^{-1} (equivalent evaporation) with pan evaporation of 0.4 mm hr^{-1}) to 35.8% (at air temperature of 28.0°C and relative humidity of 38.0%, at a wind speed of 5.7 m s^{-1} with vapor pressure deficit of 2.3 kpa, at a solar radiation of 10.7 mm day^{-1} with pan evaporation of 0.4 mm hr^{-1}). Air temperature ranged from a low of 21°C to a high of 33°C; relative humidity varied from 30 to 52.5%; wind speed ranged from 1.1 to 5.7 ms^{-1}; vapor pressure deficit varied from 1.3 to 3.5 kpa; solar radiation ranged from 7.9 to 11.5 mm day^{-1}; pan evaporation varied from 0.3 to 0.6 mm hr^{-1} and nozzle diameter varied from 11/64 to 15/64 in (Table 23.3; Table 23.4). WDEL changes with meteorological variables and nozzle diameter are depicted in Figure 23.3. Also, Pearson product-moment correlations between each pair of variables are presented in Table 23.5. Results showed that the correlation between WDEL with air temperature, wind speed, vapor pressure deficit, and pan evaporation was statistically significant (P$\leq$ 5%). Also, the correlation between evaporation with air temperature, wind speed, vapor pressure deficit, and relative humidity was statistically significant (P$\leq$ 5%).

23.3.2 WDEL Modeling by Multiple Regression Technique

The WDEL values were applied to develop model(s) relating WDEL as a function of nozzle size (D) and meteorological variables of air temperature (T), relative humidity (RH), wind speed (WS), vapor pressure deficit (es-ea), solar radiation (Rs), and pan evaporation €. The resulting best models are presented in Table 23.6.

TABLE 23.3

Max., Min., and Average of Meteorological Variables and Nozzle Diameter and Wind Drift and Evaporation Losses

Variables	Unit	Max.	Min	Average
T	(°C)	33.0	21.0	27.0
RH	(%)	52.5	30.0	41.5
WS	(m s^{-1})	5.7	1.1	3.5
es-ea	(kpa)	3.5	1.3	2.1
Rs	(mm day^{-1})	11.5	7.9	10.3
E	(mm hr^{-1})	0.6	0.3	0.4
D	(in)	15/64	11/64	13/64
WDEL	(%)	35.8	23.1	29.0

WDEL= Wind drift and evaporation losses (%), T= air temperature (°C), Ws= wind speed (m s^{-1}), es-ea= vapor pressure deficit (kpa), RH= relative humidity (%), E= pan evaporation (mm hr^{-1}), Rs= solar radiation (mm day^{-1}, equivalent evaporation), D= nozzle diameter (in).

TABLE 23.4

Wind Drift and Evaporation Losses (WDEL) from a Sprinkler under Different Nozzle Sizes and Meteorological Variables

D	T	RH	WS	es-ea	Rs	E	EWL
(in)	(°C)	(%)	(m s⁻¹)	(kpa)	(mm day⁻¹)	(mm hr⁻¹)	(%)
11/64	30.0	33.0	1.1	2.8	10.7	0.5	26.7
	31.5	32.0	1.3	3.1	9.9	0.5	30.9
	24.5	50.0	5.7	1.5	11.5	0.3	29.3
	26.5	42.5	4.3	2.0	11.1	0.4	29.8
	25.5	42.0	5.7	1.9	10.7	0.4	31.1
	22.0	47.0	4.2	1.4	8.3	0.3	24.1
	28.0	41.0	3.6	2.2	7.9	0.4	31.0
	26.5	46.0	3.0	1.9	11.5	0.4	26.0
	28.0	44.0	3.0	2.1	11.1	0.4	28.7
13/64	28.0	35.0	1.1	2.5	10.7	0.5	25.4
	33.0	30.0	1.4	3.5	9.9	0.6	32.7
	22.0	52.5	4.4	1.3	11.5	0.3	27.7
	28.0	39.0	5.7	2.3	11.1	0.4	35.2
	28.0	38.0	5.7	2.3	10.7	0.4	35.8
	23.5	45.0	4.2	1.6	8.3	0.3	26.2
	27.0	41.0	4.2	2.1	7.9	0.4	30.8
	25.0	48.0	3.2	1.6	11.5	0.3	24.1
	30.0	40.0	3.2	2.5	11.1	0.5	31.6
15/64	27.0	38.0	1.1	2.2	10.7	0.4	23.1
	32.0	30.5	1.4	3.3	9.9	0.6	31.3
	26.0	47.0	5.1	1.8	11.5	0.3	30.2
	29.0	40.0	3.6	2.4	10.7	0.4	30.3
	27.0	40.0	5.7	2.1	10.7	0.4	34.4
	21.0	48.0	4.2	1.3	8.3	0.3	25.2
	25.0	43.5	3.6	1.8	7.9	0.4	25.6
	27.0	45.0	3.0	2.0	11.5	0.4	27.4
	28.5	43.5	3.1	2.2	11.1	0.4	28.9

WDEL= Wind drift and evaporation losses (%), T= air temperature (°C), Ws= wind speed (m s⁻¹), es-ea= vapor pressure deficit (kpa), RH= relative humidity (%), E= pan evaporation (mm hr⁻¹), Rs= solar radiation (mm day⁻¹, equivalent evaporation), D= nozzle diameter (in).

The three-variable model was considered suitable to estimate WDEL from a sprinkler based on the evaluation indices viz. R^2 (=99.9%), Cp (=1.5), and MSE (=0.90).

$$WDEL = (0.44 \times T) + (2.30 \times WS) + (4.22 \times es - es) \tag{23.7}$$

$$R^2 = 0.999 \ MSE = 0.90 \ Cp = 1.5$$

TABLE 23.5
Correlation Coefficients between Dependent and Independent Variables

	T	RH	WS	es-ea	RS	E	D	WDEL
T		-0.86	-0.55	0.96	0.19	0.96	0.00	0.50
P-value		0.00	0.00	0.00	0.33	0.00	1.00	0.01
RH			0.59	-0.95	0.09	-0.96	-0.02	-0.38
P-value			0.00	0.00	0.64	0.00	0.94	0.05
WS				-0.61	0.02	-0.61	-0.03	0.40
P-value				0.00	0.93	0.00	0.88	0.04
es-ea					0.06	1.00	0.01	0.45
P-value					0.76	0.00	0.97	0.02
RS						0.06	-0.01	0.14
P-value						0.76	0.94	0.48
E							-0.01	0.45
P-value							0.98	0.02
D								-0.01
P-value								0.94
WDEL								

WDEL= Wind drift and evaporation losses (%), T= air temperature (°C), Ws= wind speed (m s^{-1}), es-ea= vapor pressure deficit (kpa), RH= relative humidity (%), E= pan evaporation (mm hr^{-1}), Rs= solar radiation (mm day^{-1}, equivalent evaporation), D= nozzle diameter (in).

TABLE 23.6
Multiple Regression Models for WDEL

Number of variables	Included variables	R2	Adj. R2	Cp	MSE
1	T	25.0	22.0	199.2	9.25
2	WS, es-ea	92.3	91.7	1.7	0.98
3	T, WS, es-ea	99.9	98.0	1.5	0.90
4	T, WS, es-ea, E	93.5	92.4	2.1	0.91
5	T, WS, es-ea, Rs, E	93.6	92.1	4.0	0.94
6	T, RH, WS, es-ea, Rs, E	93.6	92	6.0	0.99
7	T, RH, WS, es-ea, Rs, E, D	93.6	91.2	8.0	1.04

T= air temperature (°C), Ws= wind speed (m s^{-1}), es-ea= vapor pressure deficit (kpa), RH= relative humidity (%), E= pan evaporation (mm hr^{-1}), Rs= solar radiation (mm day^{-1}, equivalent evaporation), D= nozzle diameter (in).

This model indicated that WDEL increased with an increase in T, WS, and es-ea. Similar results were reported by Frost and Schwalen (1955), Yazar (1984), Trimmer (1987), Tarjuelo et al. (2000), Faci et al. (2001), and other researchers [7, 10, 27, 28, 30]. Good agreement was observed between measured and predicted WDEL by the three-variable model (Figure 23.4). Therefore, the results of this model would be valid and acceptable for prediction purposes.

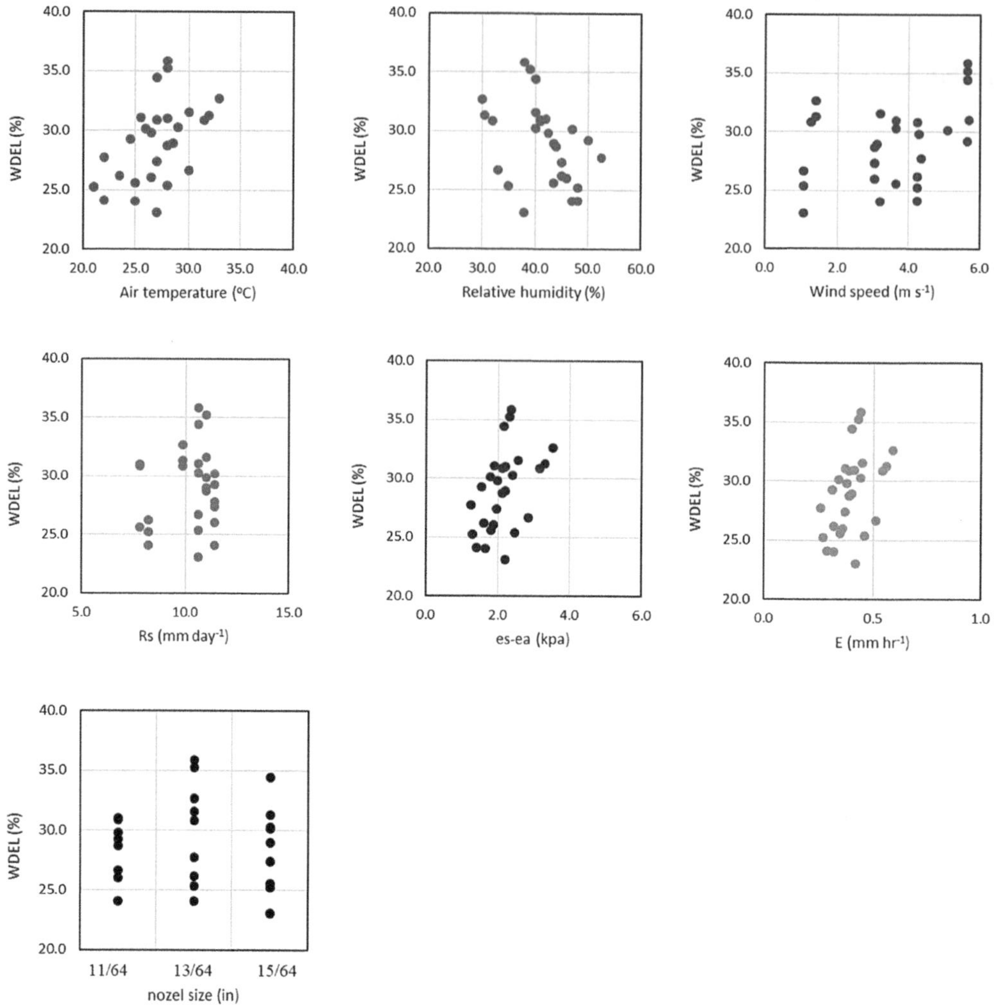

FIGURE 23.4 Relationship between WDEL; and nozzle sizes and meteorological variables.

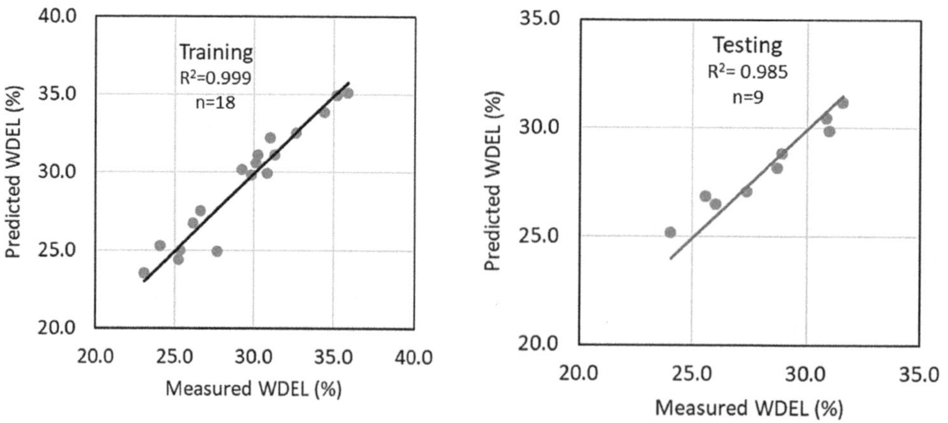

FIGURE 23.5 Predicted WDEL versus measured WDEL for training and testing stages.

23.4 CONCLUSIONS

This study on wind drift and evaporation losses (WDEL) from a sprinkler under the semiarid climate environment supports the following conclusions. The WDEL from the field experiments ranged from 23.1% to 35.8% with an average of 29.0% under the semiarid climate conditions. The WDEL model indicted that WDEL increased with increase in air temperature, wind speed and vapor pressure deficit. With the current findings such as irrigation under low temperature and low wind speed conditions, farmers could save water consumption by sprinkler irrigation. By improving irrigation efficiency, they could meet the crop water requirements in water deficit conditions.

BIBLIOGRAPHY

Al-Ghobari, H. M., El-Marazky, M. S., Dewidar, A. Z., and Mattar, M. A. (2018). Prediction of wind drift and evaporation losses from sprinkler irrigation using neural network and multiple regression techniques. *Agricultural Water Management*, 195, 211–221.

Ali, S. M. A. (1977). Effect of reduced pressure on the performance of center-pivot sprinkler irrigation systems. Doctoral dissertation, Oklahoma State University, Stillwater, OK.

Ali, S. M. A., and Barefoot, A. D. (1981). Low trajectory sprinkler patterns and evaporation loss [Irrigation]. In *Paper-American Society of Agricultural Engineers (Microfiche Collection).* American society of Agricultural Engineers Paper No. 81-2085, ASAE, 1981, 24 p.

Christiansen, J. E. (1942). *Irrigation by Sprinkling. California Agricultural Experiment Station Bulletin 670.* University of California, Berkeley, CA, 4275.

Clark, R. N., and Finley, W. W. (1975). Sprinkler evaporation losses in the southern plains. *ASAE*, Paper no.75-2573. ASAE, St.Joseph, MI.

Dylla, A. S., and Shull, H. (1983). Estimating losses from a rotating-boom sprinkler. *Transactions of the ASAE*, 26(1), 123–0125.

Eslamian, S. S., and Amiri, M. J. (2011). Estimation of daily pan evaporation using adaptive neural-based fuzzy inference system. *International Journal of Hydrology Science and Technology*, 1(3/4), 164–175.

Faci, J. M., Salvador, R., Playán, E., and Sourell, H. (2001). Comparison of fixed and rotating spray plate sprinklers. *Journal of Irrigation and Drainage Engineering*, 127(4), 224–233.

Faria, L. C., Nörenberg, B. G., Colombo, A., Dukes, M. D., Timm, L. C., Beskow, S., and Caldeira, T. L. (2019). Irrigation distribution uniformity analysis on a lateral-move irrigation system. *Irrigation Science*, 37(2), 195–206.

Frost, K. R. (1963). Factors affecting evapotranspiration losses during sprinkling. *Transactions of the ASAE*, 6(4), 282–0283.

Frost, K. R., and Schwalen, H. C. (1955). Sprinkler evaporation losses. *Agricultural Engineering*, 36(8), 526–528.

Hermsmeier, L. F., and Whisler, F. D. (1973). *Evaporation during Sprinkler Application in a Desert Climate.* Unpublished Paper No. 73- 216, presented at the Summer Meeting. American Society of Agricultural Engineers, Lexington.

Kraus, J. H. (1966). Application efficiency of sprinkler irrigation and its effects on microclimate. *Transactions of the ASAE*, 9(5), 642–0645.

Kumar, J., Kumar, D., Kumar, D., and Mittal, H. (2013). Wind drift and evaporation losses under various operating conditions of sprinkler irrigation system. *INMS*, 12(2), 136–140.

Manke, E. B., Nörenberg, B. G., Faria, L. C., Tarjuelo, J. M., Colombo, A., Neta, M. C. C. C., and Parfitt, J. M. B. (2019). Wind drift and evaporation losses of a mechanical lateral-move irrigation system: Oscillating plate versus fixed spray plate sprinklers. *Agricultural Water Management*, 225, 105759.

Mclean, R. K., Ranjan, R. S., and Klassen, G. (1994). Measuring spray evaporation losses from sprinkler irrigation systems. *Canadian Agricultural Engineering*, 42(1), 1–8.

Murray, F. W. (1967). On the computation of saturation vapor pressure. *Journal of Applied Meteorology and Climatology*, 6, 203–204.

Nasseri, A., and Lotfi-Sadigh, A. (1999). The effect of nozzle diameter on wind drift and evaporation losses. *Journal of Iranian Agricultural Engineering Research Institute*, 4(5), 1–11.

Ortiz, J. N., Tarjuelo, J. M., and De Juan, J. A. (2009). Characterization of evaporation and drift losses with centre pivots. *Agricultural Water Management*, 96(11), 1541–1546.

Playán, E., Garrido, S., Faci, J. M., and Galán, A. (2004). Characterizing pivot sprinklers using an experimental irrigation machine. *Agricultural Water Management*, 70(3), 177–193.

Playán, E., Salvador, R., Faci, J. M., Zapata, N., Martínez-Cob, A., and Sánchez, I. (2005). Day and night wind drift and evaporation losses in sprinkler solid-sets and moving laterals. *Agricultural Water Management*, 76(3), 139–159.

Sadeghi, S. H., Peters, T. R., Amini, M. Z., Malone, S. L., and Loescher, H. W. (2015). Novel approach to evaluate the dynamic variation of wind drift and evaporation losses under moving irrigation systems. *Biosystems Engineering*, 135, 44–53.

Sadeghi, S. H., Peters, T., Shafii, B., Amini, M. Z., and Stöckle, C. (2017). Continuous variation of wind drift and evaporation losses under a linear move irrigation system. *Agricultural Water Management*, 182, 39–54.

Seginer, I., and Kostrinsky, M. (1975). Wind, sprinkler patterns, and system design. *Journal of the Irrigation and Drainage Division*, 101(4), 251–264.

Spurgeon, W. E., Thompson, T. L., and Gilley, J. R. (1983). Irrigation management using hourly spray evaporation loss estimates. Microfiche collection. ASAE Paper 1983, No. 83-2591. ASAE, St. Joseph, MI, 22 p.

Steiner, J. L., Kanemasu, E. T., and Clark, R. N. (1983). Spray losses and partitioning of water under a center pivot sprinkler system. *Transactions of the ASAE*, 26(4), 1128–1134.

Sternberg, Y. M. (1967). Analysis of sprinkler irrigation losses. *Journal of the Irrigation and Drainage Division*, 93(4), 111–124.

Tarjuelo, J. M., Ortega, J. F., Montero, J., and De Juan, J. A. (2000). Modelling evaporation and drift losses in irrigation with medium size impact sprinklers under semi-arid conditions. *Agricultural Water Management*, 43(3), 263–284.

Trimmer, W. L. (1987). Sprinkler evaporation loss equation. *Journal of Irrigation and Drainage Engineering*, 113(4), 616–620.

Wiser, E. H., Van Schilfgaarde, J., and Wilson, T. V. (1961). Evapotranspiration concepts for evaluating sprinkler irrigation losses. *American Society of Agricultural and Biological Engineers*, 4(1), 0128–0130.

Yazar, A. (1984). Evaporation and drift losses from sprinkler irrigation systems under various operating conditions. *Agricultural Water Management*, 8(4), 439–449.

Zamanisepahvand, S., Torabi, H., and nasrolahi, A. (2023). Estimation of wind drift and evaporation losses of sprinkler irrigation systems under the influence of different exploitation managements (case study: Choghahoroshi Khorramabad Plain). *Iranian Journal of Irrigation & Drainage*, 17(1), 129–141.

Index